T0140469

Advances in Intelligent Systems and Computing

Volume 709

Series editor

Janusz Kacprzyk, Systems Research Institute, Polish Academy of Sciences,
Warsaw, Poland
e-mail: kacprzyk@ibspan.waw.pl

The series "Advances in Intelligent Systems and Computing" contains publications on theory, applications, and design methods of Intelligent Systems and Intelligent Computing. Virtually all disciplines such as engineering, natural sciences, computer and information science, ICT, economics, business, e-commerce, environment, healthcare, life science are covered. The list of topics spans all the areas of modern intelligent systems and computing such as: computational intelligence, soft computing including neural networks, fuzzy systems, evolutionary computing and the fusion of these paradigms, social intelligence, ambient intelligence, computational neuroscience, artificial life, virtual worlds and society, cognitive science and systems, Perception and Vision, DNA and immune based systems, self-organizing and adaptive systems, e-Learning and teaching, human-centered and human-centric computing, recommender systems, intelligent control, robotics and mechatronics including human-machine teaming, knowledge-based paradigms, learning paradigms, machine ethics, intelligent data analysis, knowledge management, intelligent agents, intelligent decision making and support, intelligent network security, trust management, interactive entertainment, Web intelligence and multimedia.

The publications within "Advances in Intelligent Systems and Computing" are primarily proceedings of important conferences, symposia and congresses. They cover significant recent developments in the field, both of a foundational and applicable character. An important characteristic feature of the series is the short publication time and world-wide distribution. This permits a rapid and broad dissemination of research results.

Advisory Board

Chairman

Nikhil R. Pal, Indian Statistical Institute, Kolkata, India
e-mail: nikhil@isical.ac.in

Members

Rafael Bello Perez, Faculty of Mathematics, Physics and Computing, Universidad Central de Las Villas, Santa Clara, Cuba
e-mail: rbellop@uclv.edu.cu

Emilio S. Corchado, University of Salamanca, Salamanca, Spain
e-mail: escorchado@usal.es

Hani Hagras, School of Computer Science and Electronic Engineering, University of Essex, Colchester, UK
e-mail: hani@essex.ac.uk

László T. Kóczy, Department of Information Technology, Faculty of Engineering Sciences, Győr, Hungary
e-mail: koczy@sze.hu

Vladik Kreinovich, Department of Computer Science, University of Texas at El Paso, El Paso, TX, USA
e-mail: vladik@utep.edu

Chin-Teng Lin, Department of Electrical Engineering, National Chiao Tung University, Hsinchu, Taiwan
e-mail: ctlin@mail.nctu.edu.tw

Jie Lu, Faculty of Engineering and Information, University of Technology Sydney, Sydney, NSW, Australia
e-mail: Jie.Lu@uts.edu.au

Patricia Melin, Graduate Program of Computer Science, Tijuana Institute of Technology, Tijuana, Mexico
e-mail: epmelin@hafsamx.org

Nadia Nedjah, Department of Electronics Engineering, University of Rio de Janeiro, Rio de Janeiro, Brazil
e-mail: nadia@eng.uerj.br

Ngoc Thanh Nguyen, Wrocław University of Technology, Wrocław, Poland
e-mail: Ngoc-Thanh.Nguyen@pwr.edu.pl

Jun Wang, Department of Mechanical and Automation Engineering, The Chinese University of Hong Kong, Shatin, Hong Kong
e-mail: jwang@mae.cuhk.edu.hk

More information about this series at http://www.springer.com/series/11156

Pankaj Kumar Sa · Sambit Bakshi
Ioannis K. Hatzilygeroudis
Manmath Narayan Sahoo
Editors

Recent Findings in Intelligent Computing Techniques

Proceedings of the 5th ICACNI 2017, Volume 3

 Springer

Editors
Pankaj Kumar Sa
Department of Computer Science
 and Engineering
National Institute of Technology, Rourkela
Rourkela, Odisha
India

Sambit Bakshi
Department of Computer Science
 and Engineering
National Institute of Technology, Rourkela
Rourkela, Odisha
India

Ioannis K. Hatzilygeroudis
Department of Computer Engineering
 and Informatics
University of Patras
Patras, Greece

Manmath Narayan Sahoo
Department of Computer Science
 and Engineering
National Institute of Technology, Rourkela
Rourkela, Odisha
India

ISSN 2194-5357   ISSN 2194-5365   (electronic)
Advances in Intelligent Systems and Computing
ISBN 978-981-10-8632-8   ISBN 978-981-10-8633-5   (eBook)
https://doi.org/10.1007/978-981-10-8633-5

Library of Congress Control Number: 2018934925

© Springer Nature Singapore Pte Ltd. 2018
This work is subject to copyright. All rights are reserved by the Publisher, whether the whole or part of the material is concerned, specifically the rights of translation, reprinting, reuse of illustrations, recitation, broadcasting, reproduction on microfilms or in any other physical way, and transmission or information storage and retrieval, electronic adaptation, computer software, or by similar or dissimilar methodology now known or hereafter developed.
The use of general descriptive names, registered names, trademarks, service marks, etc. in this publication does not imply, even in the absence of a specific statement, that such names are exempt from the relevant protective laws and regulations and therefore free for general use.
The publisher, the authors and the editors are safe to assume that the advice and information in this book are believed to be true and accurate at the date of publication. Neither the publisher nor the authors or the editors give a warranty, express or implied, with respect to the material contained herein or for any errors or omissions that may have been made. The publisher remains neutral with regard to jurisdictional claims in published maps and institutional affiliations.

This Springer imprint is published by the registered company Springer Nature Singapore Pte Ltd.
The registered company address is: 152 Beach Road, #21-01/04 Gateway East, Singapore 189721, Singapore

Foreword

Message from the General Chairs Dr. Modi Chirag Navinchandra and Dr. Pankaj Kumar Sa

Welcome to the 5th International Conference on Advanced Computing, Networking, and Informatics. The conference is hosted by the Department of Computer Science and Engineering at National Institute of Technology Goa, India, and co-organized with Centre for Computer Vision & Pattern Recognition, National Institute of Technology Rourkela, India. For this fifth event, held on June 1–3, 2017, the theme is security and privacy, which is a highly focused research area in different domains.

Having selected 185 articles from more than 500 submissions, we are glad to have the proceedings of the conference published in the *Advances in Intelligent Systems and Computing* series of Springer. We would like to acknowledge the special contribution of Prof. Udaykumar R. Yaragatti, Former Director of NIT Goa, as the chief patron for this conference.

We would like to acknowledge the support from our esteemed keynote speakers, delivering keynotes titled *"On Secret Sharing"* by Prof. Bimal Kumar Roy, Indian Statistical Institute, Kolkata, India; *"Security Issues of Software Defined Networks"* by Prof. Manoj Singh Gaur, Malaviya National Institute of Technology, Jaipur; *"Trust aware Cloud (Computing) Services"* by Prof. K. Chandrasekaran, National Institute of Technology Karnataka, Surathkal, India; and *"Self Driving Cars"* by Prof. Dhiren R. Patel, Director, VJTI, Mumbai, India. They are all highly accomplished researchers and practitioners, and we are very grateful for their time and participation.

We are grateful to advisory board members Prof. Audun Josang from Oslo University, Norway; Prof. Greg Gogolin from Ferris State University, USA; Prof. Ljiljana Brankovic from The University of Newcastle, Australia; Prof. Maode Ma, FIET, SMIEEE from Nanyang Technological University, Singapore; Prof. Rajarajan Muttukrishnan from City, University of London, UK; and Prof. Sanjeevikumar Padmanaban, SMIEEE from University of Johannesburg, South

Africa. We are thankful to technical program committee members from various countries, who have helped us to make a smooth decision of selecting best quality papers. The diversity of countries involved indicates the broad support that ICACNI 2017 has received. A number of important awards will be distributed at this year's event, including Best Paper Awards, Best Student Paper Award, Student Travel Award, and a Distinguished Women Researcher Award.

We would like to thank all of the authors and contributors for their hard work. We would especially like to thank the faculty and staff of National Institute of Technology Goa and National Institute of Technology Rourkela for giving us their constant support. We extend our heartiest thanks to Dr. Sambit Bakshi (Organizing Co-Chair) and Dr. Manmath N. Sahoo (Program Co-Chair) for the smooth conduction of this conference. We would like to specially thank Dr. Pravati Swain (Organizing Co-Chair) from NIT Goa who has supported us to smoothly conduct this conference at NIT Goa.

But the success of this event is truly down to the local organizers, volunteers, local supporters, and various chairs who have done so much work to make this a great event.

We hope you will gain much from ICACNI 2017 and will plan to submit to and participate in the 6th ICACNI 2018.

Best wishes,

Goa, India   Dr. Modi Chirag Navinchandra
Rourkela, India   Dr. Pankaj Kumar Sa
                                          General Chairs, 5th ICACNI 2017

Preface

It is indeed a pleasure to receive an overwhelming response from academicians and researchers of premier institutes and organizations of the country and abroad for participating in the 5th International Conference on Advanced Computing, Networking, and Informatics (ICACNI 2017), which makes us feel that our endeavor is successful. The conference organized by the Department of Computer Science and Engineering, National Institute of Technology Goa, and Centre for Computer Vision & Pattern Recognition, National Institute of Technology Rourkela, during June 1–3, 2017, certainly marks a success toward bringing researchers, academicians, and practitioners in the same platform. We have received more than 600 articles and very stringently have selected through peer review 185 best articles for presentation and publication. We could not accommodate many promising works as we tried to ensure the highest quality. We are thankful to have the advice of dedicated academicians and experts from industry and the eminent academicians involved in providing technical comments and quality evaluation for organizing the conference in good shape. We thank all people participating and submitting their works and having continued interest in our conference for the fifth year. The articles presented in the three volumes of the proceedings discuss the cutting-edge technologies and recent advances in the domain of the conference.

We conclude with our heartiest thanks to everyone associated with the conference and seeking their support to organize the 6th ICACNI 2018 at National Institute of Technology Silchar, India, during June 4–6, 2018.

Rourkela, India	Pankaj Kumar Sa
Rourkela, India	Sambit Bakshi
Patras, Greece	Ioannis K. Hatzilygeroudis
Rourkela, India	Manmath Narayan Sahoo

In Memoriam: Prof. S. K. Jena (1954–2017)

A man is defined by the deeds he has done and the lives he has touched; he is defined by the people who have been inspired by his actions and the hurdles he has crossed. With his deeds and service, Late Prof. Sanjay Kumar Jena, Department of Computer Science and Engineering, has always remained an epitome of inspiration for many. Born in 1954, he breathed his last on May 17, 2017, due to cardiac arrest. He left for his heavenly abode with peace while on duty. He is survived by his loving wife, beloved son, and cherished daughter.

He is known for his ardent ways of problem-solving right from his early years. Even at 62 years of age, his enthusiasm and dedication took NIT Rourkela community by surprise. From being the Superintendent of S. S. Bhatnagar Hall of Residence to Dean of SRICCE to Head of the Computer Science Department to a second term as the Head of Training and Placement Cell, he not only has contributed to the growth of the institute, but has been a wonderful teacher and researcher guiding a generation of students and scholars. Despite this stature, he was an audience when it came to hearing out problems of students, colleagues, and subordinates, which took them by surprise being unbiased in judgments. His kind and compassionate behavior added splendidly to the beloved teacher who could be approached by all. His ideas and research standards shall continue to inspire generations of students to come. He will also be remembered by the teaching community for the approach and dedication he has gifted to the NIT community.

Committee: ICACNI 2017

Advisory Board Members

Audun Josang, Oslo University, Norway
Greg Gogolin, Ferris State University, USA
Ljiljana Brankovic, The University of Newcastle, Australia
Maode Ma, FIET, SMIEEE, Nanyang Technological University, Singapore
Rajarajan Muttukrishnan, City, University of London, UK
Sanjeevikumar Padmanaban, SMIEEE, University of Johannesburg, South Africa

Chief Patron

Udaykumar Yaragatti, Director, National Institute of Technology Goa, India

Patron

C. Vyjayanthi, National Institute of Technology Goa, India

General Chairs

Chirag N. Modi, National Institute of Technology Goa, India
Pankaj K. Sa, National Institute of Technology Rourkela, India

Organizing Co-chairs

Pravati Swain, National Institute of Technology Goa, India
Sambit Bakshi, National Institute of Technology Rourkela, India

Program Co-chairs

Manmath N. Sahoo, National Institute of Technology Rourkela, India
Shashi Shekhar Jha, SMU Lab, Singapore
Lamia Atma Djoudi, Synchrone Technologies, France
B. N. Keshavamurthy, National Institute of Technology Goa, India
Badri Narayan Subudhi, National Institute of Technology Goa, India

Technical Program Committee

Adam Schmidt, Poznan University of Technology, Poland
Akbar Sheikh Akbari, Leeds Beckett University, UK
Al-Sakib Khan Pathan, SMIEEE, UAP and SEU, Bangladesh/Islamic University in Madinah, KSA
Andrey V. Savchenko, National Research University Higher School of Economics, Russia
B. Annappa, SMIEEE, National Institute of Technology Karnataka, Surathkal, India
Biju Issac, SMIEEE, FHEA, Teesside University, UK
Ediz Saykol, Beykent University, Turkey
Haoxiang Wang, GoPerception Laboratory, USA
Igor Grebennik, Kharkiv National University of Radio Electronics, Ukraine
Jagadeesh Kakarla, Central University of Rajasthan, India
Jerzy Pejas, Technical University of Szczecin, Poland
Laszlo T. Koczy, Szechenyi Istvan University, Hungary
Mithileysh Sathiyanarayanan, City, University of London, UK
Palaniappan Ramaswamy, SMIEEE, University of Kent, UK
Patrick Siarry, SMIEEE, Université de Paris, France
Prasanta K. Jana, SMIEEE, Indian Institute of Technology (ISM), Dhanbad, India
Saman K. Halgamuge, SMIEEE, The University of Melbourne, Australia
Sohail S. Chaudhry, Villanova University, USA
Sotiris Kotsiantis, University of Patras, Greece
Tienfuan Kerh, National Pingtung University of Science and Technology, Taiwan
Valentina E. Balas, SMIEEE, Aurel Vlaicu University of Arad, Romania
Xiaolong Wu, California State University, USA

Organizing Committee

Chirag N. Modi , National Institute of Technology Goa, India
Pravati Swain, National Institute of Technology Goa, India
B. N. Keshavamurthy, National Institute of Technology Goa, India
Damodar Reddy Edla, National Institute of Technology Goa, India

B. R. Purushothama, National Institute of Technology Goa, India
T. Veena, National Institute of Technology Goa, India
S. Mini, National Institute of Technology Goa, India
Venkatanareshbabu Kuppili, National Institute of Technology Goa, India

Contents

About the Editors

Pankaj Kumar Sa received his Ph.D. degree in Computer Science in 2010. He is currently serving as an assistant professor in the Department of Computer Science and Engineering, National Institute of Technology Rourkela, India. His research interests include computer vision, biometrics, visual surveillance, and robotic perception. He has co-authored a number of research articles in various journals, conferences, and chapters. He has co-investigated some research and development projects that are funded by SERB, DRDOPXE, DeitY, and ISRO. He has received several prestigious awards and honors for his excellence in academics and research. Apart from research and teaching, he conceptualizes and engineers the process of institutional automation.

Sambit Bakshi is currently with Centre for Computer Vision & Pattern Recognition of National Institute of Technology Rourkela, India. He also serves as an assistant professor in the Department of Computer Science and Engineering of the institute. He earned his Ph.D. degree in Computer Science and Engineering. He serves as an associate editor of *International Journal of Biometrics* (2013–), *IEEE Access* (2016–), *Innovations in Systems and Software Engineering* (2016–), *Plos One* (2017–), and *Expert Systems* (2018–). He is a technical committee member of IEEE Computer Society Technical Committee on Pattern Analysis and Machine Intelligence. He received the prestigious Innovative Student Projects Award 2011 from the Indian National Academy of Engineering (INAE) for his master's thesis. He has more than 50 publications in journals, reports, and conferences.

Ioannis K. Hatzilygeroudis is an associate professor in the Department of Computer Engineering and Informatics, University of Patras, Greece. His research interests include knowledge representation (KR) with an emphasis on integrated KR languages/systems; knowledge-based systems, expert systems; theorem proving with an emphasis on classical methods; intelligent tutoring systems; intelligent e-learning; natural language generation; and Semantic Web. He has several papers published in journals, contributed books, and conference proceedings. He has over 25 years of teaching experience. He is an associate editor of *International Journal on AI Tools* (IJAIT), published by World Scientific Publishing Company, and also serving as an editorial board member to *International Journal of Hybrid Intelligent Systems* (IJHIS), IOS Press, and *International Journal of Web-Based Communities* (IJWBC), Inderscience Enterprises Ltd.

Manmath Narayan Sahoo is an assistant professor in the Department of Computer Science and Engineering, National Institute of Technology Rourkela, Rourkela, India. His research interest areas are fault-tolerant systems, operating systems, distributed computing, and networking. He is the member of IEEE, Computer Society of India, and The Institution of Engineers, India. He has published several papers in national and international journals.

Part I
Data Mining, NLP, Text Mining, Social Media Analysis

Ashtadhyayi—An Experimental Approach to Enhance Programming Languages and Compiler Design Using Panini's Grammar

A. Soumya Mahalakshmi and Minal Moharir

Abstract The astonishing foresight and vision of ancient scientists remain unparalleled. They augmented the conversations and thoughts about philosophy and science, often revealing structured explanations for phenomena that were way ahead of their time. One such instance is the emergence of the Sanskrit language which was characteristic of the literature during the Vedic period in Ancient India, in third century BC. Sanskrit has been widely accepted as an extremely logical language and the sole credit for this goes to Panini, Sanskrit Grammarian and vastly regarded as the first programmer of the world. In his work *Ashtadhyayi*, he summarizes the logic and grammar for Sanskrit in the form of 4000 *Sutras*, effectively building a machine that generates thousands of Sanskrit words and sentences. While linguists around the world have begun realizing similarities to Backus-Naur form in *Ashtadhyayi*, Sanskrit is being claimed as the best language for Artificial Intelligence and Natural Language Processing. The aim of this project is to identify and understand the *Sutra* style of Panini Grammar, and to exploit the optimizations in language for better programming languages. The effectiveness of applying a similar optimized grammar for use in C programming language has been explored. Further, the results have been extrapolated to understand the advantage of using this grammar in CUDA-C in a graphical processing unit. The performed experiments validate the efficiency and pronounced the enhancement of using a Panini-inspired compiler for C as well as CUDA-C, which can lead to path-breaking speed and a new paradigm to approach fast data technologies using next generation GPU systems.

Keywords Panini · Ashtadhyayi · Programming languages
C · CUDA-C · Compiler design · Grammar · Sanskrit · Sutras

A. Soumya Mahalakshmi (✉) · M. Moharir
Department of Computer Science and Engineering, R V College of Engineering,
Bengaluru, Karnataka, India
e-mail: a.soumyamahalakshmi@gmail.com

© Springer Nature Singapore Pte Ltd. 2018
P. K. Sa et al. (eds.), *Recent Findings in Intelligent Computing Techniques*,
Advances in Intelligent Systems and Computing 709,
https://doi.org/10.1007/978-981-10-8633-5_1

1 Introduction

The Indian Civilization was an ardent contributor to the ancient understanding of science and technology. The astonishing foresight and vision of ancient scientists remain unparalleled. They augmented the conversations and thoughts about philosophy and science, often revealing structured explanations for phenomena that were way ahead of their time. Ancient India was a land of scholars, sages, seers, and scientists. Research has revealed outstanding examples of genius inventions such as the mathematical digit "zero". Aryabhata created a symbol for zero and its integration into the place-value system enabled one to write large numbers using only ten symbols. Binary numbers involve the most rudimentary understanding of computer science, and it was first described by the Pingala in Chandahśāstra, a Sanskrit treatise on prosody. Another instance in relation to computer science is the chakravala method, which evolved as a cyclic algorithm to obtain integer solutions for intermediate quadratic equations. Research has suggested the presence of explanations and origins of theories such as the atomic theory and heliocentric theory in Ancient Indian texts long before the rest of the world documented it. A similar observation can be made regarding the knowledge of metallurgy, surgery, rocket science, and medicine.

In this regard, it is worth mentioning the achievement of Panini, who has gained the reputation of being the pioneer in grammars and languages. He was hailed as a Sanskrit grammarian who worked on phonetics, phonology, and morphology, thereby providing a complete and comprehensive grammar for Sanskrit. Incidentally, the word "Sanskrit" means "complete" or "perfect" and it was thought of as the divine language, or language of the gods. The Sanskrit language was indicative of the Vedic period in India, and was used in nearly all scientific and literary documents of the time. In several contexts, Panini has been identified as the world's first programmer, owing to his outstanding contribution in defining a structured grammar for languages. Today, linguists and programmers around the world have identified the merits of his grammar used in the Sanskrit language, which in many ways helps present day understanding of Natural Language Processing and Artificial Intelligence (Fig. 1).

Panini's work is largely compiled in his treatise called *Astadhyayi*, whose name indicates the presence of eight chapters, each subdivided into quarter chapters. Panini gives formal production rules and definitions to describe Sanskrit grammar. Panini's constructions of the Sanskrit language using a grammar were vastly similar to a mathematical function, which integrated 1700 basic elements such as nouns, verbs, etc. [1].

There has been considerable speculation about the possibility of the presence of rules in ancient Indian logic and grammar that can aid advancement in cognitive and computer science. The significance of the context-sensitive grammars of Panini was understood only when the Chomsky normal form was introduced in the nineteenth century. Formally, Panini's grammar, Ashtadhyayi is studied together with the dhatupatha (a list of verbal roots arranged into sublists), and the ganapatha (a list of various classes of morphs). Ashtadhyayi provides a structure for the

Fig. 1 An artist's
interpretation of Panini

Courtesy : India's Postal Stamp of Panini, 2004

analysis of sentences, which has been described as a machine generating words and sentences of Sanskrit [2].

Sanskrit's phonology, morphology, and syntax are described in Ashtadhayayi in a collection of 4000 sutras, which define its structure as a rule-based system. Each sutra is a reference to a rule, which can be definitions, theorems (linguistic facts), and meta-theorems (rules regarding rules). Since Ancient India followed a mechanism of oral propagation of tradition, it was important to make the 4000 sutras as concise as possible, consisting of only three words each, achieved through various optimizations. Hence, it is generally agreed that the Paninian system is based on a principle of economy, an Occam's razor. This makes the structure to be of special interest to cognitive scientists [2]. Traditionally, a sutra is defined as the most concise of statements which uses as few letters as possible. The Grammar follows a principle of the following form:

iko yan aci

Each sutra can be analogous to a production rule that defines a grammar. Therefore, it is only fitting to observe that the Panini system advocates the use of only three tokens in a production rule, achieved through employing several algebraic devices such as prefixes and suffixes. Panini took the idea of providing a context for action in terms of its relations to agents and situation, given by the karaka theory. The following semantic notions capture the various aspects of action, which have been used for optimizations [3]:

a. Apadana: that which is fixed when departure takes place
b. Sampradana: the recipient of the object
c. Karana: the main cause of the effect; instrument
d. Adhikarana: the basis, location
e. Karman: what the agent seeks most to attain; deed, object
f. Kartr: one who is independent; the agent

Here, any sentence is optimized to only three tokens, using prefixes, suffixes, recursion, and context. However, in a regular C compiler, there is no upper limit on the number of tokens in a statement. Therefore, in this study, an attempt has been made to employ these ideas to reduce the number of tokens in a production rule that defines a grammar for programming languages. Hence, LEX and YACC tools were used to define a limited functionality compiler for C programming language, which was designed using grammar production rules influenced by these optimizations.

2 Methodology

A. *Understanding*

1. To analyze the sutra style of grammar which is highly optimized using prefixes and suffixes
2. To ponder over the present grammar of programming languages such as C and CUDA-C

B. *Design*

1. To develop a C compiler using YACC for the new Panini inspired grammar
2. To develop a matrix multiplication program in C, with the new compiler

C. *Validation*

1. To compile and run the program against a conventional compiler and the Panini compiler
2. To compare, prorate, and analyze the results obtained
   (Fig. 2).

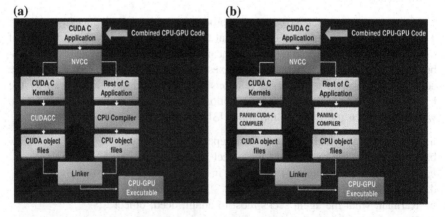

Fig. 2 Design of compiler design of **a** conventional CUDA-C compiler **b** optimized Panini CUDA-C compiler

Table 1 Experimental results for optimization using Panini compiler

CUDA-C language (s)	C language (s)	Type of compiler
1.01	44.1537	Conventional compiler
0.6059 (Prorated)	26.4922	Optimized compiler

3 Experimental Results

A C program for matrix multiplication was compiled and executed over a conventional C compiler, and then with a compiler optimized using Panini grammar and time taken was 44.1537 s and 26.4922 s, respectively, for input size 2000. When executed on NVIDIA Tesla K40, the time taken to execute matrix multiplication on a CUDA compiler was 1.01 s. Through proration, we can reduce the time taken to 0.6059 s (Table 1).

4 Conclusion

i. It can be concluded that using Panini's optimizations in the grammar of a compiler can bring about a marked speed enhancement of nearly 40–50%.
ii. When implemented for CUDA compilers, it can revolutionize fast data technologies for next generation GPU systems.
iii. Panini has indeed proved himself to be the world's first programmer.

Acknowledgements We gratefully acknowledge the support of NVIDIA Corporation with the donation of the Tesla K40 GPU as a part of GPU Research Center, used for this research.

References

1. O'Connor, J.J., Robertson, E.F.: "Panini", School of Mathematics and Statistics, University of St Andrews, Scotland
2. Bhate, S., Kak, S.: Panini's grammar and computer science. Ann. Bhandarkar Orient. Res. Inst. **72**, 79–94 (1993)
3. Kak, Subhash C.: The Paninian approach to natural language processing. Int. J. Approximate Reasoning **1**(1), 117–130 (1987)

Movie Box-Office Gross Revenue Estimation

**Shaiwal Sachdev, Abhishek Agrawal, Shubham Bhendarkar,
Bakshi Rohit Prasad and Sonali Agarwal**

Abstract In this research work, movie box-office gross revenue estimation has been performed using machine learning techniques to effectively estimate the amount of gross revenue a movie will be able to collect using the public information available after its first weekend of release. Here, first weekend refers to first three days of release namely Friday, Saturday, and Sunday. This research work has been done only for the movies released in USA. It was assumed that gross revenue is equal to the amount of money that is collected by the sale of movie tickets. Data collected has been collected from IMDB and Rotten Tomatoes for movies released from the year 2000–2015 only. Multiple linear regression and genre-based analysis was used to effectively estimate the gross revenue. Finally, Local regression methods namely local linear regression, and local decision tree regression were used to get a better estimate.

Keywords Gross revenue · Box-office · Linear regression · Decision tree regression · IMDB · Rotten tomatoes · Machine learning

S. Sachdev (✉) · A. Agrawal · S. Bhendarkar · B. R. Prasad · S. Agarwal
e-mail: sonali@iiita.ac.in
Indian Institute of Information Technology Allahabad, Allahabad, India
e-mail: iit2013196@iiita.ac.in

A. Agrawal
e-mail: iit2013128@iiita.ac.in

S. Bhendarkar
e-mail: iit2013172@iiita.ac.in

B. R. Prasad
e-mail: rs151@iiita.ac.in

S. Agarwal
e-mail: sonali@iiita.ac.in

© Springer Nature Singapore Pte Ltd. 2018
P. K. Sa et al. (eds.), *Recent Findings in Intelligent Computing Techniques*,
Advances in Intelligent Systems and Computing 709,
https://doi.org/10.1007/978-981-10-8633-5_2

1   Introduction

Film industry is a big business in United States. It is one of the biggest players in the entertainment industry. Predicting the gross revenue of a movie has become need of the hour. Lot of researchers have used different models but still there is no computational model that can effectively predict the box-office gross revenue movie will be able to collect. This depends on a lot of factors like number of available theater screens, budget of the film, star cast, genre of the movie, MPAA rating (Motion Picture Association of America film rating system), and release year. This paper uses a combination of regression methods and specific genre-based method using the revenue data collected from IMDB and Rotten Tomatoes of past 15 years movies to estimate the gross revenue after its first weekend of release. This method can be used by movie producers and production studios who by looking at estimated values of revenue can take different steps on deciding the budget for things like marketing and promotion. Also can be used by movie theaters as they can also estimate their sale of tickets. If the estimated revenue is lower than expected, studios may increase their promotion budget and may even think of releasing the movie outside the domestic space. In these cases, studios may try to release it in more theaters and invest more in advertisement. The goal of this research paper is to propose a method that will be able to effectively estimate the box-office gross revenue for a movie using the public information available after its first weekend of release.

2   Related Work

A lot of factors that affect the revenue prediction model have been studied and used by different researchers in combination with multiple machine learning algorithms. Do Critics Reviews Really Matter? As mentioned in [1], Eliashberg and Shugan analyzed the impact of critic reviews on box-office success. They divided the role of critic into two dimensions, influencer, and predictor. Critic is said to be an influencer, when his or her reviews influence the box-office results. And to be a predictor when he or she is able to predict the success of movie based on reviews. Ravid in [2] concluded that box-office revenue increases with increase in positive reviews by reviewers. However, according to Reinstein and Snyder in [3], few critics have the power to manipulate the box-office revenue hence, only critic ratings cannot judge the financial success. Research work of Litman [4] used multiple linear regression to predict the box-office success. He concluded that star cast, genre of movie, MPAA rating, and release date are all determinants of the financial success of the film. Forswell in their work [5] used linear regression model on features like Opening Weekend revenue, budget of the movie and number of theater screens. Robert in [6] divided the feature set into three types; simple which is numeric only, complex that is numeric, and text and sentiment which includes all possible types. They used Logistic regression for classification. Simonoff in [7] used different parameters like

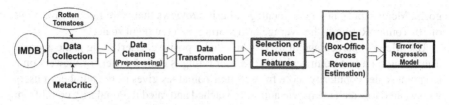

Fig. 1 Proposed methodology

production budget, whether movie is a sequel, star power, etc., that can predict the box-office revenue before release. Marton and Taha in [8] tried to do early prediction of movie success using the Wikipedia activity Big Data. Big Data analytics and computing is one of the current focus of research these days in multiple domains [9, 10]. Vitelli in [11] tried to create a set of features and did extract values from graphical properties of the actor–actor graph, actor–movie graph, and movie–movie graph relationships. Dursun Delen in [12] used neural network with features like MPAA rating, genre of movie, star value, and sequel importance for prediction of prerelease revenue. Work by Anast [13] related genre to movie attendance and concluded that violence and eroticism have a positive correlation with movie attendance. Unlike previous works Ryan as in [14] tried to estimate the foreign box-office revenue which depends on domestic success, language adaptability, cultural differences, MPAA rating, etc. Thorsten in [15] did a study whether the success of one feature can affect the effect of other on revenue or not.

Considering related research, we find that none of them did the genre-based analysis. Moreover, all features cannot be a determinant for all types of movies. So, this paper applies a genre-specific approach along with regression models to effectively estimate the box-office gross revenue. The proposed methodology is shown in Fig. 1.

3 Proposed Methodology

3.1 Data Collection

Movie related data was collected from the two most popular websites IMDB and Rotten Tomatoes for the past 15 years (year 2000–2015). Movies collected consisted of many genres namely Action, Adventure, Animation, Thriller, Horror, Sci-Fi, etc.

Field Information. Within the collected data, some of the general features were the title of movie, genre of movie, release date, budget of the movie(USD), number of screens it was released in the opening weekend, opening weekend revenue(USD), IMDB user rating, TomatoMeter, TomatoRating, UserMeter, UserRating, Popularity from IMDB, Rotten Tomatoes, and gross domestic revenue for the movie in USA.

Field Description. TomatoMeter is based on the published opinions of hundreds of film critics and is a trusted measurement of movie quality for millions of moviegoers. Meter represents the percentage of critic reviews that were rated above 5 out of 10. TomatoRating is the average critic rating on a ten point scale.

UserMeter is percentage of users who rated it positive. UserRating is the average user rating on a five point scale. IMDB user rating is done by users or viewers on a ten point scale. Popularity score from Rotten Tomatoes gives us the number of users who wanted to watch the movie and who watched and rated it. Popularity count from IMDB is the number of movie page views for that week.

3.2 Data Cleaning

Movies with insufficient information were removed from the dataset. Budget of the movies was in different currencies. Currency Conversions was done to convert all budget values into USD.

3.3 Data Transformation

Box-Office gross revenue is simply the total sum of the money collected by the sale of movie tickets. Average movie ticket prices have increased rapidly each year rising from 5.39 USD in 2000 to around 8.66 USD in 2016. To adjust the revenue data collected namely budget of the movie, opening weekend revenue, and box-office gross revenue, the change in ticket price was considered as an inflation parameter and all the revenue was changed according to ticket prices in 2016.

Standardization. Various features in the dataset are brought together to the same level or scale. This will make all the different features having different ranges to contribute proportionally to the final gross revenue. In fact, the ranges are unbounded, budget of the movie or box-office gross revenue has no maximum or minimum value. After performing standardization, all the features will have zero mean and unit variance. Each sample and feature is standardized using Eq. 1.

$$x_1 = \frac{x - mean}{Standard deviation} \tag{1}$$

4 Estimation Using Regression-Based Modeling

Mean absolute percentage error (MAPE), also known as mean absolute percentage deviation (MAPD) is the mean of the percentage error for each sample. Let A denote the actual value and F denote the predicted value. Let n number of test movies, then Eq. 2 specifies the calculation of MAPE.

$$MAPE = \frac{100}{n} \sum_{t=1}^{n} \left| \frac{A_t - F_t}{A_t} \right| \tag{2}$$

Assuming linear relationship between gross revenue and all the other features, multiple linear regression was used. Using this model and all the features, error (MAPE) was around 110%. Equation 3 describes the linear model where W is the weight and b is the bias.

$$Y = W1 * Budget + W2 * OpenWeek + W3 * Screens + W4 * userRating + \cdots + b \tag{3}$$

4.1 Training Split on the Basis of Genre

Each genre of movie shares a different relationship with each of the features. For example, budget of the movie maybe important for Action or Animation movies but not for a Horror movie. UserRatings or UserReviews has more role to play in the prediction model for movies which are of documentary or Horror type and lesser on Adventure or Comedy movies. Whole training set is split on the basis of genre of movie. Then feature selection algorithms are used to find the best set of features for each genre. This way, model will adapt to different genres of movies and use the features accordingly.

4.2 Feature Selection

Feature selection is used basically for two reasons:

- To avoid over fitting by reducing number of features and to improve generalization of model.
- To gain better understanding of features and their relationship to response variable.

Among various techniques best subsets regression guaranteed the best result in our work. It selects best set of features based on statistical criteria; Mean absolute percentage error (MAPE). Result obtained after best subsets regression method is given in Table 1. Here, OpenWeek refers to the opening weekend revenue and Popul refers to popularity. Result shows that the budget of the movie does not affect the gross revenue of horror, documentary, Sci-Fi, and history movies. Opening weekend is the most important feature that is present in all of the genres. Critic rating and popularity do not seem to affect the performance of a lot of movies. Only documentary, drama, history, and music were affected a bit. Multiple linear regression model is used for each genre of the movie. Here, training data is the set of movies released between the year 2000–2013 and test data is of the movies released from the year 2014–2015. Test set consists of 400 movies.

Table 1 Best combination and MAPE for each genre

Genre	Best combination	MAPE
Action	OpenWeek, Popul(IMDB)	46.45
Adventure	OpenWeek, budget	49.16
Animation	OpenWeek, budget	36.69
Drama	OpenWeek, budget, Popul(IMDB)	95.45
Comedy	OpenWeek, budget	50.26
Sci-Fi	OpenWeek	24.20
Romance	OpenWeek, budget, Popul(Rotten)	90.45
Music	OpenWeek, budget, UserMeter, Popul(Rotten)	60.70
Fantasy	OpenWeek, budget, screens, Popul(IMDB), UserRating	47.90
History	OpenWeek, Popul(IMDB), TomatoRating	62.38
Documentary	OpenWeek, Popul(IMDB), TomatoRating, UserRating	42.57
Horror	OpenWeek, Popul(Rotten)	23.47
Mystery	OpenWeek, budget, Popularity(IMDB), Popularity(Rotten)	49.89

5 Multi-genre Analysis

The movies in the test set were having more than one genre. To handle such type of data, linear model was run for each genre using the best set of features found using best subsets regression. Then final estimated box-office gross revenue is the arithmetic mean of gross revenue estimated by the model for each genre. Error (MAPE) was found to be 53.96%.

5.1 Training Split on the Basis of TomatoRating and Screens

After splitting the training set on the basis of genre, then it was split on the basis of TomatoRating and number of screens as shown in Fig. 2. Then linear regression model is run and error (MAPE) is calculated as shown in Table 2.

For TomatoRating, we observed there are two type of movies
Type1, High Critic Rating > 5.
Type2, Low Critic Rating ≤ 5.

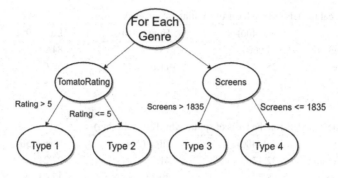

Fig. 2 Training split on the basis of TomatoRating and screens

Table 2 Linear regression result after split for test set

Split	All	High	Low
Rating	34.18 (400)	42.6 (244)	24.25 (156)
Screens	39.92 (400)	26.96 (210)	127.32 (190)

For Number Of Screens, we observed there are two type of movies:

Mean value of number of screens taken from the Training Set is 1835.

Type3, High Critic Rating >1835.

Type4, Low Critic Rating ≤ 1835.

5.2 Local Regression Models

Neighbor Search. For each test movie, nearest samples in the training set are found out. Two tuples are made on the basis of genre of movie like for Adventure movies, (Opening Weekend, Budget) will be the tuple. Now, Euclidean distance between the test tuple and each of the training data tuple is calculated and 50 nearest ones are considered to be the training set. Multiple linear regression and decision tree regression were run using these 50 neighbors as the training set. In Table 3, movies with higher number of screens (that is less than 1835) gave better result whereas in Table 4, movies with lower rating gave better result. But, on comparing the two tables, movies with higher rating performed better than movies with lower number of Screens.

On observing the two tables, model performs poorly on movies with low number of screens (that is less than 1835). To overcome this, two-way split is used. When the movie was released on screens > 1835, model should use the Type3 training set. When the movie is released on screens ≤ 1835, model should use split on the basis of TomatoRating and then use Type1 or Type2 training Set accordingly. This approach is shown in Fig. 3.

Table 3 Local models (on the basis of splitting by screens)

Algorithm	All (400)	High (210)	Low (190)
Linear regression	37.966	18.6	84.9
Decision tree regression	25.77	11.8	50.6

Table 4 Local models (on the basis of splitting by TomatoRating)

Algorithm	All (400)	High (244)	Low (156)
Linear regression	32.47	41.38	23.21
Decision tree regression	28.78	29.17	17.04

Fig. 3 Final approach

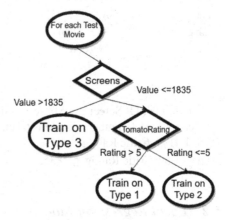

Using the approach in Fig. 3, local decision tree regression gave the best result. Error (MAPE) for the test set of 400 movies was found to be 24.76%.

6 Conclusion

Presented work performs genre-based splitting of training set and further division on the basis of number of screens. If the number of screens is less than mean value 1835, then we use Type 1 or Type 2 training set. Error is minimum when movies with higher number of screens (210 movies) were tested. This was as low as 11.8% Using this combined type of approach (Fig. 3) and Local decision tree regression algorithm, error (MAPE) for 400 test movies was 24.76%.

Table 5 Test on recent 5 releases of 2016

Name	Real gross	Predicted gross	MAPE (Percent)
Conjuring 2	102,461, 593 USD	117,240,783 USD	14.42
The Angry Birds	107,506,776 USD	145,956,013 USD	35.5
Deadpool	363,024,263 USD	358,836,741 USD	1.15
The Legend of Tarzan	126,585,313 USD	113,305,012 USD	10.49
The Jungle Book	363,995,937 USD	425,317,712 USD	16.8

7 Test on Recent Releases of 2016

Some recent releases of 2016 were tested and results are shown in Table 5.

References

1. Eliashberg, J., Shugan, S.M.: Film critics: influencers or predictors? J. Mark. 68–78 (1997)
2. Ravid, S.A.: Information, blockbusters, and stars: a study of the film industry. J. Bus. **72**(4), 463–492 (1999)
3. Reinstein, D.A., Snyder, C.M.: The influence of expert reviews on consumer demand for experience goods: a case study of movie critics. Working Paper, University of California-Berkeley and George Washington University (2000)
4. Litman, B.R.: Predicting success of theatrical movies: an empirical study. J. Pop. Cult. **16**(4), 159–175 (1983)
5. Apte, N., Forssell, M., Sidhwa, A.: Predicting Movie Revenue. CS229, Stanford University (2011)
6. Yoo, S., Kanter, R., Cummings, D.: Predicting Movie Revenue from IMDb Data. Stanford University (2011)
7. Simonoff, J.S., Sparrow, I.R.: Predicting movie grosses: winners and losers, blockbusters and sleepers. Chance **13**(3), 15–24 (2000)
8. Mestyn, M., Yasseri, T., Kertsz, J.: Early prediction of movie box office success based on Wikipedia activity big data. PloS One **8**(8), e71226 (2013)
9. Prasad, B.R., Agarwal, S.: Comparative study of big data computing and storage tools: a review. Int. J. Database Theory Appl. **9**(1), 45–66 (2016)
10. Prasad, B.R., Agarwal, S.: Stream data mining: platforms, algorithms, performance evaluators and research trends. Int. J. Database Theory Appl. **9**(9), 201–218 (2016)
11. Predicting Box Office Revenue for Movies: Matt Vitelli (2015)
12. Sharda, R., Delen, D.: Predicting box-office success of motion pictures with neural networks. Expert Syst. Appl. **30**(2), 243–254 (2006)
13. Anast, P.: Differential movie appeals as correlates of attendance. J. Q. **44**(1), 86–90 (1967)
14. de Silva, B., Compton, R.: Prediction of foreign box office revenues based on wikipedia page activity (2014). arXiv:1405.5924
15. Hennig-Thurau, T., Houston, M.B., Walsh, G.: Determinants of motion picture box office and profitability: an interrelationship approach. Rev. Manag. Sci. **1**(1), 65–92 (2007)
16. Internet Movie DataBase: http://www.imdb.com/
17. Rotten Tomatoes: https://www.rottentomatoes.com/

Classification of Short Text Using Various Preprocessing Techniques: An Empirical Evaluation

H. M. Keerthi Kumar and B. S. Harish

Abstract In recent decades, microblogs generate large volumes of data in the form of short text. Twitter has been one of the most widely used microblogging sites. Twitter data consist of noise due to shortness, which need to be preprocessed to find the accurate sentiment expressed by the user. The major challenges in short texts are the presence of noisy data like URLs, misspelling, slang words, repeated characters, punctuation, etc. To handle these challenges, this paper proposes to combine various preprocessing techniques with different classification methods as a tool for Twitter sentiment analysis. We evaluated the effect of noisy data like URLs, hashtags, negations, repeated characters, punctuations, stopwords and stemming. We use n-gram representation model to find the bindings and further applied support vector machine (SVM) and K-nearest neighbors (KNN) multi-class classifiers for sentiment classification. Experiments are conducted to observe the effect of various preprocessing techniques on Stanford Twitter Sentiment Dataset. The extensive experimental results are presented to show the effect of various preprocessing techniques to classify short texts.

Keywords Short text · Preprocessing · Support vector machine
K-Nearest neighbor · Classification

1 Introduction

In recent years, microblogs play a vital role in information sharing and communication. Microblog is a form of multimedia blogging that allows users to send brief text updates or micro-media such as photos or audio clips and publish them.

H. M. Keerthi Kumar (✉)
JSS Research Foundation, Mysuru, Karnataka, India
e-mail: hmkeerthikumar@gmail.com

B. S. Harish
Sri Jayachamarajendra College of Engineering, Mysuru, Karnataka, India
e-mail: bsharish@sjce.ac.in

© Springer Nature Singapore Pte Ltd. 2018
P. K. Sa et al. (eds.), *Recent Findings in Intelligent Computing Techniques*,
Advances in Intelligent Systems and Computing 709,
https://doi.org/10.1007/978-981-10-8633-5_3

Many micro-blog sites like Tumblr, Twitter, Posterous, FriendFeed, etc., are used for information sharing and communication. Twitter is one of the most popular and commonly used microblogging services. Twitters are accessible through website interface and numerous mobile devices. Millions of users are sharing information on various topics ranging from political debate, products, stock market, etc. On Twitter, users post and read messages are restricted to 140-characters, which are called "tweets" [5]. In tweets, users share their views and opinions known as "sentiment", in the form of text, photos, and audio clips, where text shares a major part in communication. These tweets hold the key for determining sentiment of a population. By analyzing sentiment on tweets, we can identify the kind of emotions, mainly as positive, negative or neutral.

Sentiment analysis is treated as a classification task as it classifies the orientation of tweets into different classes or polarity [3, 21]. Sentiment classification methods can be classified into machine learning, lexicon based methods, and linguistic methods [18]. Many researchers [7, 12, 13] have claimed that lexicon-based methods and linguistic methods do not perform well on sentiment classification, due to nature of an opinionated text which requires more understanding of text. However, the occurrence of some keywords could be the key for an accurate classification [10]. In sentiment analysis, machine learning methods are used to train an algorithm based on a set of keywords or features, which describes the polarity and then test on another set whether it is able to detect the keywords and give the accurate classification. Machine learning classifiers such as Naive Bayes (NB), maximum entropy (ME), and support vector machine (SVM) are used in [13] for sentiment classification.

Twitter sentiment analysis using machine learning techniques encompasses tasks such as preprocessing, feature extraction and selection, representation, classification or clustering and evaluation. In tweets, users make spelling mistakes, slang words and use emoticons for expressing their views. Moreover, tweets contain a large amount of noise data, such as URLs, punctuation, etc. [3], which need to be preprocessed. In this paper, we are exhibiting the impact of preprocessing in determining sentiment on tweets. However, this work concentrates more on conventional preprocessing techniques used to eliminate noisy data which do not contribute enough to Twitter sentiment classification. Although few works concentrate on twitter preprocessing techniques, still generic solution needs to be developed for efficient classification. In this work, we focus on exploring preprocessing techniques to uplift the performance of sentiment classification, including the effect of URLs, usernames, hashtags, negations, repeater characters normalization, punctuation, stopword removal and stemming. The experimentations are performed rigorously on Stanford Twitter Sentiment Dataset [1] to show that sentiment accuracy increases when URLs are removed, username elimination, Hashtag content retained, negation transformation and repeated character normalization. We also represented feature space in unigram, bigram, and trigram representation and further applied support vector machine (SVM) and K-nearest neighbors (KNN) multi-class classifiers for sentiment classification.

The rest of the paper is organized as follows: Sect. 2 explains related work on sentiment analysis on Twitter data. Section 3 portrays the methodology used in this paper. Section 4 contains experimentation and related results along with the discussion. Finally, we conclude our work with outlining future work in Sect. 5.

2 Related Work

Over the time, microblogs are used for expressing sentiments on an event or topic. Twitter is one of the most commonly used micro-blog to express sentiment over the current issues. Many researchers concentrated their studies to understand sentiments expressed in twitter. Twitter contain a large amount of noisy data, such as URLs, user names, punctuations symbols, etc. These characters make sentiment classification a bit difficult and challenging and thus preprocessing plays a vital role in Twitter sentiment analysis.

Based on the research work by many researchers, it has been proved that the preprocessing is the main aspect in sentiment analysis [4, 5, 10, 16, 20]. To deal with these, many researchers proposed various preprocessing techniques along with the algorithm based on supervised, semi-supervised, and unsupervised machine learning approaches with lexicon-based approaches. Bao et al. [4] described the effectiveness of preprocessing techniques on Twitter data. The method uses unigram, bigram representation with Liblinear classifier to classify the data into positive and negative classes. The experiment shows that noisy data like URLs, negation transformation and repeated characters normalization have a positive impact on classification accuracy while stemming has a negative impact. Singh et al. [16] brief the role of text preprocessing in Twitter sentiment analysis. The method explains text normalization as the process of purification of tweets, where each step eliminates the noise data. This method defines the significance and sentiment strength of slang and unidentified words in tweets. Support vector machine (SVM) classifier is used to evaluate and measure the impact of preprocessing on sentiment classification.

In [10], Haddi et al. describe the sentiment analysis on online movie reviews. The different preprocessing methods are used to reduce noise in the text. The results of preprocessing techniques show that data transformation and filtering can significantly enhance the performance of SVM classifier on sentiment identification. Uysal and Gunal [20] explore the impact of preprocessing on text classification. Sentiment analysis was conducted on Turkish and English languages by choosing appropriate preprocessing task such as tokenization, stopword removal, lowercase conversion, and stemming. By employing preprocessing, a significant improvement was found on classification accuracy whereas inappropriate combinations resulted in degrading the accuracy. Tripathy et al. [19] used machine learning techniques such as naive Bayes (NB), maximum entropy (ME), SVM, and stochastic gradient descent (SGD) classification using n-gram approach. Unigram, bigram, trigram models and their combinations are used for classification on IMDb movie review dataset. The accuracy of different methods is examined in order to access their performance on the basis of parameters such as precision, recall, f-measure, and accuracy. Agarwal et al. [3] applied novel approaches to preprocess tweets. The methods replace URL, target name, negations, and repeated characters with appropriate terms. The results of their experiment illustrated that appropriate text preprocessing methods can significantly increase the accuracy of the classifier. Saif et al. [15] described the role of preprocessing to reduce sparsity issue in twitter sentiment analysis. Experiment results illustrated that appropriate text preprocessing techniques can significantly reduce sparsity and increases the classification accuracy.

In literature, many researchers [3, 10, 15, 16, 19, 20] described the role of preprocessing techniques by selecting the appropriate combination of techniques to improve the classification performance. The Twitter data consists of URLs, slang words, misspellings, punctuation, and abbreviations which make preprocessing a challenging task. By eliminating the above noisy data, we can reduce misclassification in sentiment analysis. Although various preprocessing techniques exist in the literature, the problem of sentiment classification on short text is still challenging, with no generic solution and remains an open research area. In this work, we perform extensive experimentation to show the impact of preprocessing techniques on Stanford Twitter Sentiment Dataset.

3 Methodology

The text preprocessing is the initial step in sentiment analysis, where noisy data are eliminated from the dataset. Here, we apply various preprocessing techniques to reduce the noisy data. We adopt the following process for Twitter sentiment analysis.

3.1 Tweet Preprocessing

In this step, we are removing or replacing noisy data in each tweet, which do not contribute much for sentiment classification. We are using eight traits to process tweets, namely URL removal, username replace with white space, hashtag removal, handle negation, characters normalization, punctuation removal, stopword elimination and stemming.

URL removal: In tweets, the user posts URL along with text to provide supporting information about the text like "http://bit.ly/IMXUM". These URL links become noisy data during sentiment analysis. We are eliminating URL links in each tweet and replacing it with a space.

Username: There are usernames like "@LATimesautos", "@XPhile1908", that start with symbol "@", the symbols indicate the username or target person. Here, we are concentrating our work toward finding the sentiment on each tweet and not on any targeted persons. The contribution of username is less on sentiment analysis, so we replace all username with a white space.

Hashtags: Hashtags marked with the symbol "#", which means that tweets are associated with the particular topic and also consists opinion expressed in the tweets. We removed only symbol "#", retaining the contents.

Handling negation: Negations play a vital role in sentiment classification, the negative word, e.g., "not", "n't", etc., in which co-occurrence with other word changes the orientation of text into different polarity. Considering the effect of negation, we applied abbreviations for short terms like "don't", "can't", "n't", etc., terms to "do not", "cannot", "not", etc., words, respectively, which changes the sentiment of the tweet.

Character normalization: Words with consecutive characters, e.g., "looovvv-veee", are more common in tweets and users tend to use this way to express their opinion or sentiments. Thus, it is necessary to deal with these words to make them more formal. Here, consecutive characters mean repeated characters more than 3 times in a word. This needs to be normalized to give a formal representation. Here, we replace repeated characters more than three times to single characters.

Punctuation: We removed all the punctuations symbols like ",", " ' ", "$", "?", "!", etc., from the dataset, which does not contribute to the sentiment of tweets.

Stopwords: Stopword refer to most common words used in the tweets like a, all, am, an, and, any, are, etc. [9]. We eliminated these English stopwords, which contribute less to the sentiment of tweet.

Stemming: Stemming is used to achieve feature reduction. Stemming is the process of bend words to their root or base form, by eliminating words ending with "er", "ing", "ed", etc. [16]. Here, we apply stemming to reduce the feature space.

Emotion symbol is used to express sentiment in tweets, e.g., ":(, =]" means sad or negative emotion. Here in our work, we are considering only the sentiment related to text so we replace these emotions symbol with white space. We eliminated all symbols, digits, single character, and other nonalphabetic symbols.

In this work, we are demonstrating the role of preprocessing in Twitter sentiment analysis by observing the impact on sentiment classification accuracy. The combinations of techniques are applied to identify the impact of methods in sentiment classification.

3.2 Representation

Preprocessed tweets are represented using n-gram representation model [8]. N-gram is a contiguous sequence of n number of words. In this case, each n-gram is one features space whose dimension is equal to the number of n-grams [11]. When $n = 1$, it represents unigram, where each word represents a feature. Similarly, for $n = 2$, 3 represent bigram, trigram, respectively. Weight values are associated with each pair using the term frequency (TF) scheme. The unigram, bigram, and trigram feature representations are used to represent the preprocessed tweets. We have used TF to find the number of terms appeared in each tweets.

3.3 Classification

In this work, we employ multi-class support vector machine (SVM) and K-nearest neighbor (KNN) to train the preprocessed tweets. The SVM algorithm has several advantages, which are important for learning a sentiment classifier from a large Twitter dataset [6, 17]. SVM is a widely used classifier in sentiment analysis tasks. It can effectively conduct classification task in high-dimensional feature space [14]. In this paper, we used SVM for multi-class problem to classify the tweet into positive, negative or neutral. On the other hand, KNN is also used as nonparametric method used for pattern classification. KNN classification is based on the class of their closest neighbors, most often, more than one neighbor is taken into consideration, and here K denotes the number of neighbors taken into account in determining the class [2].

4 Experimental Results and Discussion

In this section, we explore the results obtained when various types of preprocessing techniques are applied to Stanford Twitter datasets.

4.1 Dataset Description

The experiment was carried out on Stanford Twitter Sentiment Dataset [16]. The dataset is in English language which consists of 498 tweets, where 182 positive, 177 negative, and 139 neutral tweets. Each tweet comes with the labels: positive, negative, and neutral.

4.2 Experimentation

In the experiment, we carried out step-by-step process to evaluate the impact of preprocessing methods on sentiment classification. The Stanford Twitter Sentiment Dataset of labeled 498 tweets are taken randomly for training and testing purpose. The equal propositions of positive, negative, and neutral tweets are taken for training and testing data. Experimentation was conducted using the combination of 50 training and 50 testing, 60 training and 40 testing, and 70 training and 30 testing bases, respectively. For classification purpose, we used multi-class SVM and KNN classifier for multi-class problem. The classification accuracy is considered as a metric to evaluate the individual and combination of various preprocessing techniques. The experimentation was conducted using R Studio Version 0.99.903 and R-3.1.3 Language to perform sentiment analysis on Stanford Twitter Sentiment Dataset.

In the first step, we removed all the URLs in each tweet and represented using n-gram feature representation. The representation includes unigram, bigram and trigram on each process. Table 1 shows the effect of URLs on the classification accuracy. The result shows KNN classifier on unigram provide better accuracy when compared to bigram and trigram feature representation. The SVM classifier gives better accuracy in bigram when training and testing ratio is increased.

In next method, we removed username which starts with "@" symbol. From Table 2, we can see the effect of username (@) removal from the dataset. The results show the accuracy of SVM classifier increases when the username is removed. The results are obtained for unigram, bigram, and trigram feature representation.

Table 1 Classification accuracy for URL removal

Classifiers	Training: testing	Unigram	Bigram	Trigram
SVM	50:50	**68.80**	63.60	36.40
	60:40	**73.00**	68.50	47.00
	70:30	75.40	**85.40**	35.70
KNN	50:50	**77.60**	66.40	65.20
	60:40	**87.00**	78.50	77.50
	70:30	**88.00**	76.10	75.40

Table 2 Classification accuracy for username removal

Classifiers	Training: testing	Unigram	Bigram	Trigram
SVM	50:50	**76.00**	74.40	50.80
	60:40	68.50	**78.00**	37.50
	70:30	64.90	**73.50**	72.80
KNN	50:50	**76.50**	65.60	62.80
	60:40	**83.50**	71.00	69.00
	70:30	**86.09**	76.80	76.15

Continuing the process of preprocessing, we removed Hashtag "#" symbol and retained the content. Table 3 shows the impact of hashtags (#) removal on classification accuracy. The result increases in the accuracy using KNN classifier for unigram and bigram feature representation.

Further, we carried out by combining first two techniques, i.e., URL and username removal. Table 4 presents the results of URLs and username removal from the dataset. We get better results after bigram features are affiliated with feature space.

To continue experimentations by reducing features from the original feature space, we applied three methods jointly, i.e., URL, username, and hashtags. Table 5 shows the result of a combination of techniques, which gives average accuracy when compared to Tables 1, 2 and 3.

Table 3 Classification accuracy for Hashtag (#) removal

Classifiers	Training: Testing	Unigram	Bigram	Trigram
SVM	50:50	66.00	**70.80**	54.00
	60:40	67.00	**69.00**	50.50
	70:30	**68.80**	68.20	65.50
KNN	50:50	**77.60**	72.40	71.60
	60:40	**79.00**	67.00	67.00
	70:30	**88.70**	84.10	80.39

Table 4 Classification accuracy for URL and username removal

Classifiers	Training: testing	Unigram	Bigram	Trigram
SVM	50:50	**74.00**	69.20	37.60
	60:40	69.00	**77.00**	36.50
	70:30	70.10	**81.40**	72.84
KNN	50:50	**77.20**	64.40	60.80
	60:40	**81.00**	70.00	68.00
	70:30	**80.70**	74.83	76.15

Table 5 Classification accuracy for URL and username removal and hashtags

Classifiers	Training: testing	Unigram	Bigram	Trigram
SVM	50:50	**72.00**	69.20	57.20
	60:40	74.00	**75.00**	63.00
	70:30	74.10	**79.40**	70.86
KNN	50:50	**75.60**	65.20	64.00
	60:40	**79.50**	76.00	73.00
	70:30	**86.09**	77.40	77.48

In consideration of the effect of negation, we applied abbreviations for short terms like "don't", "can't", "n't", etc., terms to "do not", "cannot", "not", etc., words respectively, which changes the sentiment of the tweet. Table 6 shows the results of handling negations with other three methods jointly. The experimental results increases when negations are applied on unigram, bigram representation of dataset.

Words with consecutive characters, e.g., "looooooooovvvvvveee", are more common in tweets, and users tend to use this way to express their opinion or sentiments. Thus, it is necessary to deal with these words to make them more formal. Here consecutive character means repeated characters more than three times in a word. This needs to be normalized to give formal representation. Table 7 shows the result after performing normalization of characters with other methods jointly.

In the next set of experiments, we eliminated the punctuation's present in the corpus and combined it with previous methods. Table 8 presents the results after removing punctuations in the dataset. The punctuations are used in tweets to express the strong feeling toward the polarity but here we are considering tweets related to positive, negative, or neutral. In our work, punctuation becomes noisy data, after eliminating punctuation there is an increment in the accuracy of unigram and bigram representation.

To continue with reducing features from the original feature space, we introduced stopword removal to the dataset. Table 9 shows the effect of stop words removal along with previous techniques. There is a sharp decline in classification accuracy of KNN classifier. The results illustrate that stop words contribute less toward sentiment classification, when applied jointly with other preprocessing methods.

Table 6 Classification accuracy for URL and username removal, hashtags, and negation

Classifiers	Training: testing	Unigram	Bigram	Trigram
SVM	50:50	**74.80**	69.20	38.00
	60:40	**78.50**	74.50	45.00
	70:30	80.79	**86.75**	54.30
KNN	50:50	**74.00**	64.00	63.60
	60:40	**86.50**	70.50	67.50
	70:30	**90.00**	79.47	78.80

Table 7 Classification accuracy for URL and username removal, hashtags, negation, and character normalization

Classifiers	Training: testing	Unigram	Bigram	Trigram
SVM	50:50	**66.80**	54.40	64.00
	60:40	74.50	**78.50**	62.00
	70:30	80.13	**81.45**	79.40
KNN	50:50	**73.60**	62.00	61.20
	60:40	**81.00**	72.50	70.50
	70:30	**82.78**	76.15	76.80

Table 8 Classification accuracy obtained for URL and username removal, hashtags, negation, and character normalization, punctuation

Classifiers	Training: testing	Unigram	Bigram	Trigram
SVM	50:50	**72.40**	71.20	52.00
	60:40	**80.00**	79.50	68.50
	70:30	74.83	**80.13**	76.82
KNN	50:50	**76.80**	65.60	63.60
	60:40	**82.00**	74.00	73.50
	70:30	**90.00**	80.79	80.13

Table 9 Classification accuracy for URL and username removal, hashtags, negation and character normalization, punctuation, and stopwords removal

Classifiers	Training: testing	Unigram	Bigram	Trigram
SVM	50:50	**74.40**	60.40	36.00
	60:40	**79.00**	76.00	56.50
	70:30	**82.11**	80.13	65.56
KNN	50:50	**75.60**	68.00	68.00
	60:40	**80.50**	75.50	75.00
	70:30	**88.07**	85.43	76.80

Table 10 Classification Accuracy for URL and username removal, hashtags, negation and character normalization, punctuation, stopwords removal and steaming

Classifiers	Training: testing	Unigram	Bigram	Trigram
SVM	50:50	**80.00**	68.00	20.40
	60:40	**83.50**	61.00	36.00
	70:30	**88.07**	82.78	36.40
KNN	50:50	**76.50**	70.40	68.00
	60:40	**87.50**	77.00	76.00
	70:30	**84.70**	82.11	79.47

Stemming is also used to achieve feature reduction. Table 10 reveals the result of stemming with other processes jointly. The result shows increase in accuracy in both unigram and bigram representation. The experimental results indicate that the preprocessing is a basic step for sentiment classification. The step-by-step processes of applying preprocessing methods increases the accuracy of classification.

4.3 Discussion

The preprocessing techniques are applied to eliminate noisy data and normalize the dataset. In this work, we applied eight text preprocessing methods to normalize the corpus. Each method exhibits a substantial amount of effects on the classification accuracy. Tables 1, 2, 3 shows the results related to URL, username, and hashtag removal, respectively. The result shows that there is an increase in accuracy when we apply unigram, bigram and classify using SVM and KNN classifiers. From Tables 4, 5, 6, 7, 8 and 9, we applied the combination of various preprocessing techniques to normalize the dataset. The results show a slight increase as well as a decrease in the classification accuracy on the classifiers with respective n-gram representation. Table 10 gives the overall result of various preprocessing techniques applied on the dataset. The overall result illustrates the performance of sentiment classification increases when URL removal, username replace with white space, hashtag removal, negation, character normalization, punctuation removal, stopword elimination and stemming are applied. We observe that various preprocessing techniques clearly indicate an increase in performance of the classifiers with unigram and bigram representations.

5 Conclusion and Future Work

In Twitter, sentiment analysis has become a recent and meaningful topic for researchers. The length limitation, various topic discussion, informal language, slang words and rich in symbols, all these characters of tweet make sentiment analysis a challenging. In this paper, we conducted a series of experimentation to verify the effectiveness of various preprocessing techniques on Stanford Twitter Sentiment Dataset. We used preprocessing techniques like URL removal, username replace with white space, hashtag removal, negation handling, character normalization, punctuation removal, stopword elimination and stemming. We demonstrated the role of various preprocessing techniques in Twitter sentiment classification.

In future, we would like to incorporate natural language processing (NLP) based text preprocessing techniques like lemmatization techniques, part-of-speech (POS) tags, etc., for topic-based sentiment analysis and also include different feature selection methods which enhance the classification performance.

References

1. http://help.sentiment140.com/for-students/
2. Adeniyi, D., Wei, Z., Yongquan, Y.: Automated web usage data mining and recommendation system using k-nearest neighbor (knn) classification method. Appl. Comput. Inform. **12**(1), 90–108 (2016)
3. Agarwal, A., Xie, B., Vovsha, I., Rambow, O., Passonneau, R.: Sentiment analysis of twitter data. In: Proceedings of the Workshop on Languages in Social Media, pp. 30–38. Association for Computational Linguistics (2011)
4. Bao, Y., Quan, C., Wang, L., Ren, F.: The role of pre-processing in twitter sentiment analysis. In: International Conference on Intelligent Computing, pp. 615–624. Springer (2014)

5. Bhuta, S., Doshi, A., Doshi, U., Narvekar, M.: A review of techniques for sentiment analysis of twitter data. In: 2014 International Conference on Issues and Challenges in Intelligent Computing Techniques (ICICT), pp. 583–591. IEEE (2014)
6. Chang, C.C., Lin, C.J.: LibSVM: a library for support vector machines. ACM Trans. Intell. Syst. (TIST) **2**(3), 27 (2011)
7. Ding, X., Liu, B., Yu, P.S.: A holistic lexicon-based approach to opinion mining. In: Proceedings of the 2008 International Conference on Web Search and Data Mining, pp. 231–240. ACM (2008)
8. Fusilier, D.H., Montes-y Gomez, M., Rosso, P., Cabrera, R.G.: Detecting positive and negative deceptive opinions using pu-learning. Inf. Process. Manage. **51**(4), 433–443 (2015)
9. Ghag, K.V., Shah, K.: Comparative analysis of effect of stopwords removal on sentiment classification. In: 2015 International Conference on Computer, Communication and Control (IC4), pp. 1–6. IEEE (2015)
10. Haddi, E., Liu, X., Shi, Y.: The role of text pre-processing in sentiment analysis. Procedia Comput. Sci. **17**, 26–32 (2013)
11. Lima, A.C.E., de Castro, L.N., Corchado, J.M.: A polarity analysis framework for twitter messages. Appl. Math. Comput. **270**, 756–767 (2015)
12. Melville, P., Gryc, W., Lawrence, R.D.: Sentiment analysis of blogs by combining lexical knowledge with text classification. In: Proceedings of the 15th ACM SIGKDD International Conference on Knowledge Discovery and Data Mining, pp. 1275–1284. ACM (2009)
13. Pang, B., Lee, L., Vaithyanathan, S.: Thumbs up?: sentiment classification using machine learning techniques. In: Proceedings of the ACL-02 Conference on Empirical methods in Natural Language Processing-Volume 10, pp. 79–86. Association for Computational Linguistics (2002)
14. Ren, Y., Wang, R., Ji, D.: A topic-enhanced word embedding for twitter sentiment classification. Inf. Sci. **369**, 188–198 (2016)
15. Saif, H., He, Y., Alani, H.: Alleviating data sparsity for twitter sentiment analysis. In: CEUR Workshop Proceedings (CEUR-WS. org) (2012)
16. Singh, T., Kumari, M.: Role of text pre-processing in twitter sentiment analysis. Proced. Comput. Sci. **89**, 549–554 (2016)
17. Smailovi_c, J., Gr_car, M., Lavra_c, N., _Znidar_si_c, M.: Stream-based active learning for sentiment analysis in the _nancial domain. Information Sciences 285, 181–203 (2014)
18. Thelwall, M., Buckley, K., Paltoglou, G.: Sentiment in twitter events. J. Am. Soc. Inform. Sci. Technol. **62**(2), 406–418 (2011)
19. Tripathy, A., Agrawal, A., Rath, S.K.: Classification of sentiment reviews using n-gram machine learning approach. Expert Syst. Appl. **57**, 117–126 (2016)
20. Uysal, A.K., Gunal, S.: The impact of preprocessing on text classification. Inf. Process. Manage. **50**(1), 104–112 (2014)
21. Zainuddin, N., Selamat, A.: Sentiment analysis using support vector machine. In: 2014 International Conference on Computer, Communications, and Control Technology (I4CT), pp. 333–337. IEEE (2014)

Named Entity Recognition in Text Documents Using a Modified Conditional Random Field

G. Veena, Deepa Gupta, S. Lakshmi and Jeenu T. Jacob

Abstract The Named Entity Recognition in documents is an active and challenging research topic in text mining. The major objective of our work is to extract a phrase from the sentence and classify this phrase to one of the predefined named entities. The proposed system works in two layers, in the first phase each and every word in the phrase is tagged using word feature extraction approaches. In the second phase the model recognizes named entities in the phrase level using Modified Conditional Random Field. This work identifies four classes of entities such as Person, Organization, Location and Other. Our algorithm first parses the text document and identifies the sentence structure. From this sentence structure concepts are extracted. In this work the feature extraction module make use of the yahoo Geoplanet Web service for identifying the location. We have created person ontology of all available Indian names to check whether a word is name or not. Inorder to check whether the word is organization or not we have used a database with company name indicators. Finally, our MCRF assign a label to the tagged phrase.

Keywords Named entity identification · Text · MCRF · NER
Geoplanet

G. Veena (✉) · S. Lakshmi · J. T. Jacob
Department of Computer Science & Applications, Amrita Vishwa Vidyapeetham,
Amrita University, Kollam, India
e-mail: veenag@am.amrita.edu

S. Lakshmi
e-mail: laxmimaaalu@gmail.com

J. T. Jacob
e-mail: jeenutjacob@gmail.com

D. Gupta
Department of Mathematics, Amrita Vishwa Vidyapeetham Amrita University,
Bengaluru, India
e-mail: deepagupta.verma@gmail.com

© Springer Nature Singapore Pte Ltd. 2018   31
P. K. Sa et al. (eds.), *Recent Findings in Intelligent Computing Techniques*,
Advances in Intelligent Systems and Computing 709,
https://doi.org/10.1007/978-981-10-8633-5_4

1   Introduction

Text Mining has a wide range of history in the area of Artificial Intelligence and Machine Learning. Named Entity Recognition is one of the text analysis processes in Text Mining. Within the past few years lots of research work has been achieved in the Named Entity Recognition category. But these formal works were focused on the statistical analysis of words using some contextual clues. The model proposes an innovative way to depict the concept in the phrase level. The Triplet Generation [1] and Belief Network generation [1] are the first phase of the proposed system. We classified each node in the Belief Network into predefined categories such as person, organization, location and other. Our algorithm first parses the text document and identifies the sentence structure. From this sentence structure concepts are extracted and each word in the concepts is tagged using the Word Feature Extraction approaches. In the proposed system makes use of the yahoo Geoplanet Web service [2] for identifying the location. It consists of a geographical hierarchy of the countries. We have created a Trie data structure called person ontology consists of all available Indian names to check whether word is name or not. The model containing a database with company name indicator inorder to check whether the word is organization or not. This database consists of company name indicators such as Ltd., Inc, Corporation etc. Next phase is the probability calculation of tagged words using a Modified Conditional Random Field and labeling a phrase with suitable Named Entity.

This paper describes the related works based on NER and proposed solutions in Sect. 2 and Sect. 3 respectively. Section 4 consists of experiments and result obtained through this research followed by a conclusion and future work.

2   Related Works

The Named Entity Recognition methodologies are specifically divided into two categories, handcrafted rules and learning based methods [3]. Techniques primarily based on handcrafted rules consist of designing and implementing lexical syntactic extraction patterns for sequence labeling. They employ existing records facts together with dictionaries that can often identify named entities [4]. The learning based approach uses machine learning techniques to perform entity identification and its classification [5, 18, 19].

In HMM [6], a model for computing the likelihood of words occurring within predefined name class called statistical bigram language model [7]. Every word is represented by a state in the bigram model, and there is a probability associated with every transition to the current word. Then the parameters are trained to maximize the joint likelihood of training sets. The limitation of this approach is that we have to train the large dataset to calculate the parameters and it is only suitable

Fig. 1 Sequence labeling problem of HMM

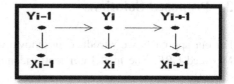

Fig. 2 Sequence labeling problem of MEM

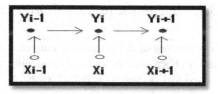

for small level sequence labeling problem. Figure 1 shows sequence labeling problem of HMM. Here Y is the label and X is the observation sequence.

Maximum Entropy Model is a learning based directed graphical model for sequence labeling For labeling a specific observation, make use of previous label and current observation [8]. The strategy concentrates on maximizing the entropy of information and generalizing the maximum amount as possible for the training data. The disadvantage of ME Model is that they probably suffer from the "label bias problem" [9] where states with low entropy transition distributions "effectively ignore their observations". Figure 2 shows sequence labeling problem of MEM.

Conditional Random Fields is a variant of Markov Random Field computing joint probability based on certain conditional approach [10]. For Named Entity Recognition problem a linear CRF is used for calculating maximum-likelihood based on the clique generated [11]. The model computes a score based on the set of feature function recognizing the label for an observation. Figure 3 shows general structure of CRF.

Now a days several open NLP tools are available such as Stanford Core NLP Suit [12], Apache OpenNLP [13], Natural Language Toolkit [14] etc. But the main limitations within these all are most of them are never identifying a local place or a local name. To overcome these all we introduce a Modified Conditional Random field for phrase labeling.

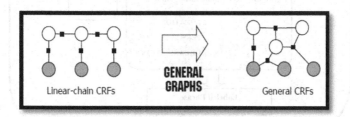

Fig. 3 Sequence labeling problem of CRF

3   Proposed Solutions

For our approach, we introduce an innovative Named Entity Recognizer based on Modified CRF. The model efficiently identifies the named entity in phrase level (Fig. 4).

3.1   Document to Text Converter

The purpose of this phase is to convert the input document into the text format. We use an open source document converter to convert the input document. The converted text document is redirected to the Pre-processing phase.

3.2   Pre-processing

The Pre-processing phase [1] describes the generation of verb-argument structure and concept extraction. This section involves document cleaning and Part of speech Tagging.

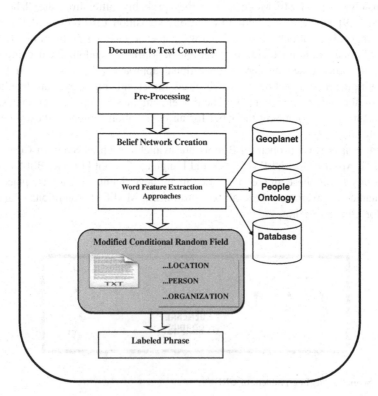

Fig. 4 System architecture

3.3 Belief Network Creation

In the Belief Network Creation [1, 15, 16] a directed acyclic graph is generated which represents subject, object and the relationship between each other. The previously generated verb-argument structure is processing in this phase. Each node in the Belief Network contains the subject or object.

3.4 Word Feature Extraction Approaches

In word feature extraction phase we have to label each word in the phrase. We introduce certain new approaches to extract the word features like Geoplanet for location tagging, people ontology for person tagging and a database containing company name indicators for organization tagging.

Each word in the phrase is first redirected to the location checking. We use Geoplanet for location checking, which is a wider system contains all the geographical ontology. Geoplanet uses a hierarchical model for locations that provides both vertical consistency and horizontal consistency of location geography. If the location is valid the algorithm labeled the word as LOC. Additional information extraction is also possible using Geoplanet such as extracting the location hierarchy and location features etc. If the identifier fails to extract the location, it is redirected to the person identifier algorithm.

The person identifier makes a look up in the knowledge base, check it is a valid name or not. We create a Trie data structure called people ontology, which can insert and find strings much faster than set and hash table. The Trie structure contains all available Indian names.

The person identifier algorithm makes a look up in the Trie data structure for a particular name. If the name is in the people ontology then the word is labeled as PER. The model uses a database for identifying the organization name. The database consists of a set of lists containing company indicators and used as background knowledge, instead of predetermined and flat set of lists. If it is a valid name, the identifier gives the organization tag ORG to the corresponding word.

3.5 Modified Conditional Random Field

In Named Entity Recognition the linear chain CRF [10] computes the joint probability based on current label, previous label and current observation.

$$P(y|x, \lambda) = \frac{1}{Z(x)} \exp\left(\sum_j \lambda_j F_j(y, x)\right) \tag{1}$$

where y is the sequence label and x is the observation sequence for define the probability of a particular label given observation. Here λ_j is the parameter estimated from the training data and F_j is the set of real valued feature functions. $Z(x)$ is the normalization factor.

In Modified Conditional Random Field, for a given phrase s the model computes score of a label l given S by adding up the weighted feature over all words in the phrase.

$$\text{Score}(l|S) = \sum_{j=1}^{m} \sum_{i=1}^{n} \lambda_j F_j(S, i, l_i, l_{i-1}) \tag{2}$$

l_i is the label in the position i and l_{i-1} is label at the previous position. The proposed model reducing the number of feature functions.

Finally, the model transforms the scores into probabilities by exponentiation and normalizing.

$$P(l|s) = \frac{\exp[\text{Score}(l|s)]}{\sum_{l'} \exp[\text{Score}(l'|s)]} = \frac{\exp\left[\sum_{j=1}^{m} \sum_{i=1}^{n} \lambda_j F_j(S, i, l_i, l_{i-1})\right]}{\sum_{l'} \exp\left[\sum_{j=1}^{m} \sum_{i=1}^{n} \lambda_j F_j(S, i, l'_i, l'_{i-1})\right]} \tag{3}$$

The example feature functions are,

To label the ORGANIZATION tag we define only one feature function.

$f_1(S,i, l_i, l_{i-1})=$ *if ORG presents anywhere in a sentence; 0 otherwise.*

To label the PERSON tag we define three feature functions.

$f_2(S, i, l_i, l_{i-1})=$ *1 if $l_{i-1}=$ PER and $l_i=$ PER; 0 otherwise.*
$f_3(S, i, l_i, l_{i-1})=$ *1 if $l_{i-1}=$ PER and $l_i=$ LOC; 0 otherwise.*

$f_4(S, i, l_i, l_{i-1})=$ *1 if $l_{i-1}=$ LOC and $l_i=$ PER; 0 otherwise.*

The model defines a feature function for label the Location tag to the phrase.

$f_5(S, i, l_i, l_{i-1})=$ *1 if $l_{i-1}=$ LOC and $l_i = $ LOC; 0 otherwise.*

Figure 5 Shows MCRF which is characterized by the following,

Fig. 5 Modified conditional random field

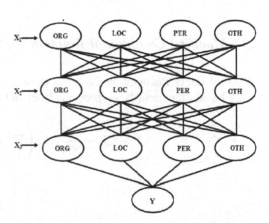

For each and every words present in the observation the model assigns a tag based on its features. Our model creates a clique [11] to represent this tagged phrase. Then the score of this tagged phrase is computed using Eq. (2).Then we have used Eq. (3) for computing the probability.

Algorithm 1 Modified CRF Algorithm.

```
Input: A phrase S.
Output: Labelled phrase.
 Let X₁, X₂, X₃..... Xₙ be the words present in S
      for each word X present in S
            if
                  Isloc(X)==true;
                    y =  LOC;
                     else if
                                Isper(X)==true;
                    y =  PER;
                     else if
                                Isorg(X)==true;
                    y =  ORG;
            else
                    y = OTH;
 Let Y₁, Y₂, Y₃..... Yₙ be the label associated with S.
      Calculate Score_LOC(L);Calculate Score_PER(L);
      Calculate Score_ORG(L);
      if
      Score_LOC(L)!= 0
      Print "Location";
      else if
      Score_PER(L)!= 0
                        . Print "Person";
      else if
       Score_ORG(L)!= 0
                           Print "Organization";
      else
                           print "Other";
            end if
             end if
        end for
 end
```

---

4   Experiments and Results

The experimental setup we collected 100 articles from the site https://www. sciencedaily.com and generated a Probabilistic Graphical Model and a Historical Dataset. The model consists of certain word feature extraction approaches such as

Fig. 6 Belief network

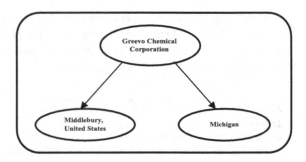

Geoplanet, people ontology, and an Org_Dictionary. Our experimental set up uses the site https://developer.yahoo.com/geo/geoplanet/ for Location name tagging. Yahoo! Geoplanet helps bridge the gap between the real and virtual worlds by providing an open, permanent, and intelligent infrastructure for geo-referencing data on the Internet.

Example **1: Greevo Chemical Corporation is located in Middlebury, United States. The company was founded by Michigan** From the above example first the conceptual structure extracted then Belief network is generated using Extended DB scan Algorithm [1] (Fig. 6).

Here the node contains the word **"Greevo Chemical Corporation"**. The model tagged each word in the phrase with Feature tags.

Greevo   Chemical   Corporation

↓   ↓   ↓

PER   ORG   ORG

Finally the phrase is labeled as ORGANIZATION tag using the implemented algorithm.

4.1 Named Entity Recognition Efficiency

The efficiency of named entity recognition is evaluated using F-measure and Entropy calculation. The Entropy is measured for evaluating the consistency of the system. The accurate system has the lower entropy and higher F-measure. The entropy value is measured as:

$$E = \sum_{i=1}^{n} p(x_i)\log_2 p(x_i)$$

Table 1 Output of stanford parser

Phrase	Named entity
University	Organization
Institute	Organization
of	Organization
Technology	Organization

E   Entropy value.

$p(x_i)$ probability of the incident.

4.2 Accuracy Measure

An accuracy measure is calculated for the comparison of our model by formal models. Our model is suitable for simple set of sequences and avoids repetitive iterations. Formal models like HMM and Bayesian network, each variable in these models depends directly only on its parents in the network. The Conditional random field defines a large set of feature function which made the system as a complex one. Proposed system reduce the number of feature function and consider the whole concept to label a phrase. The proposed system provides an accurate solution for the capitalization problem of Stanford NER. Our system correctly identifies the named entities of capitalized and non-capitalized words using the modified CRF algorithm.

We give the input **'University Institute of Technology'** to the Stanford NER, the classifier classified it correctly.If the input is: **'university institute of technology'**.

Then the Output become (Table 1).

The main difference between the Conditional Random Feild and the proposed model is that, the proposed model reducing the number of feature functions. Figures 7 and 8 show the comparison between the existing CRF models and proposed MCRF.

Fig. 7 Comparison of feature functions

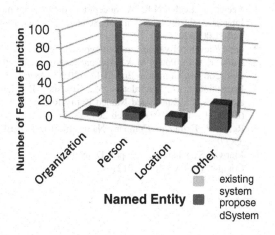

Fig. 8 Accuracy analysis

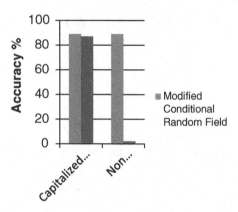

The NER fails to classify the input without proper capitalization of words. Proposed algorithm solves this problem by using Modified Conditional Random Field.

5 Conclusion

In our proposed solution, for a sentence a triplet is generated. We introduced a method for Named Entity Recognition using modified CRF and certain word feature extraction approaches. Our future work includes Word Sense Disambiguation module for identifying the correct sense associated with the triplet.

Acknowledgements We are thankful to Dr. M. R. Kaimal, Chairman Department of Computer Science, Amrita University for his valuable feedback and suggestions.

References

1. Veena, G., Lekha, N.K.: A concept based clustering model for document similarity. In: 2014 International Conference on Data Science & Engineering (ICDSE). IEEE. 978-1-4799-5461-2/ 14/$31.00 ©2014
2. https://developer.yahoo.com/geo/geoplanet/guide/
3. Patawar, M.L., Potey, M.A.: Approaches to named entity recognition: a survey. Int. J. Innov. Res. Comput. Commun. Eng. **3**(12) (2015) (*An ISO 3297: 2007 Certified Organization*)
4. Farmakiotou, D., Karkaletsis, V., Koutsias, J., Sigletos, G., Spyropoulos, C.D., Stamatopoulos, P.: Rule based Named Entity Recognition For Greek financial texts. In: MITOS (EPET II – 1.3 – 102) 28 Feb 2013
5. Rau, L.F.: Extracting Company Names from Text. IEEE (1991). CH2967-8/91/0000/0029$01 .OO *0*
6. Morwal, S., Jahan, N., Chopra, D.: Named entity recognition using HMM. Int. J. Nat. Lang. Comput. (IJNLC) **1**(4) (2012)

7. DANIEL M. BIKEL, RICHARD SCHWARTZ and RALPH M. WEISCHEDEL." An Algorithm that Learns What's in a Name." ACM Digital Library 1999
8. Hui, N., Hua, Y., Ya-zhou, T., Hao, W.: A method of Chinese named entity recognition based on maximum entropy model. In: Proceedings ofthe 2009 IEEE International Conference on Mechatronics and Automation 9–12 Aug, Changchun, China
9. https://en.wikipedia.org/wiki/Maximum-entropy_Markov_model
10. Hanna M. Wallach." Conditional Random Fields: An Introduction." Eighteenth International conference on Machine Learning February 24, 2004
11. Sutton, C., McCallum, A.: An introduction to conditional random fields. Found. TrendsR_ Mach. Learn. 4(4), 267–373 (2012).https://doi.org/10.1561/2200000013
12. http://stanfordnlp.github.io/CoreNLP/
13. https://opennlp.apache.org
14. http://www.nltk.org
15. Veena, G., Krishnan, S.: A concept based graph model for document representation using coreference resolution. In: (ISTA'15), 10–13 Aug 2015, Kochi, Kerala, India. Springer
16. Veena, G., Unmesha Sreedevi, U.B.: Improving the Accuracy of Document Similarity Approach using Word Sense Disambiguation (WCI-'15), 10–13 Aug 2015, Kochi, Kerala, India(ACM)ISBN No. 978-1-4503-3361-0
17. https://en.wikipedia.org/wiki/Conditional_random_field
18. Prasad, G., Fousiya, K.K., Kumar, M.A., and Soman, K.P.: Named entity recognition for malayalam language: a CRF based approach. In: 2015 International Conference on (ICSTM), Chennai (2015)
19. Bharath Ganesh, H.B., Abinaya, N., Kumar, A.M., Vinayakumar, R., Soman K.P.: AMRITA-CEN@NEEL: identification and linking of Twitter Entities, Named entity extraction & linking challenge. In: Proceedings of the 5th Workshop on Making Sense of Microposts (#Microposts2015), vol. 1395. CEUR, Florence, Italy (2015)

Knowledge-Based Text Mining in Getting Perfect Preferences in Job Finding

Shaziya Islam and Manpreet Kaur

Abstract Although we know that finding the most suitable job using the Internet takes various hours because the job portal gives us the preferences of jobs based on some particular keyword stored in their database but it may not be the preferences you want, so in order to remove wrong relevancy of job preferences and to be appropriate, we have the concept of text mining with knowledge-based in order to filter out most suitable preferences based on our searching criteria, so the problem is to give 100% accuracy using knowledge-based text mining using some technique, so in order to give accurate means 100% results, we have used concept of knowledge-based text mining using R studio by which we get only the job preferences which we want according to our criteria, hence this paper gives 100% accuracy in finding that.

Keywords Text mining · TM · Knowledge-based test mining

1 Introduction

"No one is satisfied with their job in today's scenario" this quotation is based on the survey done on various people working in various organization this means that in every organization, employees want to change their job for betterment of their position, salary, working environment, or the location so each and everyone is registered in some job portals for getting good opportunities and every morning they open their mail for finding some, and they get the mail "10 jobs matching your profile and when they open the mail 9 out of 10 jobs are not the preferences of jobs which they want and 1 out of 10 jobs are the actual job preferences Which we want or to which we can look up for…". And due to this we are wasting our Time in

S. Islam (✉) · M. Kaur
Bhilai, India
e-mail: Shaziya.islam26@gmail.com

M. Kaur
e-mail: Preet.kaur963@gmail.com

© Springer Nature Singapore Pte Ltd. 2018
P. K. Sa et al. (eds.), *Recent Findings in Intelligent Computing Techniques*,
Advances in Intelligent Systems and Computing 709,
https://doi.org/10.1007/978-981-10-8633-5_5

reading all the preferences and also we are wasting the memory of our mail account like for example, Someone registered on shine job portal for the job of assistant professor in computer science and engineering department and the job preferences are given in below image.

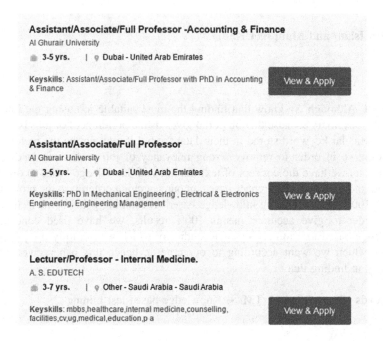

None of the job preferences matches his/her criteria because this suggestion is based on your current position, or it can be based on your profession it is not filtering or refining the appropriate criteria which we want, sometimes it suggests us the job lower than the salary which we are getting presently so that is also not good, hence in order to overcome this problem, there is concept "TEXT MINING" and particularly knowledge-based text mining and in order to proceed with let us see what is text mining and what is knowledge-based text mining.

2 Text Mining

Text mining is also known as *text data mining*, which is equivalent to analysis of text in order to have some good or valuable text from the bulk of text, the process of deriving high-quality information from text or extracting the information according to our need [2]. Quality information is typically derived through the patterns such as statistical pattern learning [2]. Text mining usually involves the process of structuring the input text (usually parsing, along with the addition of some derived linguistic features and the removal of others, and subsequent insertion into a database), deriving patterns within the structured data, and finally evaluation and interpretation of the output. "High quality" in text mining usually refers to some combination of relevance, novelty, and interestingness. Typical text mining tasks include text categorization, text clustering, concept/entity extraction, production of granular taxonomies, sentiment analysis, document summarization, and entity relation modeling (i.e., learning relations between named entities).

Text mining also referred as the text analytics or data text mining helps us in finding some particular words in some particular text or words based on the structure of some particular data, highest quality data can be found from bags of words or big data or from any ordinary document generally text mining includes text categorization, text clustering, and text concept [1–3].

2.1 Knowledge-Based Text Mining

Text mining is the actual technique of refinement of text from bulk of text, but in deep knowledge of text mining, there are types of text mining applied for different criteria for practical examples and in that, there are various techniques and among them there is knowledge-based text mining too and to define this technique, we can divide the word into two words, i.e., **knowledge + based** which means it is dealing with the concept of knowledge so there is some type of requirement of knowledge before processing means if we are dealing with some data online or offline then before processing that data to for getting useful result, we have to go through some knowledge of that data, and consider the data frame having the information about the job finder in below image.

	A	B	C	D	E
1	job position	qualificaton	location	branch	salary
2	Assistant professor	PHD	saudi	computer sceince & engineering	50000
3	Assistant professor	PHD	saudi	mechanical engineering	40000
4	lecturer	m.tech	saudi	mechanical engineering	35000
5	lecturer	m.tech	saudi	mechanical engineering	35000

So this is the data frame which we have in excel sheet and saved with .csv extension, .csv stands for comma separated values because in computing theory comma separated values is a text having numbers and text separated by comma in plain text so if we can use the csv file, it will be better for our project to mine the text in a better way.

Hence in order to understand the data we have to first load it, in R studio which we are using in mining text in a better way suppose for the above data frame we have to find the details of the data hence we can find it as:

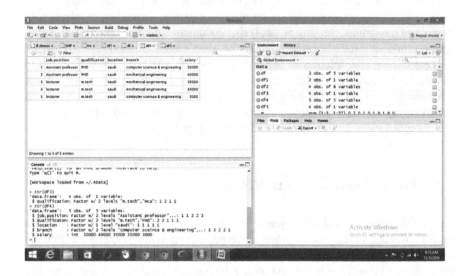

As we can see in the above figure that all the observations made by the **str()** function in order to enquire about the text is made. Hence in order to proceed further, we should find out some knowledge of the data, as this is the small data so it is not required but in the large database, this task can be completed within seconds. We can also use the plot function in R to get the data frame information graphically which can help us to understand the data frame or data set more accurately and for large amount of data. We can use the plot for getting the information about the data frame as easily as writing:

plot(data frame name), like for example in my case the data frame name is df4 so I have written:

plot(df4)and the result is

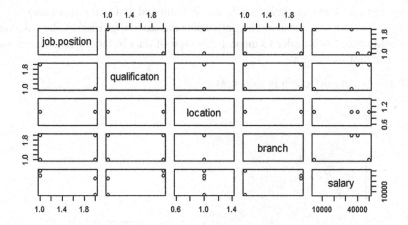

Also, I have another data set like only for educational qualification needed by the companies in my case its educational institutions hence it is data frame as **df3** and the information in that data frame is like:

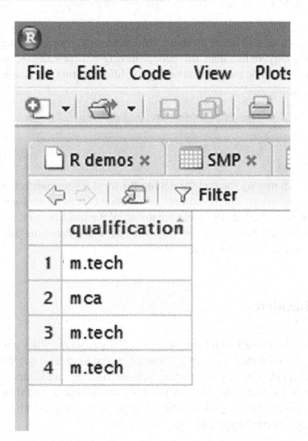

As seen above in df3, there is only one column and that is for qualification this example is set for the recruiter who only wants the searching criteria to be based on the qualification so in order to inspect this type of data in large quantity, we can again use plot function as:

Plot(df3)-> And the graph is shown as:

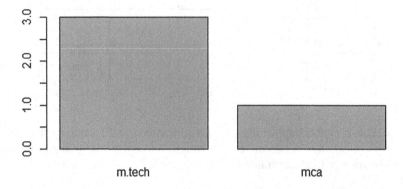

It is clear from the picture that there is more requirement of people having master's degree in technology rather than the people having master's degree in computer application so this will be the great advantage of differentiating the people in terms of their qualification.

2.2 *R Tool*

Tool which we have used in mining the text for getting the most appropriate result is the R tool which gives us various functionalities and functions for inspecting and accessing the text, so we have used the R studio for doing all the tasks here taking two data frames namely df2, df4 having the required data and manipulated the data frames using various functions available in R studio.

3 Problem Identification

As said before that no one is satisfied with their job hence 95% people are willing to change their job and that is why they register in some job portal to at least get the job preferences according to the job they want which is based on their qualification, location, and some specific keyword but this is not fulfilled at all, people start their work by opening their mails to have a look for the preferences and they got about 10 jobs preferences or 20 job preferences like shown below:

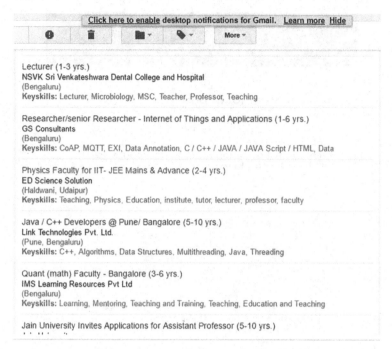

As shown above, this is the job preferences got for any particular registration in some of the job portal and the criterion was to have the job in Gulf countries in academics for the post of lecturer in computer science and engineering department but as shown in the picture, these are all irrelevant jobs preferences got against the job search criteria, which will not be seen further this is one situation now coming to another situation like shown below:

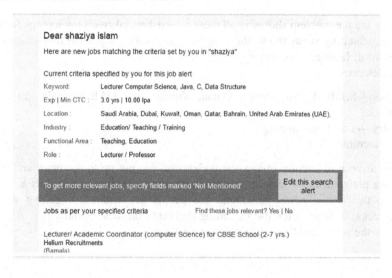

Now see the problem shown in all these screenshots taken on my laptops for the criteria, searching job as shown the criteria is specified clearly in the first image that is keyword: lecturer—computer science, java, c, c++.

Preferences:

Location—Saudi Arabia, Dubai, Oman, Kuwait, and UAE basically all Gulf countries.
Industry—education/teaching
Role—lecturer

But as seen in another image, the preferences are not up to the mark and they give the preferences for lecturer in math's, physics, and chemistry so this is of no use, so this is one of the situations that people get irritated of getting wrong preferences, so what is the use of getting registered on these portals.

Now the most irritating situation shown below:

In above picture, the job search criteria are based on the preferred location, and the criteria but when going through this link the specifications are as follows:

Branch—computer science and engineering
Post—Assistant professor
Qualification—PHD

Now after having a good watch in this criteria again, the eligibility criteria are not fulfilled because the criteria of education were set to master's degree in computer science and engineering and the result is for PhD, so now this is again the most irritating situation because time has been given for registering to the portal and this is the result which is not at all matching the specified criteria, this is only one case if there are 10 cases like this then this is waste of time in accessing these links and knowing about the job, so there should be some criteria for this so that if getting some of the preferences then it should be valid and based on specific criteria only, there should be no other preferences given other than the specific criteria.

4 Solution to the Problem

If we use knowledge-based text mining in finding appropriate job preferences, then we can get 100% result of getting only specific criteria-related job preferences this will be made on large amount of data set but for showing the result I have made two data frames:

df2-> data frame having the information of the candidates registered on some of the job portals and it can be loaded to R studio as:

Reading jobdir.csv (discussed above)
df2 <-read.csv("C:/Users/admin/Documents/texts/jobdir.csv")

And now to see whether the data frame is loaded successfully or not, we have to write the data frame name and press enter.

df2

	Name	qualification	branch	location	salary	position
1	shaziya islam	m.tech	computer science & engineering	INDIA	50000 inr	Assistant professor
2	nadim nizami	mca		india	15000	lecturer
3	manpreet kaur	mtech	computer science & engineering	INDIA	20000	Assistant professor
4	neha chaubey	mtech	electronics engineering	india	30000	lecturer

As shown in the above image, there are four candidates with the information—qualification, branch, location, position, and expected salary which are the basic information needed for registering on some job portal.

Now coming on the jobs directory coming from some employer to some job portal say:

df4-the data frame having the jobs for the candidates it can be loaded to the R studio as:

df4 <-read.csv ("C:/Users/admin/Documents/texts/job.csv")

And now to see whether the data frame is loaded successfully or not, we have to write the data frame name and press enter.

df4

	job.position	qualificaton	location	branch	salary
1	Assistant professor	PHD	saudi	computer sceince & engineering	50000
2	Assistant professor	PHD	saudi	mechanical engineering	40000
3	lecturer	m.tech	saudi	mechanical engineering	35000
4	lecturer	m.tech	saudi	mechanical engineering	35000
5	lecturer	m.tech	saudi	computer sceince & engineering	3000

So here is the jobs directory data frame loaded into df4 having fields job position, qualification, location, branch, and salary so now we want that according to the qualification, branch, and location I should get the accurate result for the position otherwise NA for not applicable position so in R studio we can do it in seconds using the simple sql queries available in R studio but for using these functions, we have to load the sqldf package available in R studio as:

library (sqldf)

And it will be loaded as shown below:

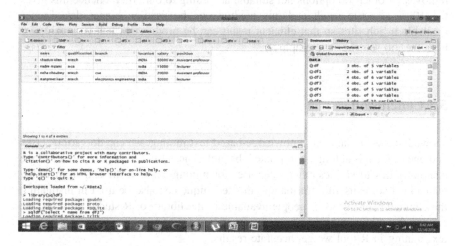

And after loading sqldf package, we can use various queries available in sql to be used in data frames.

So with our problem as said before we have two data frames df2 and df4, df2 is containing the information of the candidate registered in job portal and df4 is containing the information from the employer regarding the job posting so the query will be

sqldf("select name, df4.position, df4.branch, df4.salary, df4.location from df2 join df4 where df2.qualification = df4.qualification and df2.branch = df4.branch")

and the result will be as shown below:

Name	Position	Branch	Salary	Location
1 Shaziya Islam	Lecturer	Cse	2000	Saudi
2 Neha Choubey	Lecturer	Cse	2000	Saudi

As shown in the screenshot taken from R studio working in windows

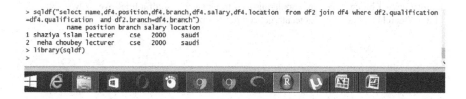

As shown in the image, only two names are selected from the two data frames which are exactly suitable for the job position asked by the recruiter hence this removes all other job options not suitable according to our criteria hence this gives the 100% accurate result and this method can be used in bigger data frames available in job sites so that efficient result can be produced which is suitable for both the job searcher and employer.

5 Conclusion

As we have seen that using knowledge-based text mining, we can do various tasks and one of them is getting appropriate jobs preferences from some job portal and we can apply this to various other categories for mining some of the text from which we can get some useful information, this technique can also be used in instigating the comments or tweets of people available in the library of R studio, in short, we can say that all the things are available for our accessing but using knowledge-based text mining in R tool we get accurate results.

References

1. https://www.gettinggeneticsdone.com/2010/05/use-sql-queries-to-manipulate-data.html
2. Shehata, S., Karray, F., Kamel, M.: A concept-based model for enhancing text categorization. In: Proceedings of the 13th International Conference on Knowledge Discovery and Data Mining (KDD '07), pp. 629–637 (2007)
3. Gaikwad, S.V., Chaugule, A., Patil, P.: Text mining methods and techniques. Int. J. Comput. Appl. **85**(17), 0975–8887 (2014)
4. Wu, S.-T., Li, Y., Xu, Y.: Deploying approaches for pattern refinement in text mining. In: Proceedings of the IEEE Sixth International Conference on Data Mining (ICDM '06), pp. 1157–1161 (2006)
5. Mooney, R.J., Nahm, U.Y.: Text mining withi nformation extraction. In: Daelemans, W., du Plessis, T., Snyman, C., Teck, L. (eds.) Proceedings of the 4th International MIDP Colloquium, Sept 2003, Bloemfontein, South Africa, pp. 141–160. Van Schaik Publishers, South Africa (2005)
6. Rathor, A.S., Garg, D.P.: Analysis on text mining techniques. Int. J. Adv. Res. Comput. Sci. Softw. Eng. ISSN 2277 128X
7. Shinde, M.R., Gill, P.C.: Pattern discovery techniques for the text mining and its applications. Int. J. Sci. Res. (IJSR) ISSN (Online) 2319-7064 Impact Factor (2012), 3.358 **3**(5) (2014)
8. https://www.r-bloggers.com/manipulating-data-frames-using-sqldf-a-brief-overview/
9. https://searchbusinessanalytics.techtarget.com/definition/text-mining

A Distributed Framework for Real-Time Twitter Sentiment Analysis and Visualization

Jamuna S. Murthy, G. M. Siddesh and K. G. Srinivasa

Abstract Social networking site such as Twitter contributes to the huge amount of text data every day and hence provides the best opportunity for sentiment analysis (SA). Existing systems for SA used Hadoop and MySQL and hence lacked real-time analysis. Thus, this research work aims at introducing a distributed framework that processes and analyzes tweets in real time using apache storm and Redis database. Proposed system focuses on key aspects needed for SA in terms of data collection, data parsing, and data visualization. A novel algorithm called Emoticon-Polarity-SentiWordNet (EPS) is used for Twitter sentiment analysis and the classification results are visualized on a real-time web application. Evaluation results proved that our system outperforms the existing system.

Keywords Real-time data analytics · Hadoop · Apache storm
Redis · D3

1 Introduction

Nowadays, people use Twitter for posting short messages called tweets to express their thoughts and opinions on different aspects ranging from simple ones such as "Hey @xyz!!Gunnyt☺sleep" to themes such as "#IPL-2017". Tweets are undoubtedly rich with user information and thus performing SA on the tweets offers companies and organizations a fast and effective way of planning marketing strategies and helps in quick decision-making [1]. But due to the tweet length

J. S. Murthy · G. M. Siddesh
Department of ISE, Ramaiah Institute of Technology, Bengaluru, India
e-mail: jamunamurthy.s@gmail.com

G. M. Siddesh
e-mail: siddeshgm@gmail.com

K. G. Srinivasa (✉)
Ch. Brahm Prakash Government Engineering College, New Delhi, India
e-mail: kgsrinivasa@gmail.com

© Springer Nature Singapore Pte Ltd. 2018
P. K. Sa et al. (eds.), *Recent Findings in Intelligent Computing Techniques*,
Advances in Intelligent Systems and Computing 709,
https://doi.org/10.1007/978-981-10-8633-5_6

restriction of 140 characters, many people use irregular expressions, emoticons, smiles, and abbreviations for saving the room of space for their tweets. This has increased tweet sentiment analysis problems to greater extent showing data sparsity and sarcasm. Over the years, many classifiers were trained by the researchers for sentiment classification of tweets but most of the algorithms used the traditional approaches such as bag of words model, unigrams–bigrams model, POS Tagging, etc., which lead to less classification accuracy. Also, the existing framework for SA used Hadoop for processing a massive set of tweets and used MySQL database for querying. But Hadoop and MySQL are traditional approaches and do not support analysis over real-time. Hence, the proposed system aims at implementing a distributed real-time framework for Twitter sentiment analysis with a novel classification algorithm called EPS. The keys aspects of the framework lie in implementing three major modules called the data collection, data parsing, and data visualization.

2 Literature Review

Sentiment analysis deals with analyzing the user-generated data and hence many researchers have worked on the same and proposed different techniques which are discussed in this section. Name entity recognition (NER) was implemented by Alan Ritter et al. [2] which tried to rebuild natural language processing (NLP) pipeline using POS tagging and polarity-based sentiment classification with an accuracy of 72%. Vinh et al. [3] used a combination of lexicon-based approach and naïve Bayes classifier for sentiment classification using MapReduce and increased the classification accuracy up to 73.7%. Pang et al. [4] reported an accuracy of movie review tweets with 81%, 80.4%, and 82.9% for naive Bayes, MaxEnt, and SVM-based classifier, respectively. Balamurali et al. [5] proved that unigram and trigram model outperforms the trigram model when used with naive Bayes classification for tweet sentiment analysis. On the other hand, the reverse was true in case of SVM and MaxEnt which was proved by Chang et al. [6]. Although the above methods tried to increase the classification accuracy to some extent, the proposed EPS algorithm outperforms all the existing classifiers by increasing the classification accuracy to 85%.

3 Proposed Work

The proposed framework for real-time SA consists of three major modules called data collection, data parsing, and data visualization as shown in Fig. 1 The main objective of "Data Collection" module is to capture the sparse tweets continuously over real time. The tweets are polled using Twitter streaming API with the help of java library Twitter4j and are consumed using Kafka [7], a distributed message queuing system. Thus, tweets collected using kafka are injected to Apache Storm [8] for data parsing. The "Data Parsing" module includes two components data

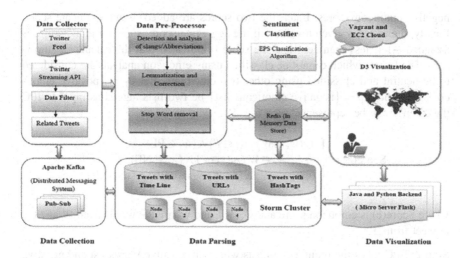

Fig. 1 Proposed architecture

preprocessor and sentiment classifier. Data preprocessor acts as a natural language processing pipeline to extract the important features from the tweet using dictionary based approach. For each tweet, at first, detection and analysis of slangs and abbreviations are done to correct the irregular spells. Later, the longer words are stemmed to short words and are corrected. Next, the skip words such as "http", "to", etc., are removed. Finally, the preprocessed tweet is sent to sentiment classifier for further classification. The sentiment classifier implements proposed EPS algorithm in terms of three classifiers called improved emoticon classifier (IEC), enhanced polarity classifier (EPC) and SentiWordNet Classifier (SNC) and the algorithm description is given below.

The set of tweets T is defined as

$$T = \{t_1, t_2, \ldots, t_n\}$$

If each tweet t contains w words then set of words W is defined as

$$W = \{w_1, w_2, \ldots, w_m\}$$

Then the sentiment score S, calculated for each word is given as

$$\text{Score} = \sum_{i=1}^{n} \sum_{j=1}^{m} S_{t_i w_j}$$

Step 1: IEC does classification of emoticons. First, the regular expression is used to detect the presence of emoticons in tweets. A manually tagged rich list of emoticons with positive and negative is initialized. The emoticon in the tweet is matched against the defined positive and negative list to obtain matched positive and negative emoticon scores. Later, the two scores are added to get aggregate score. Finally, the tweet is given a value 1 if the aggregate score is greater than zero, it is assigned −1 if less than zero and the score 0 indicated that the calculated sum is zero. The tweets which cannot be classified using emoticon analysis are considered to be neutral and classified using other two classifiers. Let $P_E = \{$positive emoticons list$\}$ and $N_E = \{$negative emoticons list$\}$ be two lists tagged as input to IPC then step 1 can be represented as:

$$Score(e) = \begin{cases} 1, (w_x \in W) \wedge (t \in T) \wedge (w_x \in PE) \\ -1, (w_y \in W) \wedge (t \in T) \wedge (w_y \in NE) \\ 0, (w_2 \in W) \wedge (t \in T) \wedge (w_2 \notin PE) \wedge (w_2 \notin NE) \end{cases}$$

where Score(e) is emoticon score and w_x, w_y, and w_z are the words from W and t is a tweet from T.

Step 2: EPC takes the input as two lists which are positive words list and negative words list this is known as "bag of words tagging". Generally, words are domain independent. IPC works only with words with correct spelling and if there is a combination of positive or negative in tweet it is classified as neutral and addressed using SNC. Words are identified based on splitting the words in tweets with delimiter. The words list used are collected from [9] and both the lists are combined to obtain roughly around 9500 words. EPC is an enhancement from [10] and hence the name. It works similar to IEC except that their emoticons are used and here words are used. Let $P_W = \{$positive words list$\}$ and $N_W = \{$negative words list$\}$ then Step 2 is simply represented:

$$Score(w) = \begin{cases} 1, (w_x \in W) \wedge (t \in T) \wedge (w_x \in PW) \\ -1, (w_y \in W) \wedge (t \in T) \wedge (w_y \in NW) \\ 0, (w_2 \in W) \wedge (t \in T) \wedge (w_2 \notin PW) \wedge (w_2 \notin NW) \end{cases}$$

where Score(w) word score of IPC, w_x, w_y, and w_z are the words from W and t is a tweet from T.

Step 3: SNC is based on the SentiWordNet dictionary and the tweets are classified based on sentiment identified from the dictionary and it assigns different weights for the words based on the type of sentiment using parts of speech. Similar to the previous step in IPC, here too the words are delimited and the sentiment value is calculated using SentiWordNet library. The aggregate sentiment score is calculated by adding the weight of each word which was assigned using SentiWordNet. Finally, the tweet is given the value 1 if the aggregate score is greater than zero, it is assigned −1 if less than zero and the score 0 indicated that the calculated sum is zero. Step 3 can be represented in a simple way as:

$$\text{Score}(s) = \begin{cases} 1, (w_x \in W) \wedge (t \in T) \wedge (\text{weight}(w_x) > 0) \\ -1, (w_y \in W) \wedge (t \in T) \wedge (\text{weight}(w_y) < 0) \\ 0, (w_2 \in W) \wedge (t \in T) \wedge (\text{weight}(w_2) < 0) \end{cases}$$

where Score(s) is sentiment score of SNC and weight (w_x), weight (w_y), weight (w_z) are the words from W and t is a tweet from T.

Step 4: For the refined tweets from the preprocessor first IEC in Step 1 is applied, next EPC and finally SNC. If IPC classifies the tweet as neutral, it goes to Step 2 and if the tweet is classified as neutral at this step too, it moves to Step 3. The tweets which are not classified using any classifier are considered to be neutral. Final classification step can be expressed as

$$\text{Class} = \begin{cases} \text{Positive}, (S_e > 0) \vee (S_e = 0 \wedge S_w > 0) \vee (S_e = 0 \wedge S_w = 0 \wedge S_s > 0) \\ \text{Negative}, (S_e < 0) \vee (S_e = 0 \wedge S_w < 0) \vee (S_e = 0 \wedge S_w = 0 \wedge S_s < 0) \\ \text{Neutral}, (S_e = 0) \wedge (S_w = 0) \wedge S_w = 0 \end{cases}$$

The tweets parsed during the data parsing module are stored in the Redis database and are visualized using "Data Visualization" module. The visualization module is built with the help of javascript library D3 and using web application library Flask.

4 Evaluation Results

The results are evaluated as throughput and scalability to check the performance of data collection and data parsing modules. Throughput is evaluated as the number of messages collected using the data collection module. Series of three tests were conducted on three different days and the average is taken to discuss the results. From Fig. 2, we observe that the average rate of input messages increases when the

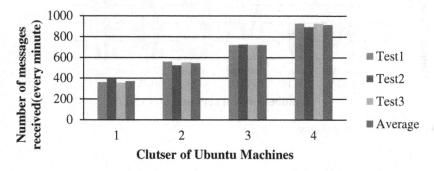

Fig. 2 Evaluation of throughput

number of worker nodes increase. Thus, by implementing a minimal of 4 node cluster for our framework we can collect a variety of tweets from each node and reach up to the benchmark of collecting millions of tweets per day.

Here, the scalability of the framework is evaluated by altering the number of nodes and vCPUs. Series of five tests were conducted and the average is taken to discuss results. As the results in Fig. 3 show raising the number of workers increases the performance of sentiment component dramatically. When the number of virtual CPUs cannot satisfy the computing requirements of the component, increasing number of workers cannot enhance the performance of the component, it may even decrease the number of output messages.

For example, comparing the first two rows of Fig. 3, the number of workers is added from 1 to 2, the average outputs decrease from 589 to 381 messages per minute. By comparing the first row to the third row, which keeps the same workers and increases the number of virtual CPUs, the average message output increases from 589 to 748. We can conclude that one virtual CPU cannot cover the computing resources of one Storm sentiment worker. In the third and fourth row, we increase the number of workers, with 4 virtual CPUs, the average output increases from 748 to 1184 per minute. Likewise, in rows five and six, with six virtual CPUs, increasing worker number raises the output from 1454 to 1654 per minute. When computing resources are adequate, increasing the number of workers leads to increased output as well. As the system applies Twitter 4j to collect data, whose average data fetching rate is 1071 messages per minute, our system can completely process all collected messages, with more than four virtual CPUs and two worker threads.

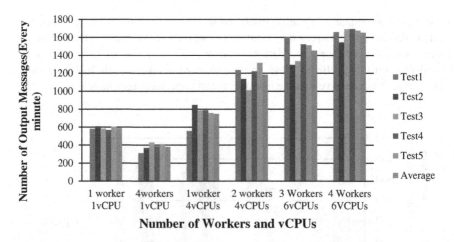

Fig. 3 Evaluation of scalability

5 Conclusion

This research work has proposed a distributed real-time Twitter sentiment analysis and visualization framework by implementing a novel algorithm for SA called EPS. The whole framework is implemented in the form of three major modules which are easily reusable. Apache Storm is a core part of the framework which makes our system distributed and provides real-time stream processing capabilities. Evaluation results prove that our framework provides best results of throughput and scalability. This proves that our system can scale well with real-time big data analytics. Future research directions include experimenting with other data resources like Facebook, Instagram, etc.

References

1. Agarwal, A., et al.: Sentiment analysis of twitter data. In: Proceedings of the Workshop on Languages in Social Media. Association for Computational Linguistics (2011)
2. Ritter, A., Clark, S., Etzioni, O.: Named entity recognition in tweets: an experimental study. In: Proceedings of the Conference on Empirical Methods in Natural Language Processing. Association for Computational Linguistics (2011)
3. Khuc, V.N., et al.: Towards building large-scale distributed systems for twitter sentiment analysis. In: Proceedings of the 27th Annual ACM Symposium on Applied Computing. ACM (2012)
4. Pang, B., Lee, L., Vaithyanathan, S.: Thumbs up?: sentiment classification using machine learning techniques. In: Proceedings of the ACL-02 Conference on Empirical Methods in Natural Language Processing, vol. 10. Association for Computational Linguistics (2012)
5. Balamurali, A.R., Joshi, A., Bhattacharyya, P.: Harnessing wordnet senses for supervised sentiment classification. In: Proceedings of the Conference on Empirical Methods in Natural Language Processing. Association for Computational Linguistics (2011)
6. Chang, C.C., Lin, C.J.: LibSVM: a library for support vector machines. ACM Trans. Intell. Syst. Technol. (TIST) 2(3), 27 (2011)
7. Garg, N.: Apache Kafka. Packt Publishing Ltd (2013)
8. Jain, A., Nalya, A.: Learning Storm. Packt Publishing (2014)
9. http://www3.nd.edu/mcdonald/Word_Lists.html
10. http://www.cs.uic.edu/

Building an Information Retrieval Apparatus for Kannada Language: A Rule-Based Machine Interpretation System with Noise Suppression

Shivani Kulkarni and R. H. Goudar

Abstract India is a home to various vernaculars. Due to the widespread usage of the Internet, the amount of information available is huge, thus people tend to bend toward these available resources for the information retrieval for various vernaculars. Indian dialects give charming four particular vernacular families, to be specific; those Indo-Aryan, those Austro-Asiatic, those Tibeto-Burman, and the Dravidian. Kannada belongs to the Dravidian family. The main focus of this paper is interpretation of Kannada language in particular for the people familiar in English to get access to the required information without much difficulty. Suitably, it is fundamental to build an information retrieval (IR) instrument or device for the Kannada composite. Keeping in mind the end goal to build up an IR apparatus encompassing the Kannada dialect, the specifics of the dialect are to be recognized. This paper gives a rule-based approach which additionally strips fourteen diverse significant portrayals of postfixes (pratyaya in Kannada), along with the subclasses by interpreting morphophonemic assurance and involving a zero-shot noise suppressing algorithm for the texts or document(s) encompassing images along with them.

Keywords Information retrieval · Kannada · Rule-based approach
Dialect · Suffixes (pratyaya) · Zero-shot noise suppressing algorithm

1 Introduction

India is a home to different vernaculars and comprises of 122 unique dialects [1] and out of which 12 languages are considered to be official by the constitution. Indian vernaculars give four distinctive lingo families, to be specific; those

S. Kulkarni (✉) · R. H. Goudar
Department of Computer Network Engineering, Visvesvaraya Technological University, Belgaum 590018, Karnataka, India
e-mail: kulkarnishivani024@gmail.com

R. H. Goudar
e-mail: rhgoudar.vtu@gmail.com

© Springer Nature Singapore Pte Ltd. 2018   63
P. K. Sa et al. (eds.), *Recent Findings in Intelligent Computing Techniques*,
Advances in Intelligent Systems and Computing 709,
https://doi.org/10.1007/978-981-10-8633-5_7

Indo-Aryan, the Austro-Asiatic, those Tibeto-Burman, and the Dravidian. Kannada has a place in the Dravidian family. For building up an IR apparatus for the Kannada dialect specifically, the specifics of the dialect will be seen. This creation accommodates a decision-based drive that strips 14 particular critical portrayals of postfixes (pratyaya in Kannada) and furthermore the subclasses.

1.1 Approaches to Machine Interpretation

IR is the technique to get the data from customer request. These days, the search for information in different dialects is extended rather than in the native lingos. Interpretation is classified into three ways, they are; (a) Dictionary-based interpretation (b) Machine interpretation (c) Corpus-based interpretation [2].

- Dictionary-based Interpretation—Here, a bilingual word reference will be used for playing out the understanding or interpretation.
- Machine interpretation—In machine interpretation [3], no mankind's mediation will be included and the interpretation is performed toward the machine naturally.
- Corpus-based Interpretation—Here, the information will be gotten from those tantamount, furthermore parallel corpus.

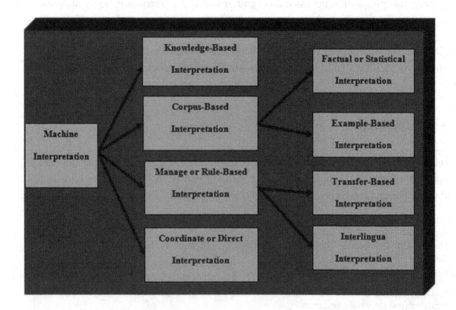

Fig. 1 Approaches to machine interpretation

MT is divided into four major approaches by means of interpretation [4]. Figure 1 illustrates these approaches.

Coordinate or direct interpretation wherein, each expression's interchanges character over a sentence will be deciphered by facilitating a look in the bilingual word reference. The inclination here is that the deciphered brisk will be direct to the observer.

Manage or rule-based interpretation incorporates transfer-based and interlingua interpretation. It might be in a general sense a get together of models that hold extra information with respect to the semantics of the wellspring. Thus, interpretation might be achieved by coordinating those outlines from asserting choices.

Corpus-based interpretation incorporates statistical and example-based interpretation.

The *knowledge-based interpretation* incorporates both straight and orthological data. It is used in artificial intelligence.

The following is the purpose of introducing this paper:

a. To assist people to get the glimpse about the Kannada language thoroughly.
b. For the developers who are not well versed with the language can get the required interpretations accordingly by eliminating the lingual barrier.
c. To develop a simple system that could be used by anybody who wishes to gain the required outcome of interpretations.

2 Literature Survey

In this section, the work done by different authors is looked into. The different ways of machine interpretation proposed by them are given.

B. N. V. Narasimha Raju et al. [2] gave the classification for the machine interpretation and the explanation for each approach involved is provided. A statistical machine interpretation tool is implemented in this paper with the aid of parallel corpus. T. Siddiqui and U. S. Tiwary [3], discussed the natural languages versus the machine learning and the grammar-based language model, along with the syntactic and semantic analysis. D. D. Rao [4] discussed the machine interpretation with a proper introduction to it. K. Narayana Murthy [5] gave the specifics about the Kannada language and how it can be processed for the purpose of interpretation. Knowles F. [6] discussed the linguistics and the creation of the lexical database and how word references play a vital role in the interpretation process. Mallamma V. Reddy, Dr. M. Hanumanthappa [7] discussed root word identifier for different lingos and their respective characteristics and how they are obtained and stored.

The above-discussed papers lack the involvement of different techniques such as formatting (which includes deformatting and reformatting), noise suppression and morphological analysis into one single experiment. This proposed paper considers these methods into one single information retrieval apparatus that aids in having an advanced system. Section 3 below gives a precise and an appropriate system for better machine interpretation by combining these techniques.

3   Proposed System

In this section, a machine interpretation gadget is designed which can spare an impressive measure of human power and also provides an effective understanding of how a specific case lingo is interpreted to another record. The uncovered square diagram of the suggested system is illustrated in Fig. 2.

3.1   Dialect Identification

It is the process of perceiving the dialect of the document(s) by transferring the records by entering the brisk. The distinguishing proof will be refined for the two lingos English and Kannada. Those reports are iteratively checked for their nearness.

The expressions from that corpus would be sustained and those postfixes and the expressions are recognized iteratively in the end by the framework illustrated in the Fig. 3. Around the twist of the assignment, the root word [7] might be obtained.

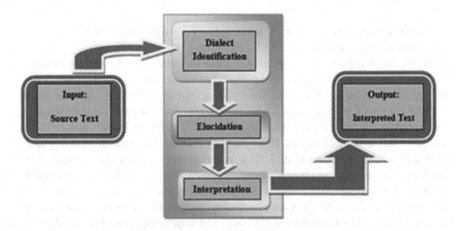

Fig. 2 Structure of the proposed system

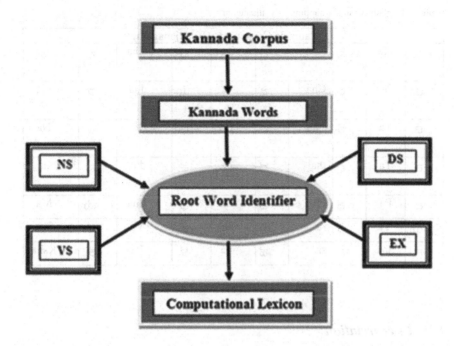

Fig. 3 Block diagram showing the input and output to the root word identifier. NS: noun suffix; VS: verb suffix; DS: dual functional suffix

3.2 Elucidation

Machine elucidation module considers character by character the words from one lingo and converts it to another lingo without losing its phonological characteristics. The process here is simple and straightforward and considers and maps characters in Kannada to English and vice versa. Tables 1 and 2 illustrate this mapping.

Table 1 English-Kannada mapping (Swaragalu)

Kannada Vowels (Swaragalu)									
ಅ	A	ಆ	aa,A	ಇ	I	ಈ	ee, I	ಉ	U	ಊ	oo,U
ಋ	Ru	ಎ	E	ಐ	ai,Eai	ಏ	ai	ಒ	O	ಔ	oa, O
ಒ	Ou	ಅಂ	Um	ಅ:	Ah				

Table 2 English-Kannada napping (Vyanjanagalu)

Kannada Consonants (Vyanjanagalu)									
ಕ	Ka	ಖ	Kha	ಗ	ga	ಘ	Gha	ಙ	~ga
ಚ	Cha	ಛ	Cha	ಜ	ja	ಝ	Jha	ಞ	~ja
ಟ	Ta	ಠ	Tha	ಡ	Da	ಢ	Dha	ಣ	Na
ತ	Ta	ಥ	tha	ದ	da	ಧ	dha	ನ	na
ಪ	Pa	ಫ	Pha	ಬ	ba	ಭ	Bha	ಮ	Ma
ಯ	Ya	ರ	ra	ಲ	la	ವ	Va	ಶ	Sha
ಷ	Sha	ಸ	sa	ಹ	ha	ಳ	La	ಕ್ಷ	Ksha

3.3   Interpretation

Interpretation basically is the process of converting or translating the written text in one lingo to another while preserving its genuine and precise meaning. Machine interpretation [5, 6] structures that produced elucidations between the two best particular dialects would be called "bilingual frameworks". In order to design a prototype for the Interpretation system, following steps are to be taken. These steps are highlighted in the Fig. 4 below.

Deformatting, besides Reformatting, is making that machine elucidation procedure less complex and moreover subjective. The record that holds figures, flowcharts, etc., does not require elucidation.

Morphological principle might be the examination of the expressions besides, investigation. It breaks down those structures of expression, for example, root, stems, words, and prefixes, moreover additions.

Syntactic examination or parsing is the technique wherein the strings about characters or pictures in the substance are to be investigated, conceivably beforehand, normal lingo then again over machine lingos that fit in with the formal linguistic principles.

Commotion (i.e., Noise) Suppression will be utilized to diminish the essentialness of the lesquerella germane expressions (i.e., less applicable words). The noise suppression regularizer in the understanding technique that will actually cover the impact of lesquerella relevant expressions. A noise smothering zero-shot learning algorithm is used to decreasing the less applicable words.

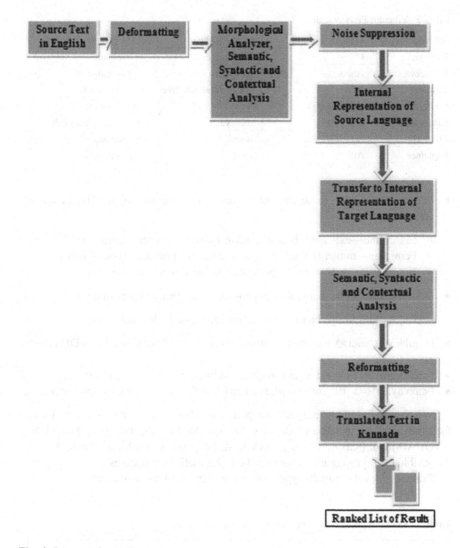

Fig. 4 Interpretation system

4 Kannada Classification of Suffixes

Kannada is a South Indian, Dravidian lingo. Kannada expressions could be requested correspondingly as [8], nouns, verbs, adjectives, articles, and stop words explained below:

Table 3 Vibhakti Pratyayagalu

Vibhakti	Markers	Inflection type	Example
Prathama	U	Nominative	ramanu
Dweteeya	Annu	Accusative	ramanannu
Truteeya	iMda	Instrumental/Ablative	ramaninda
Chaturthi	ige/ge/kke/akke	Dative	ramanige
paMchami	daseyiMda	Ablative	ramanadaseyinda
ShashThi	A	Genetive	ramana
Saptamee	Alli	Locative	ramanalli

- (<nAmapada>) => these are the words that name the items. Things are of sorts;

  - Masculine—ex; rama, bharat, naman (names), master (instructor)
  - Feminine—ramaa (name), geetaa, sita (names) huDugi (young lady)
  - Neutral—ex; mara (tree), kai (hand), mane (house) and so on,.

- (<kriyApada>) => words that portray the activities being performed

  - Ex; hattu (climb), baa (come), hogu (go) kare (call), and so on,

- (<guNavAchaka>) => the describing words. Ex; Chikka (small), doDDa (big), etc.,.
- (<saRvanAma>) => Ex; avanu (he), avalu (she), adu (that), idu (this).
- (<avyaya>) => Ex; mEle (up), keLage (down), yaake (why), ityaadi (etc.) etc.,.

In Kannada, additions are only the pratyaya. There are around 14 assorted sorts for pratyayas or postfixes. Besides vowels to be specific; (<akArAtMa>), (<ikArAtMa>), (<ukArAtMa>), (<ekArAtMa>), (<aikArAtMa>). Table 3 gives the <vibhakti pryayagalu> alongside their articulations and cases.

Table 4 gives the Sandhi types and the examples of the word connection.

Table 4 Sandhi types and examples of word connection

SANDHI TYPE	COMPLEX WORD	SIMPLE/INFLECTED WORD
ಲೋಪ	ಮಾತಿಲ್ಲ	ಮಾತು+ಇಲ್ಲ
ಆಗಮ	ಮನೆಯನ್ನು	ಮನೆ+ಅನ್ನು
ಆದೇಶ	ಮರಗಾಲು	ಮರ+ಕಾಲು

5 Conclusion

In this paper, a machine interpretation apparatus is prescribed for the English to Kannada, what's more, the other way around understanding. The model means amid interpreting the records effectively using the rule-based approach, morpho-phonemic assurance, and the commotion (noise) smothering regularizer. New principles have been incorporated which will make the structure of a more prominent adequacy.

References

1. Baldonado Census of India: Abstract of speakers' strength of language and mother dialect-2001. www.censusindia.gov.in
2. Narasimha Raju, B.N.V. et.al.: Interpretation approaches in cross language information retrieval. In: Proceedings of International Conference on Computer and Communications Technologies (ICCCT). IEEE (2014)
3. Siddiqui, T., Tiwary, U.S.: Natural Language Processing and Information Retrieval. Oxford Press (2008)
4. Rao, D.D.: Machine interpretation a gentle introduction. Resonance (1998)
5. Narayana Murthy, K.: Computer processing of Kannada language, University of Hyderabad
6. Knowles, F.: The pivotal role of the dictionaries in a machine interpretation system. In: Lawson, V. (ed.) Practical experience of machine interpretation. North Holland, Amsterdam (1982)
7. Reddy, M.V., Hsanumanthappass, M.: Machine translation for English to Kannada
8. http://www.nammakannadanaadu.com/vyakarana/index.php

A Context-Aware Fuzzy Classification Technique for OLAP Text Analysis

Anirban Chakrabarty, Santanu Roy and Sudipta Roy

Abstract Online Analytical processing (OLAP) has been a major area of interest for different organizations since the last decade. Traditionally OLAP has been carried on numeric data and there has been relatively limited research in the domain of OLAP for text data. The work suggests context-aware classification using fuzzy logic which considers several co-related concepts for multi- dimensional OLAP text analysis. In the proposed model, each dimension is related to some contextual factors where a context propagation technique and a look-up table have been suggested for efficient and fast searching on OLAP dimensions. To validate the proposed model, experiments have been performed using several publicly available resumes collected from different job portals, considering Job Specialization, Region of person, and Experience as dimensions. The experimental results confirm the importance of the approach in efficient searching of diverse resume by specifying various OLAP operations at different levels of granularity for improved decision-making.

Keywords Fuzzy classification · Context-awareness · Online analytical processing · Text analysis · Concept hierarchy · Look-up table

1 Introduction

The incessant growth of text related content in various kinds of business applications and web based communication has spurred the need to analyze both structured data in the form of files and unstructured text such as blogs, tweets, emails etc.

A. Chakrabarty (✉) · S. Roy
Assam University, Silchar, India
e-mail: chakrabarty.anirban@rediffmail.com

S. Roy
e-mail: sudipta.it@gmail.com

S. Roy
Future Institute of Engineering and Management, Kolkata, India
e-mail: santanuroy84@gmail.com

© Springer Nature Singapore Pte Ltd. 2018
P. K. Sa et al. (eds.), *Recent Findings in Intelligent Computing Techniques*,
Advances in Intelligent Systems and Computing 709,
https://doi.org/10.1007/978-981-10-8633-5_8

While online analytical processing (OLAP) techniques have been proven very useful for analyzing and mining structured data, they face challenges in handling text data. Typically, OLAP queries are unable to evaluate text data that are usually related to some contextual information but which must be considered to retrieve accurate results.

Navigation is supported in OLAP systems through multiple dimensions from one view to another which can be effectively used to mine OLAP data. Usually summarized information required for decision making can be retrieved from OLAP systems using well defined aggregation operator and complex queries which generally take long response time.

In this work, we propose a fuzzy context classification technique for OLAP text data analysis. The text documents are represented by vectors of weighted concepts and the dimensions are related to different contextual factors. Further, it integrates a context navigation technique on the dimensions represented by concept hierarchies to better co-relate the concepts of textual data. The work suggests a new data structure—a look-up table, for faster processing of OLAP queries instead of the consolidated fact table.

The rest of this paper is organized as follows: Section 2 presents an overview of related work. Section 3 conveys the methodology. Section 4 presents a context navigation method and a look-up table for a multi-dimension system. Section 5 provides the proposed algorithm with time complexity analysis. The experimental metrics and results are discussed in Sect. 6. The conclusion in Sect. 7 summarizes the findings and contributions of this work.

2   Related Work

In the era of Data warehouse and data mining technology, OLAP has been used for interactive analysis of multidimensional data at different levels through various operations like drill down, rollup, pivoting [1]. Usually data warehouses support OLAP data cube which can represent summarized information using dimension and fact tables. Further a data cube facilitates data navigation at different levels of granularity [2]. Text can be in the form of data records or character fields and also as links to records thru joinable common attributes, thus conceptually text can be in structured data and unstructured text data [3]. It has been observed from related study that traditional data cube has been extended to support text analysis in OLAP domain for e.g. Text Cube [4], Topic Cube [5]. More recent work has used data cubes in new domains, such as OLAP on sequences, spatial data and mobile data [6]. However keeping with the boom of Internet and the ever increasing business intelligence applications, the most important domain of interest is that of text data [7].

A study of recent work in this direction proposed a blend of keyword search and OLAP technology in order to efficiently evaluate the content of a multidimensional text database [8]. Further study in this direction revealed the use of a Topic Cube which allows users to effectively explore the unstructured text data from different semantic topics [5]. More Recent work in this direction describes an effective solution for recommending textual OLAP over data warehousing environments where the recommendation process is based on text semantics and query personalisation to improve the relevance of the retrieved results [9]. The suggested work in this chapter is different from other methods discussed above. While data cubes suggested earlier were represented by text fragments in a user query, the proposed model uses a context navigation technique to extract meaningful information stored in a concept hierarchy. A related study on Fuzzy logic reveals Fuzzy C-Means (FCM) is one of the basic unsupervised methods which finds the centroid of every cluster and assigns membership (μ) for each object whose summation for all clusters is unity [10].

$$\sum_{i=1}^{k} \mu_{ij} = 1 \tag{1}$$

here μ_{ij} represents the membership value of the jth data in the ith cluster and k indicates the number of clusters. FCM uses an objective function OJ_m with the aim to minimize it on constrained and unconstrained basis to yield Eq. (2)

$$OJ_m = \sum_{i=1}^{k} \sum_{j=1}^{n} (\mu_{ij})^m |x_j - c_i|^2 \tag{2}$$

A novel validity index was recommended by Fukuyama-Sugeno involving both the membership values and separation between cluster centroids was optimized [11] as in Eq. (3) shown below.

$$FS = \sum_{i=1}^{k} \sum_{j=1}^{n} (\mu_{ij})^m [|x_j - v_i|^2 - |v_i - v'|^2] \tag{3}$$

In the above equation the first term measures the compactness among clusters and the second term is a degree of separation between each cluster and the mean vector (v') of cluster centroids.

3 Methodology

The work proposes a fuzzy context classification technique for OLAP environment on text data where contextual factors are represented by concept hierarchy of dimensions. A context navigation method computes the concept weights and disseminates the weight from the leaf nodes to ancestor nodes to better co-relate the text. A new data structure, called Look-up table has been proposed which allows efficient searching of the desired documents by quantifying their occurrences and also enables OLAP operations at a faster rate than querying from the fact table. The initial pre-processing of raw text involves stop word removal, stemming, and part of speech tagging has been used for extracting the relevant concepts present in the form of noun (NN), adjective (ADJ) and verb (VBG). The notion of fuzzy context classification has been used to identify persons with two or more specialization. A visual representation of methodology is provided in Fig. 1.

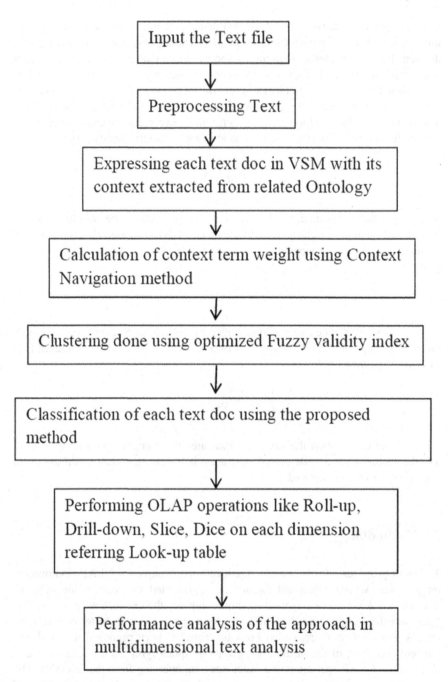

Fig. 1 Methodology

4 Proposed Context Navigation and Lookup Table in OLAP

This section discusses in detail the two novel ideas used in the work (a) Context navigation (b) Look-up table data structure.

4.1 Context Navigation

This work defines context as any information that can be used to characterize the situation of an entity—which can be person, a place, or an object that is considered relevant to the user and the application. A system is context-aware if it uses context to provide relevant information to its users [12]. The objective of our context navigation method is to circulate the scores apportioned to leaf nodes (n_L) to its ancestor nodes (n_A) for weight calculation through a tree structure. The distance (n_A, n_L) is the number of edges between them and can be computed as in Eq. (4).

$$\text{Weight}(n_A, n_L) = \text{Weight}(n_A) + \text{Weight}(n_L)^{\text{distance}(n_A, n_L)+1} \qquad (4)$$

4.2 Look-up Table

The Look-up table is an m × n matrix where m is the number of text documents that have been classified by the proposed classifier and n is the number of columns in the matrix in which the 1st column is the document id and rest (n − 1) columns are total number of contexts present in different levels defined in the concept hierarchy of any particular dimension. Any value *Val[m, $n_{n \neq 1}$]* corresponding to *n*th context of the *m*th document in the Look-up table can be either '0' or '1', where '0' means the document does not contain that particular context whereas value '1' confirms the presence of the context in that document instead of searching the entire fact table.

5 Proposed Context Based Fuzzy Classification Technique

The classification method starts from the highest level of hierarchy and compares the test document v with the symbolic vector of each node, if the two vectors are close enough based on similarity measure it decides that the test document belongs to the class the node represents, if no such close match is found based on threshold determined experimentally, it can be concluded that the document does not belong to this class and so it stops. The proposed model allows users to search a set of

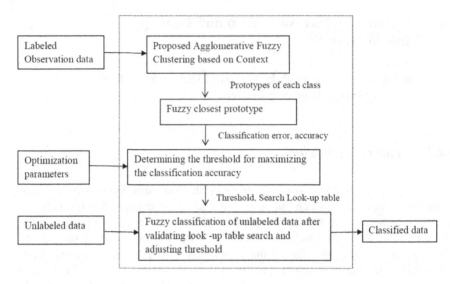

Fig. 2 Steps of the proposed fuzzy hierarchical classification

documents by mentioning desired contexts by referring from the look-up table. After consulting the look-up table, the documents with id value = '1' for searched contexts are retrieved. The steps of the proposed model are shown in Fig. 2.

Generally in fuzzy based classification, a function is used to calculate the membership vector $(m_1, \ldots, m_n)$ for the object O, where m_i is the degree of membership of O to class c_i, mapping a feature vector to a membership vector for the n classes. In data mining, this function is unknown and it is the task to find this function. In this work, a fuzzy validity function has been derived for clustering data shown in Eq. (5).

$$FI_{new} = \sum_{i=1}^{k} \sum_{j=1}^{n} \frac{(\mu_{ij})^m |x_j - c_i|^2}{n} - \frac{1}{k} \sum_{i=1}^{k} |c_i - c'|^2 \qquad (5)$$

Here the aim is to minimize the function by increasing the separation between cluster centroids and minimizing the global compactness among clusters. The first factor in Eq. (5) measures the overall compactness of the fuzzy partitions defined as the ratio between variance and cardinality while the second term indicates the average separation between cluster centroids. Since a good cluster partition produce low global compactness and high value of separation among partitions, the appropriate number of clusters was found by minimizing the objective function FI_{new} for over C_i for fixed partitions and μ_{ij} for fixed C_i subject to the constraint that $\sum_{i=1}^{k} \mu_{ij} = 1$.

5.1 Proposed Algorithm Fuzzy Classification in a Concept Hierarchy

Let v be the vector representation of the text document to be classified.
Let Class = NULL;
NodeSet = {set of concepts or nodes inserted for a particular dimension};
While (NodeSet ! = NULL) do
Retrieve a $node_i$ from NodeSet
NodeSet = NodeSet - $node_i$;
Compute the prototype (or symbolic vector) for $node_i$: sv = Wcom ($node_i$)
Calculate the similarity between v and sv : sim = cosine(v,sv);
If(sim > T) then
Class = Class U {v};
Look-Up table updated with '1' for matched contexts.
Update fuzzy membership (μ) for the class assigned
NodeSet = NodeSet U {getChildren($node_i$)}
Else print 'Test document not similar to the class'
End
Return Class

5.2 Time Complexity Analysis of Proposed Classifier

The extraction of features from a document and training the features in concept hierarchy takes a linear time of $O(n)$. The factors involved for time complexity computation: Computation of weight using term frequency and inverse document frequency is $O(n)$ again weight updation using context navigation technique involves $O(n \log n)$ time. The calculation of threshold takes $O(n \log n)$ time which takes into consideration the sorting cost of edges. Querying from Lookup table takes $O(n)$ time, thus the time complexity remains bounded by $O(n \log n)$.

6 Experiment and Result Analysis

6.1 Data Set Used for Experiment

The data set used for experimentation purpose was obtained from different resume portals from where 782 resumes of persons belonging to various skill sets were collected from different publicly accessible job portals available online [13–15]. Wikipedia and WordNet has been used as external online knowledge sources to construct the concept hierarchy [16, 17].

6.2 Experimental Metric Used

(a) **Classification accuracy**: Classification accuracy that refers to the percentage of documents correctly classified to a class and is defined as below:

$$\text{Accuracy (Acc)} = (\ C1\ + C4)/(C1 + C2 + C3 + C4) \tag{6}$$

C1 No. of documents selected in the class which are correct
C2 No of documents which are selected but should not belong to the class
C3 No of documents which should have been in the class but are rejected.
C4 No. of docs which should not be in class and not considered.

(b) **Searching time for various OLAP operations**

Actual searching time = Time taken to find out the document-ids containing the desired contexts + Retrieval Time of documents with the corresponding document-ids.

6.3 Results and Discussion

6.3.1 Experiment-1

The first experiment was devised to perform a comparison of accuracy with other established classification methods like Fuzzy C-means, Fuzzy nearest neighbour and Fuzzy k-medoids. The dimensions considered were: Dimension-1: Skill Specialization; Dimension-2: Region of person, Dimension-3: Experience, Dimension-4 Salary Range. The comparison results are shown in Fig. 3.

Results in Table 1 suggest that Fuzzy Nearest neighbour and fuzzy k-medoids perform significantly better in classification when compared with Fuzzy C-means, however the proposed classifier outperforms other classifiers for all dimensions by reducing the number of false negative cases for concept specific text dimensions.

6.3.2 Experiment-2

The objective of this section was to study the benefit of applying our context-aware fuzzy classification technique in various OLAP operations like-rollup, drill-down, slice and dice using the proposed look-up table in comparison to search from the fact table.

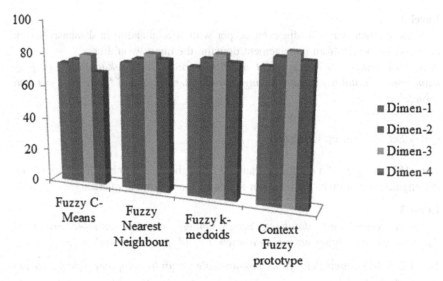

Fig. 3 Comparison of classification accuracy dimension wise

Table 1 Comparison of accuracy values with other classifiers

Classification method	Average accuracy (%) of classification dimension wise using data of resumes			
	Dimen1	Dimen2	Dimen3	Dimen4
Fuzzy C-means	73.7	76.4	79.7	69.6
Fuzzy-nearest neighbour	76.9	79.6	83.1	80.3
Fuzzy k-medoids	77.3	82.5	85.7	81.2
Context based fuzzy prototype	80.5	86.4	89.9	85.1

Queries for Drill-Down

At Level 1:
If the HR Manager of a company wants to count the total number of resumes from the management domain
Select count(fileid) from look-up-table group by management having 'management' = 1;

Level 2:
If the manger wants to further drill down and counts the number of resumes with specialization in database from management domain, the query would be:
Select count(fileid), management, database from look-up-table group by management, database having 'management' = 1 and 'database' = 1;

Level 3:

If the manger wants to dig even deeper with specialization in database having expertise on oracle from management domain, the query would be:

Select count(fileid), management, database, oracle from look-up-table group by management, database, oracle having 'management' = 1 and 'database' = 1 and 'oracle' = 1;

Query Set for Roll-up Operation

If the HR Manager of a company wants to count the total number of resumes from the engineering domain then we can represent the following queries:

Level 3:

Select count(fileid), database, programming from look-up-table group by database, programming having database='1' and 'programming' = 1;

Level 2: *Select count(fileid) from look-up-table group by computer_science having computer_science='1';*

Level 1:

Select count(fileid) from look-up-table group by engineering having engineering='1';

Slice Operation

The slice operation selects one particular dimension from a given cube and provides a new sub-cube.

Select count(fileid) from look_up_table_region where Location='Kolkata' group by location;

Dice Operation

Dice selects two or more dimensions from a given cube and provides a new sub-cube.

Dice Query

Select count(file_id) from look_up_table_topic as t_topic, look_up_table_location as t_location where t_topic.file_id = t_location.file_id and (topic = 'autocad' or topic = 'autodesk') and location='Kolkata' group by topic, location;

Discussion of Results

The experimental results evidently show that with increasing level of concept hierarchy the execution time for drill-down increases substantially and for slice and

Table 2 Operation time in wall clock (milliseconds) for the dimension domain specialization using proposed classification method

Levels	Rollup	Drill down	Slice	Dice
Level-1	448.71	93.12	14.89	176.42
Level-2	272.46	218.67	16.17	251.40
Level-3	196.72	396.84	19.06	370.72

Table 3 Operation time in wall clock (milliseconds) for dimension skill specialization using search from fact table by aggregation

Levels	Rollup	Drill down	Slice	Dice
Level-1	1105.8	864.8	56.3	472.6
Level-2	865.2	1346.2	62.5	906.3
Level-3	347.4	2046.7	76.9	1674.4

dice it increases but less progressively. This is due larger number of concept nodes lower down the hierarchy and more specificity of information. While the roll-up operation for both the data sets take maximum time at the topmost level due to larger time for aggregation up the concept hierarchy. Table 2 and Table 3 above shows the operation time of OLAP operations for the proposed model and fact table respectively (Fig. 4).

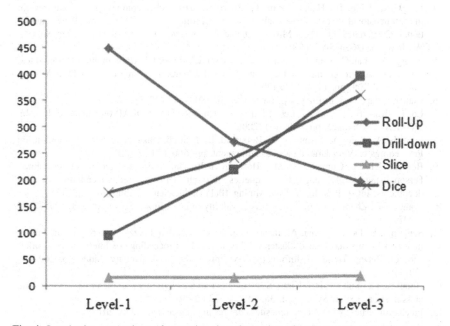

Fig. 4 Level wise comparison of operation time for various OLAP operations using proposed context based classification for the dimension job specialization

7 Conclusion and Future Scope

In this work a context based fuzzy classification technique has been proposed on OLAP dimensions to analyse unstructured text in a multidimensional database. The work integrated a context navigation technique on the dimensions represented by concept hierarchies to propagate the concepts of text data. Experiments were performed using resume data to validate the model and comparative results with other contemporary classification methods based on both accuracy and query execution time for OLAP operations revealed the superiority of our approach. In future the work will apply the same approach on a much larger and wider data set apt for OLAP domain to explore its application on broader perspectives.

References

1. Bansal, N., Koudas, N.: Blogscope a system for online analysis of high volume text streams. In: Proceedings of the 33rd International Conference on Very Large Data Bases, VLDB Endowment, VLDB'07, pp. 1410–1413 (2007)
2. Jiang, F., Pei, F., Chee Fu, A.W.: 'Ix-cubes: iceberg cubes for data warehousing and olap on xml data. In: Proceedings of the Sixteenth ACM Conference on Information and Knowledge Management, CIKM'07, pp. 905–908. ACM, USA (2007)
3. Chakravarthy, V., Gupta, H., Roy, P., Mohania, M.: Efficiently linking text documents with relevant structured information. In: Proceedings of the 32nd International Conference on Very Large Data Bases, VLDB Endowment, VLDB'06, pp. 667–678 (2006)
4. Lin, C.X., Ding, B., Han, J., Zhu, F., Zhao, B.: Text cube: computing IR measures for multidimensional text database analysis. In: Proceedings of the 2008 Eighth IEEE International Conference on Data Mining, ICDM'08, pp. 905–910. IEEE Computer Society, Washington, DC, USA (2008)
5. Zhang, D., Zhai, C., Han, J.: Topic cube: topic modelling for OLAP on multidimensional text databases. In: Proceedings of the Ninth SIAM International Conference on Data Mining, SIAM, SDM'09, pp. 1124–1135 (2009)
6. Han, J.: Olap, Spatial. Encyclopedia of GIS, pp. 809–812 (2008)
7. Lo, E., Kao, B., Ho, S., D. Lee, Chui, C.K., Cheung, D.W.: OLAP on sequence data. In: SIGMOD Conference, pp. 649–660 (2008)
8. Simitsis, A., Baid, A., Sismanis, Y., Reinwald, B.: Multidimensional content exploration. In: Proceedings of Very large database endowment, pp. 660–671 (2008)
9. Berbel, T.R.L., Gonzàlez, S.M.: How to help end users to get better decisions? Personalising OLAP aggregation queries through semantic recommendation of text documents. Int. J. Bus. Intell. Data Mining 10(1) (Inderscience Publications) (2015)
10. Wanga, W., Zhanga, Y.: On fuzzy cluster validity indices. Fuzzy Sets Syst. 158, 2095–2117 (2007)
11. Sengupta, S., De, S., Konar, A., Janarthanan, R.: An improved fuzzy clustering method using modified Fukuyama-Sugeno cluster validity index. In: Proceedings of International Conference on Recent Trends in Information Systems, Jadavpur University, India, pp. 269–274 (2011)
12. Hong, J.Y., Suh, E.H., Kim, S.J.: Context-aware systems: a literature review and classification. Expert Syst. Appl. 36, 8509–8522 (2009)
13. FreeResumesites. www.freeresumesites.com/resume. Accessed 2 July 2016

14. Free-resume-database. http://www.9amjobs.com/free-resumedatabase.aspx. Accessed 5 July 2016
15. Eresumex. http://www.eresumex.com. Accessed 3 July 2016
16. Wikipedia. http://en.wikipedia.org. Accessed 20 Aug 2016
17. Wordnet. https://wordnet.princeton.edu. Accessed 21 Aug 2016

A Survey on Event Detection Methods on Various Social Media

Madichetty Sreenivasulu and M. Sridevi

Abstract Today, social media play a very important role in the world to share the real-world information, everyday existence stories, and thoughts through the virtual communities or networks. Different types of social media's such as Twitter, Blogs, News Archieve, etc., have heterogeneous information with various formats. This information is useful to the real time events such as disasters, power outage, traffic, etc. Analyzing and understanding such information on different social media are a challenging task due to the presence of noisy data, unrelated data, and data with different formats. Hence, this paper focuses on various event detection methods in different types of social media and categorizes them according to the media type. Moreover, features and data sets of various social media are also explained in this paper.

Keywords Social media · Twitter · Event detection · User generated content
YouTube · News archive

1 Introduction

Nowadays, social media play a very important role in the world. Social media can be used for sharing information's, ideas, and interests among the people throughout the world. Lot of people communicated through social media by sending and receiving messages. Social media can be different types such as Twitter, news archive, multimedia, BlogPosts, web pages, Facebook, etc. Social media generate huge amount of information it will be very much useful to the people. The information can be described as something happened outside is known as event. Social media contains

M. Sreenivasulu (✉) · M. Sridevi
Department of Computer Science & Engineering, National Institute of Technology,
Tiruchirappalli, Tiruchirappalli, Tamil Nadu, India
e-mail: sreea568@gmail.com

M. Sridevi
e-mail: msridevi@nitt.edu

© Springer Nature Singapore Pte Ltd. 2018
P. K. Sa et al. (eds.), *Recent Findings in Intelligent Computing Techniques*,
Advances in Intelligent Systems and Computing 709,
https://doi.org/10.1007/978-981-10-8633-5_9

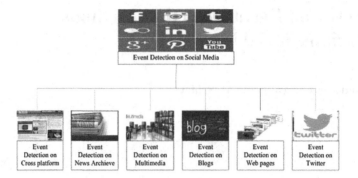

Fig. 1 Event detection on different types of social media

information about related to be event and not related to event. During a crisis, colossal information can be posted in social media. In that, determining the information related to event is challenging task on social media. Different types of social media have the different characteristics. Figure 1 describes about event detection on social media.

Whenever social event occurs, traditional media take 2 h or even days for reporting the news of a particular event, at the same time corresponding information may begin to spread promptly in the social media like Twitter. Twitter is a free social networking microblogging service which allows users for registering and post short texts (messages) on microblog called tweets. Each tweet allows 140 characters which may be text images, hyperlink, videos, etc. As per the statistics given in [1], average of 41 million users have been registered in July 2009 which was drastically increased to 317 million users, in August 2016. Whenever the user follows any another user, user is able to read the tweets and share the same tweet, called retweet. A user has the ability to read the tweet which is related to other users. In twitter, the user-generated contents are updated each and every time which will be useful to the users in the twitter [2]. During disasters, large colossal information can be generated in the microblog. At that time, it is very important to know where the disasters occur and where can be disseminated? And there is a need to know what is happening? in those locations. Extraction of the event is a challenging task. By entering pertinent word to search engine (Google, Bing, Yahoo! etc) which gives very large roll of URL which are related to some concepts of interests. At the present, researchers mainly focus on ranking of the list of documents based on the weight [3].

Users easily access the news, web, videos by leverage the search engines and video sharing websites such as Google, YouTube, Baidu and YouKu. Newswires like CNN, BBC, and CCTV also publish news videos. According to the 2014 YouTube data, for every minute, 100 h of videos are uploaded to YouTube, and each month six billion hours of videos are watched on the YouTube. These factualities describe a new challenge for the users to clench the great events available from searching video databases [4]. In general, belief of the users is initially user-generated content (UGC) and got huge popularity and then gradually dim into oblivion. Whenever analyzing more than 350k YouTube videos, it needs at least 1 year time for uploading videos to get huge popularity. These videos are called sleeping beauties. Sleeping beauties will get huge popularity after one year which is challenging a problem. Identifying such videos will not only comfort the advertisers, but also the designers of recommendation systems who search for to maximum user satisfaction [5].

Due to ubiquitous of the We-media, information can be received and published into numerous forms at anywhere at any time with leverage the Internet.The rich cross-media information contains multimodal data in multiple media and has many audiences. The large information available on different events which have great impact compare to single media information. Hence, detection of the event in cross-media is a challenging problem in the following aspects (1) It contains the different media data which has different characteristics (2) Topics are presented among the noisy web data. Detection of the event in cross-media is very helpful for the government organization and advertising agencies [6].

Today, there is huge growth in web technology, social websites have become the flexible platform for users to access the world and exchanging the ideas (opinions). However, prodigious explosion in the extensiveness of web pages from the social media has made the strenuous problem for the users to access the hot topics and web administrators to detect the web activities properly. The content of social media is less restricted and less predictable compared to others (news articles) and also it is short, noisy and sparse [7].

2 Literature Survey

First story detection (FSD) or new event detection can be defined as identifying the event which is unseen or unknown previously. In [8], the author proposed a method which is called Inverted Index for FSD, which is efficient and it is helpful for large data streams. This approach was used by the Umass FSD system. The limitation of the proposed method is that it is not scalable to unbounded streams and processing of one document will take longer time. Hence, it will not provide good performance for Twitter due to continuous tweets, it is clear as some document takes long time for processing each document than arrival of tweets. In [9], the author proposed a method called probabilistic approaches for FSD. The main advantage of this method is that it presents the data in structured way but it is more expensive and also very less suitable for processing large amount of data.

In [10], the author proposed a method for event detection in blogs called temporal random indexing. Blogs are less structured compared to newswire and it can be updated very fast. The main limitation of this method is that the events based around keywords need to specify the keywords explicitly and require lot of volume for semantic shift. It is not suitable for a large-scale event detection.

Support Vector Machine (SVM) for event detection in the Twitter during disasters is used in [11]. It detects specific types of events and it is more attractive for small sets of events with high precision. It is unfeasible for any type of events. If large number of events are there then it requires separate classifier which needs to be trained for each event and also requires labeled data.

In [12], the author proposed a hybrid method to detect the events from the noisy and fragmented tweet. In this paper, the author considered power outage as event and power outage detection as an event detection. Hybrid method is a combination of heterogeneous information network and supervised topic model. Heterogeneous information network contains both temporal and spatial information of tweets. Supervised topic model used heterogeneous information network for improving the accuracy of event detection compared to existing methods like SVM, Bayesian model, etc.

A method was developed for extracting the traffic events like traffic, public transport, water supply, weather, sewage, and public safety from the tweets which are posted by the observations [17]. In general, for detecting the events, there is a need to understand the tweet stream which is a challenging task in social media. In this paper, hypothesize the problem of annotating the tweet stream as a sequence labeling problem. For the sequence labeling models, there is a need for manual labeled training data. The proposed method is a novel approach for creating the automatic training label based on the instance-level domain knowledge and it indicates locations in a city and possible event occurrences. It gives comparable performance to annotated tasks which is more advantageous for training and avoids the need of the manual labeled training process.

A novel framework is proposed in [4] to better the group associated web videos to events. First, the data preprocessing is performed to select features and tag relevance learning. Next, multiple correspondence analysis is applied to explore the correlations between terms and events with the assistance of visual information. Co-occurrence and visual near duplicate feature trajectory induced from near duplicate key frames (NDKs) are combined to calculate the similarity between NDKs and events. Finally, a probabilistic model is proposed for news web video event mining, where both visual temporal information and textual distribution information are integrated.

In [7], a method is explored in the view of similarity diffusion. The method uses clustering-like pattern across similarity cascades (SCs). SCs are chain of subgraphs generated by pruning the similarity graphs with a set of thresholds and capturing the topics with the help of the maximum cliques. At last, they discovered the real topics in effective manner from huge number of candidates with the use of topic-restricted similarity diffusion method. This method is experimented on three public datasets such as MCG-WEBV, YKS and Social Event Detection 2012 (SED2012). In this experiment, they considered only social event as web topics and ignored the geographical information. It gives better performance. The summary of the methods discussed for event detection on various social media is tabulated in Table 1.

Table 1 Summaries of event detection on various social media

Sl. No.	Author and reference	Media	Method	Features	Datasets	Advantages
1	Wayne Xin Zhao et al. [13]	News articles	Burst VSM	Bursty features	68 millon web pages	• Effective for large corpus of news archieve
2	Thorsten Brants et al. [14]	News stories	IncrementalTF-IDF model	Hellinger distance document similarity normalization	TDT2, TDT3 andTDT4 corpora	• Gives good results in news stories. • Reduces the cost
3	Ting Hua et al. [15]	Twitter	Targeted-domain spatio-temporal event detection (ATSED)	Twitter location information	July 2012 to May 2013 of both twitter and GSR data	• Automatic assignment of labels to the tweets
4	Maia Zaharieva et al. [16]	MultiMedia	An unsupervised, multistaged approach	Time and location information	306, 159 Flickr photos and 969 YouTube videos	• More robust grouping of media for different platforms
5	Lingyang Chu et al. [6]	MultiMedia	Multimodality fusion framework and a topic recovery (TR)	Multimodality graph	MCG-WEBV and YouKu Sina dataset (YKS)	• Very flexible for different platforms
6	Haifeng Sun et al. [12]	Twitter	Supervised topic model with heterogeneous information	Temporal, spatial and information	10,000 labeled tweets related to power outage detection	• locate events without any additional measurements
7	Pramod Anantharam et al. [17]	Twitter	Automatic labeling to training data	Locations of city and event term	4 months tweets of San Francisco Bay area	• Uses best classification model • Suitable for traffic agents
8	Wei Wang et al. [18]	News articles	A multiple instance learning based on CNN	Labels at the document level	GSR 19795 news article related to event and 3759 articles (not related)	• Works very fast and feasible for news articles

3 Conclusion

The enormous amount of user-generated content (UGC) can be disseminated in social media. Analyzing and understanding the UGC are very helpful to the users at the time of disasters, power outage, traffic, etc. Event detection is one of the important tasks and targets at identifying real-world occurrences. However, event detection on social media must be efficient and accurate. In this paper, survey of different methods is explained based on event detection with various types of social media. In future, A new approach for event detection in social media will be proposed to improve the accuracy and speed.

References

1. Kwak, H., Lee, C., Park, H., Moon, S.: What is twitter, a social network or a news media? In: Proceedings of the 19th International Conference on World wide web, pp. 591–600. ACM (2010)
2. Pervin, N., Fang, F., Datta, A., Dutta, K., Vandermeer, D.: Fast, scalable, and context-sensitive detection of trending topics in microblog post streams. ACM Trans. Manag. Inf. Syst. (TMIS) 3(4), 19 (2013)
3. Mori, M., Miura, T., Shioya, I.: Topic detection and tracking for news web pages. In: Proceedings of the 2006 IEEE/WIC/ACM International Conference on Web Intelligence, pp. 338–342. IEEE Computer Society (2006)
4. Zhang, C., Xiao, W., Shyu, M.-L., Peng, Q.: Integration of visual temporal information and textual distribution information for news web video event mining. IEEE Trans. Hum.-Mach. Syst. 46(1), 124–135 (2016)
5. Sikdar, S., Chaudhary, A., Kumar, S., Ganguly, N., Chakraborty, A., Kumar, G., Patil, A., Mukherjee, A.: Identifying and characterizing sleeping beauties on youtube. In: Proceedings of the 19th ACM Conference on Computer Supported Cooperative Work and Social Computing Companion, pp. 405–408. ACM (2016)
6. Chu, L., Zhang, Y., Li, G., Wang, S., Zhang, W., Huang, Q.: Effective multimodality fusion framework for cross-media topic detection. IEEE Trans. Circuits Syst. Video Technol. 26(3), 556–569 (2016)
7. Pang, J., Jia, F., Zhang, C., Zhang, W., Huang, Q., Yin, B.: Unsupervised web topic detection using a ranked clustering-like pattern across similarity cascades. IEEE Trans. Multimed. 17(6), 843–853 (2015)
8. Allan, J., Lavrenko, V., Malin, D., Swan, R.: Detections, bounds, and timelines: Umass and tdt-3. In: Proceedings of Topic Detection and Tracking Workshop, pp. 167–174 (2000)
9. Ahmed, A., Ho, Q., Eisenstein, J., Xing, E., Smola, A.J., Teo, C.H.: Unified analysis of streaming news. In: Proceedings of the 20th International Conference on World Wide Web, pp. 267–276. ACM (2011)
10. Jurgens, D., Stevens, K.: Event detection in blogs using temporal random indexing. In: Proceedings of the Workshop on Events in Emerging Text Types, pp. 9–16. Association for Computational Linguistics (2009)
11. Sakaki, T., Okazaki, M., Matsuo, Y.: Tweet analysis for real-time event detection and earthquake reporting system development. IEEE Trans. Knowl. Data Eng. 25(4), 919–931 (2013)
12. Sun, H., Wang, Z., Wang, J., Huang, Z., Carrington, N., Liao, J.: Data-driven power outage detection by social sensors. IEEE Trans. Smart Grid 7(5), 2516–2524 (2016)
13. Zhao, W.X., Chen, R., Fan, K., Yan, H., Li, X.: A novel burst-based text representation model for scalable event detection. In: Proceedings of the 50th Annual Meeting of the Association for

Computational Linguistics: Short Papers-Volume 2, pp. 43–47. Association for Computational Linguistics (2012)

14. Brants, T., Chen, F., Farahat, A.: A system for new event detection. In: Proceedings of the 26th Annual International ACM SIGIR Conference on Research and Development in Informaion Retrieval, pp. 330–337. ACM (2003)

15. Hua, T., Chen, F., Zhao, L., Chang-Tien, L., Ramakrishnan, N.: Automatic targeted-domain spatiotemporal event detection in twitter. GeoInformatica **20**(4), 765–795 (2016)

16. Zaharieva, M., Del Fabro, M., Zeppelzauer, M.: Cross-platform social event detection. IEEE MultiMed. (2015)

17. Anantharam, P., Barnaghi, P., Thirunarayan, K., Sheth, A.: Extracting city traffic events from social streams. ACM Trans. Intell. Syst. Technol. (TIST) **6**(4), 43 (2015)

18. Wang, W., Ning, Y., Rangwala, H., Ramakrishnan, N.: A multiple instance learning framework for identifying key sentences and detecting events. In: Proceedings of the 25th ACM International on Conference on Information and Knowledge Management, pp. 509–518. ACM (2016)

A Comparative Study of Online Resources for Extracting Target Language Translation

Vijay Kumar Sharma and Namita Mittal

Abstract Today, online resources are effectively used for identifying target language translations. This study belongs to Hindi to English translations resources, which will be very helpful for Machine Translation (MT) systems or Cross-Lingual Information Retrieval (CLIR). A number of online translation resources are studied and their efficiency and effectiveness are discussed in this paper. Wikipedia, Indo WordNet, and ConceptNet usability for translation are discussed. Different online dictionaries are evaluated and their translation effectiveness are discussed. The Google and Bing online translation systems are also analyzed.

Keywords Indo WordNet · ConceptNet · Dictionary · Wikipedia
Translation

1 Introduction

Magnification of multilingual content on the web needs some translation mechanism which would be able to cross the language barriers. The numbers of non-English users are increased tremendously, and all of them are unable to express their queries in English. The translation mechanism is useful for Cross-Lingual Information Retrieval (CLIR), where a user can search in his native language without caring the target documents language. Dictionaries and Corpora are the most widely used resources for translation. These approaches are simple and efficient but not so effective because they suffered from Out Of Vocabulary (OOV) issues, i.e., words which are not found in the dictionaries or corpora [1].

V. K. Sharma (✉) · N. Mittal
Malaviya National Institute of Technology Jaipur, Jaipur, India
e-mail: sharmavijaykumar55@gmail.com

N. Mittal
e-mail: mittalnamita@gmail.com

© Springer Nature Singapore Pte Ltd. 2018
P. K. Sa et al. (eds.), *Recent Findings in Intelligent Computing Techniques*,
Advances in Intelligent Systems and Computing 709,
https://doi.org/10.1007/978-981-10-8633-5_10

In the case of CLIR, Latent Semantic Indexing (LSI), Latent Dirichlet Allocation (LDA), and Explicit Semantic Indexing (ESI) approaches are effective but so complex, and they process a corpus with high computation cost. A word may have multiple translations in the case of a dictionary or corpora that come in Word Translation Disambiguation (WTD) issue. Co-occurrence statistics and Point-wise Mutual Information (PMI) are the most widely used WTD models, but their performance depends on a corpus quality and coverage. Word Embeddings (WE) bring a revolutionary change in WTD [2, 3].

Nowadays, online resources are going to be used for finding a word translation. Researcher uses online dictionaries, Wikipedia and online translation system for the foreign language. Our contribution in this work for the Hindi language is to: (1) Analysis of WordNet, Indo WordNet, and ConceptNet. Are we able to use them for translation? (2) Analysis of available Hindi-English online dictionaries. Is it beneficial to use online dictionaries instead of off-line dictionaries? (3) Is there any significant difference between Google and Bing translation systems? (4) Is Wikipedia being useful as a translation resource? Section 2 represents related work. Different online translation resources are discussed in Sect. 3. Section 4 represents a detailed discussion and conclusion is discussed in Sect. 5.

2 Related Work

Vahid et al. analyzed Google and Bing translator system for foreign language [4]. Sharma et al. used an off-line Hindi-English dictionary for translation and transliteration mining algorithms were used for handling OOV words [5]. The bilingual word embeddings are used to select the best translation and Wikipedia knowledge base for finding word translation dynamically [2, 3]. Sorg et al. used explicit semantic indexing where Wikipedia articles are used as external concepts [6]. Vulic et al. showed that dual semantic space based translation models CL-LSI, CL-LDA are effective but not efficient [7].

Jagarlamudi et al. prepared a Statistical Machine Translation (SMT) system which trained on aligned parallel sentences [8]. Mahapatra et al. used the sentence word overlap score and WordNet similarity score to select the best translation [9]. Saravanan et al. used the transliteration generation or mining technique for OOV words [10]. Surya et al. and Shishtla et al. used the CRF model for OOV word transliteration [11]. Larkey et al. and Nie et al. used the probabilistic dictionary for query translation [12, 13]. Bradford et al. used the SMT to create parallel corpora and Cross-Lingual LSI method was used for CLIR [14]. Udupa et al. used the GIZA++ tool to create a probabilistic lexicon and machine transliteration for OOV words [15].

3   Online Translation Resources

Manually constructed dictionaries and corpora are the traditional sources of translation, but they suffered from OOV issues, so researchers are moving on to online resources because they are dynamic in nature. Various online resources are analyzed and used for foreign languages but not analyzed for the Hindi language. WordNet, Indo WordNet, ConceptNet, Online dictionaries, Online translation system, and Wikipedia are analyzed in next subsections.

3.1   Hindi WordNet, Indo WordNet and Concept Net

The Hindi WordNet[1] system brings different lexical and semantic relations between the Hindi words. Word meanings are represented as lexical information and can be termed as a lexicon based on psycholinguistic principles. It is developed by Centre for Indian Language Technology (CFILT), IIT Bombay.

Indo WordNet is an advanced version of WordNet[2] having some advanced features. It includes different lexical and semantic relations in Hindi, English, and 17 other Indian languages. Synonymy and Gloss features can be useful for translation; as Hindi word synonymy features, gloss sentences and English synonymy features, gloss sentences are the mutual translations.

Concept Net is a semantic network containing lots of things computers should know about the world, especially when understanding text written by people. It contains *concept nodes* and *relationship set* for every word. Concept nodes are the words, or short phrases and relationship set shows the existing relation between concepts. It is available in world's different languages including the Hindi language. Two ConceptNet[3] features can be used for translation, i.e., "TranslationOf" and "RelatedTo".

3.2   Online Dictionaries and Wikipedia Based Translation

Various online dictionaries are available for translations; many of them are stable and not updated automatically. Some dictionaries are timely dynamic; updated after a certain period. Automatically constructed dictionaries are suffered from a lot of ambiguities than the manually constructed dictionaries, but dictionary coverage capacities of automatically constructed dictionaries are greater. Different Hindi to English online dictionaries are analyzed and presented in Table 1.

---

[1]http://www.cfilt.iitb.ac.in/wordnet/webhwn/.

[2]http://www.cfilt.iitb.ac.in/indowordnet/index.jsp.

[3]http://conceptnet5.media.mit.edu/.

Table 1 Comparison of different online dictionaries

Dictionary	Creator	Length	Searching	Weak point
Universal word (UW)	CFILT, IIT Bombay	136109	Exact and max match	OOV issue, Inclusion of irrelevant and wrong translations
Shabdkosh (SK)	Student IIT Delhi	100000	Exact match	OOV issue, Inclusion of unnecessary relevant translations
Raftar (RF)	rafter.in	_	Exact match	Roman transliterated searching, few relevant translations found
HinKhoj (HK)	hindkhoj.com	_	Exact match	Roman transliterated searching, few relevant translations found
Tamilcube (TC)	tamilcube.com	200000	Max match	Inclusion of unnecessary relevant translations
Indiatyping (IT)	indiatyping.com	_	Exact and max match	Inclusion of unnecessary relevant translations
Online translation	Google and Bing	_	Semantic, syntactic	Handle OOV, MWE, NE. no significant difference identified

The Wikipedia[4] is an online encyclopedia which is editable by users across the World Wide Web. It is very useful for resource-poor languages like Hindi, and its structure and content make it amenable to linguistic research. Every Wikipedia article has the *title* and *inter-wiki link* features which are useful for extracting target language translations.

4 Results and Discussion

The online translation resources are analyzed with the help of FIRE 2010 and 2011 topic sets which contain 50 queries. The evaluation matrices are not available for analyzing online translation resources, so all resources are analyzed manually. A comparative analysis of WordNet, Indo WordNet, and ConceptNet is presented in Table 2. Each row represents a query where the first column shows a query word, and next column shows number of senses on WordNet (W), Indo WordNet (I), and ConceptNet (C). WordNet shows only senses, but the translation is achieved by Indo WordNet because Indo WordNet contains Hindi synonym and English synonym mutually while WordNet does not have English synonyms. Sometimes, a word has multiple WordNet senses but do not have a single sense on the Indo WordNet. Even the Indo WordNet shows empty linked pages for the pages that are suggested by itself, while those pages have multiple senses on WordNet. The Indo

---

[4]https://hi.wikipedia.org/wiki/Special:Search?search=&go=Go.

Table 2 Comparative analysis among WordNet, Indo WordNet, and ConceptNet

S.no.	Word	W,I,C	Word	W,I,C	Word	W,I,C	Word	W,I,C
				FIRE 2010 Queries				
1	गुज्जरों	2,0,0	मीणा	0,0,0	समुदाय	3,3,1	संघर्ष	1,0,1
2	राम	2,2,0	मंदिर	2,2,0	आडवाणी	0,0,0	सिंघल	0,0,0
				FIRE 2011 Queries				
3	स्वाइन	0,0,0	फ्लू	1,0,0	टीके	0,0,0		
4	माइकल	0,0,0	जैक्सन	0,0,1	अचानक	1,0,1	मृत्यु	1,2,1

Table 3 Comparative analysis of online dictionaries

S.no.	Word	NOT	Word	NOT	Word	NOT	Word	NOT
				FIRE 2010 Queries				
1.	रिश्वत	$$2$$$	बदले	$10$$0	संसद	$21$$3	प्रश्न	$$$$$$
2.	गुटखा	000000	मालिकों	000$00	अन्डरवर्ल्ड	000000	उलझाव	3413$4
				FIRE 2011 Queries				
3.	अबू	$00$$0	गरीब	$$$$$$	जेल	$$$$$$	अत्याचार	$$$$$$
4.	भोपाल	000001	गैस	$$1$$$	दुर्घटना	$$$$$$		

WordNet has fewer entries than the WordNet. So it cannot be useful for extracting target language translation. The ConceptNet features *TranslationOf* and *RelatedTo* render target language translation. A Hindi query word is searched for Hindi version of ConceptNet. The numbers of ConceptNet features are 0 for most of the words and 1 for few words. So the ConceptNet cannot be used for extracting target language translation.

Six online dictionaries are evaluated which are briefly discussed in the previous section. A comparative analysis is presented in Table 3. Each query word's next column has a six digit entry. Each digit represents the number of translations that are found in each online dictionary for a query word. The $ symbol is used in case of finding more than five translations. The order of digits is according to the dictionaries order that is [UW, SK, RF, HK, TC, IT]. Most of the entries are $ that again put a WTD issue. Many dictionaries show entries with digit 0 for the Named Entity (NE) terms or Newly Identified (NI) terms which is an issue of OOV words or dictionary coverage. Offline dictionaries are also useful with transliteration mining algorithm [5]. Offline dictionaries are same as stable online dictionaries. Online dynamic dictionaries achieve better performance than the offline dictionaries.

Two online translation systems Google and Bing are analyzed manually. There is not any significant difference between them but few statistics are noted as (i) Mostly Google and Bing both give the same translations, if different then the translations are synonyms of each other (ii) If a word is not translated and it requires

Table 4 Wikipedia online and off-line Hindi data statistics

Dataset	Available Hindi articles	Available English inter-wiki link articles	Percentage of availability of inter-wiki link over the total available article (%)
FIRE 2010	4185	3244	77.51
FIRE 2011	2460	1849	75.16
Hindi Wikipedia dumps	140k	37k	26.42

transliteration, then transliterations variation can occur in both of the transliteration. Wikipedia online encyclopedia and off-line dumps[5] are also analyzed for translation. Online Wikipedia analysis for FIRE 2010 and 2011 datasets are represented in Table 4. Around 75% online Wikipedia Hindi titles have English inter-wiki links for the FIRE 2010 and 2011 query set, and around 26% off-line Hindi Wikipedia articles from the offline Hindi Wikipedia dumps have English inter-wiki links.

5 Conclusion

The online translation resources are analyzed for Hindi to English translation. The WordNet, Indo WordNet, and ConceptNet are manually analyzed for Hindi language and concluded, that these resources are not rich and have poor vocabulary coverage, so cannot be used for translation at this stage. Both offline and online dictionaries need WTD. The Online dictionaries will be more effective than the offline dictionaries due to their dynamic nature. There is not any significant difference between Google and Bing. Only synonym and transliteration differences are occurred during analysis. The significant results are achieved during online Wikipedia analysis [3]. Off-line Wikipedia may also be useful, but it needs a large preprocessing and refining time because of only 26% availability of English inter-wiki links.

---

[5]https://dumps.wikimedia.org/.

References

1. Sharma, V.K, Mittal, N.: Cross lingual information retrieval (CLIR): review of tools, challenges and translation approaches. In Information System Design and Intelligent Application, pp. 699–708 (2016)
2. Sharma, V.K., Mittal, N.: Exploring bilingual word vectors for Hindi-English cross-language information retrieval. In: Proceedings of the International Conference on Informatics and Analytics. ACM (2016)
3. Sharma, V.K., Mittal, N.: Exploiting Wikipedia API for Hindi-English cross-language information retrieval. Procedia Comput. Sci. **89**, 434–440 (2016)
4. Hosseinzadeh Vahid, A.: A comparative study of online translation services for cross language information retrieval. In: Proceedings of the 24th International Conference on World Wide Web. ACM (2015)
5. Sharma, V.K., Mittal, N.: Cross lingual information retrieval: a dictionary based query translation approach. In: Advances in Intelligent Systems and Computing (2016)
6. Sorg, P., Cimiano, P.: Exploiting Wikipedia for cross-lingual and multilingual information retrieval. Data Knowl. Eng. **74**(2012), 26–45 (2012)
7. Vulić, Ivan, De Smet, Wim, Moens, Marie-Francine: Cross-language information retrieval models based on latent topic models trained with document-aligned comparable corpora. Inf. Retr. **16**(3), 331–368 (2013)
8. Jagarlamudi, J., Kumaran, A.: Cross-lingual information retrieval system for indian languages. Advances in Multilingual and Multimodal Information Retrieval, pp. 80–87. Springer, Berlin Heidelberg (2007)
9. Mahapatra, L., Mohan, M., Khapra, M.M., Bhattacharyya, P.: OWNS: cross-lingual word sense disambiguation using weighted overlap counts and wordnet based similarity measures. In: Proceedings of the 5th International Workshop on Semantic Evaluation. Association for Computational Linguistics, pp. 138–141 (2010)
10. Saravanan, K., Udupa, R., Kumaran, A.: Crosslingual information retrieval system enhanced with transliteration generation and mining. In: Forum for Information Retrieval Evaluation (FIRE-2010) Workshop (2010)
11. Surya, G, Harsha, S, Pingali, P, Verma, V.: Statistical transliteration for cross language information retrieval using HMM alignment model and CRF. In: Proceedings of the 2nd Workshop on Cross Lingual Information Access (2008)
12. Larkey, L.S., Connell, M.E., Abduljaleel, N.: Hindi CLIR in thirty days. ACM Trans. Asian Lang. Inf. Process. (TALIP) **2**(2), 130–142 (2003)
13. Nie, J., Simard, M., Isabelle, P., Durand, R.: Cross-language information retrieval based on parallel texts and automatic mining of parallel texts from the Web. In: Proceedings of the 22nd Annual International ACM SIGIR Conference on Research and Development in Information Retrieval. ACM, pp. 74–81 (1999)
14. Bradford, R., Pozniak, J.: Combining modern machine translation software with LSI for cross-lingual information processing. In: 2014 11th International Conference on Information Technology: New Generations (ITNG). IEEE, pp. 65–72 (2014)
15. Udupa, R., Jagarlamudi, J., Saravanan, K.: Microsoft research india at fire2008: Hindi-english cross-language information retrieval. In: Working Notes for Forum for Information Retrieval Evaluation (FIRE) Workshop (2008)

A Fuzzy Document Clustering Model Based on Relevant Ranked Terms

K. Sreelekshmi and R. Remya

Abstract The web today is a growing universe of vast amounts of documents. Clustering techniques help to enhance information retrieval and processing huge volume of data, as it groups similar documents into one group. The relevant feature identification from a high-dimensional data is one of the challenges in text document clustering. We propose a sentence ranking approach which finds out the relevant terms in the documents so as to improve the feature identification and selection. Preserving the correlation between terms in the document, the document vectors are mapped into a lower dimensional concept space. We used k-rank approximation method which minimizes the error between the original term-document matrix and its map in the concept space. The similarity matrix is converted into a fuzzy equivalence relation by calculating the max-min transitive closure. On this, we applied fuzzy rules to efficiently cluster the documents. Our proposed method has shown good accuracy than previously known techniques.

Keywords Feature identification · Sentence ranking · Dimension reduction
Fuzzy document clustering

1 Introduction

A knowledge repository typically has vast and huge amounts of formal data elements, which are generally available as documents. With the rapid growth of text documents in knowledge repositories over time, textual data have become high-dimensional,

K. Sreelekshmi (✉) · R. Remya
Department of Computer Science and Engineering, Amrita School of Engineering,
Amritapuri, India
e-mail: sreewarrior20@gmail.com

R. Remya
e-mail: remyar@am.amrita.edu

K. Sreelekshmi · R. Remya
Amrita Vishwa Vidyapeetham Amrita University, Coimbatore, India

© Springer Nature Singapore Pte Ltd. 2018   103
P. K. Sa et al. (eds.), *Recent Findings in Intelligent Computing Techniques*,
Advances in Intelligent Systems and Computing 709,
https://doi.org/10.1007/978-981-10-8633-5_11

which increases the processing time, and thereby diminishing the performance of the system. Thus, effective management of this ever-increasing volume of documents is essential for fast information retrieval, browsing, sharing, and comprehension. Text clustering is useful in organizing large document collections into smaller meaningful groups. Clustering aims at grouping objects having similar properties into the same group and others into separate groups based on the information patterns enclosed in it [1–4]. Numerous methods have been proposed and implemented to solve the clustering problem. The clustered documents will have highest similarity to the ones in the same cluster and least similarity to the ones in the other clusters.

According to the literature, clustering algorithms use either top-down or bottom-up approaches. Those which use top-down algorithms [5–7] work with the value of k which needs to be fixed and known in advance. Eventhough bottom-up approaches are effective, they lack performance in certain datasets. Recently, several types of biologically inspired algorithms have been proposed for clustering [8].

Despite of all these approaches, however, document clustering still presents certain challenges. This includes optimizing feature selection for document representation, reduction of this representation into a lower dimensional pseudo-space, with less information loss and incorporating mutual information between the documents into a clustering algorithm. Our work proposes an approach which extracts the relevant terms by a sentence ranking approach. The work also proposes a dynamic rank reduction method to map the documents into a low dimensional concept space. Finally, using the similarity measures, documents are clustered based on fuzzy rules.

2 Related Work

A method proposed in [9] uses sentence ranking and clustering based summarization approach to extract essential sentences from the document. A weighted undirected graph is constructed according to the order of sentences in a document, to discover central sentences. The weights are assigned to edges by using sentence similarity and discourse relationship between sentences. The scores of sentences are obtained by a graph ranking algorithm. Sentences in the document are clustered by using a Sparse Nonnegative Matrix Factorization. Tian et al. [10] use a page ranking approach to rank the sentences in a document. Li et al. [11] perform constrained reinforcements on a sentence graph, which unifies previous and current documents to determine importance of sentences. The constraints ensure that the most important sentences in current documents are updates to previous documents.

In [12], the authors proposed a fuzzy clustering framework which uses statistical measures to form sentence clusters. Related sentences are grouped by applying an enhanced fuzzy clustering algorithm. Semantically similar sentences are identified using Expectation Maximization (EM) framework and Page Rank score of an object. The work [13] focuses mainly on three issues, namely, optimizing feature selection, a low-dimensional document representation, and then incorporating these information into a clustering algorithm. To address the issue of feature extraction, it considers a domain-specific ontology that provides a controlled vocabulary. The original feature space is mapped into a lower dimensional concept space, using Singular Value Decomposition (SVD). The relationship between documents is modeled using the fuzzy compatibility relation. With the help of a cluster validation index, all the data sequences are allocated into clusters. The drawback of the work is that it is restricted only to a particular domain with a controlled vocabulary. A fuzzy controlled genetic algorithm, combined with semantic similarity measure is used for text clustering in [14]. In this work, a hybrid model that combines the thesaurus based and the Semantic Space Model (SSM) approach is proposed. The biologically inspired principles of genetic algorithm with the help of fuzzy controllers significantly improved clustering accuracy.

Even though all these works address the problem with feature identification and extraction, accuracy still remains as an open problem to be considered. In our work, features are identified by a sentence ranking approach. Rather than fixing the dimension using a cluster validation index, we proposed a dynamic dimension-fixing approach which retains almost all of the information content (i.e., minimizes the error). We build a fuzzy controlled system where the clusters are determined automatically, based on the information patterns formed.

3 Proposed Approach

The work mainly focuses on three major phases in text document clustering.

1. Relevant feature selection and extraction.
2. Document representation so as to retrieve mutual information between documents.
3. Efficient clustering method.

Solution approach is shown in Fig. 1. Each module is explained in detail in the following subsections.

Fig. 1 Block diagram of the proposed work

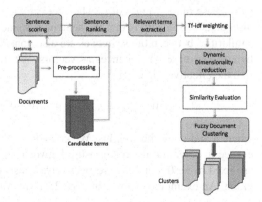

Preprocessing of the data is done by stop word removal, lemmatization, and specific pattern removal to obtain the candidate terms. They are then written along with its frequencies to a temporary file, which is fed to the proposed system for further processing and clustering.

3.1 Feature Identification and Extraction

Feature identification is a challenging task in document clustering. A method to efficiently acquire the relevant terms in a document is one of the motives of the work. To reveal the relevance of the terms in the context, we put forth a method, which considers the *importance value* of sentences. For each document, we build a sentence vocabulary by considering the terms in the temporary file and the sentences in that document. Here, we introduce a new weighting measure to the terms, as the product of term frequency in a particular sentence and logarithmic ratio of total sentences to the number of sentences in which the term appears. Using (1), we obtain the weight of each term, where tf and sf represent the frequency of the term in the particular sentence and number of sentences in which the term appears, respectively.

$$W_{term} = tf_{term}. \log_{10} \frac{N}{sf} \tag{1}$$

In the news corpus, introduction sentences convey the entire information in the article, and thus will have more significance compared to the sentences that follows. To accommodate this feature, we assign positional weights to the sentences in our document collection. The relationship between sentences is modeled as an undirected graph with sentences as nodes and, edges show which all sentences are related to each other. The relationship between sentences are found by taking advantage of the discourse relations [9], which reveals how much a sentence is dependent on another, by context. There is a set of predefined discourse terms. Let s_i and s_j be the two sentences where, s_j follows s_i in sequence. If s_j starts with a discourse term, then s_j is dependent on s_i. Thus, s_j imposes a weight on to s_i. This results in an increase in the weight of s_i. The necessary condition for two sentences to have a discourse relationship is that the sentences s_i and s_j should be adjacent. The cosine similarities between the sentences are also considered to improve the edge weights, by considering each sentences as vectors. We then combine all these measures to find the effective edge weights.

$$W_{effective} = a.W_{VSM} + b.W_{dis} + (1 - a - b).W_{pos} \tag{2}$$

where a and b are the weights. We considered the discourse relation has more relevance in revealing the relationship between sentences than cosine similarity, and are fixed to be 0.3 and 0.5, respectively. Thus, we get a fully connected weighted graph. An example is shown in (Fig. 2). Here, s_1 to s_6 are the sentences and edges

Fig. 2 Sentence weighting

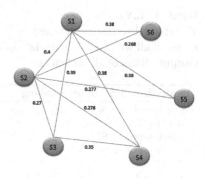

show the weights assigned. The weights are normalized and given to the sentence ranking module. We then ranked the sentences by the Page Ranking Algorithm [9]. Thus, the top ranked k sentences are retrieved. With the assumption that all the terms in the important sentences will also be important, the relevant terms are extracted from the top ranked sentences. This is our relevant term collection. We then used the traditional tf-idf scoring to assign weights to the retrieved relevant terms in the document. Let G be the graph with E and V, the edges and vertices, respectively. Sentences are modeled as nodes of the graph. The algorithm returns the weights for all the sentences in the document.

Input : G(E,V): The graph
E : set of all edges.
V : set of all vertices.
N : Total number of lines in Di.
K : A multiplication constant.
Output : Edge Weights.

Algorithm 1 Edge weighting
___

for each i ϵ 1...N :
for each j ϵ 1...N :
if i==j :
set W_{ij}=0
else :
foreach : e_{ij} ϵ G(E,V)   // for i,j ϵ 1...N
Wpos(S_i)=N*K/Line number
if $S_j[0]$ $\in$ 'discourse terms' && i and j are adjacent :
Improve the weight W_{ij}.
endIf
endFor
endIf
endFor
endFor
Normalize the weight: $W_{ij} = W_{ij} / \sum_{i=1}^{n} W_{ik}$.   //k: number of edges from node i.
return weight.
___

Input : G(E,V): The graph
V : set of all vertices(sentences).
E : set of all weighted edges(relationship between sentences).
Output : Rank of all sentences.

Algorithm 2 Sentence Ranking

for each sentence $S_i \in$ doc D_i :	// for i $\in$ 1...N
add v_i to G.V	
for each sentence S_i :	$i\epsilon$ 1...N:
for each sentence S_j :	$j\epsilon$ 1...N:
if Si $\sim$ Sj :	
Add an undirected edge e_{ij} to G.E.	
Add weight w_{ij} to the edge e_{ij}.	
endIf	
Compute rank of v_i	
endFor	
endFor	
R(v_i)= d.$\sum R_j$ x w_{ij}+(1 - d)	//d: constant
endFor	
return rank, R	

3.2 Dimensionality Reduction

The term-document matrix, having the tf-idf score, is fed to the decomposition module. The original term-document matrix is mapped to a low dimensional concept space, using SVD. The drawback in using SVD for decomposition is that the reduced rank, k is fixed by inspection of the singular values. In the proposed method, k is considered dynamically by converting this into an energy conservation problem. Let λ_1, $\lambda_2 \ldots \lambda_n$ be the singular values [15]. The energy conserved for n-dimension will be,

$$E_n = \frac{\lambda_1^2 + \lambda_2^2 + \cdots + \lambda_i^2}{\lambda_1^2 + \lambda_2^2 + \cdots + \lambda_n^2} \tag{3}$$

where n is the total number of documents and i = 1, 2, ..., n. Here, we consider 80% energy (information) is conserved. i.e., $E_n \geq 80\%$.
Input : A; *Term − document matrixofsize* $m \times n$.
Diagonal values of Sigma ($\lambda_1, \lambda_2, \ldots \lambda_n$)
Output: Reduced Dimension.

Algorithm 3 Dimension Reduction

---

for each λ_i :   // i = 1,2,...,n

if $\lambda_i{}^2 / \sum_1^n \lambda_i{}^2 \leq 80\%$:

continue

else :

dimension=i

break

endIf

endFor

return dimension.

3.3 Similarity Evaluation

The document similarity is evaluated with the vector space model, in the reduced concept space. Each document is considered as a vector and the similarity is found [13].

3.4 Fuzzy Document Clustering

The fuzzy set theory is an extension of the classical set theory with degrees of membership for its elements [16]. In this work, we used the documents' similarity as the membership degrees to form a fuzzy equivalence relation. A fuzzy equivalence relation can be defined using the mathematical preliminaries defined in [13]. To cluster the documents, the following steps are done:

- To apply fuzzy rules, the similarity matrix is to be normalized to bring the values in the range [0, 1] .
- Considered max-min closure to find transitive closure R_T.
- Chose proper λ (cut-set) to find all feasible clusters.

Input : The similarity matrix R'
Output : Clusters.

Algorithm 4 Fuzzy Clustering

Perform a normalization computation on R'
$R_{ij}=(R_{ij}'+1)/2$
Find a transitive closure $R^T=R^{n-1}$
Find a suitable λ-cutset$\in [0,1]$
for all $R^T[i,j] < \lambda$:
if i==j :
set $R^T[i,j]=0$
else :
set $R^T[i,j]=1$
endIf
Select the docs corresponding to 1's to fall in the same cluster
return clusters

---

4   Experimentation

To evaluate the proposed approach, we considered 406 documents from the 20-newsgroups collection, available in UCI Machine Learning Repository. The dataset includes documents from four different topics.

4.1   SStress Criteria

The dissimilarity between two matrices is given by the SStress criteria. SStress is defined as (4).

$$SStress = \sum_{i=1}^{n} \sum_{j=1}^{n} \left(S_{ij}^2 - \widehat{S_{ij}^2} \right) \tag{4}$$

S_{ij} and $\widehat{S}_{ij}$ refers to the elements in the original term-document matrix and the matrix in the reduced concept space, respectively. From Fig. 3, we can see that SStress decreases with the increase of dimension k. To fix an optimal dimension retaining 80% of the information, we fix the dimension at 247 which is shown in Fig. 4. With this dimension, we could map the original term-document matrix into a concept space without much information loss, thereby reducing the error.

4.2   Proper Cutoff for Clusters

Considering the accuracy of clusters formed against different λ-cut values, we chose a proper λ value by inspection. Figure 5 shows the plot of number of clusters versus

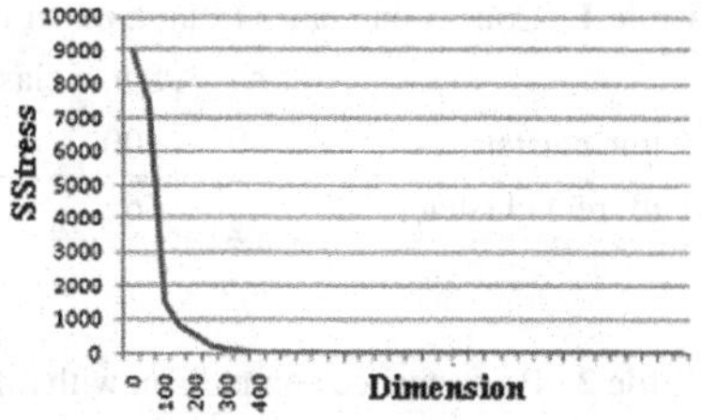

Fig. 3 The SStress between
original matrix and the
matrix in reduced concept
space

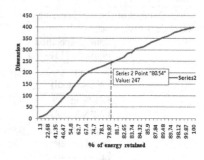

Fig. 4 Choosing optimal
rank, k

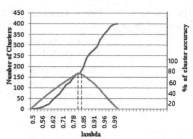

Fig. 5 Choosing a cutoff for
fuzzy clustering

λ value for the dataset under consideration and we chose optimal λ value between
0.8 and 0.85 as shown.

5 Result and Analysis

We used the metrics; precision, recall and F1 score to compare our work with existing
method (5).

$$Precision(P) = \frac{TP}{TP + FP}, Recall(R) = \frac{TP}{TP + FN}, F1Score = \frac{2.PR}{P + R} \quad (5)$$

TP, FP, and FN represents the true positive, false positive, and false negative rates,
respectively. The proposed method correctly clustered majority of the documents.
The confusion matrix is obtained as shown in Table 1. We used the selected subset of

Table 1 Confusion matrix (total number of documents = 406)

	Same class	Different class
Same cluster	300	40
Different cluster	36	30

Table 2 Performance comparison with different datasets

Dataset	Method	Precision	Recall	F-measure
Selected subset of 406 documents	Fuzzy clustering on domain-specified ontology [13]	0.77	0.82	0.79
	K-Means (WEKA tool)	0.668	0.60	0.632
	DB-Scan (WEKA tool)	0.72	0.75	0.73
	Fuzzy clustering based on relevant ranked terms	0.88	0.89	0.83
Document collection from [13]	Fuzzy clustering on domain-specified ontology [13]	0.78	0.80	0.78
	Fuzzy clustering based on relevant ranked terms	0.82	0.76	0.79

406 documents and those used in [13] to compare the performance of each method. Also, traditional approaches like K-means and DB-SCAN are run on our dataset using WEKA tool. Performance analysis of different methods is shown in Table 2.

6 Conclusion

We propose a method to select and extract relevant terms from documents by a sentence ranking approach. Also, a dynamic method has been proposed to choose a suitable rank in the lower dimensional space without losing relevant information. Fuzzy clustering applied to the similarity relation gives good quality clusters. From the experiment results obtained, it is evident that our system performs better than the traditional approaches and the one proposed in [13].

References

1. Gurrutxaga, I., Albisua, I., Arbelaitz, O., Martin, J.I., Muguerza, J., Perez, J.M., Perona, I.: An efficient method to find the best partition in hierarchical clustering based on a new cluster validity index. Pattern Recogn. **43**(10), 3364–3373 (2010)
2. Nguyen, C.D., Krzysztof, J.C.: GAKREM: a novel hybrid clustering algorithm. Inf. Sci. **178**, 4205–4227 (2008)
3. Saha, S., Bandyopadh yay, S.: A symmetry based multi objective clustering technique for automatic evolution of clusters. Pattern Recogn. **43**(3), 738–751 (2010)
4. Menon, R.R.K., Aswathi, P.: Document classification with hierarchically structured dictionaries. Adv. Intell. Syst. Comput. **385**, 387–397 (2016)
5. Selim, S., Ismail, M.: K-means-type algorithm: generalized convergence theorem and characterization of local optimality. IEEE Trans. Pattern Anal. Mach. Intell. **6**, 81–87 (1984)
6. Bandyopadhyay, S., Maulik, U.: An evolutionary technique based on K-means algorithm for optimal clustering in R. Inf. Sci. **146**, 221–237 (2002)
7. Harikumar, S., Surya, P.V.: K-medoid clustering for heterogeneous datasets. Proc. Comput. Sci. **70**, 226–237 (2015)
8. Song, W., et al.: Genetic algorithm for text clustering using domain-specified ontology and evaluating various semantic similarity measures. Expert Syst. Appl. **36**, 9014–9095 (2009)
9. Ge, S.S., Zhang, Z., He, H.: Weighted graph model based sentence clustering and ranking for document summarization, Singapore National Research Foundation, Interactive Digital Media R&D Program, pp. 90–95 (2010)
10. Tian, J., et al.: Ranking sentences in scientific literatures. In: 11th International Conference on Semantics, Knowledge and Grids, pp. 275–282. IEEE (2015)
11. Li, Xuan, et al.: Update summarization via graph-based sentence ranking. IEEE Trans. Knowl. Data Eng. **25**(5), 1162–1174 (2013)
12. Uma Devi, M., et al.: An enhanced fuzzy clustering and expectation maximization framework based matching semantically similar sentences. In: 3rd International Conference on Recent Trends in Computing, Procedia Computer Science, vol. 57, pp. 1149–1159 (2015)
13. Yue, Lin, et al.: A fuzzy document clustering approach based on domain-specified ontology. Data Knowl. Eng. **100**, 148–166 (2015)
14. Song, Wei, et al.: Fuzzy control GA with a novel hybrid semantic similarity strategy for text clustering. Inf. Sci. **273**, 156–170 (2014)
15. Grewal, B.S.: Higher Engineering Mathematics. Khanna Publishers
16. Zadeh, L.A.: Fuzzy sets. Inf. Control **8**, 338–353 (1965)

Euclidean Distance Based Particle Swarm Optimization

Ankit Agrawal and Sarsij Tripathi

Abstract This paper proposes a technique for improving the convergence speed and the final accuracy of the Particle Swarm Optimization (PSO) by introducing a new adaptive inertia weight strategy based on Euclidean distance. This change does not inflict any major modifications to the basic algorithm. The proposed technique has shown significantly better performance as compared to other PSO variants on a test suite of ten optimization test functions evaluated on following performance metrics: time to locate the solution, scalability, quality of the final solution, and frequency of hitting the optima.

Keywords Swarm intelligence · Particle swarm optimization (PSO)
Inertia weight · Convergence · Exploration and exploitation

1 Introduction

Particle Swarm Optimization (PSO) is a stochastic, population-based search strategy initially presented by Kennedy and Eberhart in 1995 [1]. For optimization of the function, any of its gradient information is not required in PSO [1, 2]. It is conceptually easy and can be done using only primitive mathematical operators. PSO is also inspired from socio-cognition and the boid's method of Craig Reynolds [2]. The swarming behavior of the animals herding, insects, fish schooling and birds flocking is emulated by PSO, where the swarm cooperatively search for food.

The basic entity in PSO is the particle. The set of particles which conduct the search in the multi-dimensional search space is known as a swarm. Each particle has a velocity and a position at any particular instant of time. The position of every

A. Agrawal (✉) · S. Tripathi
National Institute of Technology Raipur, Raipur 492010, Chhattisgarh, India
e-mail: ankitagrawal648@gmail.com

S. Tripathi
e-mail: stripathi.cs@nitrr.ac.in

© Springer Nature Singapore Pte Ltd. 2018
P. K. Sa et al. (eds.), *Recent Findings in Intelligent Computing Techniques*,
Advances in Intelligent Systems and Computing 709,
https://doi.org/10.1007/978-981-10-8633-5_12

individual or particle in the population represents a candidate solution to the optimization problem.

Each particle moves a step at a time through the search space looking for the optimal solution. Every particle remembers their respective best coordinate position, also known as the 'pbest' position which corresponds to the best solution (fitness) achieved by the particle so far. Similarly, the overall best coordinate position which corresponds to the entire swarm's best position, also known as the 'gbest' position is also remembered and all the particles of the swarm are aware of this position. To determine the optimal solution, all the particles of the swarm move with their respective velocity towards the current optimum solution i.e. towards the pbest and the gbest position.

In this paper, we propose a new variant of PSO i.e. EDPSO which improves the performance of the algorithm. EDPSO has adaptive inertia weight strategy based on Euclidean distance, which uses change in the mean of Euclidean distance between particles and the gbest, as a feedback parameter.

2 Background

The PSO is a cooperative method, in which N number of particles moves through the search space of D dimensions. The particle's position depends on its neighbor's experience and its own experience. The position of the ith particle can be written as: $x_i = (x_{i1}, x_{i2}, x_{i3}, \ldots, x_{iD})$, where $x_{id} \in [L_d, U_d], d \in [1, D]$, and U_d and L_d are the upper and lower bound of the dth dimension of the search space. Similarly, the velocity of the ith particle can be written as: $v_i = (v_{i1}, v_{i2}, v_{i3}, \ldots, v_{iD})$. At each time step t, the swarm particles update their position and velocity according to following equations:

$$v_{id}(t+1) = \omega * v_{id}(t) + c_1 * R_{1id} * (pbest_{id} - x_{id}(t)) + c_2 * R_{2id} * (gbest_d - x_{id}(t))$$

(1)

$$x_{id}(t+1) = x_{id}(t) + v_{id}(t+1)$$ (2)

where ω is the inertia weight, R_{1id} and R_{2id} are independent and uniformly distributed random positive numbers in range (0, 1), c_1 and c_2 are cognitive and social learning parameters respectively (mostly $c_1 = c_2$), $pbest_{id}$ is ith particle's best previous position, $gbest_d$ is the best previous position of particle among all particles of the swarm.

Inertia weight maintains the balance between the ability of PSO to explore the search space and ability to converge on a specific region of the search space. All the inertia weight strategies proposed by researchers can be classified into four main groups: constant [3], random [4], time varying and adaptive inertia weight strategy.

Some famous time varying inertia weight law includes linearly decreasing inertia weight [5], sigmoid [6], Sugeno function [7], simulated annealing [8], logarithmic decreasing law [9], and exponential decreasing law [10]. Then various adaptive methods were introduced that adjusts inertia weight using the process feedback to gain better control over population diversity. Some of the feedback parameters include the number of updated best positions [11], the best fitness achieved [12], the distance between particles [13], or the standard deviation in components of all particles [14].

3 Proposed Inertia Weight Law

In this paper, we have applied a new inertia weight technique in which Euclidean distance is utilized for determining a suitable inertia weight in each iteration such that it maintains the balance between the global and local search capability of PSO. The change in the arithmetic mean of Euclidean distance between the particles' position and the gbest position is taken into account for this purpose.

First, we calculate the Euclidean distance between all the particles and the global best position as:

$$Euc\_dist_i = \sqrt{\sum_{d=1}^{d=D} (gbest_d - x_{id})^2}. \tag{3}$$

where i denotes the ith particle. Then we calculate the arithmetic mean of the above Euclidean distances as:

$$\mu(t) = \overline{Euc\_dist} = \sum_{i=1}^{i=N} Euc\_dist_i/N. \tag{4}$$

To use the mean Euclidean distance in calculation of the inertia weight, the mean value is normalized between 0 and 1 as:

$$\mu_{new}(t) = \frac{\mu(t) - \mu_{\min}(t)}{\mu_{\max}(t) - \mu_{\min}(t)}. \tag{5}$$

Then inertia weight is taken as:

$$\omega = \omega_{\min} + (\omega_{\max} - \omega_{\min}) \times |\mu_{new}(t) - \mu_{new}(t-1)|. \tag{6}$$

The arithmetic mean of Euclidean distance between particles' position and the gbest position represents how far the particles are from the gbest position with a single value. Exploration capability of PSO increases when the change in above mean value is significant, and the exploitation capability enhances when the change in mean value is not significant. This property can be seen in Fig. 1 which shows how inertia weight varies with time when EDPSO applied on the sphere function with dimension 30. Initially, the values of inertia weight are large in the first few iterations which indicate that EDPSO is performing exploration in the beginning. After this short period, inertia weight converges to 0.427 (approx.), which is a suitable value for the Sphere function and oscillates around it.

4 Experimental Setup

4.1 Benchmark Functions

Table 1 contains the list of ten well-known benchmark functions [15] we have used to evaluate the performance of EDPSO and compare with other inertia weight strategies. The first five functions are unimodal functions while the rest are multimodal functions.

4.2 Algorithms Compared

The performance of the proposed EDPSO algorithm have been compared with following algorithms: (a) GPSO [5] (b) Sugeno [7] (c) AIWPSO [11], and (d) w-PSO [14].

Fig. 1 Variations of the inertia weight as a function of the step number t

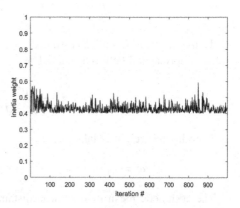

Table 1 Benchmark functions used in experiments

Function	Mathematical representation
Sphere function	$f_1 = \sum_{i=1}^{D} x_i^2$
Rotated hyper-ellipsoid function	$f_2 = \sum_{i=1}^{D} \sum_{j=1}^{i} x_j^2$
Rosenbrock function	$f_3 = \sum_{i=1}^{D-1} \left[100(x_{i+1}^2 - x_i^2) + (x_i - 1)^2 \right]$
Sum squares function	$f_4 = \sum_{i=1}^{D} i x_i^2$
Rastrigin function	$f_5 = 10D + \sum_{i=1}^{D} \left[x_i^2 - 10\cos(2\pi x_i) \right]$
Ackley function	$f_6 = -20\exp\left(-0.2\sqrt{\frac{1}{D}\sum_{i=1}^{D}\cos(x_i^2)} \right) - \exp\left(-\frac{1}{D}\sum_{i=1}^{D}\cos(2\pi x_i) \right) + 20 + \exp(1)$
Griewank function	$f_7 = \sum_{i=1}^{D} \frac{x_i^2}{4000} - \prod_{i=1}^{D} \cos\left(\frac{x_i}{\sqrt{i}} \right) + 1$
Powell function	$f_8 = \sum_{i=1}^{D/4} \left[(x_{4i-3} + 10x_{4i-2})^2 + 5(x_{4i-1} - x_{4i})^2 + (x_{4i-2} - 2x_{4i-1})^4 + 10(x_{4i-3} - x_{4i})^4 \right]$
Schwefel function	$f_9 = 418.9829 - \sum_{i=1}^{D} x_i \sin(\sqrt{|x_i|})$
Levy function	$f_{10} = \sin^2(\pi w_1) + \sum_{i=1}^{D-1} (w_i - 1)^2 [1 + 10\sin^2(\pi w_i + 1)] + (w_D - 1)^2 [1 + \sin^2(2\pi w_D)],\ where$ $w_i = 1 + \frac{x_i - 1}{4}$

4.3 Evaluation and Comparison Criteria

We conducted two distinct sets of experiments for better judgment of the performance and comparison of algorithms over ten benchmark functions in dimensions 10 and 30 on Matlab. Each algorithm runs 50 times with swarm size 40 and $c_1 = c_2 = 2$ [16]. In the first experiment, the mean and the standard deviation of best solution are recorded after $4 * 10^5$ functional evaluations (FEs). This experiment determines the accuracy of all PSO variants. In the second experiment, the algorithm terminates either when fixed number of FEs ($4 * 10^5$) is carried or when algorithm achieves the solution with the specified accuracy. Here we record the number of FEs required to obtain the solution with specified accuracy by the algorithm. Also in this experiment the results are recorded only when algorithm finds the solution in at least 15 out of 50 runs. This experiment determines which PSO variant converge quickly towards the global optimum.

4.4 Results and Discussion

In both set of experiments, recently reported algorithms in literature [11, 14], mostly found the best solutions. In first experiment (Table 2), for dimension 10, EDPSO outperformed the other algorithms for functions f_1, f_2, f_4, f_8 and yield

Table 2 Mean and standard deviation of the best solutions in 50 runs for 5 PSO variants

Fun	Dim	Mean best fitness (Standard deviation)				
		EDPSO	GPSO	Sugeno	AIWPSO	w-PSO
f_1	10	0(0)	5.73235e−262(0)	0(0)	0(0)	8.51272e−13 (2.00072e−12)
	30	4.00436e−190(0)	2.67101e−65 (7.70833e−65)	2.02217e−158 (1.04454e−157)	4.78818e−148 (1.87727e−147)	1.42636e−04 (1.64748e−04)
f_2	10	0(0)	6.23702e−264(0)	0(0)	0(0)	2.77554e−12 (5.49616e−12)
	30	7.40000e+03 (1.72390e+04)	1.98000e+04 (2.22683e+04)	3.38000e+04 (3.31287e+04)	1.12000e+04 (1.80295e+04)	8.40000e+03 (1.60814e+04)
f_3	10	6.04416e+04 (2.39795e+05)	1.40231e+05 (3.50419e+05)	1.40466e+05 (3.50326e+05)	2.00122e+04 (1.41420e+05)	1.00625e+05 (3.02846e+05)
	30	1.41076e+05 (3.50091e+05)	3.21310e+05 (4.70334e+05)	3.21744e+05 (4.70029e+05)	1.60689e+05 (3.70045e+05)	2.01146e+05 (4.03516e+05)
f_4	10	0(0)	2.00000e+02 (1.41421e+03)	0(0)	0(0)	6.88126e−12 (2.92691e−11)
	30	1.14000e+04 (1.73805e+04)	2.98000e+04 (3.19751e+04)	2.98000e+04 (3.37754e+04)	1.22000e+04 (2.31490e+04)	1.40000e+04 (2.29463e+04)
f_5	10	5.25338e+00 (3.27248e+00)	1.35314e+00 (9.58977e−01)	2.38790e+00 (1.33336e+00)	4.51711e+00 (2.54429e+00)	5.17383e−01 (9.04333e−01)
	30	7.20547e+01 (2.39575e+01)	2.28615e+02 (1.41384e+03)	3.84651e+01 (1.00297e+01)	5.63145e+01 (1.66366e+01)	2.23760e+01 (8.25177e+00)

(continued)

Table 2 (continued)

Fun	Dim	Mean best fitness (Standard deviation)				
		EDPSO	GPSO	Sugeno	AIWPSO	w-PSO
f_6	10	2.05220e+01 (5.59168e−01)	2.00000e+01 (2.12186e−07)	2.00000e+01 (2.16358e−07)	1.96000e+01 (2.82843e+00)	1.96000e+01 (2.82843e+00)
	30	2.09752e+01 (6.48274e−01)	2.00000e+01 (1.82413e−08)	2.00000e+01 (1.80902e−08)	2.00000e+01 (1.65441e−07)	2.00000e+01 (2.30803e−05)
f_7	10	7.33122e−02 (3.03813e−02)	5.20452e−02 (2.36817e−02)	5.69722e−02 (2.70632e−02)	6.69027e−02 (3.65682e−02)	5.36209e−02 (2.10820e−02)
	30	1.42082e−02 (1.66270e−02)	1.79006e−02 (1.68284e−02)	6.97821e−02 (3.88676e−01)	1.09687e−02 (1.36424e−02)	2.16604e−02 (2.49721e−02)
f_8	10	3.49383e+04 (1.30264e+05)	5.82391e+04 (4.16578e+04)	1.27129e+05 (2.94012e+05)	6.15645e+04 (1.79862e+05)	6.67114e+04 (2.17449e+05)
	30	1.39691e+06 (1.01078e+06)	3.68100e+06 (1.46334e+06)	5.08370e+06 (1.43249e+07)	1.07770e+06 (8.97018e+05)	1.78647e+06 (1.35512e+06)
f_9	10	3.59576e+03 (3.14166e+01)	3.58155e+03 (1.48008e+01)	3.58542e+03 (1.41088e+01)	3.58138e+03 (2.02207e+01)	3.57490e+03 (1.21518e+01)
	30	1.08826e+04 (7.83317e+01)	1.08663e+04 (1.02545e+02)	1.08635e+04 (1.07744e+02)	1.08405e+04 (7.78811e+01)	1.07682e+04 (5.79738e+01)
f_{10}	10	6.18144e−02 (2.63930e−01)	1.49976e−32 (1.38235e−47)	1.49976e−32 (1.38235e−47)	9.08648e−03 (6.42511e−02)	1.65127e−13 (5.95327e−13)
	30	1.03523e+03 (8.68263e+02)	1.34956e+03 (1.05363e+03)	1.73842e+03 (1.09297e+03)	9.22079e+02 (9.14339e+02)	9.94359e+02 (9.53816e+02)

competitive result for functions f_3, f_7, and f_9. For other functions the performance was not up to the mark when compared with other existing algorithms. Similarly, for dimension 30, EDPSO outperformed others, for functions f_1, f_2, f_3, f_4 and yield competitive result for functions f_5, f_7, f_8, f_9, and f_{10}. The algorithms are scored depending on number of times they yield the best and comparable solutions. Scores of EDPSO, w-PSO, AIWPSO, Sugeno, GPSO are 16, 14, 16, 12 and 11 respectively.

In the second experiment (Table 3), following the criteria mentioned in Sect. 4.3, functions which were not able to locate the solution are: function f_3, f_5, f_6, f_9 for both the dimensions, function f_7 for dimension 10 and function f_8 and f_{10} for dimension 30. In remaining cases, EDPSO and AIWPSO outperformed other PSO variants. The performance of EDPSO is slightly better than AIWPSO.

The insights of the searching behavior of PSO variants can be seen in their convergence curves in Fig. 2. In some cases, convergence graph of EDPSO is a straight line indicating a constant convergence rate during the whole course of the run. Also, the performance of EDPSO doesn't get affected by increase in the dimension. By looking at result of the second experiment, we found that in comparison to other PSO variants, EDPSO converges quickly towards the best solution.

Above experiment shows, that the EDPSO is significantly better and highly comparable to other PSO variants in most of the cases. Also it has the capability to converge quickly towards the best solution.

Table 3 Mean and standard deviation of FEs required out of 50 runs to find the solution with specified accuracy for 5 PSO variants

Fun	Dim	Accuracy	Mean of functional evaluations (Standard deviation)				
			EDPSO	GPSO	Sugeno	AIWPSO	wPSO
f_1	10	1.00E−03	3682.40 (356.53)	137201.60 (3810.12)	30449.60 (1286.33)	5347.20 (351.09)	57181.60 (7206.86)
	30	1.00E−02	12945.31 (931.51)	212293.88 (3820.03)	57906.40 (1889.77)	16422.40 (1044.49)	188860.00 (28901.48)
f_2	10	1.00E−03	4157.60 (327.70)	140771.20 (3604.32)	31558.40 (1376.87)	5942.40 (414.68)	65744.80 (9622.04)
	30	1.00E−02	15647.50 (1433.19)	0	64308.00 (1191.09)	19926.45 (1292.61)	268598.57 (38652.01)
f_3	10	1.00E−03	0	0	0	0	0
	30	1.00E−02	0	0	0	0	0
f_4	10	1.00E−03	4166.40 (302.63)	141072.80 (3989.77)	31388.80 (1229.86)	5909.60 (372.86)	65205.60 (8403.53)
	30	1.00E−02	15870.86 (1468.19)	0	63152.50 (3195.20)	19571.43 (988.16)	286177.60 (46969.32)
f_5	10	1.00E−03	0	0	0	0	188857.65 (57786.29)
	30	1.00E−02	0	0	0	0	0
f_6	10	1.00E−03	0	0	0	0	0
	30	1.00E−02	0	0	0	0	0
f_7	10	1.00E−03	0	0	0	0	0
	30	1.00E−02	11318.71 (1785.74)	208656.00 (8616.31)	57069.63 (12291.37)	13908.57 (1866.07)	121723.81 (15694.62)
f_8	10	1.00E−03	10371.61 (2004.32)	161774.12 (5981.47)	46215.65 (12703.95)	11264.52 (1998.92)	42045.16 (27923.88)
	30	1.00E−02	0	0	0	0	0
f_9	10	1.00E−03	0	0	0	0	0
	30	1.00E−02	0	0	0	0	0
f_{10}	10	1.00E−03	4927.50 (607.06)	140716.80 (4438.35)	32612.00 (1592.09)	32085.60 (1479.44)	54917.60 (5128.99)
	30	1.00E−02	0	0	0	0	0

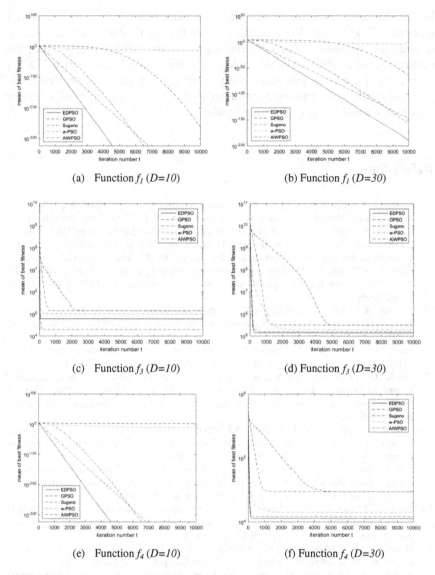

Fig. 2 Mean of best fitness for 50 independent runs as a function of step number

5 Conclusions

In this paper, a new PSO variant (EDPSO) with adaptive inertia weight strategy is introduced for the global optimization. The main aim of the study is to balance the exploration-exploitation trade-off effectively. The inertia weight is based on the Euclidean distance and varies in the range [0.4, 0.9]. The proposed PSO variant i.e.,

EDPSO has been tested on ten standard benchmark functions and compared with four other PSO variants. The experiment shows that the EDPSO converges quickly towards the optimum solution in comparison to other PSO variants. Also, the performance of EDPSO doesn't degrade in high-dimensional search space. EDPSO is simple and can be applied successfully to solve the optimization problems. The future research work will be focused on exploring the capability of proposed algorithm for real world engineering problems.

References

1. Kennedy, J., Eberhart, R.C.: Particle swarm optimization. In: Proceedings of IEEE International Conference on Neural Networks, vol. 4, pp. 1942–1948, Perth, Australia (1995)
2. Kennedy, J., Eberhart, R.C., Shi, Y.: Swarm Intelligence. Morgan Kaufman, USA (2001)
3. Eberhart, R.C., Shi, Y.: A modified particle swarm optimizer. In: Proceedings of the IEEE Congress on Evolutionary Computation (CEC'98), pp. 69–73, Anchorage, AK (1998)
4. Eberhart, R.C., Shi, Y.: Tracking and optimizing dynamic systems with particle swarms. In: Proceedings of the IEEE Congress on Evolutionary Computation (CEC'01), pp. 94–100, Seoul, South Korea (2001)
5. Eberhart, R.C., Shi, Y.: Empirical study of particle swarm optimization. In: Proceedings of the IEEE Congress on Evolutionary Computation CEC'99), pp. 1945–1950, Washington, DC, (1999)
6. Malik, R.F., Rahman, T.A., Hashim, S.Z.M., Ngah, R.: New particle swarm optimizer with sigmoid increasing inertia weight. Int. J. Comput. Sci. Secur. 1, 35–44 (2007)
7. Lei, K., Qiu, Y., He, Y.: A new adaptive well-chosen inertia weight strategy to automatically harmonize global and local search ability in particle swarm optimization. In: Proceedings of the First International Symposium on Systems and Control in Aerospace and Astronautics, pp. 977–980, Harbin (2006)
8. Hassan, W.A., Fayek, M.B., Shaheen, S.I.: PSOSA: an optimized particle swarm technique for solving the urban planning problem. In: Proceedings of the International Conference on Computer Engineering and Systems, pp. 401–405 (2006)
9. Gao, Y., An, X., Liu, J.: A particle swarm optimization algorithm with logarithm decreasing inertia weight and chaos mutation. In: Proceedings of the International Conference on Computational Intelligence and Security, vol. 1, pp. 61–65 (2008)
10. Chen, G., Huang, X., Jia, J., Min, Z.: Natural exponential inertia weight strategy in particle swarm optimization. In: Proceedings of the Sixth World Congress on Intelligent Control and Automation (WCICA), vol. 1, pp. 3672–3675 (2006)
11. Nickabadi, A., Ebadzadeh, M.M., Safabakhsh, R.: A novel particle swarm optimization algorithm with adaptive inertia weight. Appl. Soft Comput. 11, 3658–3670 (2011)
12. Nikabadi, A., Ebadzadeh, M.: Particle swarm optimization algorithms with adaptive inertia weight: a survey of the state of the art and a novel method. IEEE J. Evol. Comput. (2008)
13. Zhan, Z.-H., Zhang, J., Li, Y., Chung, H.S.-H.: Adaptive particle swarm optimization. IEEE Trans. Syst. Man Cybern. Part B Cybern. 39, 1362–1381 (2009)
14. Kessentini, S., Barchiesi, D.: Particle swarm optimization with adaptive inertia weight. Int. J. Mach. Learn. Comput. 5(5), 368–373 (2015)
15. Surjanovic, S., Bingham, D.: Virtual Library of Simulation Experiments: Test Functions and Datasets. http://www.sfu.ca/~ssurjano/ (2013)
16. Suresh, K., Ghosh, S., Kundu, D., Sen, A., Das, S., Abraham, A.: Inertia-adaptive particle swarm optimizer for improved global search. In: Eighth International Conference on Intelligent Systems Design and Applications, ISDA'08, vol. 2, pp. 253–258. IEEE (2008)

Clustering and Visualization on Web Search Results: A Survey

Shefali Kedia, Kishor Wagh and Prashant Chatur

Abstract A query fired on web search engines provide snippets in a ranked order of most visited once at the top. The results obtained are easy and understandable for the user. Certain queries those are ambiguous in nature fail to provide best match results. Clustering can solve this problem to a certain extent. The use of tf-idf vector followed by clustering through k-means++ and reorganization of the snippets will make user find the search more easy.

Keywords Web search result · K-means++ · Clustering

1 Introduction

Internet is taking over everything. A man's daily needs now include Internet. Any information needed is just a click away. Search engines play an important role in today's era of technology. Information required on any topic is searched and within seconds a lot of data is obtained. A query is fired and a ranked list of results also known as snippets is obtained. These snippets are short and can provide little information about the document inside. They are straightforward for queries that are unambiguous. The limitation is that a user has to explore the results to get desired domain. The results are ranked according to the most visited links. A user searching for fruit when types "apple" does not get the fruit link even in the first 100 links of result. Users then need to go through the links to find the desired result or give detailed query resulting in time consumption.

S. Kedia (✉) · K. Wagh · P. Chatur
Department of Computer Science and Engineering, Government College of Engineering,
Amravati 444604, Maharashtra, India
e-mail: kediashefali@gmail.com

K. Wagh
e-mail: kishorwagh2000@gmail.com

P. Chatur
e-mail: prashant_chatur@rediffmail.com

© Springer Nature Singapore Pte Ltd. 2018
P. K. Sa et al. (eds.), *Recent Findings in Intelligent Computing Techniques*,
Advances in Intelligent Systems and Computing 709,
https://doi.org/10.1007/978-981-10-8633-5_13

Clustering of web snippets [1] makes the searching for the user efficient and user friendly. The existing technologies improve snippets by using color glyphs, tag clouds, which add information about the documents.

Vector representation of snippets helps in knowing the word frequency of the unique words. The snippets are chosen as centers that are far away from each other. K-means++ is then applied to these snippets for clustering. After clusterization of these snippets, they are again made the tag clouds. This is done using different colors for different clusters that now becomes easy for the user to obtain desired result.

2 Related Work

2.1 Query Occurrence Visualization

Heimonen and Jhaveri [2] stated a scheme in which the query results are shown in a small icon which is placed beside the web search link. This scheme neither reduced time consumption of the task nor gave efficient results. Also, there was not any change in the number of users that use web search engines. Users said that it was unobtrusive and easy.

2.2 TileBars

Hearst [3] introduced a visualization technique TileBars, used for a Boolean query. It also provides the data on term distribution of the Boolean queries in the results. It gives the length of the web page, the term frequency, and query distribution. They are displayed on a tilebar column beside each search result. It helps in making the judgement about the web page easy.

2.3 Tag Clouds

Kuo et al. [4] replaced snippets by tag clouds and each tag's size, color, indexing is based on the snippet rank.

2.4 Hotmap

Hoeber and Yang [5] provide a heat scale to state the term frequency of the query term. The more frequency is given by red color and the lesser ones by the orange color. These are placed beside the snippet.

2.5 Rank Spiral

Spoerri [6] presented a rankspiral which scans large documents and the titles in a single window. It also uses spiral mapping that reduces occlusions and maximizes density. It solves the problem of labeling.

2.6 Resultmaps

Clarkson et al. [7] stated a hierarchical method that uses tree representation to represent the search results. The information is highlighted and indexed according to the ranks.

2.7 Clustering on Similarity Measure

Wagh [8] proposed a method of similarity measuring which measures the similarity of snippets to make clusters.

3 Design Rationale

Traditional list based results are not useful for ambiguous queries such as apple, jaguar, etc. or a broad topic. Visualization makes it more user friendly, easy to navigate the web search results. It could be an add-on on the traditional list-based technique than to be its replacement. It gives a more comprehensive view of the snippets.

3.1 Arrangement Based on Similarity

The tf-idf vector representation is made of each obtained snippet by calculating unique words and their frequency. The snippets are then grouped on similarity according to the keyword of query.

3.2 Ranking of Documents

The highest visited documents are ranked first in the clusters. Thus, visualization on web search results preserves the rankings of the results. The size of these link results can be resized and colored according to their ranks and clusters for better understanding.

3.3 Uncluttered Layout

The snippets those are similar to each other are grouped and may overlap when are projected on the graph. It can be difficult to find the relevant result the user is looking for. This can be solved by reorganizing using colors and ranks.

4 System Architecture

The steps of the proposed method are:

1. Preprocessing the obtained snippets and creating tf-idf vector for each snippet.
2. Plotting the vectors on graph and applying k-means++ algorithm to form clusters.
3. Organization and visualization of clusters according to the ranking by grouping it in different colors to get a comprehensive view and easy to understand (Fig. 1).

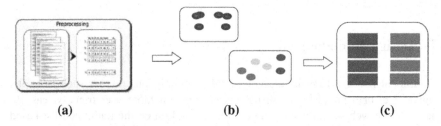

(a)   **(b)**   **(c)**

Fig. 1 **a** Preprocessing query. **b** Clustering snippets. **c** Organization of results

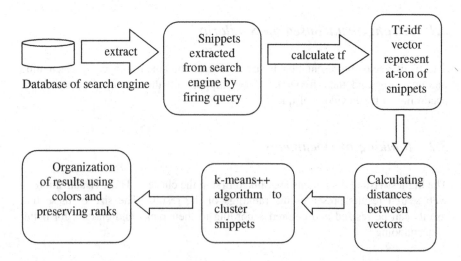

Fig. 2 Architecture flow diagram

First, preprocessing of the snippets is done. The results are obtained through API of a web search engine by firing a query and passing it. The results are then extracted and stopword removal and stemming are done. Then, the tf-idf vector representation of each snippet is obtained.

Further, each vector of snippet is treated as a point and distances between each is calculated. Following this step, the vectors are then clustered by using a k-means ++ [9] algorithm that uses an automatically detects the centers unlike k-means.

Lastly, these clustered snippets are reprojected by giving colors to them according to the clusters to which they belong (Fig. 2).

5 Conclusion

We have proposed a technique that uses clustering on the snippets to group them according to a certain similarity based on the keywords of the query. Clustering and visualization on the snippets make it user friendly, easy, simple, and efficient. The use of k-means++ instead of k-means helps in improvising efficiency. The limitation is that it is useful for ambiguous queries rather than unambiguous ones.

References

1. Gomez-Nieto, E., San Roman, F., Pagliosa, P., Casaca, W., Helou, E.S., de Oliveira, M.C.F.: Similarity preserving snippet-based visualization of web search results. IEEE Trans. Vis. Comput. Graph. **20**, 457–463 (2014)
2. Heimonen, T., Jhaveri, N.: Visualizing query occurrence in search result lists. In: Proceedings of the Ninth International Conference on Information Visualisation, pp. 877–882 (2005)
3. Hearst, M.A.: TileBars: visualization of term distribution information in full text information access. In: Proceedings of the ACM SIGCHI Conference on Human Factors in Computing Systems, pp. 59–66 (1995)
4. Kuo, B.Y.-L., Hentrich, T., Good, B.M., Wilkinson, M.D.: Tag clouds for summarizing web search results. In: Proceedings of the 16th International Conference on World Wide Web (WWW), pp. 1203–1204 (2007)
5. Hoeber, O., Yang, X.D.: The visual exploration of web search results using Hotmap. In: Proceedings of the 10th International Conference on Information Visualization, pp. 157–165 (2006)
6. Spoerri, A.: RankSpiral: toward enhancing search results visualization. In: Proceedings of the IEEE Symposium on Information Visualization, pp. 208–214 (2004)
7. Clarkson, E., Desai, K., Foley, J.: Resultmaps: visualization for search interfaces. IEEE Trans. Vis. Comput. Graph. **15**(6), 1057–1064 (2009)
8. Govardhan, P.H., Wagh, K.P., Chatur, P.N.: Web document clustering using proposed similarity measure. In: National Conference on Emerging Trends in Computer Technology, pp. 15–18 (2014)
9. Arthur, D., Vassilvitskii, S.: k-Means++: the advantages of careful seeding. In: Proceedings of the 18th Annual ACM-SIAM Symposium on Discrete algorithms (SODA), pp. 1027–1035 (2007)

10. Nizamee, M., Shojib, M.: Visualizing the web search results with web search visualization using scatter plot. In: Proceedings of the IEEE Second Symposium on Web Society, pp. 5–10 (2010)
11. Sawaitul, S.D., Wagh, K.P., Chatur, P.N.: Classification and prediction of future weather by using back propagation algorithm-an approach. Int. J. Emerg. Technol. Adv. Eng. **2**, 110–113 (2012)
12. Dehankar, S., Wagh, K.P.: Web page classification using apriori algorithm and Naive Bayes classifier **3** (2015)
13. Charate, R.G., Chatur, P.N., Wagh, K.P.: Document filtering: intelligent inference system for web. Int. J. Manag. IT Eng. **3**, 207–222 (2013)
14. Kolhe, S., Wagh, K.P.: Semantic similarity based on information content. Int. J. Comput. Sci. Appl. 82–86 (2010)

Sentiment Analysis Using Weight Model Based on SentiWordNet 3.0

Jitendra Kumar, Jitendra Kumar Rout, Anshu katiyar and Sanjay Kumar Jena

Abstract Sentiment analysis also is known as opinion mining, is the process of analyzing the sentiment of public opinions, attitudes, comments, etc. In this paper, we present experimental results of NB, SVM, Random forest Machine Learning classifier using Weight model on three different datasets. The contributions of this paper are: (a) Generate feature weight model using SentiWordNet 3.0, (b) assign feature weight to every feature according to POS tag, and (c) suitable combination selection of adj, adv, verb, and noun to give better results.

Keywords SentiWordNet 3.0 · Naive Bayes · Support vector machine
Logistic regression · Random forest · Pointwise mutual information

1 Introduction

Sentiment analysis consists of data mining, data warehousing, and Natural Language Processing (NLP) which basically uses supervised techniques, unsupervised techniques and semi-supervised techniques Rout et al. [1] to classify the sentiments. Supervised techniques require some domain-specific knowledge to train the system whereas unsupervised techniques do not require any information. In this paper document, level supervised sentiment analysis is performed on the movie review website "www.imdb.com", online shopping website "www.amazon.com" and one other

J. Kumar (✉) · J. K. Rout · A. katiyar · S. K. Jena
National Institute of Technology, Rourkela, Rourkela 769008, India
e-mail: jtndrkumar969@gmail.com
URL: http://www.springer.com/lncs

J. K. Rout
e-mail: jiturout@gmail.com

A. katiyar
e-mail: anshukiot09@gmail.com

S. K. Jena
e-mail: skjena@nitrkl.ac.in

© Springer Nature Singapore Pte Ltd. 2018   131
P. K. Sa et al. (eds.), *Recent Findings in Intelligent Computing Techniques*,
Advances in Intelligent Systems and Computing 709,
https://doi.org/10.1007/978-981-10-8633-5_14

which provides data about businesses, online food delivery service Eat24 provider website www.yelp.com, datasets are using the different machine learning techniques which are Support Vector Machine (SVM), Naive Bayes (NB) and Random forest. The experiment uses a unigram model with different folds and shows the respective results.

The major contribution of the work is as follows:

- Select the dataset and perform preprocessing in order to get the better result.
- Perform lexical based analysis using SentiWordNet and assign the weight using Pointwise Mutual Information (PMI) to each word.
- Assign different combinations of input to the classifier and find out the best combination for better results.

2 Literature Survey

Sentiment analysis can be used in the variety of fields. E-commerce sites, social media sites like Amazon, Facebook, Twitter, etc., generate a huge amount of data in unstructured formats. To find out meaningful information from unstructured data, we need NLP. Dash et al. [2] perform on Twitter data sets and Hamouda and Rohaim [3] performed on e-commerce sites to find the positive and negative sentiment of review using polarity score of each word with the help of SentiWordNet. Rout et al. [4], Singh et al. [5] worked on aspect level analysis using POS tags. Denecke [6], Singh et al. [5], Medagoda et al. [7] perform on different datasets using different combination of lexicons from noun, adjective, adverb, and verb and obtained better results.

Medagoda et al. [7] performed his work on the Sinhala languages as less work is available in that language. He considered only two type of lexicons that is adjective and adverb. Initially, the English SentiWordNet mapped the Sinhala dictionary to the English dictionary and then sentiment score is calculated. Finally, the machine learning techniques such as SVM, Naive Bayes, and J48 are used. Among these methods, Naive Bayes gave better result.

Denecke [6] proposed multilingual framework support, through which reviews in any language can be analyzed. Initially, the language is mapped to English dictionary. Three types of lexicons are used adjective, verb, and noun and weight of each word are assigned using SentiWordNet.

3 Data Description

To evaluate performance, we used different datasets in the field of sentiment analysis which are easily available online [8]. In this work, we used three datasets. (i) Internet Movie 'IMDB' [9] (sentence polarity dataset). (ii) 'Yelp' data[1] are collected from www.yelp.com. and (iii) 'Amazon' data [10] are collected from www.amazon.com.

---

[1]https://www.yelp.com/dataset_challenge.

All datasets contain 1000 reviews and its corresponding label '0' or '1' (0/1 represent negative/positive review respectively).

4 Methodology

The supervised machine learning techniques require some training knowledge as a properly labeled data to learn after that it gives a result to the test data accordingly. This is the constraint with supervised learning but it is used to give better result [11]. In this work, three classifiers are used to classify which are Support Vector Machine (SVM), Naive Bayes (NB), Logistic regression, and Random forest. To evaluate the obtained result four types of parameters are used which are accuracy, precision, recall, and F-score. For more details, refer Tripathy et al. [11].

5 Our Approach

5.1 Preprocessing

Preprocessing plays the vital role to improve accuracy. Real word data are generally inconsistency, incomplete, noise. The collected data are passed through preprocessing model so that they give more accurate results and reduce computational overhead. In preprocessing model, the following steps are performed.

- Expand abbreviations and replace slangs: Expand all abbreviation and replace all slangs with correct words/phrases. For example, C.V. replace with Curriculum Vitae, a.m replace with Ante Meridian, p.m replace with Post meridian, Prof. replace with Professor, etc.
- Filter English language: Filtering of English word performed with the help of WordNet. It insures that data contains only those words which are available in WordNet.
- Stemming: Find the root word from a given word for example play is the root word of playing, played, player, players etc.
- Spelling Correction: To obtain better results, text data should be free from spelling mistakes.
- Remove stop words, URLs, numeric and special characters etc. All these words are used most commonly and do not affect sentiment.
- Part of speech (POS) tagging: The process of assigning a part of speech to each word in a sentence. POS information play an important role in sentence detection because any word with different POS tag have a different meaning. In this work, Stanford POS tagger has been used.

Table 1 Subset of SentiWordNet3.0

POS	PosScore	NegScore	SynsetTerms
a	0.25	0	dissilient#1
n	0.5	0.25	coup#2
n	0	0.125	screamer#4 scorcher#2
r	0	0.25	noxiously#1 harmfully#1 detrimentally#1
v	0.375	0	lilt#1

Table 2 term#POS frequency in SentiWordNet3.0

term#POS	Positive frequency	Negative frequency
meticulous#a	2	0
unsound#a	0	5
scorch#n	0	1
attentiveness#n	2	0
precious#r	0	1
haltingly#r	1	0
convert#v	4	1
accuse#v	1	0

5.2 Weight Model

SentiWordNet[2] is a lexical resource for opinion mining. SentiWordNet assigns to each synset of WordNet three sentiment scores: positivity, negativity, and objectivity. In this work, SentiWordNet 3.0 has been used to calculate the weight of each term. SentiWordNet 3.0 look like Table 1.

On the basis of how often it is used every synset term ranked, therefore term#POS pair can appear multiple times with same or different polarity score. PolarityScore of each term calculated with the help of Eq. 1.

$$PolarityScore = PosScore - NegScore \qquad (1)$$

Every term#POS pair of SentiWordNet classified into two classes, i.e., positive or negative. If PolarityScore of term#POS pair has positive then pair considered in positive class if PolarityScore has negative then pair considered in negative class if

---

[2]http://sentiwordnet.isti.cnr.it/.

PolarityScore is equal to zero then that pair has been discarded. In Table 2 represent the sample of term#POS pair with its frequency present in SentiWordNet3.0. After finding the negative and positive frequency of all term#POS pair calculate Point-wise Mutual Information (PMI) of every pair with both class (negative and positive) Durme and Lall [12]. To calculate PMI in this work Eq. 2 has been used.

$$PMI(w, c) = max\{log_2 \frac{P(w, c)}{P(w) * P(c)}, 0\} \tag{2}$$

where P(w, c) is the probability of occurrence w in class 'c', P(w) is the probability of occurrence 'w' independent of class. P(c) is the probability of occurrence class 'c'. And PMI(w, c) is the PMI between 'w' and class 'c'.

Every term#POS pair assigns the weight that is calculated by Eq. 3.

$$W(w) = PMI(w, positive) - PMI(w, negative) \tag{3}$$

where W(w) is the weight of w(term#POS), PMI(w, positive) is PMI score of 'w' with class positive. PMI(w, negative) is PMI score of 'w' with the class negative.

Finally, in weight model, the weight of every term#POS pair normalized in the range [−1, 1] using Eq. 4.

$$x'_i = \frac{x_i - X_{min}}{X_{max} - X_{min}}(new\_max - new\_min) + new\_min \tag{4}$$

where x_i is the instant of X, x_i' is the normalized valued of x_i. X_{min} and X_{max} are the minimum and maximum value of X, respectively. And new_min and new_max are the minimum and maximum value of normalized X (−1, 1).

5.3 Weight Assignment and Feature Combination Selection Model

Preprocessed data has sixteen tags, however, these sixteen tags mapping with four tags (i.e., adjective, adverb, noun, verb) because in weight model SentiWordNet3.0 has been used which is based on four POS tags. The mapping process is done by Table 3. Each term#POS pair of dataset assigns its corresponding weight using weight model. Note: only subjective(PolarityScore ≠ 0) features are selected from dataset those are present in weight model.

Table 3 Penn Treebank tags set to SentiWordNet POS tags conversion

Wquivalent	Penn Treebank tag set	SentiWordNet POS tag
Adjective	JJ, JJR, JJS	a
Adverb	RB, RBR, RBS	r
Noun	NN, NNS, NNP, NNPS	n
Verb	VB, VBD, VBG, VBN, VBP, VBZ	v

Here, the main problem is data transformation into a format so that machine learning classifier can process. And another problem is that find out the best combination of adj, adv, verb, and noun. Those problems are solved using following steps:

- Format of data for classifier

  R1: $\lambda 1_1, \lambda 2_1, \lambda 3_1, \lambda 4_1, \lambda 1_1 + \lambda 2_1, \lambda 1_1 + \lambda 3_1, \lambda 1_1 + \lambda 4_1, \lambda 2_1 + \lambda 3_1, \lambda 2_1 + \lambda 4_1, \lambda 3_1 + \lambda 4_1 \cdots$

  R2: $\lambda 1_2, \lambda 2_2, \lambda 3_2, \lambda 4_2, \lambda 1_2 + \lambda 2_2, \lambda 1_2 + \lambda 3_2, \lambda 1_2 + \lambda 4_2, \lambda 2_2 + \lambda 3_2, \lambda 2_2 + \lambda 4_2, \lambda 3_2 + \lambda 4_2 \cdots$

  .

  .

  .

  Rn: $\lambda 1_n, \lambda 2_n, \lambda 3_n, \lambda 4_n, \lambda 1_n + \lambda 2_n, \lambda 1_n + \lambda 3_n, \lambda 1_n + \lambda 4_n, \lambda 2_n + \lambda 3_n, \lambda 2_n + \lambda 4_n, \lambda 3_n + \lambda 4_n \cdots$

- Where: $\lambda 1_i$ is the sum of weight of all adj feature present in the ith review, $\lambda 2_i$ is the sum of weight of all adv feature presents in the ith review, $\lambda 3_i$ is the sum of weight of all verb feature presents in the ith review and $\lambda 4_i$ is the sum of weight of all noun feature presents in the ith review.

6 Results

Table 4 represents the fourfold, fivefold, and tenfold cross validation results (accuracy, precision, recall and F1-score values) on three different datasets (IMDB, Yelp, and Amazon) using four classifiers (SVM, NB, Logistic regression and Random forest). In fourfold cross validation, NB classifier obtains best result on Amazon dataset which is 69.2% accuracy, in fivefold cross validation Random forest classifier obtains the best result on Amazon dataset which is 72% accuracy. And in tenfold cross validation, SVM and Random forest classifiers obtain the best results on Yelp and Amazon datasets which are 74% accuracy. Comparison with other existing work is shown in Table 5.

Table 4 4, 5, and tenfold cross validation results on different datasets

Classifiers	Datasets	4-fold				5-fold				10-fold			
		Accuracy	Precision	Recall	F1-score	Accuracy	Precision	Recall	F1-score	Accuracy	Precision	Recall	F1-score
SVM	Amazon	62.4	69.51	44.53	54.28	66.5	72.13	43.13	53.98	67	78.12	49.012	60.24
	imdb	63.6	48.19	80.8	60.37	66	50.78	76.47	61.03	68	65.21	73.77	69.23
	Yelp	64.4	79.74	46.32	58.6	64	80.59	48.21	60.33	74	80	47.45	59.57
NB	Amazon	69.2	70.1	53.12	60.44	68.5	68.38	52.94	59.66	70	70.73	56.86	63.04
	imdb	66	59.03	49.49	53.84	65.5	64.06	49.39	55.78	73	70	45.9	55.44
	Yelp	65.2	80.51	45.25	57.94	63	80	46.42	58.75	69	74.19	37.7	50
Logistic regression	Amazon	63.2	67.7	50.78	58.03	67	66.23	50	56.98	68	68.18	58.85	63.15
	imdb	66.4	65	39.39	49.05	66.5	66.66	43.37	52.55	69	74.19	37.7	50
	Yelp	64.4	61.57	86.02	71.77	63	62.98	86.6	72.93	72	75	55.93	64.07
Random forest	Amazon	68.4	57.33	67.18	61.87	72	67.44	56.86	61.7	74	68.88	60.78	64.58
	imdb	65.2	50	61.61	55.2	65	51.54	60.24	55.55	70	71.42	57.37	63.63
	Yelp	66.4	67.59	53.67	59.83	65.55	68.37	71.42	69.86	65	76.74	55.93	64.7

Table 5 Result comparison with other results

Author	Dataset used	Method	Accuracy (%)
Hamouda and Rohaim [3]	Amazon product review	Average on sentence and Average on review	68.63
Medagoda et al. [7]	Sinhala opinions	NB	60
Denecke [6]	IMDB in German	Simple Logistic Classifier	66
Ohana and Tierney [13]	IMDB	SWN used as features	69.35
Goel et al. [14]	Twitter	NB	58.4
Proposed work	Yelp	SVM	74
Proposed work	Amazon	Random forest	74

7 Conclusion

We performed sentiments analysis on movie review data, e-commerce website Amazon product reviews dataset and www.yelp.com reviews dataset. SentiWordNet approach is used as a baseline to assign the weight to each sentence. Wight Assignment model used the PMI to improve accuracy. Different combinations of noun, adjective, verb, and adverb are implemented. Machine learning techniques such as SVM, NB, Random forest, Logistic regression are used to classify the sentiments. 74.00% accuracy is achieved using SVM with 10 folds Yelp datasets.

References

1. Rout, J.K., Dalmia, A., Choo, K.-K.R., Bakshi, S., Jena, S.K.: Revisiting semi-supervised learning for online deceptive review detection. IEEE Access 5(1), 1319–1327 (2017)
2. Dash, A.K., Rout, J.K., Jena, S.K.: Harnessing twitter for automatic sentiment identification using machine learning techniques. In: Proceedings of 3rd International Conference on Advanced Computing, Networking and Informatics, pp. 507–514. Springer (2016)
3. Hamouda, A., Rohaim, M.: Reviews classification using SentiWordNet lexicon. In: World Congress on Computer Science and Information Technology. IAENG (2011)
4. Rout, J.K., Singh, S., Jena, S.K., Bakshi, S.: Deceptive review detection using labeled and unlabeled data. Multimed. Tools Appl. 1–25 (2016)
5. Singh, V.K., Piryani, R., Uddin, A., Waila, P.: Sentiment analysis of movie reviews: a new feature-based heuristic for aspect-level sentiment classification. In: 2013 International Multi-Conference on Automation, Computing, Communication, Control and Compressed Sensing (iMac4s), pp. 712–717. IEEE (2013)
6. Denecke, K.: Using sentiwordnet for multilingual sentiment analysis. In: IEEE 24th International Conference on Data Engineering Workshop, 2008 (ICDEW 2008), pp. 507–512. IEEE (2008)
7. Medagoda, N., Shanmuganathan, S., Whalley, J.: Sentiment lexicon construction using SentiWordNet 3.0. In: 2015 11th International Conference on Natural Computation (ICNC), pp. 802–807. IEEE 2015

8. Kotzias, D., Denil, M., De Freitas, N., Smyth, P.: From group to individual labels using deep features. In: Proceedings of the 21th ACM SIGKDD International Conference on Knowledge Discovery and Data Mining, pp. 597–606. ACM (2015)

9. Maas, A.L., Daly, R.E., Pham, P.T., Huang, D., Ng, A.Y., Potts, C.: Learning word vectors for sentiment analysis. In: Proceedings of the 49th Annual Meeting of the Association for Computational Linguistics: Human Language Technologies, vol. 1, pp. 142–150. Association for Computational Linguistics (2011)

10. McAuley, J., Leskovec, J.: Hidden factors and hidden topics: understanding rating dimensions with review text. In: Proceedings of the 7th ACM Conference on Recommender Systems, pp. 165–172. ACM (2013)

11. Tripathy, A., Agrawal, A., Rath, S.K.: Classification of sentiment reviews using n-gram machine learning approach. Expert Syst. Appl. **57**, 117–126 (2016)

12. Durme, B.V., Lall, A.: Streaming pointwise mutual information. Adv. Neural Inf. Process. Syst. 1892–1900 (2009)

13. Ohana, B., Tierney, B.: Sentiment classification of reviews using SentiWordNet. In: 9th. IT & T Conference, p. 13 (2009)

14. Goel, A., Gautam, J., Kumar, S.: Real time sentiment analysis of tweets using naive Bayes. In: *2016 2nd International Conference on Next Generation Computing Technologies (NGCT)*, pp. 257–261. IEEE (2016)

Privacy of Organization in Online Social Networks

Priyanja Singh and Sarang Shrivastava

Abstract With the rise in the participation of people on Online Social Networks (OSN), there has been a significant increase in reports of privacy leakage, both from users, as well as organizations. Although there are numerous solution frameworks for protecting user privacy, there are very few effective online tools to safeguard the privacy of organizations, most of which are based on manual altering of content. A lot of sensitive information of an organization can go public, intentionally or unintentionally. Analyzing and classifying the publicly available information will be very valuable to organizations. In this research, we were able to generate a privacy score for a social media post about a large organization. We utilized the data from internal Wiki, as well as public blogs of the organization, and evaluated the publicly available information on online social networks, especially, Twitter. We have used a Semantic Web representation for storing the data. The results show that a private information leak can be detected by our solution framework.

Keywords Experimentation · Security · Privacy · Semantic Web
Online social networks

1 Introduction

In the past few years, there has been a tremendous rise in the usage of online social networking platforms. This has been accompanied by a considerable rise in privacy leaks and related issues reported by individual users. The data collection techniques used by the operators are, more often than naught, non-privacy preserving. This implies that the data which a user is providing voluntarily or is being leaked through cookies can be used to identify the user from a group of users. Not only is the privacy of user under threat on OSN, the private and confidential information of organizations have also been shared on these platforms. The OSNs these days are used by

P. Singh (✉) · S. Shrivastava
Motilal Nehru National Institute of Technology Allahabad, Allahabad 211004,
Uttar Pradesh, India
e-mail: priyanjasingh15@gmail.com

© Springer Nature Singapore Pte Ltd. 2018   141
P. K. Sa et al. (eds.), *Recent Findings in Intelligent Computing Techniques*,
Advances in Intelligent Systems and Computing 709,
https://doi.org/10.1007/978-981-10-8633-5_15

some employees to discuss social work, which can put their employers in a spot if proprietary or confidential content is discussed. Several examples can be taken from the past, Israel Military officer sharing upcoming raid information on Facebook [1], a woman's health reports shared on Facebook, Microsoft employee sharing trade secrets with a blogger [2], etc. Most of these instances were a result of a casual attitude of the employees of an organization regarding a piece of critical information. Sometimes, organizations are also the target of hackers, who utilize such mistakes from employees and use the leaked information to gather intelligence about organization [1].

2   Related Work

Although there have been many case studies related to user privacy on OSN and some solutions proposed to safeguard user privacy on OSN, there has been little work done with regard to a solution framework to contain the damage caused by a leakage of confidential information of an organization. Our area of study, thus, has been to propose a solution framework and implement a working system that can detect leakage of privacy-sensitive content about an organization on social networking platforms.

- Solutions for user privacy protection: The k anonymity model [3] is a method that can be used by OSNs to implement privacy preserving techniques for sharing user data with third parties. Another model suggested in this regard is the differential privacy model [4]. Any database, containing user information, created using differential privacy model, provides similar analytic results if any one data set is removed from this database. This means that a user can afford to share her information and not face a threat of identity revelation. For the mobile application domain, there is a study by Lin et al. [2].
- Bag of Words model: This approach can be utilized for classifying the information revealed on OSN, about an organization, as confidential or nonconfidential. However, such a naive approach would demand a lot of data, since this model fails to capture the essential relation between terms, or in other words, ignores the semantics of the information.
- Online solutions: Almost all organizations have signed Nondisclosure Agreements (NDAs) with their employees. Many organizations also provide security awareness and training for their employees. However, these o-line solutions can only provide a deterrent against sharing critical information by employees.

3   DataSet

For the study, we have implemented our solution for Adobe Systems India. The framework that we propose hinges on two sets of data, internal repositories and publicly available information.

Table 1 List of node categories for word classification

Category	Represented as
Employee names	emp
Product names	prd
Places/Office locations	plc
Marketing words	mkt
Technical terms	tec
IP addresses	ips
Customers	cst
Phone numbers	phn
URLs	url
Dates and time	dat
Alphanumeric entries	aln
Other English words	ews

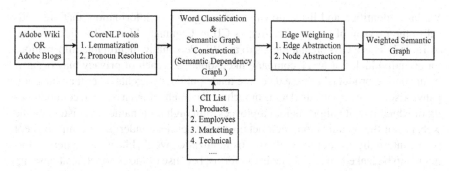

Fig. 1 Constructing a weighted semantic graph

- Internal Repository. The data in this category is very dependent on the business orientation of the organization which uses this technology. For an IT company like Adobe, information related to the list of items in Table 1 will form the part of internal repository. We obtained the web pages from the internal wiki of Adobe for the private sources.
- Publicly available blogs. The amount of data in this category can be huge. However, for efficient working of the algorithms for privacy leak detection, it is sufficient to have information related to the topics from the Internal Repository (Fig. 1).

The data from these two sources was split into two parts, one for graph construction and one for graph weighing. The organization was also asked to provide a list of its employees, office locations, and product names.

4 Solution Architecture

We have used the semantic graph representation of the organization's database (both public and private). This model of information storage will take care of the "independence" assumption made by a bag of words model. The solution architecture can be divided as

1. Identifying the features of confidential information
2. Preprocessing the data from the sources
3. Semantic Graph Construction
4. Quantify and identifying Patterns corresponding to Private and Public
5. Scoring the social stream.

4.1 Identifying Features of Confidential Information

We have identified and listed down some categories of information, that are critical to the privacy of an organization, as CII or Company Identifiable Information, in Table 1. These CIIs are analogous to the PII (Personally Identifiable Information) for an individual. For an IT company like Adobe, a breach of privacy may occur if its upcoming products' release dates are made public, by agents other than the company. Also, there are certain details or technical content about a project, confidentiality of which is quite high. As for employees, although their names are visible on the websites of the organization, their details such as project undertakings must be kept secure internally. In addition to these, no organization would like its customers' information to be leaked on social media platforms, because of the contractual anonymity agreements.

4.2 Preprocessing the Data from the Sources

After obtaining the information from the organization, such as employee list, product names, etc., we construct a dictionary, which contains the above information under the appropriate headers, such as employee names, product names, locations, etc. We train a naive Bayes classier, to classify all words from the information sources into these categories. If a classier fails to identify the right basket for a word, it puts the word into 'Other English words' category.

For dates, IP addresses, URL, phone numbers, and alphanumeric entries, we have used regular expressions to put them in their respective categories. The final list of categories is listed in Table 2.

Special characters were removed from the text and lemmatization and pronoun resolution was done using CoreNLP. These two tasks for a sample sentence are depicted in Fig. 2.

Table 2 Criticality of relations

Category of node1	Category of node2	Criticality[a]
Date	Product	0.9
Product	URL	0.9
IP address	Product	0.8
Customer	Product	0.7
Employee	Product	0.6
Product	Technical terms	0.5
Phone Numbers	Employees	0.4
Phone	Place	0.2

[a]The other remaining relations have a default criticality of 0.1

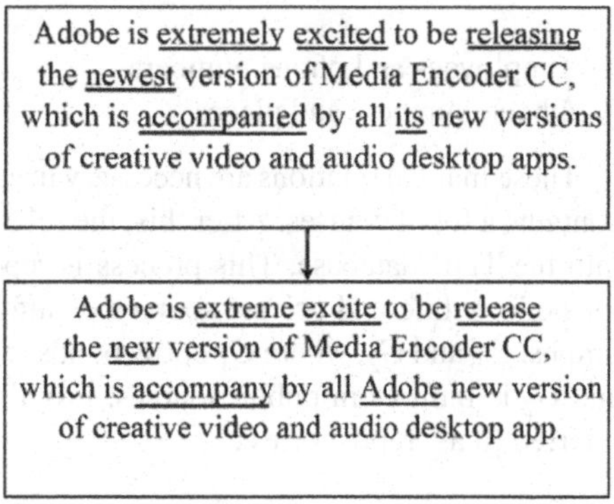

Fig. 2 Lemmatization and pronoun resolution. The words 'extremely', 'excited', etc., are replaced by 'extreme', 'excite', etc. while 'its' has been replaced by 'Adobe'

Adobe is extremely excited to be releasing the newest version of Media Encoder CC, which is accompanied by all its new versions of creative video and audio desktop apps.

Adobe is extreme excite to be release the new version of Media Encoder CC, which is accompany by all Adobe new version of creative video and audio desktop app.

4.3 Semantic Graph Construction

We have used Apache Jena TDB libraries (Java based) to store the RDF triples that will form the basis of the semantic graph. This framework supports subgraph querying by SPARQL commands. To create the semantic graph, a document is preprocessed as in the previous section. So, every sentence is sent to the NLP tool for POS tagging and then we obtain the preliminary graph from the collapsed dependencies. A semantic graph represents data as RDF (Resource Description Framework) triples, "subject predicate object". For every RDF triple, the concatenation of the categories of the nodes is used as the predicate value. This means that a relation between an employee and product will be called 'rel#empprd'. Note that the concatenation is in lexicographic order, since we are implementing an undirected graph for this study. After getting a preliminary set of edges, there are some important relationships which we add that may be skipped by NLP tool. Dependencies manually (Fig. 3)

1. Employees and Projects
2. Customer and Products
3. Products and technical words
4. Dates and Products

Fig. 3 Semantic graph of
sample sentences. Large
nodes are of higher criticality

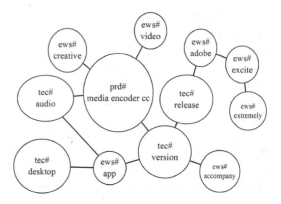

5. Employees and phone numbers
6. Phone numbers and places.

These manual relations are necessary in cases where the sentence is quite long and contains a lot of clauses. After this, the relations (edges or RDF triples) are inserted into the TDB database. This process is repeated for every sentence in the content for both, public and private sources of information. Thus we obtain two semantic graphs, G_1 and G_2, where G_1 corresponds to the graph built from internal repository and G_2 is built from public sources. From now on, G_1 and G_2 will be collectively referred to as 'repositories'.

4.4 Edge Weighing Mechanisms

We utilized two approximate methods to weigh the edges by the number of patterns that matched—
Edge Abstraction In the edge abstraction method, we pick an edge and replace the values of the nodes of this edge by their categories and obtain the query graphs. These partly abstracted graphs are then queried in the stored repositories. Then, all the results (here, results are edges present in subgraphs, as defined by the query) are compiled together and duplicates (duplicates occur in cases where an edge is present in two or more matching subgraphs) are removed. The weights of all these edges are then increased by a value corresponding to the criticality of that relation. We have defined the criticality of relations as in. Thus, if we have an edge whose predicate is rel#empprd' that matches the pattern, then its weight will be incremented by 0.6 (Fig. 4).

Node Abstraction

Node Addition For node addition, we remove all the values of nodes and replace by their categories. One by one for all categories, we substitute back the actual values

Fig. 4 Edge abstraction weighing: The nodes with single alphabets are searched by value, while the ones with 'type' prefix are searched by category

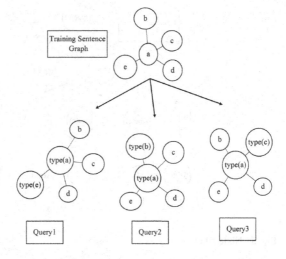

and observe the change in the number of results. The queries look like those in Fig. 5. The weights of edges of the matching results are increased in a similar way, as in Edge Abstraction, i.e., weights are increased by the criticality of the edges.

Node Removal For node addition, we remove the value of one category of nodes and replace the value by the category. One by one, this is done for all the nodes (while other nodes are restored), and the resulting graph is queried in the stored repositories. This is shown in Fig. 6. The weights of edges of the matching results are increased in a similar way, as in Edge Abstraction i.e., weights are increased by the criticality of the edges.

The nodes with single alphabets are searched by value, while the ones with 'type' prefix are searched by category.

Storing the weights We store the weights in an SQL table. The table contains the weights for edges, as required by the scoring algorithm in Sect. 4.5. Thus, the weights for both, edge abstraction and node abstraction methods, are stored as weights of edges.

4.5 Scoring the Social Stream

The following cases arise when searching for an edge of the tweet semantic graph, t_i in G_1,

1. **Both nodes found**: We proceed to find a path between these two nodes. For our study, we limited the maximum path length to 3. So if two nodes have a path length greater than 3, they carry nonsignificant semantic relation viz-a-viz the repositories. Let L_i be the length of the path, connecting the two nodes of edge

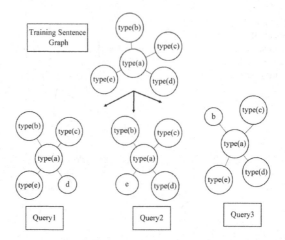

Fig. 5 Node addition

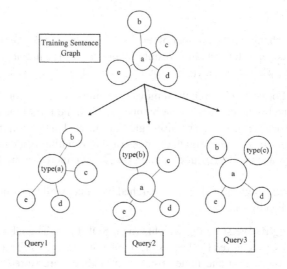

Fig. 6 Node removal

t_i, in the two repositories. We have defined L_i as the number of edges $e \in G_1$, in the repository, in the direct path from the source and destination nodes of t_i. Let these intermediate edges be $e_k \in E_1$ where k ranges from 1 to L_i. Then, weight for edge t_i, $W(t_i)$ is,

$$W(t_i) = a^{-(k-1)} * \sum_{k=1}^{L_i} W(e_k) \qquad (1)$$

If there are multiple such paths between the two nodes, we take the mean of all the scores over such paths.

2. **Only one node found**: We try to find a heuristic weight for the edge t_i. We look into the neighborhood of the node that was found, where neighborhood consists of nodes which have the same category as the missing node and are at a distance of 1 edge from the node that was found. We use the mean of the weights of such edges to evaluate a heuristic score for such cases. Let n_{i1} and n_{i2} be the nodes corresponding to the edge t_i. Let $n_{1i} \in G_1$ and $n_{2i} \notin G_1$. We denote B as the set of neighbor nodes of n_{i1} belonging to the same category as n_{2i}, such that $v_k \in B \iff isEdge(G_1, n_{i1}, v_k)$ is true, where v_k and n_{i2} belong to the same category and $v_k \in G_1$, $\forall k \in \{1, 2, 3, \ldots, N\}$. Here, N represents the cardinal number of the set B. Then, heuristic weight for the edge t_i is

$$H(t_i) = \frac{\sum_{k=1}^{N} W(n_{i1}, v_k)}{N} \tag{2}$$

   where, $W(n_{i1}, v_k)$ is the weight of the edge connecting n_{i1} and v_k.
3. **Both nodes not found**: We have assumed that such edges/relations carry very little semantic significance, so we assign zero weight to them.

$$W(t_i) = 0 \tag{3}$$

After getting the scores for every edge of the tweet graph, G_t, we need to give a final score. One possibility could be to have the final score as

$$Score = 10 * \frac{\sum_{k=1}^{|Edge|} W(t_k)_{private}}{\sum_{k=1}^{|Edge|} W(t_k)_{private} + \sum_{k=1}^{|Edge|} W(t_k)_{public}} \tag{4}$$

where $|Edge|$ is the number of edges in the tweet semantic graph. This would be the total score from the private repository over the sum of the total scores from private and public repository, multiplied by 10 to scale it from 0 to 10. However, such a score would be misleading, in the sense that even if both the terms of the denominator in Eq. 4 are low, we might end up getting a high score just because the first term is larger than the second.

So, we must have a factor to discount for this. The final scores that we provide below utilize the logistic function, with its center shifted to the right, so as to have low scores for posts having low match in both the public and private repositories.

1. The score from the repository, S_{ri}, where, i refers to the repository index, is the mean of the scores for all edges.

$$S_{ri} = \frac{\sum_{k=1}^{|Edge|} W(t_k)}{Edge_{count}} \tag{5}$$

   where $|Edge|$ is the number of edges in the post.

2. Then, we calculate the privacy score as,

$$PrivacyScore = 10 * \frac{S_{r1}}{S_{r1} + S_{r2}} \cdot \frac{1}{\exp^{-(S_{r1}+S_{r2}-10)}} \qquad (6)$$

The factor 10 in the logistic function was used to shift the logistic function toward the positive axis. This relation ensures that when both the scores, S_{r1} and S_{r2} are low, the generic score is low. Also the final privacy score between 0 and 10 is arrived at by multiplication with 10.

Thus, for every social media post, we have two scores, one corresponding to the edge abstraction weights and one for the node abstraction weights (Fig. 7).

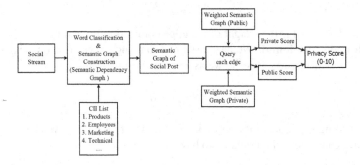

Fig. 7   Scoring the social stream

5   Results

The algorithm was tested on 21,116 tweets from February to July 2015, related to Adobe's products. The tweets were manually annotated into three categories, Very little private content, mildly private, and confidential. The annotation process was carried out by taking a survey from the employees of Adobe. The results of the scoring algorithm are classified in Figs. 8 and 9 for the two types of scores. There is a difference between the scores calculated from the two kinds of weighing mechanisms, edge abstraction and node abstraction,

1. **Edge Abstraction**: Of the 21,116 tweets, about 31% fell in the 0–4 category (very little private content), 52% were classified in 4–7 category (mildly private) and the remaining 17% were categorized in 7–10 category (confidential).
2. **Node Abstraction**: Amongst all the tweets about 42.5% fell in the 0–4 category (very little private content), 20.5% were classified in 4–7 category (mildly private) and the remaining 37% were categorized in 7–10 category (confidential).

Fig. 8 Distribution of scores of the tweets by using edge abstraction weights

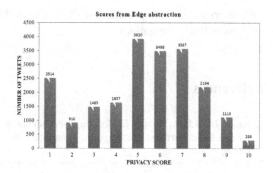

Fig. 9 Distribution of scores of the tweets by using node abstraction weights

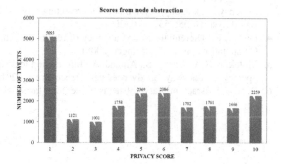

False Positives The approach to weighing the graph in Sect. 4.4 based on frequency of occurrence might give an impression that it might lead to a large number of false positives. However, this problem was avoided by having a priority list (Table 2) of all the relations and the corresponding quantum of increase in weights. This priority ensures that weights for edges of the type rel#ewsews' are low, even though such edges have a large number of occurrences. Also, weights of a particular edge are increased only if it matches the pattern of the training sentence, which was picked from a carefully chosen list of document.

6   Conclusion

The algorithm was used to design a social media management product, APrIL. There are various use cases of this. First, a watchdog for an organization to easily monitor its confidential information being shared on Online social networks. Second, a tool to monitor the activities of the employees of an organization on OSN and the information they share about the organization. Third, a utility to easily evaluate the privacy score of a public announcement, before the actual announcement, so as to have a controlled revelation of information. Fourth, effective semantic representation of the internal and public data of the organization and a query system to retrieve such information. This provides an alternative to the document search implemented by most of the organizations.

References

1. Sadeh, J.L.B.L.N., Hong, J. I.: Modeling users mobile app privacy preferences: restoring usability in a sea of permission settings. In: Symposium on Usable Privacy and Security (SOUPS) (2014)
2. Kaiser, T.: Former Microsoft employee arrested for stealing trade secrets; shared info with tech blogger (2014)
3. Sweeney, L.: k-anonymity: a model for protecting privacy. Int. J. Uncertainty Fuzziness Knowl. Based Syst. **10**(05), 557–570 (2002)
4. Dwork, C.: Dierential privacy: a survey of results. In: Theory and Applications of Models of Computation, pp. 1–19. Springer (2008)
5. Molok, N.N.A., Chang, S., Ahmad, A.: Information leakage through online social networking: opening the doorway for advanced persistence threats (2010)
6. B. NEWS. Israeli military unfriends soldier after facebook leak (2010)

Summary-Based Document Classification

P. P. Assainar Hafnan and Anuraj Mohan

Abstract Document classification is one among the major NLP tasks that facilitate mining of text data and retrieval of relevant information. Most of the existing works use pre-computed features for building the classification model. Large-scale document classification relies on the efficiency or appropriateness of feature selection for document representation. The proposed system uses text summarization for automated feature selection to build the classification model. This work considers feature selection as a sentence extraction task which can be done using extractive text summarization. The method will have the advantage of reduced feature space, as classifier will be trained on shorter summary than the original document. Also, deep learning-based summarization generates the most relevant features resulting in improved efficiency and accuracy of the classifier. Experiments showed that classification based on features generated using deep learning provides better classification accuracy.

Keywords Deep learning · Document classification · Text summarization

1 Introduction

Availability of information in the form of documents and other forms are increasing at an exponential pace everyday making text management a tedious task. Searching for and retrieval of relevant information through this vast amount of information sources is a challenging problem. Automation of these processes is carried out using natural language processing techniques. Summarization is such an NLP task for capturing the main aspects covering a document or a collection of documents. In this work, a newer approach based on document summarization is proposed for document classification. Summarization is the process of deriving a reduced representation of

P. P. Assainar Hafnan (✉) · A. Mohan
Department of Computer Science and Engineering, NSS Engineering College,
Palakkad, India
e-mail: hafnu.hafnu@gmail.com

© Springer Nature Singapore Pte Ltd. 2018
P. K. Sa et al. (eds.), *Recent Findings in Intelligent Computing Techniques*,
Advances in Intelligent Systems and Computing 709,
https://doi.org/10.1007/978-981-10-8633-5_16

a document or a collection of documents, which conveys what is actually intended by the original document or the collection. The summary can be extracted or generated from the documents. Documents can be classified based on its content or based on its attributes such as document length, author name, etc. which facilitates users to search for information based on the category. Content-based classification assigns documents to classes based on the weights assigned to the textual units in its content.

This paper presents a supervised content-based document classification system built on summary features extracted via deep learning. Existing classification models require explicit specification of features upon which the classifier needs to be trained. The idea behind this work is to determine relevant features through summarization and use these features for document classification in order to increase the accuracy and efficiency of the classifier. This work is an attempt to automate the feature selection by summarizing the documents via deep learning. These features are then used as attributes for building the classifier.

Deep learning techniques are the most efficient methods for automatically finding features and learning higher level representations from the data provided to it. Building a summary-based classifier has several advantages. As the size of the summary is much less than the size of the corresponding document, the dataset size gets reduced. The classifier needs to be trained on a smaller dataset with lesser time. Hence efficiency is ought to be increased. There will be a reduction in the dimensionality of input representations. Also, feature space for summaries will be small as compared to the feature space for the entire document collection. Hence, a smaller feature space needs to be explored which contains the most relevant features extracted from the document. If the classifier can be built on the most relevant features extracted, accuracy can be increased.

2 Literature Survey

2.1 Text Summarization

This section discusses the various methods for implementing text summarization. Sentence ranking is the major step in selecting important sentences in the extractive summarization of a document. The most traditional method in sentence ranking is the use of statistical features derived from the document [1, 2]. The statistical features used include the position of a sentence in the document, similarity of a sentence with the title, sentence length, presence of cue words or phrases, presence of frequent words, presence of words with high Tf-idf values, presence of title words in the sentence, presence of proper nouns, etc. Graph-based methods are the next approach towards the extraction of summaries from the original document. The core of all graph-based methods is almost the same which includes preprocessing tasks, building a graph model, applying ranking algorithm and finally summary generation task. Documents are represented using simple graphs or bipartite graphs [3, 4]. Graph is constructed with nodes representing the textual units and edges are drawn between the nodes and is weighted with cosine similarity, eigenvector similarity, content-based features, other features such as cue words, length of sentences, etc. Graph-based shortest path algorithms, HITS algorithm, page rank algorithm, etc. are used for ranking graph nodes.

Another approach which gained interest in recent years towards summarization employs natural language processing and machine learning techniques. The work by [5] employs support vector machine cascaded with clustering technique to obtain much better summary. A machine learning-based single document summarization system uses a set of features based on groups of words which often co-occur with each other for obtaining sentence vectors and a classifier is trained in order to make a global combination of these scores in the vector [6]. The sentences are represented as vector of features including sentence length, sentence position, etc. A concept-based automatic multi-document summarizer is built using learning mechanism [7]. Concepts extracted using mutual information combined with statistical features derived from the document are fed to appropriate learning model.

A summarization system for news articles is implemented in three phases including neural network training, feature fusion and sentence selection [8]. A system is built that utilizes bisect K-means clustering to improve the time and neural networks to improve the accuracy of the summary generated by NEWSUM algorithm [9]. A pair-based sentence ranker for the summarization of newswire documents is built using RankNet learning algorithm to score every sentence in the document and identify the most important sentences [10].

A summarization model is built using embeddings derived for sentence level and document level using convolutional neural networks with backpropagation [11]. A newer technique for summarization employs neural network and rhetorical structure theory [12]. Neural network-based training is used to extract sentences, and these sentences are fed to rhetorical structure to find relation between sentences that facilitates generation of better summary. In addition to handcrafted feature vectors of words as input, recursive neural networks are used to automatically learn ranking features [13]. A deep learning-based approach for extractive summarization uses restricted Boltzmann machine to reduce redundancy. The four features title similarity, positional feature, term weight and concept feature computed for document is the input to the training algorithm.

All the above works require explicit specification of features to be used for building the summarizer. With the development of deep learning techniques, automated generation of features can be done. One such work performs query-oriented summarization on multiple documents [14]. It employs deep learning model with three stages: concept extraction, reconstruction validation and summary generation.

2.2 Document Classification

Similar to text summarization, text classification has also been widely explored. Traditional machine learning algorithms such as decision trees, *K*-nearest neighbour, naive Bayes, support vector machines, etc. are used for text categorization [15–17]. In recent years, neural network-based classifiers are also built [18].

Accuracy of all the above methods relies on the accuracy of the feature selection process. Dimensionality reduction, i.e. selecting only the most representative features during feature selection is also an important consideration. The method proposed here aims to use summarization to extract relevant features. The focus in this approach is to replace manual feature selection with automated feature selection using deep learning in order to increase the accuracy of classification. To the best of our knowledge, this proposal is the first attempt at using summarization to assist the classification task.

3 Proposed System

This section gives the detailed description of summary-based document classifier. Figure 1 gives the architecture of the proposed system with two important modules: summarizer and classifier.

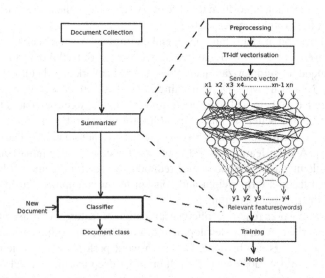

Fig. 1 Summary-based classifier

3.1 Summarizer

Summarization module works as the feature extractor for building the classifier. Documents are represented using the bag-of-words model. Deep learning technique is the core of the summarization process presented here.

Text Preprocessing. The text document is initially preprocessed to clean the textual content and produce preprocessed sentences. Preprocessing involves removal of punctuations, stop words, special characters, etc.

Tf-idf Vectorizer. Vectorizer then converts these textual sentences from preprocessing stage into corresponding tf-idf vectors. The vectorizer first builds a local vocabulary for the input document and creates a representation vector with size same as that of the vocabulary, i.e. $x_1, x_2, x_3, \ldots, x_n$, where each x_i denotes the $tf - idf$ value of the word i in the vocabulary built over the document and n denotes the vocabulary size.

Deep Architecture. The deep architecture uses restricted Boltzmann machines as its basic building block. RBM has a visible and hidden layer with weighted connections between nodes in each layer. Each possible joint configuration of the visible and hidden units has an energy determined by weights and biases. The learning procedure starts by setting the states of visible units to a training vector. Each possible visible layer and hidden layer configurations are reconstructed based on the joint probabilities. The probability computation is based on the logistic sigmoid function. Each time the configurations are updated, the weighted connections are also updated. Training vector is input to the visible layer and activations for reduced representation of document is produced as output at the hidden layer.

Summarizer is built by stacking RBMs one over the other. Figure 2 gives the stacked structure of multiple RBM units. Each layer performs computation over

Fig. 2 Stacked RBMs

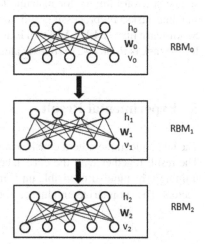

these input sentence vectors to filter out relevant features. Each RBM is fine-tuned to find the most relevant features which are fed as input to the next higher level RBM. The whole operation with RBM starts with giving the sentence vector as input. The number of nodes in the initial visible layer is determined by the length of the input sentence vector. The number of nodes is equal to the dimension of the tf-idf vector representing a document sentence. Each node in the visible layer takes as input the tf-idf value for the corresponding word or feature. With these values as input, the RBM performs its joint probability computations and computes the activations for its nodes as described previously. This processing continues until an acceptable error rate is reached. The output from the hidden layer of the last unit gives the relevant feature vector $y_1, y_2, \ldots, y_m$ with $m < n$, i.e. a vector with a lower dimension. This vector serves as the input to the classifier training. Thus, the summarizer module facilitates dimensionality reduction.

3.2 Classifier

The classification model is built using SVM classifier which takes the features extracted from the deep architecture as input. It then learns from the training samples and builds the model. After the model has been constructed, newly arriving documents can be classified based only on the known labels.

4 Experimental Setup

This work is implemented using Python language and its associated libraries upon the TensorFlow framework to implement deep learning. TensorFlow is an open source software library for machine learning in various kinds of language understanding tasks. Preprocessing and vectorization are carried out using NLTK and Scikit-learn library. LIBSVM tool is used for building the classifier. The dataset used for the task is Reuters-21578 text categorization test collection.

5 Experimental Result

The text collection without considering the label is first fed through the summarizer. The result together with the class labels is then given as input to the classifier for training. The number of visible units in RBM is set equal to the dimension of input vectors. The summarizer module can be optimized by adjusting the learning rate and

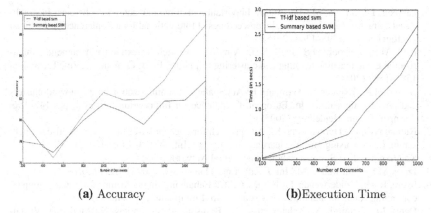

(**a**) Accuracy   (**b**)Execution Time

Fig. 3 Experimental results

number of hidden units for each RBM. Optimization aims at minimizing the error rate. The system shows better results than using normal tf-idf vectors for representing the document text. The system has an improved classification time and accuracy when compared to tf-idf-based SVM document classifier and is illustrated in Fig. 3.

6 Conclusion

The work proposed here is a supervised content-based approach for text classification. This work utilizes summarization to improve the accuracy of the text classification task. The proposed method uses deep learning which is the most recent approach for extracting the important features from the data. Feature extraction is done using the summarizer module, and these features are further utilized in text classification. Since the classifier is built on a reduced feature space, the time for building the classifier gets reduced and efficiency is improved. Accuracy is also increased as the classification model is built by training upon the most relevant features of the training sample. Hence, the accuracy of classification is limited only to the seen data. This approach can be improved further by implementing both feature extraction and classification using deep architecture.

References

1. Edmundson, H.P.: New methods in automatic extracting. J. ACM (JACM) **16**, 264–285 (1969)
2. Luhn, H.P.: The automatic creation of literature abstracts. IBM J. Res Dev. **2**, 159–165 (1958)
3. Mihalcea, R., Tarau, P.: TextRank: bringing order into texts. Assoc. Comput. Linguist. (2004)

4. Parveen, D., Strube, M.: Integrating importance, non-redundancy and coherence in graph-based extractive summarization. In: proceedings of International Joint Conference on Artificial Intelligence, pp. 1298–1304 (2015)
5. Patil, M.S., Bewoor, M.S., Patil, S.H.: A hybrid approach for extractive document summarization using machine learning and clustering technique. Int. J. Comput. Sci. Inf. Technol. **5**, 1584–1586 (2014)
6. Amini, M.R., Usunier, N., Gallinari, P.: Automatic text summarization based on word-clusters and ranking algorithms. In: European Conference on Information Retrieval, pp. 142–156. Springer, Berlin, Heidelberg (2005)
7. PadmaPriya, G., Duraiswamy, K.: An approach for concept-based automatic multi-document summarization using machine learning. Int. J. Appl. Inf. Syst. **3**, 49–55 (2012)
8. Kaikhah, K.: Text Summarization using Neural Networks (2004)
9. Igave, M.S., Gaikwad, C.M.: Int. J. Adv. Eng. Manag. Sci. **2**, 0952–0957 (2016)
10. Svore, K.M., Vanderwende, L., Burges, C.J.C.: Enhancing single-document summarization by combining RankNet and third-party sources. In: Emnlp-conll, pp. 448–457 (2007)
11. Denil, M., Demiraj, A., Kalchbrenner, N., Blunsom, P., de Freitas, N.: Modeling, Visualizing and Summarizing Documents with a Single Convolutional Neural Network (2014). arXiv:1406.3830
12. Kulkarni, A.R., Sarda, A.: Text summarization using neural networks and rhetorical structure theory. Int. J. Adv. Res. Comput. Commun. Eng. **4**, 49–52 (2015)
13. Cao, Z., Wei, F., Dong, L., Li, S., Zhou, M.: Ranking with recursive neural networks and its application to multi-document summarization. In: proceedings of the Association for the Advancement of Artificial Intelligence conference, pp. 2153–2159 (2015)
14. Zhong, S.-H., Liu, Y., Li, B., Long, J.: Query-oriented unsupervised multi-document summarization via deep learning model. Expert Syst. Appl. **42**, 8146–8155 (2015)
15. Basu, A., Watters, C., Shepherd, M.: Support vector machines for text categorization. In: Proceedings of the 36th Hawaii International Conference on System Sciences (2002)
16. kim, S.-B., Han, K.-S. Rim, H.-C., Myaeng, S.H.: Some effective techniques for naive Bayes text classification. IEEE Trans. Knowl. Data Eng. (2006)
17. Bijalwan, V., Kumar, V., Kumari, P., Pascual, J.: KNN based machine learning approach for text and document mining. J. Database Theory Appl. **7**, 61–70 (2014)
18. Zhang, X., Zhao, J., LeCun, Y.: Character-level convolutional networks for text classification. In: Advances in Neural Information Processing Systems, pp. 649–657 (2015)
19. Balaji, J., Geetha, T.V., Parthasarathi, R.: A Graph based query focused multi-document summarization. Int. J. Intell. Inf. Technol. (IJIIT) **10**, 16–41 (2014)
20. Jeonghun, Y.O.O.N., Dae-Won, K.: Classification based on predictive association rules of incomplete data. IEICE Trans. Inf. Syst. **95**, 1531–1535 (2012)
21. Kim, Y.: Convolutional Neural Networks for Sentence Classification (2014). arXiv:1408.5882
22. Lertnattee, V., Theeramunkong, T.: Class normalization in centroid-based text categorization. Inf. Sci. **176**, 1712–1738 (2006)
23. Li, W., Han, J., Pei, J.: CMAR: accurate and efficient classification based on multiple class-association rules. In: Proceedings of the IEEE International Conference on Data Mining series, pp. 369–376 (2001)
24. PadmaPriya, G., Duraiswamy, K.: An approach for text summarization using deep learning algorithm. J. Comput. Sci. **10**, 1–9 (2014)
25. Rahman, C.M., Sohel, F.A., Naushad, P., Kamruzzaman, S.M.: Text classification using the concept of association rule of data mining (2010). arXiv:1009.4582
26. Tan, S.: An improved centroid classifier for text categorization. Expert Syst. Appl. **35**, 279–285 (2008)
27. Thakkar, K.S., Dharaskar, R.V., Chandak, M.B.: Graph-based algorithms for text summarization. In: Proceedings of the 3rd International Conference on Emerging Trends in Engineering and Technology, pp. 516–519 (2010)

A Document Similarity Computation Method Based on Word Embedding and Citation Analysis

K. Lamiya and Anuraj Mohan

Abstract Document similarity is one among the most significant problems in knowledge discovery and information retrieval. Most of the works in document similarity only focus on textual content of the documents. However, these similarity measures do not provide an accurate measure. An alternative is to incorporate citation information into similarity measure. The content of a document can be improved by considering the content of cited documents, which is the key behind this alternative. In this work, citation network analysis is used to expand the content of citing document by including the information given in cited documents. The next issue is the representation of documents. A commonly used document representation is bag-of-words model. But it does not capture the meaning or semantics of the text as well as the ordering of the words. Hence, this proposed work uses word embedding representation. Word embedding represents a word as a dense vector with low dimensionality. Word2vec model is used to generate word embedding which can capture contextual similarity between words. The similarity between documents is measured using word mover's distance, which is based on the word embedding representation of words. The proposed work takes advantage of both textual similarity and contextual similarity. Experiments showed that the proposed method provides better results compared to other state-of-the-art methods.

Keywords Citation network · Word embedding · Word mover's distance
Word2vec

1 Introduction

Document represents thoughts and ideas either in written or recorded form. They are preserved in order to serve as a reference for future purposes. These documents include sales invoices, newspaper issues, executive orders, product specifications,

K. Lamiya (✉) · A. Mohan
Department of Computer Science and Engineering, NSS College of Engineering,
Palakkad, Kerala, India
e-mail: lamiyalami04@gmail.com

© Springer Nature Singapore Pte Ltd. 2018
P. K. Sa et al. (eds.), *Recent Findings in Intelligent Computing Techniques*,
Advances in Intelligent Systems and Computing 709,
https://doi.org/10.1007/978-981-10-8633-5_17

academic papers, and so on. Let us focus only on the scientific papers. Scientific papers are used for sharing research work with other researchers and also for reviewing the research done by others. Hence, it is considered as one of the primary sources for sharing information and knowledge among researchers. Basically, a scientific paper contains two interrelated aspects: content as well as citation. The main aspect of a paper is its content which demonstrates the context of that paper. The content of a paper refers to the actual textual content of it. This content can be represented using any representation models such as vector space model. The second aspect is citations, which refer to the published or unpublished sources. Citations are carefully selected by authors and are cited according to the content of the paper. Citations are normally organized into a network structure called citation network/citation graph. It is a social network which can be utilized to study patterns of relationships between papers.

The proposed work exploits citation network analysis to compute the similarity between scientific papers. Considering citations of a paper helps to improve its own content and thus to make it more understandable for readers. Also, the documents are represented using word embedding model, rather than bag-of-words-based models. This word representation can capture both textual and contextual similarities, which improves the accuracy of similarity measure.

2   Related Works

Document similarity is a most significant problem in information retrieval and knowledge discovery. Hence, many works had been conducted in this area. But most of these works only captures textual similarity between documents. Survey [1] describes several text-based similarity measures that are commonly used in information retrieval. These similarity measures only focus on the content of the document. The idea is that two documents are more similar if they share more common terms. Since these measures only focus on content of the papers, the similarity measure will not be an accurate one.

Meanwhile, many other works were introduced which rely only on citation analysis to compute document similarity measures. These measures make use of citation information extracted from citation network analysis. A link based similarity measure called SimRank [2] considers the in-links recursively and similarity between two documents is based only on the number of documents that cite both of them. P-Rank [3] take both the in-links and out-links recursively to compute similarity measure. The similarity between two documents is based on the number of documents that cite both of them and are cited by both.

Later the idea to bind both content and citation information was proposed to compute document similarity. These measures are called hybrid measures. One such method compute similarity score as a weighted linear combination of both text and link based similarity scores [4]. Another method enriches the content of a paper by considering the content of the related papers [5]. These related papers can be either cited paper or citing paper or both. Keyword extension method [6] extends the keyword set of a paper by incorporating the keywords from related papers. The idea of key extension is to find additional good terms that represent the textual content of a paper.

2.1 Word Embedding

Word embeddings [7] was formerly introduced by Bengio et al. He proposed a distributed representation for word that can overcome the issues of dimensionality. This representation was later named as Word embedding. However, it was Collobert and Weston who first demonstrated the power of pre-trained word embeddings [8]. Word embedding is considered as an interesting strand of deep learning research. A word embedding is a representation of a word in which a word is represented as a dense vector of real numbers. Word embedding builds a low-dimensional vector representation from corpus of text, which preserves the semantic meanings of words. This transformation is necessary because machine learning algorithms need their input to be vectors of continuous values rather than strings of text.

2.2 Word2vec

Mikolov et al. created word2vec model [9] for generating word embedding which enables the training and use of pre-trained embedding. Word2vec is modeled as a two-layer shallow neural network that is trained to reconstruct linguistic contexts of words. It takes input as text corpus and generates the word vectors as output. Rong explains parameter learning process of word2vec model [10]. The training results are sensitive to parameters such as architecture, training algorithm, subsampling, context window size, and dimensionality. The proposed work uses skip-gram architecture and hierarchical softmax training algorithm. The skip-gram model learns the word vector representations that are good at predicting context words. Each word embedding is trained to maximize the probability of neighboring words for a given sequence of words. The model is trained using hierarchical softmax training algorithm which helps to limit the number of output vectors that need to be updated per training instance.

2.3 Word Mover's Distance

Word Mover's Distance (WMD) [11] is a distance function used to compute distance between documents. WMD is a special case of Earth Movers Distance [12], which is used to measure the dissimilarity between multi-dimensional distributions.

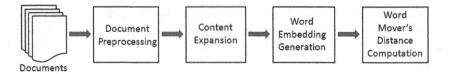

Documents

Fig. 1 Proposed system model

WMD is based on word embedding representation of words. Word vectors with same context are located closer to each other when placed in a vector space. WMD takes this advantage and compute the document distance as a function of distance between word vectors. Thus, the distance between the two documents is computed as the minimum cumulative cost required for moving all words from one document to another.

3 Proposed Methodology

This work proposes a novel similarity measure which binds content as well as citation of a document. The content of a document can be improved by considering its citation information. This work only focuses on research papers. Figure 1 explains the proposed system design. Before all computation, the documents need to be preprocessed to remove stop words and special characters. As next citation, information of a document is combined with its content and thereby improves the original content of the document. The content expansion procedure is performed for all documents in the collection. Then, word vectors are generated using Word2vec model. These word vectors capture textual and contextual similarities. At last, the distance computation is carried out using word mover's distance function.

3.1 Document Preprocessing

The system performs common text preprocessing methods to extract out meaningless words in the raw text. These methods include removal of stop words, special characters, punctuation marks, etc., and conversion to lowercase.

3.2 Content Expansion

Content expansion is the technique by which the content of each document is expanded by adding additional information. These informations are gathered from the documents that are related to it. Rather than computing citation score separately,

Fig. 2 A sample citation network

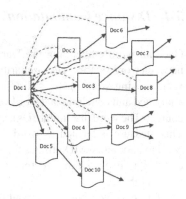

this method combines the content of a paper and the content of related papers. The related documents correspond to the cited or citing documents. The citation information is gathered by analyzing citation network.

In a citation network, each document is represented by a vertex and citation relationship by an edge between vertices. Figure 2 demonstrates a sample citation network where dashed edge represents the extraction of content from cited documents to citing documents. A document can be better represented using both directly and indirectly cited documents. Since considering all the indirectly cited papers will overwhelm the content only a subset of the same is examined. For that, a parameter is set such as number of levels that need to be considered for retrieving indirectly cited papers. Using the cited documents, the content of the document is expanded by incorporating content of cited documents.

3.3 Word Embedding Generation

Before similarity computation, documents need to be represented in some form for processing. A commonly used document representation is bag-of-words model, where the text is represented as a bag of its words. But it does not consider the grammar and even word order and only the counts the words appear in the document. This problem can be alleviated using word embedding representation where each word is represented as word vector. This transformation is necessary because machine learning algorithms need their input to be vectors of continuous values rather than strings of text. These word embeddings are generated using word2vec model with skip-gram architecture and hierarchical softmax training algorithm.

3.4 Distance Computation

Once all word embedding are learned, the next step is to compute the similarity between word vectors. The similarity is computed by measuring the distance between vectors using word mover's distance. Thus in WMD, the word dissimilarities are computed and conclude that word vectors with smaller dissimilarities share the same context. The word dissimilarities are better captured using Euclidean distance metric. This measure is termed as word travel cost, i.e., cost required to travel from one word to another. WMD utilizes the spatial property of word embedding, i.e., if word embedding is placed in a vector space, then words with same context will be closer to each other.

Thus, each word in a document is transformed into any word in other documents. Next, flow matrix is learned which express how much a word travels to other words. To transform a document entirely into another document the entire outgoing flow from each word should be equal to its normalized term frequency in the document. Similarly, the amount of incoming flow to each word must match its normalized term frequency in other documents. Finally, the distance between the two documents is computed as the minimum cumulative cost required for moving all words from one document to another, i.e.,

$$\min_{T \geq 0} \sum_{i,j=1}^{n} T_{i,j} c(i,j)$$

subject to

$$\sum_{j=1}^{n} T_{i,j} = d_i \ \forall \ i \epsilon \{1, \dots, n\}$$

$$\sum_{i=1}^{n} T_{i,j} = d'_j \ \forall \ j \epsilon \{1, \dots, n\}$$

where T represents the flow matrix and $c(i,j)$ represents word travel cost from word i to word j. Figure 3 illustrates the WMD metric on two documents $D1$ and $D2$.

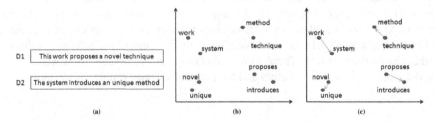

Fig. 3 An illustration of WMD metric: **a** represents two documents, **b** represents vector space, and **c** represents word travel cost associated between words

4 Experimental Setup and Results

The system was implemented using Python programming language. For document preprocessing, NLTK tool was used which is a platform to work with language processing. Also to get word vectors Word2vec toolkit in Gensim framework was made use of. It can model vector space faster than any other approaches.

For this experiment, scientific papers on different domains were collected from ScienceDirect, which is a leading source of research articles. The synthetic dataset was thus constructed with the help of domain experts. The dataset consists of papers from different areas of computer science and electronics. For each document, its cited documents were collected by exploiting corresponding citation information. Thus, each document in the dataset consists of its own abstract as well as the abstract of its cited documents. The word2vec model was trained using the synthetic dataset, and the vector representation of words was learned. Using the test set, the trained word2vec model was tested and the document similarity was computed. The ground truth for the experiment was gathered from citation information. Higher score was assigned to directly cited documents and lesser to indirectly cited documents. For a query document, k highest scoring papers were obtained, and the precision of similarity measure was computed by comparing the k highest scoring documents to those in the reference list of the query document. The proposed system was examined in two scenarios. In the first scenario, the precision of three systems, namely, basic text similarity measure, text similarity measure with citation analysis, and similarity measure with citation analysis and WMD, was analyzed. It is clear that the similarity measure with citation analysis and WMD perform better compared to other measures. In the second scenario, the system was evaluated using three distance measures: Euclidean distance, word centroid distance, and word mover's distance. Among these measures, Euclidean distance is an asymmetric one and others are symmetric measure. It was observed that the word mover's distance outperform other measures. Among the documents, the documents under similar domain have greater similarity and of different domains have lesser similarity. The results of these systems were demonstrated in Fig. 4.

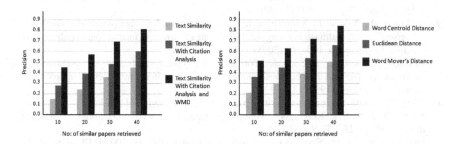

Fig. 4 Accuracy of similarities using different methods

5 Conclusion

This work combines both content and citation information to compute the similarity between documents. The citation information is incorporated in order to improve the accuracy of the system. Thus, the content of a document is expanded by including the content of related documents. The next step is to use text-based method to compute the similarity between documents. In order to represent the textual content, a distributed representation of words called word embedding is used. As next, the similarity between documents is computed using word mover's distance. The proposed method captures both textual and contextual similarities. As future work, a novel distance computation method can be developed rather than WMD so as to improve the results.

References

1. Barrn-Cedeno, A., Eiselt, A., Rosso, P.: Monolingual text similarity measures: a comparison of models over wikipedia articles revisions. In: Proceedings of the ICON: 7th International Conference on NLP, pp. 29–38 (2009)
2. Jeh, G., Widom, J.: SimRank: a measure of structural context similarity. In: Proceedings of the Eighth ACM SIGKDD International Conference on Knowledge Discovery and Data Mining (2002)
3. Zhao, P., Han, J., Sun, Y.: P-Rank: a comprehensive structural similarity measure over information networks. In: Proceedings of the 18th ACM Conference on Information and Knowledge Management (2009)
4. Chikhi, N.F., Rothenburger, B., Aussenac-Gilles, N.: Combining link and content information for scientific topics discovery. In 20th IEEE International Conference on Tools with Artificial Intelligence ICTAI'08, vol. 2, pp. 211–214 (2008)
5. Hamedani, M.R., Lee, S.C. and Kim, S.W.: On combining text-based and link-based similarity measures for scientific papers. In: Proceedings of the 2013 Research in Adaptive and Convergent Systems, pp. 111–115 (2013)
6. Yoon, S.H., Kim, S.W., Kim, J.S., Hwang, W.S.: On computing text-based similarity in scientific literature. In: Proceedings of the 20th International Conference Companion on World Wide Web, pp. 169–170. ACM (2011)
7. Bengio, Y., Ducharme, R., Vincent, P. and Jauvin, C.: A neural probabilistic language model. J. Mach. Learn. Res. 1137–1155 (2003)
8. Collobert, R., Weston, J.: A unified architecture for natural language processing: Deep neural networks with multitask learning. In: Proceedings of the 25th international conference on Machine Learning. pp. 160–167 (2008)
9. Mikolov, T., Sutskever, I., Chen, K., Corrado, G.S., Dean, J.: Distributed representations of words and phrases and their compositionality. In: Advances in neural information processing systems, pp. 3111–3119 (2013)
10. Rong, X.: Word2vec parameter learning explained (2014). arXiv:1411.2738
11. Kusner, M.J., Sun, Y., Kolkin, N.I., Weinberger, K.Q.: From word embeddings to document distance. ICML **15**, 957–966 (2015)
12. Pele, O., Werman, M.: Fast and robust earth mover's distances. In: 2009 IEEE 12th international conference on Computer vision, pp. 460–467 (2009)

Extractive Text Summarization Using Deep Auto-encoders

**K. Arjun, M. Hariharan, Pooja Anand, V. Pradeep,
Reshma Raj and Anuraj Mohan**

Abstract Finding the relevant information from a given document is one of the
major problems in today's information world. Text summarization is the process
of reducing the size of a text document in order to generate a summary that contains the salient points of the original document. The paper proposes an extractive
text summarization framework which uses Deep Neural Network (DNN) to obtain a
representative subset of the input document by selecting those sentences which contribute the most to the entire content of the document. The major advantage of the
proposed framework is its ability to discover the intrinsic semantic space that enables
the extraction of semantically relevant sentences. Hence, the information coverage
can be increased without contributing to the redundancy in the summary. The qualitative analysis in the experiments on the datasets of Multiling-2015 showed that the
proposed system produces summaries of good virtue.

Keywords Deep learning · Text summarization · Auto-encoder

1 Introduction

With the advent of Internet, the information that is available to a user is in abundance;
so abundant that sometimes the exact information that is actually required may be
scattered among multiple sources. Going through all available sources and extracting
the useful information, if done manually, is a tiresome and time-consuming task. It is
one of the important concerns in the area of Natural Language Processing (NLP) and
has many applications. Although many frameworks and algorithms have achieved
improvement in many task-specific applications, it is still a challenging job.

The main challenge in using the existing statistical methods is to keep the redundancy low in the generated summary, but cover the maximum possible information of

K. Arjun (✉) · M. Hariharan · P. Anand · V. Pradeep · R. Raj · A. Mohan
Department of Computer Science and Engineering, NSS College of Engineering,
Palakkad, India
e-mail: arjun.kay.5@gmail.com

© Springer Nature Singapore Pte Ltd. 2018
P. K. Sa et al. (eds.), *Recent Findings in Intelligent Computing Techniques*,
Advances in Intelligent Systems and Computing 709,
https://doi.org/10.1007/978-981-10-8633-5_18

the document. This is considered to be difficult since these state-of-the-art methods are based on the statistical features of the given document like sentence position, word count, etc., rather than the meaning that it conveys. Hence, there is a need for a methodology that tries to extract those sentences from the document that may cover the entire content of the document [1].

Inspired from the successful advent of deep learning due to the invention of faster GPUs, it is being applied to several NLP tasks. Deep learning is a subclass of machine learning that exploits multiple layers of nonlinear information processing for feature extraction and transformation. By using deep architecture, feature extraction from sentences can be enhanced, which would result in producing more meaningful summaries [2, 3].

This paper proposes a framework that uses deep neural networks for automatic text summarization. In the proposed method, the preprocessed document is converted into sentence vectors that are fed as input to an auto-encoder with four hidden layers which contain 1000, 750, 500, and 128 stochastic binary units, respectively. After reducing the dimensionality of the given sentence, the sentence code obtained is again used to reconstruct the given sentence vector and thereby reducing the reconstruction error and improving accuracy. Finally, the sentence codes obtained from the 128 unit hidden layer form the semantic space where similar sentences stay close. Any cluster analysis algorithm can then be applied to extract sentences that form the summary.

The remainder of this paper is organized as follows. We start by reviewing the previous study in deep learning and text summarization in Sect. 2. This is followed by a detailed description of the proposed method in Sect. 3. Section 4 sheds light on some of the experiments and evaluation conducted on the proposed model. Finally, Sect. 5 concludes this paper and discusses potential avenues for future work.

2 Literature Survey

Automatic text summarization aims to generate a summary from a given document by extracting most relevant sentences from it. A summary should not only be nonredundant but also it should keep a balance between information coverage and semantic representation. This section discusses about various text summarization techniques [4, 5]. A Machine Learning (ML) approach can be considered if we have a collection of documents and their corresponding reference extractive summaries [6]. In this, the summarization task can be seen as a two-class classification problem, where a sentence is labeled as correct if it belongs to the extractive reference summary, or as incorrect otherwise. Another effective method is using deep learning, which can guarantee the intrinsic semantic representation. One such method is based on RBM [7–9]. The features of higher dimensional space can be summarized into the following three aspects: sparsity, phenomena of empty space, and dimension effect. Many clustering algorithms have been proposed based on dimensionality reduction, some of which are Self-Organized Feature Maps (SOM), Principal Component Analysis (PCA), Multidimensional Scaling (MDS), and Fractal Dimensionality

Reduction (FDR). Another approach is based on auto-encoder [10]. It is very effective and convenient thanks to the dimensionality reduction ability of auto-encoders. The auto-encoder learns by minimizing the reconstruction error producing output and compares it with the input. When the number of hidden nodes is less than the number of input nodes, the model can be used for dimensionality reduction.

Distributed vectorial representation has become a common strategy for text representation. This type of representation can actually be used to represent the meaning of words. One such model is a skip gram model [11]. Skip gram model is an efficient method for learning distributed vector representations which represent the syntactic and semantic relationship between words. The basic objective of skip gram model is to maximize the surrounding probability given the probability of word. By using deep learning and word representation models described above, the summarization task is done by grouping the similar sentences as clusters and select the sentences from each cluster so that it constitutes a brief summary. The selection can be done by many algorithms. One method is k-means algorithm-based clustering. There are many improved K-means algorithms are available, for reducing complexity at the clustering phase [12]. In K-mean, each cluster is represented by the mean value of objects in the cluster. In clustering, the dissimilarity between data objects by measuring the distance between the pair of objects.

3 Methodology

Automatic text summarization aims at extracting meaningful sentences from a text corpus to create a short comprehensive summary of the given document. The paper proposes a framework that uses deep learning architectures for achieving this task. The existing shallow architectures are incapable of extracting certain types of complex structure from input, and hence, they can consider only the statistical features of the given document. The proposed method has the capability to generate an intrinsic semantic representation of the sentences from which salient points can be extracted. Also, it has the capability to handle the recursivity of human languages in an efficient manner.

The design mainly consists of two components:

1. Preprocessing
2. Summary generation.

3.1 Preprocessing

Initially, the input document is provided to a preprocessing unit so as to tokenize it into sentences, build a dictionary for the document, perform preprocessing, and generate the sentence vectors. The preprocessing techniques used are stemming and stop-word removal. Stemming is the process of converting or reducing words into stem. Stop-word removal is done based on predefined set of stop-words available in the NLTK library. Now to generate the sentence vectors, each sentence is compared with the word dictionary created based on the input document, and its corresponding sentence vector is created based on word count.

3.2 Summary Generation

Auto-Encoders: Auto-encoders are used in this model as the deep neural network so as to generate the summary. Auto-encoders encode the input x using an encoder function when data is passed from visible to hidden layer. A compressed representation z is obtained at the hidden layer. Reconstruction is performed when data is passed from hidden layer to output layer. Learning is accomplished in an auto-encoder by minimizing the reconstruction error

$$z = \rho_1(Wx + b) \tag{1}$$

$$x' = \rho_2(W'z + b') \tag{2}$$

The reconstruction error is

$$\delta(x, x') = ||x - x'||^2 = ||x - \rho_2(W'(\rho_1(Wx + b)) + b')||^2 \tag{3}$$

Auto-encoder performs dimensionality reduction by encoding the input. Training of auto-encoder can be divided into two stages mainly: (1) learning features (2) fine tuning. In the first stage auto-encoder takes input x and produces output x'. The reconstruction error is computed as Mean Squared Error (MSE), and the error is backpropagated through the network in the second stage. The framework of the proposed model is as follows.

1. Sentence vectors are provided as input to the DNN (as shown in Fig. 1). The size of sentence vectors is same as that of number of words of the dictionary.
2. Four hidden layers of 1000, 750, 500, and 128 units are created.
3. The output of layer with 128 units is sentence codes, which is the lower dimensional representation of sentences. Then, this output is provided as input to the reconstruction network.
4. The reconstruction network consists of layers with 128, 500, 750, and 1000 units, respectively, and they are connected in to and fro fashion to form the reconstruction network.
5. The sentence vectors are the output of the reconstruction network and they are compared with input for training.

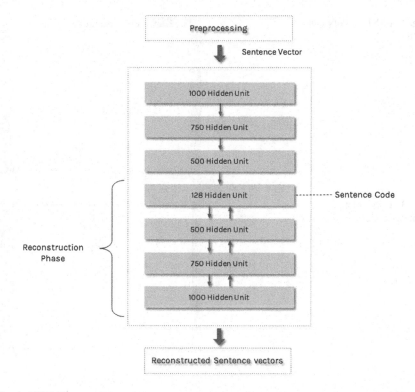

Fig. 1 DNN Framework

After the network attains convergence, the sentence codes are analyzed, and k-medoids algorithm can be applied so as to select a sentence code from each cluster. Then from sentence codes, sentences are generated and presented together as summary. The sentence code before and after training can be visualized by using t-SNE toolkit as shown in Figs. 2 and 3.

3.3 Training and Testing

Training is the process of updating weights of the network with respect to the obtained output. In this system, the sentence codes are passed to the reconstruction network so as to generate sentence vectors and compare it with the sentence vectors which are provided as the input. Then, the weights are updated according to the deviation of the output of reconstruction network from the input vectors provided.

Fig. 2 t-SNE plot of input vectors

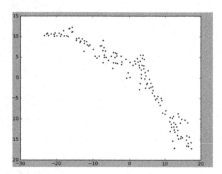

Fig. 3 t-SNE plot of sentence codes

4 Experiments and Results

Recall-Oriented Understudy for Gisting Evaluation (ROUGE) toolkit is used for the evaluation of the summary generated. ROUGE compares the generated summaries against a set of reference summaries. It measures the quality by counting the overlapping units. It requires a number of sample input documents. It also requires some sources of reference summaries to compare with.

The main dataset used is Multiling-2015, multilingual summarization of multiple documents. The dataset consists of documents belonging to various genres and of different languages. The dataset provided consists of a large number of documents collected from Internet from which 30 English documents could be used for providing input to the summarizer.

Six documents of the dataset were considered for evaluation. For getting reference summaries to compare with, an online summary generator, namely, *Sumplify* was used. *Sumplify* is an online summarizer which provides the summary of input documents with required size as specified by the user. The input size was provided based on the number of sentences that are present in the summary generated by our summarizer.

The ROUGE toolkit returns the precision, recall, and F-Score in the form of a table. We used ROUGE-1, 2, 3—which comes under ROUGE-N, an N-gram-based evaluation technique—for our evaluation, and the results are as shown in Table 1.

Table 1 Average of ROUGE-1, 2, 3 results

System summary	Avg. recall	Avg. precision	Avg F-score
TEXT1	0.49840	0.57355	0.53334
TEXT2	0.48154	0.66182	0.55764
TEXT3	0.42147	0.55038	0.47737
TEXT4	0.41843	0.67758	0.52308
TEXT5	0.44299	0.55571	0.49299
TEXT6	0.37412	0.52500	0.43689

5   Conclusion

Automatic text summarization provides the user with a shortened version of the entire content by extracting the salient sentences from the original document. There are many methods to summarize text documents that use statistical features of the document, but lack in their ability to prioritize sentences based on their contribution to the overall content. The deep learning approach proposed in this paper overcomes this by discovering the intrinsic semantic representation of the sentences using the stacked auto-encoders. The experiments conducted on the Multiling-2015 dataset shows that the proposed method produces good quality summaries. The futuristic enhancement to the proposed method can be done by incorporating the ability to summarize multiple documents.

References

1. Wong, K., Wu, M., Li, W.: Extractive summarization using supervised and semi-supervised learning. In: Proceedings of the 22nd International Conference on Computational Linguistics—COLING '08 (2008)
2. Deng. L.: A tutorial survey of architectures, algorithms, and applications for deep learning. In: APSIPA Transactions on Signal and Information Processing, vol. 3 (2014)
3. Liu, W., Wang, Z., Liu, X., Zeng, N., Liu, Y., Alsaadi, F.E.: A survey of deep neural network architectures and their applications. Neurocomputing (2016)
4. Radev, D.R., Hovy, E., Mckeown, K.: Introduction to the special issue on summarization. Comput. Ling. **28**(4), 399–408 (2002)
5. Fiori, A.: Innovative document summarization techniques: revolutionizing knowledge understanding. Information Science Reference, Hershey, PA (2013)
6. Neto, I.L., Freitas, A.A., Kaestner, C.A.A.: Automatic Text Summarization Using a Machine Learning Approach. Advances in Artificial Intelligence, Lecture Notes in Computer Science, pp. 205–215 (2002)
7. Zhong, S., Liu, Y., Li, B., Long, J.: Query-oriented unsupervised multi-document summarization via deep learning model. Expert Syst. Appl. **42**(21), 8146–8155 (2015)
8. Yao, C., Shen, J., Chen, G.: Automatic document summarization via deep neural networks. 8th International Symposium on Computational Intelligence and Design (ISCID) (2015)

9. Zhang, K., Liu, J., Chai, Y., Qian, K.: An optimized dimensionality reduction model for high-dimensional data based on restricted boltzmann machines. In: The 27th Chinese Control and Decision Conference (2015 CCDC) (2015)

10. Wang, Y., Yao, H., Zhao, S.: Auto-encoder based dimensionality reduction. Neurocomputing **184**, 232–242 (2016)

11. Mikolov, T., Sutskever, I., Chen, K., Corrado, G. S., Dean, J.: Distributed representations of words and phrases and their compositionality. NIPS, pp. 3111–3119 (2013)

12. Yadav, J., Sharma, M.: IJEET—A review of k-mean algorithm. Int. J. Eng. Trends Technol. Seventh Sense Res. Group (2013)

Rating Prediction Model for Reviews Using a Novel Weighted Textual Feature Method

Manju Venugopalan, G. Nalayini, G. Radhakrishnan
and Deepa Gupta

Abstract Online reviews and their star ratings are an important asset for users in deciding whether to buy a product, watch a movie, or where to plan for a tour. The review text is quite explanatory and expressive where the user shares his experiences in a detailed manner. But the star rating attached to a review is often subjective and at times biased to the reviewer's personality. The priorities given to different aspects or features by a user might in turn influence his rating. This discrepancy in the review content and the star rating makes a significant impact when sentiment is extracted from large voluminous data. This emphasizes the need to build a model to predict the rating based on the textual content in the reviews rather than bluntly following the star rating attached to the review. A novel weighted textual feature method is applied to assign an appropriate score to each review. Sequential Minimal Optimization (SMO) regression is applied to the feature set to predict the review rating which is based purely on the textual content of the review. The predicted rating has been compared with the actual user rating, and an analysis has been carried out to draw insights into the discrepancy. The model has been experimented in the tourism domain, and the results are quite promising. The regression residual value obtained for hotel and destination data is 0.8772 and 0.9503.

Keywords Rating prediction · Regression · Sentiment

M. Venugopalan (✉) · G. Nalayini · G. Radhakrishnan
Department of Computer Science & Engineering, Amrita School of Engineering,
Amrita Vishwa Vidyapeetham, Amrita University, Bengaluru, India
e-mail: v_manju@blr.amrita.edu

G. Nalayini
e-mail: nalashekar@gmail.com

G. Radhakrishnan
e-mail: g_radhakrishnan@blr.amrita.edu

D. Gupta
Department of Mathematics, Amrita School of Engineering,
Amrita Vishwa Vidyapeetham, Amrita University, Bengaluru, India
e-mail: g_deepa@blr.amrita.edu

© Springer Nature Singapore Pte Ltd. 2018
P. K. Sa et al. (eds.), *Recent Findings in Intelligent Computing Techniques*,
Advances in Intelligent Systems and Computing 709,
https://doi.org/10.1007/978-981-10-8633-5_19

1   Introduction

Decision-making of an individual or group is mainly influenced by what others think. Before the World Wide Web became widespread and common in use by public, we often used to ask our friends whom they would vote for in a local election, where to go for a trip and which is the better season to visit a place and so on. The development of the Internet has made it feasible to acquire information on everything instantly. Social media such as blogs, twitter, wiki, and forum websites are sources of voluminous information where users share their experience and discuss about countless topics. With increasing data volumes, there arose the necessity for automated systems that could extract user opinion from online reviews as reading and analyzing such huge volumes manually is practically impossible. Sentiment analysis is an opinion mining task that aims to determine whether writer's attitude expressed in the comments is positive or negative by analyzing large numbers of documents. Sentiment analysis is applied in many domains such as consumer market for products, social media for finding the general opinion about topics of interest or to find whether a recently released movie is a hit or not.

Online reviews are an important asset for users in deciding whether to buy a product, watch a movie or where to plan for a tour. Many websites where users share their experience about a product, movie, or a place visited also insist the user to provide a numerical rating on a scale of 1–5 usually. The review text is quite explanatory and expressive where the user shares his experiences in a detailed manner. The star rating that is often attached to a review is expected to be a numerical interpretation of the user's opinion. But these are often subjective and at times biased to the reviewer's personality and not a true reflection of the review content. The priorities given to different aspects of travel by a user might in turn influence his rating. The reader who is not specific about reading the review would bluntly be guided by the numerical rating provided by the user. This discrepancy in the review content and the star rating makes a significant impact when sentiment is extracted from large voluminous data.

Table 1 demonstrates a good example of the rating variation in two similar reviews sourced from tripadvisor.com. These reviews reflect more or less the same sentiment intensity about "tea gardens in Munnar". Both the reviews indicate that the user is pleased with the experience that can be noticed from the positive words used in the description like "nicely manicured", "must visit", "quite beautiful", "worth visiting", etc. However, the first review is rated five stars whereas the second review rating is a three star. This indicates that the rating provided by the user is not necessarily a clear indication of the sentiment, but reviewers are experts in expressing in free text format. Decision-making based on just rating would not suffice, and hence, the focus is to predict rating based on their review content.

The proposed model has been experimented on datasets from the tourism domain. Hotel and Destination reviews have been chosen. Destination reviews are characterized by varied entities when compared to product reviews. Hence, these reviews are more influenced by the varied perspective that each traveler has toward

Table 1 An example of bias in users rating and review

Location: Tea garden, Munnar	Location: Tea garden, Munnar
Rating: 5/5	Rating: 3/5
"Tea gardens in munar is like water in sea you get to see all these gardens nicely manicured. Place must visit"	"Tea gardens in munar is a must see as they are one of the famous and quite beautiful garden worth visiting"

each aspect of travel. Sentiment analysis of tourist experiences can contribute significantly to tourism development. User reviews are very relevant especially for the destination and hotel domain because they provide updated information in a timely manner and is in a detailed format than provided in any travel brochures. A novel weighted textual feature method is applied to assign a score to each review. Sequential Minimal Optimization(SMO) regression is applied to the feature set to predict the review rating which is based purely on the textual content of the review.

The paper is organized as follows: Sect. 2 describes the works that have been surveyed to understand the effective features and the various approaches used to extract sentiment from review content. Section 3 discusses in detail the proposed approach for rating prediction. Section 4 performs an analysis of the results. Finally, in Sect. 5, we draw some conclusions and propose a future extension of this work.

2 Related Works

Many research efforts have been carried out in the field of sentiment analysis over a decade. Researchers have proposed many techniques which are supervised or unsupervised in nature. Many works focused on classifying the review into positive or negative. They used machine learning approaches or lexicon-based approaches or even hybrid approaches for better results. Most approaches use bag-of-words representations [1]. Several works have explored feature engineering to find the most effective feature combination [2]. Domain-specific approaches to encounter context specificity have been attempted and resulted in improved lexicon polarities [3, 4]. Attempts have also been made to design models for extracting sentiment from tweets [5]. A changing trend to classify a review on a larger scale on a scale of 1–5 or moderately positive, extremely positive and so on) rather than a binary classification was witnesses later.

An automated system [6] to predict the ratings for the review on predefined multi aspects like atmosphere, food, value, etc., has been attempted which applies both maximum entropy classification and ordinal regression. The features incorporated were unigrams, bigrams, and word chunks. Classification outperforms the ordinal regression and the score of each aspect revealed that using textual information for review rank prediction is better than the numerical rating provided by the users. An ad hoc and a regression-based recommendation system [7] have been explored which does a sentence-wise classification of restaurant reviews into categories like food, ambience, service, etc. An interesting observation found is that if the clients are from the same location the scaled rating is similar. The authors of [8] experimented another approach using bag-of-opinions based feature extraction, where the attitude of a review is classified into root word, modifier words, and negation words. The numeric score is allotted to each opinion by using ridge regression, and the review's rating is predicted by aggregating the scores of all opinions in the review in combination with a domain-dependent unigram model. This method is suited for capturing the expressive power of n-grams and also overcoming their sparsity bottleneck. In [9], a framework that incorporates product details, textual features, and reviewer info as a three-dimensional tensor and latent factors is generated by applying tensor factorization. The Yahoo music dataset was used in [10] to design a class of time-aware matrix, which adopts time series-based parameterizations and models item drifting behaviors at multiple temporal resolutions. The ensemble combination of LR and meta-feature results in an RMSE of 22.5.

Ye Q. et al. [11] applied sentiment classification techniques on reviews from travel blogs. Comparison of SVM classifier, Naïve Bayes, and n-gram on a smaller dataset of 40 to 100 reviews showed that SVM outperformed other methods. When dataset size was increased to a range of 300–400, all three approaches showed the same outcome, and the accuracy was approximately 80%. A generalized topic and syntax model called Part-of-Speech LDA (POSLDA) to sentiment analysis has been contributed by [12]. The different semantic classes like words such as nouns, verbs, adjectives, adverbs, and syntactic classes made of functional words such as determiners, preposition, and conjunctions are used to identify the words that modify the entity. A new feature selection method is proposed. Feature selection is based on semantic classes, semantic classes with tagging, and automatic stop word removal. Movie reviews are used for document-level sentiment analysis and TripAdvisor data for aspect level. Compared with baseline both feature selection is based on semantic classes, semantic classes with tagging achieve better results. POSLDA is implemented to handle unigrams; it could be extended to handle n-grams. The authors of [13] explore segmentation and association rule techniques to know the inbound tourists in Thailand. Tourist patterns for market planning, promotion was explored by applying two-level clustering later association rule mining was applied. These results when combined with recommendation system provide a useful knowledge for travel agencies and other tourism organizations. The BESAHOT service presented in [14] collects user reviews for hotels from various websites, analyses and statistical polarity classification was performed on the review. Sentiment dictionary was built to know the polarity value of the words. In [15], they measure the effectiveness of topic model-based approaches to multi-aspect sentence labeling and multi-aspect rating prediction. For multi-aspect rating prediction, linear Support Vector Regression (SVR) and Perceptron Ranking (Prank) are used. In [16], Pairaya Juwatta Nasamran et al. have used a questionnaire approach to collect valid and reliable data on Thai travelers. The patterns learnt from data using data mining techniques are used to predict traveler preferences.

There has been ample work on analyzing the sentiments of online reviews, implicitly expressing their opinion polarities and thereby provide numeric ratings of products. Although ratings are more informative than polarities, most prior works focused on rating prediction are considering text fragments as unigrams, bag-of-words, most frequent words, etc. These gram features sparsely occur in the training set and thus fail in a robust prediction. The proposed approach attempts a unique weighted textual feature method incorporating a weightage to each review, and the prediction classifier is modeled as an SMO regression problem over an appropriately defined feature space.

3 Proposed Approach

The proposed model is portrayed in detail in Fig. 1. The data preprocessing, weighted feature extraction, and regression phases through which the model evolves are explained in detail in the following subsections.

3.1 Preprocessing

The reviews crawled are initially preprocessed which consists of the following stages: (1) spell check (misspelt words are corrected), (2) shorthand expansion (shorthands are expanded to its original form to retain the word's relative meaning

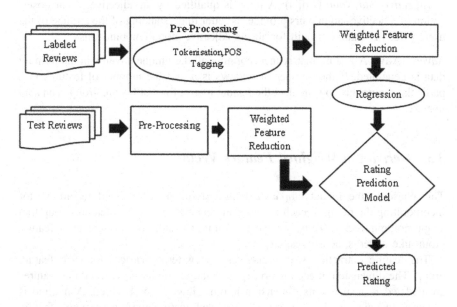

Fig. 1 Schematic diagram of proposed approach

with respect to the sentence), (3) punctuation removal, (4) tokenization (word boundaries are defined), (5) Parts-of-Speech (POS) tagging (which identifies the parts of speech tag related to every word), (6) stop word removal (non-sentimental words like or, and, the, etc., are removed), and (7) lemmatization (maps all modulated form of words into its root kind). All these preprocessing steps are performed using Stanford core NLP ToolKit. The output from this phase is a set of POS tagged opinion words.

3.2 Feature Extraction

The choice of features is finally anchored on the highly sentiment rich opinion words. Adverbs, adjectives, adjective–noun pairs (which associates the opinion and the holder), and bigrams (to extract any significant uncovered associations) are the four group of features considered for the proposed approach.

Bigrams (Bi): They have an inherent ability to capture the relationship between words (e.g., a word and its modifier or a word and its negation). A dictionary of all the bigrams that occur more than four times in each review is created. Now, a review-word matrix is constructed, where the value at the ith row and jth column is the frequency of occurrence of word j in the ith review.

Adjective (Adj): A feature matrix incorporating all the unique adjectives is built similar to that of bigrams. Here, all adjectives are considered rather than imposing any constraints.

Adjective and Noun (AdjN): A noun is qualified by an adjective. Clear observation of adjective–noun pairs provides a better understanding of the attitude of the user. A feature matrix is built for all the adjective–noun combinations.

Adverb (Adv): A feature matrix incorporating all the unique adverbs in the training data is generated. If the number of reviews is n and the number of features in a particular feature group is m, then the feature matrix for that feature group is an $n x m$ matrix.

3.3 Deriving a Weighted Feature Vector

This phase in the model emphasizes on arriving at a weighted feature vector incorporating the rating of each review and hence reducing each feature group to a single column vector. Thus, we have a column vector to represent each feature group like adverbs, adjectives, etc.

The feature extraction stage generates review-term matrices for each feature group. The dimension of each matrix is quite large; similarly, when all the features are updated in a matrix its dimension is quite huge to be handled. A method is proposed that aims at merging all features in each feature group into a single feature weight thereby reducing the count of features but retains the impact of these words in the reviews. This weighted feature method is a supervised feature reduction method as it takes into consideration the class label (rating) too for arriving at the weighted feature vector.

The approach attempts to reduce each feature group to a column vector corresponding to n reviews. Reduced feature R for each feature group is calculated using Eq. 1 where F is the feature vector of size nxm and R is a column vector of dimension $nx1$. Each element in vector R is calculated using Eq. 2 where $Rating_i$ is the rating for each review. The overall average of the review rating $AverageRating$ is calculated using Eq. 3.

$$W = F^T R \tag{1}$$

$$R = Rating_i - AverageRating \tag{2}$$

$$AverageRating = \sum_{i=1}^{n} rating[i]/n \tag{3}$$

W has an element corresponding to each feature in that feature group which represents the weightage of the feature across all reviews and has a dimension $mx1$. This weighted feature(W) is further modified as $W1$ to reflect the significance of each feature in terms of its presence across all documents using Eq. 4 where $COUNT_i$ is the number of reviews characterized by the presence of the feature corresponding to Wi.

$$W1_i = W_i/COUNT_i \tag{4}$$

The final reduced feature vector for the feature group is derived using Eq. 5

$$F_{Final} = FW_1 \tag{5}$$

F_{Final} is the reduced vector of dimension $nx1$ and has a value for each review which is the weighted representative for all the features in the feature group. This approach results in reducing each feature vector group into a column vector. This weighted feature method reduces the size of the feature set in an intellectual manner where the impact of each word on the sentence is retained. Here, the individual weights of the features are summed up based on its relation and impact on each review rating.

3.4 Regression Model for Learning and Prediction

The feature model built is input to a learning model based on regression. Regression is a function that determines the relationship between dependent variable and the independent variable; independent variable can be more than one. Dependent variable values are predicted as a function of independent variables. The independent variables in the model are bigram, adjective, adjective–noun pair, and adverb, and they are collectively denoted by F_{Final}. The dependent variable in the model is rating (RT) of each review. The rating RT is estimated by the relationship of the weighted features (F_{Final}) as shown in Eq. 6, where β is the effect of F_{Final} on the rating and ε is a measure of other terms that influence the rating.

$$RT = F_{Final}\beta + \in \qquad (6)$$

There are many types of regression model available; the proposed approach applies SMO regression for model testing. SMO regression is an iterative method that optimizes the quadratic programming problem of SVM. SMO regression allows us to understand the weights that are associated with each sentence and which directly provides the varying sentiment in the sentence. The weights are learned from a dataset which directly impact on the style in which user writes the review in a domain. Weka SMOreg function which implements the support vector machine for regression has been implemented. As explained by Smola [17], support vector machines provide the regression function estimates by covering both the quadratic programming part and its ease and advanced methods that help in dealing with large datasets. SMOreg models the star rating provided by the user as the dependent variable and the reduced features set of the opinion words extracted from the review are the independent variables. Weka SMOreg computes the SVM-based regression and provides the estimates based on the training data.

The weighted feature vector arrived at by learning the model using Eq. 5 is a representative of a weighted feature vector in that domain. Hence, it is used by the prediction model while testing for the rating of test reviews for weighted feature reduction and in turn for predicting the rating.

3.5 Corpus Statistics and Parameter Setting

The domains targeted for the experiment are hotel and destination reviews. 5000 hotel reviews have been sourced from Opinion-Based Entity ranking dataset [18]. Destination reviews have been extracted from TripAdvisor. The count of reviews from different Indian states has been listed in Table 2. The dataset is stored as a csv file which contains the reviewer name, reviewer destination, review date,

Table 2 State-wise review count

Indian state	Review count
Andhra Pradesh	451
Goa	636
Karnataka	1391
Kerala	1332
Tamil Nadu	1190

review tagline, complete review, rating, location visited_date, location, reviewer contribution, and helpful votes. The dataset contains destination rating from 1 to 5. The reviews rated 1 and 2 are considered negative, 3 as neutral, and those rated 4 and 5 as positive reviews.

Our goal is to predict the impact on the user rating after applying the weighted feature set on the dataset using regression model. The SMOreg function in Weka uses RegSMOImproved algorithm [19], which works on the multiprogramming concept where using the Karush_kuhn-Tucker (KKT) condition for dual problem, and two threshold parameters are set: they are tolerance parameter that checks for the stopping criteria and epsilon parameter—it is epsilon-insensitive loss function. The filter type selected was "no standardization" and "no normalization" and a tenfold cross-validation is performed for the proposed model.

4 Experimental Results and Analysis

A correlation coefficient of 90.25% on the train data and 88.37% on the test data is obtained. The results are tabulated in Table 3. The baseline chosen for evaluating the performance of the model is that of predicting average rating of the training data and the average of test data, though the rating is regardless of the knowledge of review text.

The regression equation outputted for destination and hotel review is shown in Eqs. 7 and 8.

$$D_{Rating} = 0.007Adj + 0.999AdN + 0.009Adv + 0.0024bi + 3.6001 \qquad (7)$$

$$H_{Rating} = 0.004Adj + 0.9977AdN + 0.016Adv + 0.0002bi + 2.999 \qquad (8)$$

Destination and hotel review ratings are calculated using the regression Eqs. 7 and 8 and are depicted in Figs. 2 and 3. In order to determine whether the regression line plotted is a good fit or bad fit, residual value is calculated for the dataset. The residuals express the difference between the data on the line and the actual data so the values of the residuals will show how well the residuals represent the data. Residual is calculated using R-square where R-square is defined as the

Table 3 Mean square error of test and training data

Data	Training correlation coefficient	Testing correlation coefficient	Training MSE	Testing MSE
SMOreg destination	0.9025	0.8837	0.28922	0.31685
SMOreg hotel	0.8731	0.8298	0.4175	0.38952
Baseline destination	–	–	1.4475	1.6444
Baseline hotel	–	–	2	1.685

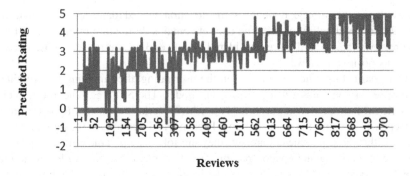

Fig. 2 Proposed model prediction for destination reviews

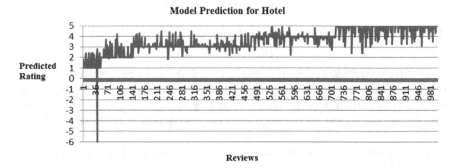

Fig. 3 Proposed model prediction for hotel review

ratio of the Sum of Squares of the Regression (*SSR*) and the Total Sum of Squares (*SST*). *SSR, SST, and R-square* calculation are shown in Eqs. 9, 10, and 11, respectively.

$$SSR = \sum_{i=1}^{n} w_i (x - y)^2 \tag{9}$$

$$SST = \sum_{i=1}^{n} w_i (x - y_{av})^2 \tag{10}$$

$$R^2 = \frac{SSR}{SST} \tag{11}$$

Residual value of the regression line in our approach is 0.9503 for destination dataset and 0.8772 for hotel dataset. These values indicate that the rating line generated satisfies the good line of fit condition. For a further understanding of our model, the destination reviews are segregated on the basis of their location and the individual average rating score of this location provided by the user is calculated using Eq. 12 and the graphical representation is shown in Fig. 4.

$$Rating = \frac{\sum_{i=1}^{5} (cnt_i * i)}{\sum_{i=1}^{5} cnt_i} \quad (12)$$

In Eq. 12, i represents the rating provided by the user for a location, and cnt_i provides the count of the distinct rating i for an individual location. When the overall rating for each location is considered, the trend of this curve indicates that there are only slight deviations; there could be changes if more reviews were introduced. It is possible that the quality of reviews used would also play a role in how much review data is actually needed for this task. Similarly, the individual hotel rating comparison is depicted in Fig. 5.

Few significantly deviating ratings are checked manually as shown in Table 4 to get an insight into the words impact and their attitude. After analysis, it was found

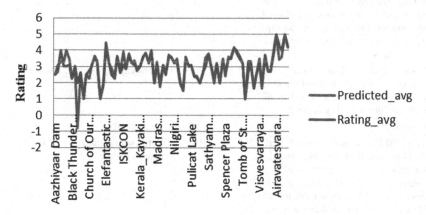

Fig. 4 Destination-specific user and proposed model average rating comparison

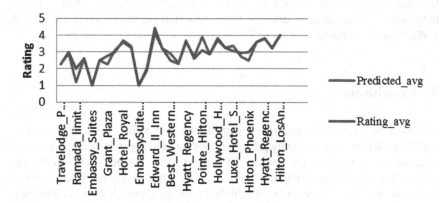

Fig. 5 Hotel-specific user and proposed model average rating comparison

Table 4 Manual comparison of user rating and proposed model prediction rating

Review	UR	MPR	Insight
I visited the temple with my spouse on may 31. We reached the west nada and parked the vehicle with great difficult as there just enough space for 1 car to move. Approached the entrance and my spouse was told to hand over the shirt to the counter outside, and i was told to wear mundu. To our dismay, we saw a board that special seva is available for an additional amount. Those who pay extra are permitted to climb the stairs and see the lord closer and they are permitted to stay for longer duration to pray. We had to wait for 2 h. When we entered kovil we were not allowed to stand for a minute. I was literally pushed. I was real annoyed by the rude behavior	1	−0.2	Here tourist completely expresses the problem faced like parking, time-consuming process
Good to find a mall and some non-veg food in the heart of the city. Otherwise a very ordinary and small mall. Multiplex quality is good	2	3.2	Words express positive attitude
Okay to Stay	1	3.2	Fewer words
TERRIBLE experience TWICE !!!!! BAD management. Bad bed coves. Floors were dirty. Smokey smell throughout. We stayed at this hotel in April 2007. We were in town for a Nascar race and arrived at our room late at night. Our room was cold	1	0.6	Words with negative score are used more in review

that users provided lower star rating based on a single incident that was bad, and they expressed the same in detail in the review. UR represents users rating, and MPR is model prediction rating in Table 4.

5 Conclusion

This work has utilized the knowledge gained from destination and hotel reviews to build a model that independently predicts the rating of a review based on the textual structure and content of a sentence. The implemented weighted feature reduction method has reduced the complexity of the model retaining the information in the feature vector. The experiment results of the proposed approach show a correlation coefficient of 90.25% for training data and 88.37% for testing data of destination reviews. An analysis of what are the scenarios where there is a notable discrepancy in the predicted rating and user rating has been carried out and meaningful justifications could be derived. The model is domain independent and can be applied in various fields to draw useful insights. The work can be extended by identifying what are the remarkable features of destinations and hotels which add more weightage with respect to a user's perspective and hence which would be more influential in rating prediction.

References

1. Pang, B., Lee, L.: Opinion mining and sentiment analysis. Found. Trends Inf. Retr. 2.1–2, 1–135 (2008)
2. Priyanka, C., Gupta D.: Identifying the best feature combination for sentiment analysis of customer reviews. ICACCI 102–108 (2013)
3. Mishra, D., Venugopalan, M., Gupta, D.: Context specific Lexicon for Hindi reviews. Procedia Comput. Sci. **93**, 554–563 (2016)
4. Sanagar, S., Gupta, D.: Adaptation of multi-domain corpus learned seeds and polarity lexicon for sentiment analysis. Comput. Netw. Commun. (CoCoNet) 50–58 (2015)
5. Venugopalan, M, Gupta D.: Exploring sentiment analysis on twitter data. In: Eighth International Conference on Contemporary Computing (IC3), pp. 241–247 (2015)
6. Gupta, N., di Fabbrizio, G., Haffner, P.: Capturing the stars: predicting ratings for service and product reviews. In: Proceedings of the NAACL HLT Workshop on Semantic Search. Association for Computational Linguistics (2010)
7. Ganu, G., Elhadad, N., Marian, A.: Beyond the stars: improving rating predictions using review text content. In: Twelfth International Workshop on the Web and Database Providence, Rhode Island, USA (2009)
8. Qu, L., Ifrim, G., Weikum, G.: The bag-of-opinions method for review rating prediction from sparse text patterns. In: Proceedings of the 23rd International Conference on Computational Linguistics, Beijing, pp. 913–921 (2010)
9. Li, F., Liu, N., Jin, H., Zhao, K., Yang, Q., Zhu, X.: Incorporating reviewer and product information for review rating prediction. In: Proceedings of the Twenty-Second International Joint Conference on Artificial Intelligence
10. Zheng, Z., Chin, T., Liu, N., Yang, Q., Yu, Y.: Rating prediction with informative ensemble of multi-resolution dynamic models. JMLR: Workshop and Conference Proceedings
11. Ye, Q., Zhang, Z.: Sentiment classification of online reviews to travel destinations by supervised machine learning approaches. Expert Syst. Appl. 6527–6535 (2009)
12. Moghaddam, S., Ester, M.: Opinion digger: an unsupervised opinion miner from unstructured product reviews. In: CIKM '10 Proceedings of the 19th ACM International Conference on Information and Knowledge Management, pp. 1825–1828
13. Zhou, H., Song, F.: Aspect-level sentiment analysis based on a generalized probabilistic topic and syntax model. In: Proceedings of the Twenty-Eighth International Florida Artificial Intelligence Research Society Conference (2015)
14. Kasper, W., Vela, M.: Sentiment analysis for hotel reviews. In: Proceedings of the Computational Linguistics-Applications Conference, Jacharanka, pp. 45–52 (2011)
15. Lu, B., Ott, M., Cardie, C., Tsou, B.: Multi-aspect sentiment analysis with topic models. In: IEEE 11th International Conference on Data Mining Workshop, pp. 81–88 (2011)

16. Juwattanasamran, P., Supattranuwong, S., Sinthupinyo, S.: Applying data mining to analyze travel pattern in searching travel destination choices. Int. J. Eng. Sci. **2** (2013)
17. Smola, A.J., Scholkopf, B.: A Tutorial on Support Vector Regression
18. Kavita, G., Zhai, C.: Opinion-based entity ranking. Inf. Retr. 116–150 (2012)
19. Shevade, S.K., Keerthi, S.S., Bhattacharyya, C., Murthy, K.R.K.: Improvements to the SMO algorithm for SVM regression. IEEE Trans. Neural Netw. 1188–1193 (1999)

RCE-OIE: Open Information Extraction Using a Rule-Based Clause Extraction Engine for Semantic Applications

D. Thenmozhi and Chandrabose Aravindan

Abstract Open Information Extraction (OIE) is a process of extracting clauses present in the text. Extraction of clauses is useful for several applications. However, the existing OIE methods do not focus on the improvement of such applications. In this paper, we present a methodology for OIE using a rule-based clause extraction engine (RCE-OIE) by considering some aspects like handling of coordinating conjunctions, negations, and relative clauses for the improvement of semantic applications. We have evaluated RCE-OIE on OIE datasets to show that our clause extraction approach is domain-independent and comparable with the state-of-the-art OIE systems. Our RCE-OIE is capable of improving the performance of downstream applications. In particular, RCE-OIE significantly improves the performance of paraphrase identification on Microsoft Research corpus when compared with the existing OIE systems.

Keywords Open Information Extraction · Rule-based clause extraction
Semantic applications · Paraphrase identification

1 Introduction

Open Information Extraction (OIE) is the process of extracting several types of clauses present in formal and informal text for all domains. Clause extraction from text is the process of syntactic simplification which focuses on extracting clauses from structurally complex sentences. This has several applications in areas such as question answering, machine translation, information retrieval, paraphrase identification, and ontology learning. Extraction of clauses can be achieved by using Open

D. Thenmozhi (✉) · C. Aravindan
Department of Computer Science and Engineering, SSN College of Engineering,
Chennai, India
e-mail: theni_d@ssn.edu.in

C. Aravindan
e-mail: aravindanc@ssn.edu.in

© Springer Nature Singapore Pte Ltd. 2018
P. K. Sa et al. (eds.), *Recent Findings in Intelligent Computing Techniques*,
Advances in Intelligent Systems and Computing 709,
https://doi.org/10.1007/978-981-10-8633-5_20

Information Extraction (OIE) methodologies. However, the focus of OIE methods is too general, and they do not possess some desirable properties, namely, handling of coordinating conjunctions, negations, and relative clauses that are helpful for improving the performance of the above applications.

Angeli et al. [2] presented an approach for OIE that uses only a few patterns for extracting clauses based on dependency parsing with improved recall. The conjunctions like "which" and "that" are not resolved by this method. For example, this method does not extract any clause from the sentence "Pesticide is a chemical which causes pollution" when executed at online CoreNLP engine.[1] However, there are two clauses present in the sentence, namely, *"Pesticide is a chemical"* and *"Pesticide causes pollution"* that contribute to ontology learning. The latter clause helps to identify the sentences "antimicrobial is a pesticide which causes pollution" and "pesticide causes pollution recently" as paraphrase in paraphrase identification. It also helps in extracting correct answer to the question "What causes pollution?". Also, this OIE approach does not extract any clause if the text contains negations which is an important aspect to be considered in many applications.

Several other approaches such as TextRunner [3], WOE [14], Reverb [6], OLLIE [11], and ClausIE [4] have been evolved for OIE. Among these methods, ClausIE outperforms the other methods [4] and extracts 12 types of clauses from the text by using several rules based on the information from dependency parsers. However, this method does not possess the properties like handling of coordinating conjunctions and relative clauses, and hence may not contribute to semantic applications. For example, from the sentence, "The laws and policies govern EPA.", ClausIE extracts the pattern ("The laws and policies", "govern", "EPA"). However, the clauses ("laws", "govern", "EPA") and ("policies", "govern", "EPA") are not extracted due to non-handling of coordinating conjunctions. These clauses may be important for semantic applications.

Hence, we propose RCE-OIE, an OIE system, based on a rule-based clause extraction engine which is domain-independent for extracting clauses from complex sentences by considering conjunctions, relative clauses, and negations that are helpful for semantic applications. This paper elaborates the design of RCE-OIE, analyzes how RCE-OIE is comparable with the existing OIE systems and what is the impact of RCE-OIE in paraphrase identification when compared with a state-of-the-art OIE system.

2 Related Work

Open Information Extraction (OIE) is the process of extracting several types of clauses present in the text for all domains. The OIE methods use information of a shallow parser or chunker to extract clauses from text. TextRunner [3] uses supervised approach by training a Bayes classifier and the triples are extracted from the

---

[1]http://corenlp.run/.

sentences by using the classifier. WOEpos [14] uses a classifier based on a labeled training corpus obtained from Wikipedia. Reverb [6] uses syntactical and lexical constraints to reduce the number of over-specified and uninformative extractions.

Several OIE methods use dependency parsing information for extracting the clauses. WOEparse [14] and OLLIE [11] use automatically generated data to learn extraction patterns on the dependency tree. Zouaq [15] and KrakeN [1] use a set of handcrafted patterns to extract clauses. ClausIE [4] is also an OIE methodology which identifies set of clauses and their types based on the dependency parsing information and a small set of domain-independent lexica. This method outperforms the other OIE systems [4]. However, many clause patterns identified by ClausIE may not be relevant for downstream applications. Angeli et al. [2] proposed an OIE methodology for relation learning. However, the method is not handling relative clauses and conjunctions [2] that may be more significant for other applications.

OIE systems may be helpful for several semantic applications. Paraphrase identification is a method of identifying a pair of sentences as paraphrase or not. It is one of the downstream applications that can make use of clauses to correctly identify the paraphrases [13]. Thenmozhi and Aravindan [12] and Angeli et al. [2] use clauses to learn relations for ontology from text. A few approaches [7] use clauses for question answering. Clauses are helpful in automatic text summarization [8] which is the process of producing a synopsis for any document.

3 RCE-OIE: A Rule-Based Clause Extraction Engine for Open Information Extraction

Our clause extraction engine (RCE-OIE) uses several domain-independent rules to extract different types of clauses from text documents. We have framed this set of rules [12, 13] to extract clauses from the sentences based on the information of a chunker. The rules 1–9 [13] are used to extract clauses that appear in the sentences as main (rule 1), dependent (rules 3 and 4), independent (rules 5, 6, and 7), relative (rule 2), and nonfinite clauses (rules 8 and 9). However, more than one rule is applicable for a sentence.

Rule 1 is the basic rule for extracting main clauses. If the sequence of chunks are in the form of $(NP_1)(VP)(NP_2)$, then NP_1 is the subject, VP is the verb, and NP_2 is the object. Rule 2 is for extracting the nonrestrictive relative clauses present in the sentences. If the sequence of chunks is in the form of $(NP_1)(PP$ having verb in gerund form$)(NP_2)$, then NP_1 is the subject, lemmatized form of PP is the verb, and NP_2 is the object. Rule 3 is for extracting dependent clauses by resolving the conjunction "which". If the sequence of chunks is in the form of $(which)(VP)(NP_2)$, then the object of the previous clause is the subject, VP is the verb, and NP_2 is the object. Rule 4 is also for extracting dependent clauses by resolving the conjunction "that, they, or it". If the sequence of chunks is in the form of $(that|they|it)(VP)(NP_2)$, then the subject of the previous clause is the subject, VP is the verb, and NP_2 is the object.

Rules 5 and 6 are for extracting multiple clauses from a single sentence. In rule 5, if the sequence of chunks is in the form of $(NP_1, NP_2, \ldots and NP_n)(VP)(NP)$, then n clauses are extracted with the subjects, namely, NP_1, NP_2 up to NP_n, verb VP, and object NP. In rule 6, if the sequence of chunks is in the form of $(NP)(VP)(NP_1, NP_2, \ldots and NP_n)$, then n clauses are extracted with the subject NP, verb VP, and objects, namely, NP_1, NP_2 up to NP_n. Rule 7 is for extracting the multiple independent clauses from a sentence. If the sequence of chunks is in the form of $(NP_1)(VP_1)(NP_2)(and|,)$ $(VP_2)(NP_3)$, then two clauses are extracted with the triples (NP_1, VP_1, NP_2), and (NP_1, VP_2, NP_3). Rules 8 and 9 will not extract the clauses in triple form. In rule 8, if the sequence of chunks is in the form of $(NP_1)(to)(VP)(NP_2)$, then the clause with no subject, verb VP, and object NP_2 is extracted. In rule 9, if the sequence of chunks are in the form of $(NP_1)(VP_1)$, then the clause with subject NP_1 and verb VP_1 is extracted with no object.

For example, for the sentence "In the United States, all pesticides, including antimicrobial are regulated under FIFRA.", the clauses extracted are "all pesticides including antimicrobial by (by rule 2)" and "all pesticides are regulated under FIFRA (by rule 1)". RCE-OIE extracts two clauses, namely, "The laws govern EPA (by rule 5)" and "policies govern EPA (by rule 5)" for the sentence "The laws and policies govern EPA".

When we compare the output of the existing OIE system, ClausIE [4] for the same examples, the clause "pesticide include antimicrobial" is not extracted by ClausIE for the first sentence due to non-handling of relative clauses. ClausIE does not extract the clauses "The laws govern EPA" and "policies govern EPA" from the second sentence due to non-handling of conjunctions.

4 Performance Comparison of RCE-OIE with State-of-the-Art OIE Systems

To show that the OIE system we have designed is general and domain-independent, we have evaluated the performance by using three OIE datasets [4], namely, Reverb dataset with 500 sentences, Wikipedia dataset with 200 sentences, and New York Times dataset (NYT) with 200 sentences. We have analyzed and studied the performance of our approach of clause extraction using the metrics mentioned by Del Corro and Gemulla (2013), namely, the number of correct extractions and precision, which is the proportion of a number of correct clauses retrieved over the number of clauses retrieved.

To show that our clause extraction approach is domain-independent, we have compared the performance with state-of-the-art methods of OIE. The results of our proposed approach and various approaches are given in Table 1. The results show that RCE-OIE is comparable with the existing OIE methods.

ClausIE performs better than all the other methods in terms of both precision and number of extractions. There are four variations of ClausIE that are reported,

Table 1 Precision comparison of RCE-OIE with state-of-the-art OIE methods

Method	Reverb	Wikipedia	NYT
ClausIE	1706/2975 (57)	598/1001 (75)	696/1303 (53)
ClausIE w/o CC	1466/2344 (62.5)	536/792 (66.7)	594/926 (64.14)
ClausIE (nonredundant)	1221/2161 (56)	424/727 (58)	508/926 (54.8)
ClausIE w/o CC (nonredundant)	1050/1707 (61.5)	381/569 (66.9)	444/685 (64.8)
OLLIE	547/1242 (44)	234/565 (41)	211/497 (42)
Reverb	388/727 (53)	165/249 (66)	149/271 (54)
WOE	447/1028 (43)	–	–
Text Runner (Reverb)	343/837 (41)	–	–
Text Runner	286/798 (36)	–	–
RCE-OIE (Our approach)	614/1226 (50.08)	224/366 (61.2)	271/490 (55.3)

namely, ClausIE with redundant clauses, ClausIE without considering conjunctions, ClausIE with no redundant clauses, and ClausIE without considering conjunctions with no redundant clauses. Among these four variations, we have compared our approach with ClausIE (no redundant), because our focus is to compare with the method that does not give redundant clauses. Our precision score is close to that of ClausIE (no redundant) approach. However, the number of extractions by ClausIE is more than twice of our method. When we compare our method with Reverb, another state-of-the-art OIE method based on shallow parsing, Reverb performs better than our method in terms of precision for all datasets. However, the number of correct extractions by our method is more than Reverb for all datasets. Further, our approach outperforms OLLIE which is a baseline for OIE. However, RCE-OIE is capable of significantly improving the performance of downstream application, namely, paraphrase identification whereas the existing OIE method ClausIE is not improving the performance significantly. This is elaborated in the next section.

5 Performance Comparison of RCE-OIE with ClausIE (An Existing OIE System) in Paraphrase Identification

Among several methods existing for paraphrase identification, a method using 15 machine translation (MT) metrics-based features outperforms the other methods [9]. Recently, we have extended this method with 4 more features, namely, concept score (CS), relation score (RS), proposition score (PS) and word score (WS) based on OIE and proposed a methodology for paraphrase identification [13] which identifies pair of sentences as paraphrase or not using 19 features. In [13], we have used RCE-

Table 2 Performance comparison between classification with 15 features, ClausIE-19 features, and RCE-OIE-19 Features

Metrics	Classification with 15 features	Classification with (ClausIE) 19 features	Classification with (RCE-OIE) 19 features
Precision	74.76	75.40	79.08
Recall	89.93	89.54	91.63
Rejection	36.94	42.04	51.90
F1-measure	81.61	81.87	84.89
f1-measure	52.23	57.22	66.27
Accuracy	72.68	73.62	78.32

OIE to extract the clauses for paraphrase identification. To show how RCE-OIE is significantly improving the performance of paraphrase identification when compared with existing OIE system, we have used the same procedure as in [13], by extracting the clauses using existing OIE approach, ClausIE.

We have measured the performance of paraphrase identification on Microsoft Research paraphrase data set [5] using the metrics namely precision, recall, rejection, F1-measure, f1-measure and accuracy as we measured in [13]. We have compared three variations, namely,

1. Classification with 15 machine translation metrics-based features.
2. Classification with ClausIE-19 features (four clause-based features using ClausIE (existing OIE system) and 15 machine translation metrics-based features).
3. Classification with RCE-OIE-19 features (four clause-based features using RCE-OIE and 15 machine translation metrics-based features).

The results of these three variations are given in Table 2. We have applied McNemar test [10] to compare the classification algorithms. When we have compared between the first two algorithms, the $McNemar_{stat} = 0.62 < McNemar_{crit} = 3.84$ ($\alpha = 0.05, df = 1$). This accepts the hypothesis that two algorithms have same error rate at significant level $\alpha = 0.05$. The *p-value* obtained from McNemar test with 95% confidence is 0.432 which is greater than 0.05, and hence, the difference is considered to be not statistically significant. In other words, ClausIE is not able to significantly improve the performance. We have compared between algorithm with ClausIE-19 features and algorithm with RCE-OIE-19 features. $McNemar_{stat} = 36.65 > McNemar_{crit} = 3.84$. This rejects the hypothesis that two algorithms have same error rate at significant level $\alpha = 0.05$. The *p-value* obtained from McNemar test with 95% confidence is 0.00 which is lesser than 0.05 and hence the difference is considered to be statistically significant. The statistical tests show that RCE-OIE significantly improves the performance of paraphrase identification whereas the existing OIE method is not improving.

6 Conclusions

The OIE system RCE-OIE we have presented in this paper is a rule-based clause extraction engine which is domain-independent. Several aspects, namely, handling of coordinating conjunctions, negations, and relative clauses are considered in the design of the engine. We have evaluated RCE-OIE on OIE datasets, namely, Reverb, NYT, and Wikipedia datasets to show that our clause extraction approach is domain-independent and comparable with the state-of-the-art OIE systems. Though our approach of clause extraction does not outperform the existing OIE approaches in general, it significantly improves the performance of paraphrase identification. We have evaluated the performance of paraphrase identification on Microsoft Research corpus and the statistical tests show that RCE-OIE significantly improved the performance of paraphrase identification while ClausIE could not. RCE-OIE may further be used in other downstream applications in future. Our RCE-OIE may further be extended with more rules to make it as a generalized OIE system.

Acknowledgements We would like to thank the management of SSN Institutions for funding the High Performance Computing (HPC) lab where this research is being carried out.

References

1. Akbik, A., Loser, A.: Kraken: N-ary facts in open information extraction. In: Proceedings of the Joint Workshop on Automatic Knowledge Base Construction and Web-scale Knowledge Extraction, pp. 52–56 (2012)
2. Angeli, G., Premkumar, M J., Manning, C.D.: Leveraging linguistic structure for open domain information extraction. In: Proceedings of the ACL, pp. 1–11 (2015)
3. Banko, M., Cafarella, M.J., Soderland, S., Broadhead, M., Etzioni, O.: Open information extraction for the web. IJCAI **7**, 2670–2676 (2007)
4. Del Corro, L., Gemulla, R.: ClausIE: clause-based open information extraction. In: Proceedings of the 22nd International Conference on World Wide Web, pp. 355–366 (2013)
5. Dolan, B., Quirk, C., Brockett, C.: Unsupervised construction of large paraphrase corpora: exploiting massively parallel news sources. In: Proceedings of the 20th International Conference on Computational Linguistics, p. 350 (2004)
6. Fader, A., Soderland, S., Etzioni, O.: Identifying relations for open information extraction. In: Proceedings of the Conference on Empirical Methods in NLP, pp. 1535–1545 (2011)
7. Furbach, U., Glockner, I., Helbig, H., Pelzer, B.: LogAnswer—a deduction-based question answering system (system description). In: International Joint Conference on Automated Reasoning, pp. 139–146. Springer (2008)
8. Hovy, E., Lin, C.: Automated text summarization and the summarist system. In: Proceedings of a Workshop on Held at Baltimore, Maryland, 13–15 Oct 1998, pp. 197–214. Association for Computational Linguistics (1998)
9. Madnani, N., Tetreault, J., Chodorow, M.: Re-examining machine translation metrics for paraphrase identification. In: Proceedings of the 2012 Conference of the North American Chapter of ACL: Human Language Technologies, pp. 182–190 (2012)
10. McNemar, Q.: Note on the sampling error of the difference between correlated proportions or percentages. Psychometrika **12**(2), 153–157 (1947)

11. Schmitz, M., Bart, R., Soderland, S., Etzioni, O.: Open language learning for information extraction. In: Proceedings of the 2012 Joint Conference on Empirical Methods in NLP and Computational Natural Language Learning, pp. 523–534 (2012)
12. Thenmozhi, D., Aravindan, C.: An automatic and clause based approach to learn relations for ontologies. Comput. J. **59**(6), 889–907 (2016). https://doi.org/10.1093/comjnl/bxv071
13. Thenmozhi, D., Aravindan, C.: Paraphrase identification by using clause based similarity features and machine translation metrics. Comput. J. **59**(9), 1289–1302 (2016). https://doi.org/10.1093/comjnl/bxv083
14. Wu, F., Weld, D.S.: Open information extraction using Wikipedia. In: Proceedings of the 48th Annual Meeting of the ACL, pp. 118–127 (2010)
15. Zouaq, A.: An overview of shallow and deep natural language processing for ontology learning. In: Ontology Learning and Knowledge Discovery Using the Web: Challenges and Recent Advances, vol. 2, pp. 16–37 (2011)

An Efficient Approach of Knowledge Representation Using Paninian Rules of Sanskrit Grammar

Bhavin Panchal, Vishvajit Bakrola and Dipak Dabhi

Abstract Computational linguistic began more than sixty years ago. The development of unambiguous knowledge representation systems for processing natural language is more fascinating and challenging task. As a being structured and word-order free language, Sanskrit seems to be a good solution. *Maharshi Panini* laid foundation of complete Sanskrit grammar in his book *Ashtadhyayi*. One of the four components of Ashtadhyayi is dhAtupatha—the verbal roots. There are more than 2000 verbal roots with sub-classification and diacritic markers encoding their morphological and syntactic properties. The usage of *dhAtupatha* gives significant contribution in knowledge representation. The work mainly focuses on the development of NLP grammar for parsing Sanskrit words and derives the meaning of unknown words using Paninian concept of dhAtupatha.

Keywords Natural language processing · Artificial intelligence · dhAtupath
Root-words · Paninian grammar · Ashtadhyayi

1 Introduction

Natural language processing (NLP) is a branch of artificial intelligence. NLP along with computational linguistics majorly focuses on the processing of natural languages (NL) to derive its meaning. Computational linguistics focuses on the analysis of language.

During processing the NL, many times word boundaries, understanding of word and Syntax of word or structure varies. At this moment, grammar of NL plays a vital

B. Panchal (✉) · V. Bakrola · D. Dabhi
C.G. Patel Institute of Technology, Uka Tarsadia University, Bardoli, India
e-mail: bpanchal083@gmail.com

V. Bakrola
e-mail: vishvajit.bakrola@utu.ac.in

D. Dabhi
e-mail: dipak.dabhi@utu.ac.in

© Springer Nature Singapore Pte Ltd. 2018
P. K. Sa et al. (eds.), *Recent Findings in Intelligent Computing Techniques*,
Advances in Intelligent Systems and Computing 709,
https://doi.org/10.1007/978-981-10-8633-5_21

199

role. This is even challenging task for NLP system to deal with such characteristics of natural language. According to grammar, language decides relations between two words. That is a main challenging task for system or machine for developing natural language processing system.

In ancient world, Sanskrit was a primary language. The vedas and all the six core fields of study to learn the vedas are written in Sanskrit. Even Sanskrit is considered mother of almost all Indo-European languages. In modern age, foundation of Sanskrit is laid by Maharshi Panini. He penned complete grammatical rules of Sanskrit in his book *Ashtadhyayi sutrapath*. This grammar is written in the form of sutras. *Ashtadhyayi* contains nearly 4000 sutras in 8 chapters of 4 sections each [1, 2]. In Sanskrit, words are not elementary parts like in other languages. Sanskrit grammar carries the whole process for generating a new word. Panini presented set of rules that focus on word formation. The process of generating new word is based on the most elementary part of the language—*dhAtu*. Sanskrit contains 2012 *dhAtu* or verbal roots. Considering these properties and features Sanskrit grammar can be stated as the best grammar for computational processing of natural languages [3].

This paper presents a novel approach for deriving the meaning of words which are unknown to the system. The meaning is derived using *dhAtu* as an elementary part of the language. In Sect. 2, we discussed knowledge representation in natural language processing. Section 3 focuses on concepts of *dhAtu*. Section 4 gives a complete description of proposed approach along with implementation and results. We have ended up with the conclusion and future work in Sect. 5.

2 Knowledge Representation in NLP

As a branch of artificial intelligence, natural language processing focuses on the processing of human spoken or natural languages, both syntactically and semantically. Basically, NLP is further classified into two broad categories—natural language understanding (NLU) and natural language generation (NLG) [4]. Up to this, many works are proposed for NLU with syntactic analysis of NL. The most natural way of encoding, transmitting and reasoning about knowledge is through NL. Determining the meaning of NL sentence requires a very large set of meanings about individual word. And the performance is dependent on the examination of different constraints related to the different searching mechanisms the find meaning of word [5]. There are several cases where a knowledge representation (KR) problem or solution has a very close parallel with NLP. Due to this, the use of KR in efficient manner becomes one of the core requirements of NLP.

Knowledge of language distinguishes language processing system from data processing. UNIX *wc* program is used to count the number of bytes, number of words and number of lines in a text file. It is a simple application when it is used for counting the number of bytes and number of lines. However, it becomes a language processing application when it counts number of words, as this requires knowledge about basics of language and what it is meant to be a word. *wc* is an example of very simple and straightforward system. Large intelligent systems like machine translation systems and question answering systems required much broader and deeper domain-specific knowledge. As in common engineering practice, the complexity of a problem can be decreased if it is decomposed into less complex sub-problems. Artificial intelligence addresses this using two principle relations as *part-of* and *is-a*. They are used to create hierarchies of constituents and classes [6].

'*The whole universe can be observed as a part-whole hierarchy where complex objects are comprised of other objects*'. To derive the meaning of an input word, we need *prior-meaning* in knowledge space, and then searching mechanism is used to find meaning from that space. The search space of meaning contains holds subspace of group of words, words and syllables. Effective KR techniques should provide a unique representation of various categories at different levels of abstraction. One of the main issues is the need of the *prior-meaning* in the search space. At this instance, unavailability of new input word requires human interaction for understanding the meaning. The system can derive meaning from similar type of previous experiences, if any. In any other cases, the system cannot generate the meaning or desired result, as the system is not having similar experiences. In this article, we have applied the same part-whole hierarchy to the sequence of characters representing sentence. The novel approach of using Sanskrit *dhAtu* in representation of knowledge seems helpful in not only understanding about words but also in deriving the meaning of words [7].

3 dhAtu Roop—*A Verbal Root*

Most of the modern languages are built upon the principle where word represents objects and entities. On a first glimpse, this seems innocent fact and absolutely harmless. But this principle is responsible for much inefficiency in the processing of natural language. Sanskrit follows different principle where word represents properties of objects or entities. In Sanskrit instead of representing an object, word *tree* becomes *taRupAdam* where it represents the property of an object tree. Here *taRu* means an object that floats and *pAdam* means an object that drinks using the feet. The word *taRupAdam* is formed using the sub-elementary components of the language—rootword. *dhAtu* is most elementary and foundational part of the language. They can be considered as the basic building blocks of the Sanskrit grammar. This foundation of the language is responsible, that being an ancient language the structure of Sanskrit is not changed since its commencement. 2012 *dhAtu* are presented in the Sanskrit grammar. Panini stated a bunch of rules for formation of new words in his book *Ashtadhyayi* [8].

As a being fundamental building blocks of the language each *dhAtu* contain its own individual meaning. *dhAtu kR* means *to do*, *to make* or *to prepare*. With the help of Paninian rules, one can form hundreds of words using *dhAtu kR*. For example, *kRit* means *is done*, *kaRm* means *work*, *kRiyA* meaning *activity*, *karmaTh* means *skillful* in doing activity. By combining more than one *dhAtu* thousands of new words can be formed unambiguously [8, 9].

kRidnt word	Word meaning
kRta	Something is completed or Finish
kRtavaana	Someone who has done something
kRti	An object as a result of doing
Karttavya	Something must be done or completed
Karma	Work
kaRa	Something is completed or Finish
Kaara	A head
kriyaA	Author, Source
kriyA	Activity
kraTu	A plan

Fig. 1 Lists of word derived from *kRidnt* and along with English meaning [8]

kRidnt word	Word meaning	dhAtu appear in words
kRtatiirtha	One who has visited by holy place	kR + tR
kRtajJa	One who acknowledges past services	kR + jJA
kRtamaarga	One who has made a path	kR + mRg
kRtadeza	One whose place is fixed	kR + diz
kRtavrata	One who has taken a resolution	kR + vR
lokakSayakRt	One who destroys the world	lok + kSii + kR

Fig. 2 Words derived using the use of *kRidnt* and multiple *dhAtu* along with English meaning [8, 9]

The words presented in Fig. 1 are formed using a single *dhAtu kR*. One of the reach approaches of Paninian grammar is the flexibility of using multiple *dhAtu* for deriving new words. Millions of new words can be formed using this approach. Figure 2 shows some of the derived words using multiple *dhAtu*.

4 Proposed Approach of Knowledge Representation Using Sanskrit Root-Words

As stated in Sect. 2, to understand the NL, we require prior-meaning in search space. This meaning can be derived to the system from the previous experiences or with the help of a human expert. With this approach, the derivation of meaning is dependent on the availability of similar meaning in the search space. As discussed in Sect. 3, *dhAtu* and the Paninian rules are useful in understanding the meaning of unknown

बालक् वारिजा रोचते

Fig. 3 Sanskrit sentence as an input

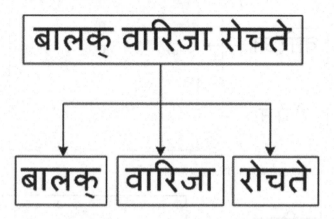

Fig. 4 Part-of-speech (POS) tagging on an input sentence

words to the system. This chapter discusses the proposed approach and the results of implementation in detail.

The input is taken in Sanskrit and with part-of-speech (POS) tagging sentence is divided into various sequences of letters. Database design is followed to store meaning of words, root-words and connectors. We have discussed various steps involved in deriving meaning of unknown words and necessary database tables. The system is working with following tables:

- Dictionary-table (contain Sanskrit words and its meaning in English language)
- Root-word table (contain Sanskrit root-word and meaning of root-word in English language)
- Connectors (contain English language connectors for connection of words)

Step I: We have taken input sentence in Sanskrit as *bAlaka vaRijA roChate* (Fig. 3).
Step II: An input sentence is a combination of three Sanskrit words. We applied part-of-speech (POS) tagging. We divided input sentence into a number of available words using POS tagging for further processing (Fig. 4).
Step III: After dividing the sentence into a number of words, the searching procedure checks individual word in Dictionary-table to derive their meaning. Here, words *bAlaka* and *roChate* found in the Dictionary-table, so we can derive their meaning easily. But our search space is lacking with meaning of the word *vaRijA* (Fig. 5).
Step IV: In Sanskrit, words are derived from the fixed set of *dhAtu*. We are having some of the root-words along with their meanings to test the algorithm. Word *vaRijA* is not available in Dictionary-table. So to derive the meaning, we applied LR parsing for searching the root in respective root-table. Hence, the sub-elementary root-words

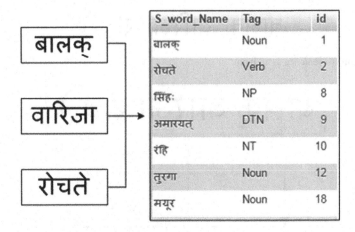

Fig. 5 Search individual word in Dictionary-table

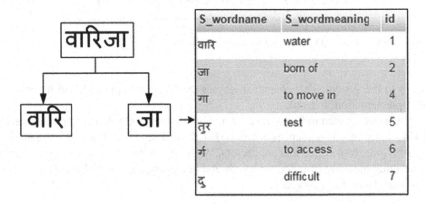

Fig. 6 Searching root-words in root-table

are available in the table along with their meaning; we derived *water* and *born of* as a meaning of *vaRi* and *jA*, respectively. At the end of the merging procedure, parsing algorithm derives the final meaning that is—born of water. This points to a characteristic of an object. From another set of prior-knowledge, we found Lotus is born of water. At the end of this step, Dictionary-table is updated with the meaning of the unknown word *vaRijA* (Fig. 6).

Step V: After the step IV, merging the meaning of the word and update the meaning in Dictionary-table (Fig. 7).

Fig. 7 Update
Dictionary-table

S_word_Name	Tag	id
बालक्	Noun	1
रोचते	Verb	2
विमान	SBJ	7
सिंहः	NP	8
अमारयत्	DTN	9
रहि	NT	10
प्रश्नपत्र	VP	11
तुरगा	Noun	12
मयूर	Noun	18
भाविन्	Sub	19
वारिजा	born of water	22

5 Conclusion

Studying Sanskrit grammar and with fundamentals of Paninian rules, we found San-skrit grammar as one of the finest NL grammar for NLP. The rules are successfully applied to derive the meaning of unknown words without any human interaction or any previous experiences. The use of *dhAtu* and grammatical rules of Sanskrit plays a significant role in deriving solution of the previous issues related under the domain of NLP based on knowledge representation.

Future Work. Future research in this area is to develop the system that gives the appropriate meaning of Sanskrit sentences and to develop a parser using Paninian rules of Sanskrit grammar. The parser should be capable to parse the input Sanskrit sentences with higher accuracy in consideration of Vibhaktis, Sandhis, Samas and Lakars.

Acknowledgements I would like to express my heart full gratitude to the department of computer engineering for providing continues support as an when needed. In addition, I also express sincere and heartfelt thanks to Dr. Madhav Gopal, JNU, Delhi for providing support in basic understating of Sanskrit language.

Last but not the least, I would like to thank Ms. Kinjal Mistree, Assistant Professor, C.G. Patel Institute of Technology and Mr. Ishank Sharma, Research Assistant, IIT-Delhi for providing necessary and fundamental knowledge about natural language processing tools and their working.

References

1. Pawan, G., Arora, V., Behera, L.: Analysis of Sanskrit text: parsing and semantic relations. In: Sanskrit Computational Linguistics, pp. 200–218. Springer, Berlin, Heidelberg (2009)
2. Kak, S.C.: The Paninian approach to natural language processing. Int. J. Approx. Reason. **1**(1), 117–130 (1987)
3. Jurafsky, D., Martin, J.H.: Speech and Language Processing (2014)
4. Kulkarni, A., Pokar, S., Shukl, D.: Designing a constraint based parser for Sanskrit. In: Sanskrit Computational Linguistics, pp. 70–90. Springer, Berlin, Heidelberg (2010)
5. Mishra, A.: Simulating the Paninian system of Sanskrit grammar. In: Sanskrit Computational Linguistics, pp. 127–138. Springer, Berlin, Heidelberg (2009)
6. Stanojevic, M., Vranes, S.: Knowledge representation for natural language understanding. Facta Universitatis Ser. Math. Inform. **21**, 93–102 (2006)
7. Teja, D., Kothuru, S.: Sanskrit in natural language processing. IJARCSSE (2015). 2277 128x
8. Inderjeet: An approach to Sanskrit as computational and natural language processing. CS J. 264–268 (2015)
9. Saxena, S., Agrawal, R.: Sanskrit as a programming language and natural language processing. Glob. J. Manag. Bus. Stud. **3**(10), 1135–1142 (2013)
10. Santos, D.: Natural language and knowledge representation. In: Proceedings of the ERCIM Workshop on Theoretical and Experimental Aspects of Knowledge Representation, pp. 195–197 (1992)

Sentence Boundary Detection for Hindi–English Social Media Text

Ajeet Singh, Bhanu Pratap Singh, Ankit Kumar Poddar and Abhishek Singh

Abstract In this paper, we present an approach of automatic sentence boundary detection for Hindi–English *Codemixed* social media texts. We develop a corpus of Hindi–English *Codemixed* posts collected from Facebook and made an in-depth study to explore the limitations of using existing rule-based sentence boundary detection systems on codemixed social media text. Our proposed approach is a rule-based sentence boundary detection approach which is tested on our developed corpus and outperforms over the existing approaches.

Keywords Social media · Codemixed · Sentence boundary detection

1 Introduction

Social Media provides a platform for people to share up-to-date information regarding any topic or current events and also to respond very fast. Since inceptions, social media like Twitter, Facebook has grown at an unprecedented rate. With the extent of ubiquitous devices, anyone can post anything he or she wants at anytime from any location. This accessing flexibility shares huge amount of data into social media communications. Among this huge amount of data, a remarkable portion of text is codemixed text.

A. Singh (✉) · B. P. Singh · A. K. Poddar · A. Singh
National Institute of Technology Agartala, Agartala, Tripura, India
e-mail: ajeets585@gmail.com
URL: http://www.nita.ac.in/

B. P. Singh
e-mail: bhanu9288@gmail.com

A. K. Poddar
e-mail: ankit4apoddar@gmail.com

A. Singh
e-mail: abhi4nita@gmail.com

© Springer Nature Singapore Pte Ltd. 2018
P. K. Sa et al. (eds.), *Recent Findings in Intelligent Computing Techniques*,
Advances in Intelligent Systems and Computing 709,
https://doi.org/10.1007/978-981-10-8633-5_22

Codemixing is the mixing of two or more languages in a single post. It is basically done when people are very much comfortable in two languages or use to speak frequently one language while another language is native language. Sometimes it is done to seek the attention of others or to improve their image in the eyes of others. Also, the probable reasons could be making people understand efficiently or it makes them sound like a person from a well to do background.

Hindi is the fourth most spoken language in the world (after Chinese, Spanish, and English) and most spoken language in India. Our current research work is detection of sentence boundary for Hindi–English (Hi-En) codemixed posts. Writing posts with Hi-En codemixed text in social media is a common practice. For example:

"plz help brothers or sisters mjhe kya karna chaihe"
"admin plz post kar dena 5th time likh rha hu..."

The posts on Facebook are quite informal, full of semi-broken words like "bro" for brother, "sis" for sister, etc., and without proper punctuations. Most of the people use this informal way of writing on social media, which consists of Hi-En codemixed data as they can express their feelings better in it.

Sentence Boundary Detection (SBD) is a very basic step in Natural Language Processing (NLP) tasks like question answering and translation. Most of the popular researches done today are based on NLP, which in turn demands SBD. SBD is a problem of deciding where one sentence ends and the other begins. This can be easily identified in formal text as some sort of punctuations are used in it. As the trend is changing on social media, so people now prefer to use Hi-En codemixed sentences, where they do not use punctuations appropriately. The codemixed sentences are full of spelling errors as well as grammatical errors. Sometimes, people merge two sentences without any marker or they end their sentences with emoticons. This makes it really very difficult to detect sentence boundaries on social media text.

Very few research papers are there on Hi-En codemixed posts except the research paper [1], which has used the rule-based approach on codemixed data (but mainly concentrated on English posts) by finding some limited patterns, based on punctuations, in the corpus like "...", "..", "!!!!", "???", etc. Most of the earlier research works are focused on formal texts. Due to the terse nature of codemixed text, SBD for Hi-En codemixed social media text is a challenging task. For example, the following Hi-En codemixed Facebook post has two sentences even though no punctuation is used.

Actual post: "mea 10 sep. ko delhi gya tha mera post off. ka xam tha"
Sentence 01: "mea 10 sep. ko delhi gaya tha"
Sentence 02: "mera post off. ka exam tha"

The rest of the paper is categorized as follows. In Sect. 2, we discuss some of the previous research work carried out on sentence boundaries, while Sect. 3 consists of the collection of corpus and preprocessing steps. Section 4 contains rule-based approach followed by its results and error analysis in Sects. 5 and 6, respectively. Lastly, Sect. 7 presents conclusion and future works.

2 Related Works

Many algorithms have been employed by various researchers for detecting sentence boundaries. Some have used the rule-based approach like by Manning et al., some have used parts of speech tagged words for detecting sentence boundaries as done by Mickheev [2], and some have used maximum entropy method as done by Reynar and Ratnaparkhi [3]. But all these researches have been carried out on English language or on a single language for formal texts.

Very few researches have been carried out on Indian languages. A method was employed by Mona et al. [4] to remove the ambiguity of a period in Kannada language. Verb suffixes and abbreviations have been used by Deepamala and Ramakanth [5] for detecting sentence boundaries in Kannada language and they have achieved an accuracy of 99.2% with this algorithm. Nagmani [6] have also used rule-based approach for Marathi language in which the rules were prepared by carefully observing the language pattern as Marathi language is just like Hindi language which most of the times ends with a verb.

But, only a handful of research can be found on Hi-En codemixed Social Media Corpus (SMC). In paper [1], a rule-based approach has been employed on it which basically contains punctuations or a group of punctuations that are usually end markers like "...", "..", "!!!!", etc. They have achieved an accuracy of 64.3%.

3 Data Collection

For detecting the accuracy of the rule-based approach for SBD on social media corpus, we have collected posts from the timeline of Facebook Pages where there are maximum chances of getting Hi-En codemixed data such as

- *Delhi Metro Confessions/Compliments*[1]
- *Resonance Confessions/Compliments*[2]
- *Jamshedpur Confessions*[3]
- *Bansal Classes Confessions*[4]

Mainly, the confessions are about the mischievous and stupid things done by the people in their childhood days or in their coaching classes. As they express their feelings on confession pages, so they are more comfortable in using Hi-En codemixed data.

From these pages, 2200 posts were collected through a third party software, Facepager.[5] Then, after manually checking the Hi-En codemixing in each post, we

---

[1]https://www.facebook.com/delhimetroconfessions/.

[2]https://www.facebook.com/RCC.KOTA/.

[3]https://www.facebook.com/aryan12121990/.

[4]https://www.facebook.com/bcplconfessions.

[5]http://facepager.software.informer.com/3.6/.

Table 1 Golden corpus size statistics

Corpus details	Count
Total posts fetched	1419
Sentences in golden corpus	16,245
Total tokens	161,960
Unique tokens	18,465

finally came to a data set of 1419 purely Hi-En codemixed post and the rest were removed.

These 1419 posts were manually broken by 3 different annotators. Due to informal nature of posts which includes lots of spelling mistakes ("dlhi" for Delhi), self-made acronyms (r8 for right, f9 for fine) as well as similar words in both Hindi and English languages ("Main" in English means chief while "Main" in Hindi means I, "Cab" in English means taxi while "cab" in Hindi means when), the annotators faced a lot of problems in manually breaking each post.

All the three annotators have Hindi as their mother tongue. The sentence breaking was agreed by annotators on 72.2% but, after mutual understanding, the inter-annotator agreement rose to 87.4%, which constitutes our golden corpus. Corpus statistics is provided in Table 1.

4 System Description

To detect the sentences in a post, we first tokenize the whole corpus using CMU Tokenizer.[6] CMU tokeniser tokenises the sentences in the following manner:

"usne mauka bhi diya...baat karne ka....bt mein nahi kar paya..."
"usne mauka bhi diya ... baat karne ka .... bt mein nahi kar paya ..."

Since we are working on Codemixed data, so people generally do not bother to put proper punctuations or spacing between two words, which leads to improper tokenisation. In few cases, CMU Tokenizer could not tokenize words correctly like

(i) If a numeral value is written along with some word.
   E.g., "5mahine", "1saal", "M21".

(ii) If is present between two words.
   E.g., "non-biharis", "RJD-JD(U)-CONGRESS", "heart-touching", "bola-ok".

(iii) If "(" or ")" is present between two words.
   E.g., "city(Delhi", "hota(matching", "NSEZ)Noida", "boys)zorr".

So, by analyzing the whole tokenized text manually, we correct the text by proper tokenization. Through experiment, we found that the accuracy of CMU tokenizer on

---

[6]http://www.cs.cmu.edu/~ark/TweetNLP/.

Table 2 Language tagging

Language tag	Tag used	Frequency
Hindi	hi	88,092
English	en	49,872
Universal	univ	20,796
Named entity	ne	2601
Acronym	acro	125
Mixed	mixed	273
Undefined	undef	201

our Hi-En codemixed coupus is 99.57%. After tokenization, we observed another important issue with some words due to its ambiguous nature of meaning like-"the", as it is an auxiliary verb in Hindi and also an article in English. So to remove ambiguity in such kind of words in the text, each token is tagged by language tagset,[7] which is shown in Table 2. One example of such ambiguity is shown below:

(i) "Mom to man gai par papa katai man ne ko taiyaar na **the**"

(ii) "Actually **the** thing is meko nahi rehna aisi duniya me jha logo ka jeena mushkil ho..."

After these preprocessing phases, we have applied our proposed rule-based approach to detect sentence boundary.

4.1 Rule-Based Approach

After in-depth study of the develop corpus, sentence boundaries are not only based on the emoticons, punctuations, or patterns as in work [1], but also sentence ends with the Hindi words which are auxiliary verbs like "tha", "thi", "hai", "hain", and "h".

Codemixed Hi-En sentences usually end with an auxiliary verb which depends on the tense whether it is past, present, or future, gender—male or female, and the numbers of persons—singular or plural we are referring to. Following are some verbs that we have encountered at the end of the sentences in our Corpus

"Kisi ka laptop ka bag chhutt gya to kisi ki pocket mar li gyi..."

"M Delhi se blng krta hu"

"m apne frnd k sath gulmohar park m btha tha"

"ladki uss ladke ko bahot pyar karti thi..."

---

[7]Language tagset is taken from [7].

Table 3 List of basic patterns denoting end of sentences

Pattern	Total occurrence	Actual line break
Single or multiple occurrences of "."	9863	7517
Multiple occurrences of ","	1913	902
Single or multiple occurrences of "?"	462	367
Single or multiple occurrences of "!"	312	219

4.2 Implementation

The code for the rule-based approach is written in Python language. All the SBD patterns in paper [1] are used as basic patterns in our work. We observed that those patterns are commonly used in Hi-En codemixed posts as sentence end. In addition to those features, we added new features (emoticons and auxiliary verbs) for our SBD task. The details of those features and their significance toward SBD are discussed in Sect. 5.

5 Experimental Evaluation

After manually breaking 1419 posts by three annotators separately and by mutual understanding, the golden corpus consists of 16,245 sentences.

Firstly, we applied regular expression only for period (.), and then it breaks our corpus into 16,148 lines, out of which 11,041 lines matched with our golden corpus, giving the accuracy of 67.96%. Similarly, regular expressions were applied only for ",", "?", and "!", which are mainly used as sentence boundaries. The accuracies for them are 36.52%, 34.32%, and 33.75%, respectively.

So, we took Period (.), ",", "?", and "!" as our base regular expression for sentence boundaries (Table 3). The accuracy that we achieved on applying them is 72.51% with 11,779 lines matching with our golden corpus and 18,760 line breaks.

As social media corpus also ends with emoticons, so after applying the base regular expression with emoticons, the accuracy that we achieved is 76.31%.

The emoticons used in this paper as sentence boundaries are :P, :\, (y), :'\, ;(, :p, -_-, :v, :-\, <3, :3, :-*, B-), =D, :*, :-@, :-C, :@, :'(, :o, :D, :O, :(, ;), :], :H.

Also, different auxiliary verbs are applied for SBD like "thi", "hain", "tha", "hu", "h", "the", "gyi", "diya", "liya", "huye", and "kiya". Due to the informal nature, sometimes "thi" is written as "thie" or "thii", "hu" is written as "hun", "hain" is written as "h", "gyi" is sometimes written as "gayi" or "gai", and "diya" as "diyaa".

Table 4 Accuracies of different auxiliary verbs as sentence boundaries

Pattern used including base pattern	Line break matched	Line break found	Accuracy (%)
"diya" or "hai"	12192	19625	75.05
"thi" or "hai"	12195	19798	75.07
"thi" or "hain" or "tha" or "hu" or "diya" or "liya"	12240	19755	75.35
"thi" or "hain" or "tha" or "hu"	12312	19587	75.79
"thi" or "tha" or "hain" or "hu"	12341	19507	75.97
"liye" or "the"	12375	19062	76.18
"thi" or "the"	12480	19096	76.82
"kiya" or "tha"	12502	18968	76.95
"thi" or "tha" or "hain"	12511	19129	77.01
"the"	12537	18796	77.17
"gyi" or "thi"	12725	18687	78.33
"thi" or "hain"	12734	18609	78.38
"thi"	12764	18548	78.57
"gyi"	12772	18417	78.62

As these all are used as auxiliary verbs while expressing on social media and hence, these are also considered as sentence boundaries.

The highest accuracy that we achieved is using the auxiliary verb "gyi" which is 78.62%, which also includes its other forms such as "gayi" and "gai". Similarly, accuracies are calculated for different sets of auxiliary verbs combined with our base regular expression and emoticons, which are tabulated in Table 4.

6 Discussion

For pure English language, period (.), "!", or "?" are generally the end of the sentences. But this is not the case in Hi-En codemixed data, here people put punctuations anywhere according to their mood instead of thinking whether it is meaningful or not. They even do not bother to correct the grammatical mistakes and spelling errors.

Due to this, our rule-based approach was unable to identify sentences breaks in some lines where no pattern was found or some line breaks occurred unnecessarily. Let us consider few sentences from our Corpus

(i) "Unhone kaha tha..k uncle aapko Google pr dekha h.."

(ii) "wo sala tha hi difaulter ...par usne kbhi mjhe serious nhi liya..."

(iii) "finally i m on my way...aur vo rithala chali gayi...."

Some single sentences are considered to be as multiple sentences due to the regular expression used for Period (.) as shown in the above examples.

(iv) "guys m Raghav Mehra nam tho suna hi hoga aap logo ne"

(v) "hua ye ki ham dono m nazre mili aur dono flow krne lage ek dusre ko"

(vi) "4 saal se ek he cell use kar raha tha samsung s3"

No unique pattern could be found out in the above sentences.

(vii) "Actually m still in metro and aaj shaam ko bhat bheed thi metro mein."

(viii) "jab meri baat hui thi tav uska breakup hua tha new new"

(ix) "meri family ne mere liye ladki dekhi thi shadi k liye,"

Hi-En codemixed sentences end with auxiliary verbs. But this is not always true. Due to breaking if sentences at 'thi', these sentences were broken from inappropriate places.

(x) "yeh bat uske phele socne chaye thi ki uske mom dad ni mangee ab bhanana bna rha h.."

(xi) "Ye chiz mujhe hurt krti thi par mujhse bolta ki mai tere ex bf jaisa kamina nahi hu."

(xii) "sb kch bdya chl rha tha then last month hmari dono ki ldai hui"

In composite sentences, an auxiliary verb is usually followed by a conjunction, which is not an ideal place to mark it as a sentence boundary.

7 Conclusion and Future Works

In this paper, we propose an automatic SBD approach for Hi-En codemixed Facebook posts. We have developed a corpus and applied rule-based approach for the task in this research work. The maximum accuracy that we have achieved on Hi-En codemixed data is 78.58%. This is an ongoing research work.

Our future works include part of speech tagging of all the tokens and then, applying machine learning algorithms including deep learning on the corpus to yield maximum accuracy.

Acknowledgements Thanks to Assistant Professor Anupam Jamatia and Assistant Professor Dwijen Rudrapal, Computer Science and Engineering Department, National Institute of Technology, Agartala for their support and guidance throughout our work.

References

1. Rudrapal, D., Jamatia, A., Chakma, K., Das, A., Gambäck, B.: Sentence boundary detection for social media text. In: ICON (2015)
2. Mikheev, A.: Tagging sentence boundaries. In: Proceedings of the NAACL, Seattle, pp. 264–271 (2000)
3. Reynar, J.C., Ratnaparkhi, A.: A maximum entropy approach to identifying sentence boundaries. In: Proceedings of the 5th Conference on Applied Natural Language Processing, Apr 1997, pp. 803–806, Washington, DC. ACL (1997)
4. Parakh, M., Rajesha, N., Ramya, M.: Sentence boundary disambiguation in Kannada texts, language in India. In: Special Volume: Problems of Parsing in Indian Languages, pp. 17–19. www.languageinindia.com. Accessed 11:5 May 2011
5. Deepamala, N., Ramakanth Kumar, P.: Sentence boundary detection in Kannada language. Int. J. Comput. Appl. (0975-8887) (2012)
6. Wanjaria, N., Dhopavkarb, G.M., Zungrec, N.B.: Sentence boundary detection for Marathi language. In: International Conference on Information Security and Privacy (ICISP2015), 11–12 Dec 2015, Nagpur, India
7. Jamatia, A., Gambäck, B., Das, A.: Part-of-speech tagging for code-mixed English-Hindi Twitter and Facebook chat messages. In: Proceedings of 10th International Conference on Recent Advances in Natural Language Processing, pp. 239-248, Hissar, Bulgaria, 7–9 Sept 2015

Text Categorization Using a Novel Feature Selection Technique Combined with ELM

Rajendra Kumar Roul and Jajati Keshari Sahoo

Abstract The rapid growth of digital documents and internet users on the web increases the searching time of a document for the end user, which affects the performance of the search engine badly. Hence, to reduce the searching time and to increase the efficiency of the search engine, text classification is the need of the day. But to do an efficient text classification, selection of good features also equally important. To address this issue, the current paper proposes an approach called Combined Correlation Discriminative Power Measure (*CCDPM*) where first the highly correlated terms (features) are removed from the corpus and then using the scores generated by discriminative power measure technique, the uncorrelated features of the corpus are ranked. Top k features are selected to generate the reduced training feature vector. For classification, Extreme Learning Machine (ELM) is used, and the empirical results on four benchmark datasets show the efficiency the proposed approach compared to other state-of-the-art feature selection techniques. Results of ELM are more promising compared to other conventional classifiers.

Keywords Classification · DPM · ELM · Feature selection · F-measure

1 Introduction

Due to the large size of the web, most of the time, the present search engine retrieves invalid links and irrelevant web pages for a submitted user query. This weakens the trust of the user on the search engine and thereby degrades its performance. Text classification, a powerful machine learning technique which categorizes an unseen document into its respective predefined class, can shed light in this direction. Two basic classifications of web pages are there: subject-based classification

R. K. Roul (✉) · J. K. Sahoo
BITS-Pilani, K.K. Birla Goa Campus, Sancoale, India
e-mail: rkroul@goa.bits-pilani.ac.in

J. K. Sahoo
e-mail: jksahoo@goa.bits-pilani.ac.in

© Springer Nature Singapore Pte Ltd. 2018   217
P. K. Sa et al. (eds.), *Recent Findings in Intelligent Computing Techniques*,
Advances in Intelligent Systems and Computing 709,
https://doi.org/10.1007/978-981-10-8633-5_23

and genre-based classification [1]. In subject-based classification, web pages are classified based on their subject or content. Topic hierarchies of web pages are build by this approach. Web pages in genre-based classification are classified into genre or functional-related factors. For example, some web pages genres are "multimedia", "home page", "online transaction", and "news headlines". This classification will help users to find their immediate interest. There are many classification techniques that exist in real and can be divided into two broad categories: eager learner and lazy learner. According to eager learner classification technique, the learner built a classi-fication model when the training dataset was given before it receives the test dataset. It can be thought as if the learning model is ready and eager to classify the new test dataset. Examples of this category are decision tree, Bayesian network, support vector machine, rule, and association-based classifier. But in lazy learner classifi-cation technique, the things are different. Here, instead of building a classification model, it simply stores the training dataset and when it sees the test dataset, it does the classification based on the similarity to the stored training dataset. It consumes extra spaces. Examples include k-nearest neighbors (k-NN) and case-based reason-ing. Much research works have been done in the field of web document classification [2–6].

Feature selection plays a major role in text classification because the selection of important features not only reduces the training time but also increases the perfor-mance of the classifier by reducing the irrelevant features from the corpus. Many researchers have worked in this domain [7–11]. In a good feature set, features should uncorrelated (not predictive of) with each other, but should be highly correlated (predictive of) with the class [12]. If a feature is highly correlated with other fea-tures, then it is said to be redundant. If two features are highly correlated, then one does not add or give any additional information (as it is determined by the other). Therefore, if the number of highly correlated features is large, then it is beneficial to reduce this number because it affects the training time as well as the accuracy of a classifier. As specified by the approach [12], each subset of highly correlated features is also uncorrelated to the other subsets. This solves the problem of choos-ing features that are uncorrelated to each other. The aim is to create a novel feature selection technique that evaluates the worth of a feature. Results of the experimental work done by Langley and Sage [13] and Kohavi and Sommerfield [14] on feature selection techniques show that redundant information should be eliminated along with the irrelevant features as well. Therefore, there must exist a balance between relevance (based on class prediction) and redundancy (correlation). Discriminating Power Measure (DPM) [15] evaluates how well a feature can help to discriminate between two classes. The DPM value for a feature is a strong measure for class pre-diction and should have considered into account during the feature selection.

Considering selection of good features and using Extreme Learning Machine (ELM) as the classifier will enhance the performance of the text classification process, this paper proposed a novel technique where initially the entire corpus is divided into different clusters to bring similar documents into one place and then from each cluster, the highly correlated terms are discarded. Finally, using DPM scores, top k features are selected from each cluster which then merged together to generate the reduced training feature vector. By this approach, we only eliminate those features that are highly correlated to other features and have a low DPM value. ELM is used for classification of documents. For experimental work, four traditional feature selection techniques (chi-square [16], binormal separation [17], information gain [18] and GINI index [19]) are used. Empirical results on four well-known machine learning datasets show that the proposed approach is promising and the performance of ELM outperforms other state-of-the-art classifiers. The paper is organized on the following lines: Sect. 2 discusses the background details of ELM briefly. In Sect. 3, the proposed approach is discussed. Section 4 discusses the experimental analysis, and finally, in Sect. 5, the proposed work is concluded.

2 Background

2.1 Extreme Learning Machine

ELM, developed by Huang et al. [20], is a single layer feed forward network. Simple to implement and extremely quick learning speed, requirement of less human intervention, good generalization capability, no adjustment of hidden layer biases and input weights, avoids local minimization, no backpropagation, etc., are some of the important features which makes ELM more popular compared to other traditional classifiers.

Brief on ELM:
Consider N different examples (x_i, y_i), where $x_i = [x_{i1}, x_{i2}, \ldots, x_{in}]^T \in R^n$ and $y_i = [y_{i1}, y_{i2}, \ldots, y_{im}]^T \in R^m$, such that $(x_i, y_i) \in R^n \times R^m$, $i = 1, 2, \ldots, N$. ELM has an activation function $g(x)$ and L hidden layer nodes (Fig. 1). Given input $\mathbf{x}$, ELM output function can be written as

$$y_j = \sum_{i=1}^{L} \beta_i g(w_i \cdot x_j + b_i) \tag{1}$$

where $j = 1, \ldots, N$, w_i and b_i are randomly generated hidden node parameters (Fig. 1). $w_i = [w_{i1}, w_{i2}, w_{i3}, \ldots, w_{in}]^T$ is the weight vector that joins the "n" input nodes to the ith hidden node. b_i is the bias of the ith hidden node. β is the weight vector that connects each hidden node to every output node and is represented as $\beta = [\beta_1, \ldots, \beta_L]^T$. $g(\mathbf{x})$ is responsible to map the input feature space of n-dimension to L-dimensional hidden layer space. The reduced form of Eq. 1 is represented in Eq. 2, where Y and H are the output and hidden layer matrix, respectively.

$$H\beta = Y \tag{2}$$

$$\beta = H^+ Y \tag{3}$$

where H^+ is the Moore–Penrose inverse [21].

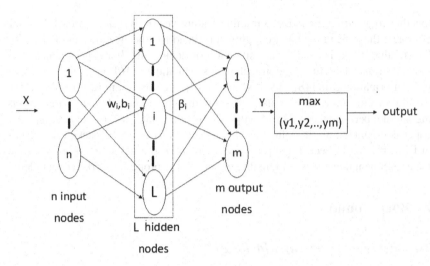

Fig. 1 Architecture of ELM

2.2 Fuzzy C-Means

Fuzzy C-Means (FCM) algorithm [22] distributes a finite collection of n documents into c clusters. It returns a list of c cluster centroids along with a matrix which shows the degree of membership of each document to different clusters. It aims to minimize the following function:

$$T_m = \sum_{i=1}^{n} \sum_{j=1}^{c} v_{ij}^m ||d_{ij}||^2$$

where distance $d_{ij} = x_i - c_j$, m generally set to 2 is the fuzzy coefficient, c_j is the centroid (vector) of cluster j, x_i is the ith document, $v_{ij} \in [0, 1]$ is the degree of membership of x_i with respect to c_j and ($\sum_{j=1}^{c} v_{ji} = 1, i = 1, \dots, n$). One can iteratively find the updated values of c_j and v_{ij} with each iteration by using the following equations:

$$c_j = \frac{\sum_{i=1}^{n} v_{ij}^m - x_i}{\sum_{i=1}^{n} v_{ij}^m} \quad (4)$$

$$v_{ij} = \frac{1}{\sum_{k=1}^{c} \left(\frac{||d_{ij}||}{||x_i - c_k||}\right)^{\frac{2}{m-1}}} \quad (5)$$

3 Proposed Approach

The proposed approach is discussed using the following steps:

Step 1 *Preprocessing of documents*:

Consider a corpus having set of classes $(C = c_1, c_2, \ldots, c_n)$ of documents $(D = d_1, d_2, \ldots, d_p)$. All documents are preprocessed which includes lexical analysis, stop-word elimination, and stemming, and then index terms are extracted. The term-document matrix is constructed using the vector space model, where TF-IDF [16] values are used to measure the weight of the terms (t) in their respective document (t_{ij}) (Table 1).

Step 2 *Clusters formation*:

The entire corpus is get clustered to discover the groups of similar documents and terms, respectively. For this purpose, traditional FCM clustering algorithm is applied to the term-document matrix of the corpus which generates "s" term-document clusters $td = \{td_1, td_2, \ldots, td_s\}$ having each td_i is of dimension $r \times b$, where b is the number of documents.

Step 3 *Important features selection from each cluster*:

Top features are selected from each term-document cluster td_i using the following steps:

i. Computing the cosine similarity for each term with the centroid of the term-document cluster td_i:

- Find the centroid (sc_i) of td_i

$$sc_i = \frac{\sum_{j=1}^{r} t_j}{r} \tag{6}$$

- Compute the cosine similarity between each term $t_j \in td_i$ and sc_i.

ii. Constructing correlation matrix:

The correlation (cr) [16] between two terms t_i and t_j (Table 2) is computed as follows:

Table 1 Term-document matrix

	d_1	d_2	d_3	...	d_p
t_1	t_{11}	t_{12}	t_{13}	...	t_{1p}
t_2	t_{21}	t_{22}	t_{23}	...	t_{2p}
t_3	t_{31}	t_{32}	t_{33}	...	t_{3p}
.	.	.	.	...	.
.	.	.	.	...	.
t_r	t_{r1}	t_{r2}	t_{r3}	...	t_{rp}

$$cr_{t_i t_j} = \frac{C_{t_i t_j}}{\sqrt{(V_{t_i} * V_{t_j})}} \tag{7}$$

where $C_{t_i t_j}$ is the covariance between t_i and t_j and V_{t_i} and V_{t_j} are their variance, respectively, as shown below.

$$V_{t_i} = \frac{1}{b-1} \sum_{m=1}^{b} (X_{im} - \overline{X}_i)^2 \tag{8}$$

$$V_{t_j} = \frac{1}{b-1} \sum_{m=1}^{b} (X_{jm} - \overline{X}_j)^2 \tag{9}$$

where $\overline{X}_i$ and $\overline{X}_j$ represent the mean of b documents of t_i and t_j, respectively. Covariance is a measure of the joint variability of two terms t_i and t_j, and is represented as follows:

$$C_{t_i t_j} = \frac{1}{b-1} \sum_{m=1}^{b} (X_{im} - \overline{X}_i)(X_{jm} - \overline{X}_j) \tag{10}$$

iii. Removing highly correlated terms from the cluster:
Select a term t_i which has highest cosine-similarity score from the cluster td_i and find a set of terms whose correlations are very high with respect to t_i. $(\leq -0.9 || \geq 0.9)$[1] from Table 2. Terms that are highly correlated do not discriminate well among classes and can be considered as sort of synonyms and therefore removed from the cluster td_i. Repeat this step for the remaining terms in the cluster td_i till the cluster get exhausted. Now the cluster td_i having terms which are uncorrelated and we named it as a new cluster td_i'.

iv. Computing discriminating power measure for each term:
The Discriminating Power Measure (DPM) is generally calculated for each term belonging to a class as a relevance measure which is able to distinguish between different categories.

- For each term $t_i \in td_i'$, the document frequency inside that cluster (DF_{in}) and document frequency outside that cluster (DF_{out}) are calculated.
- DF inside the cluster td_i':

$$DF_{in, t_i} = \frac{\#documents \; containing \; the \; term \; t_i \; and \; \in td_i'}{\#documents \; belong \; to \; cluster \; td_i'} \tag{11}$$

---

[1]Decided based on the experiment so that we should not lose more terms.

Table 2 Correlation matrix

	t_1	t_2	t_3	...	t_r
t_1	cr_{11}	cr_{12}	cr_{13}	...	cr_{1r}
t_2	cr_{21}	cr_{22}	cr_{23}	...	cr_{2r}
t_3	cr_{31}	cr_{32}	cr_{33}	...	cr_{3r}
.	.	.	.	...	.
.	.	.	.	...	.
t_r	cr_{r1}	cr_{r2}	cr_{r3}	...	cr_{rr}

- DF outside the cluster td_i':

$$DF_{out,t_i} = \frac{\#documents\ containing\ term\ t_i\ and\ \notin td_i'}{\#documents\ \notin\ cluster td_i'} \tag{12}$$

- The difference for cluster td_i for a term t_i is computed as follows:

$$DIFF_{td_i',t_i} = |DF_{in,t_i} - DF_{out,t_i}| \tag{13}$$

- For each term, the DPM is calculated by the following equation

$$DPM(td_i', t_i) = \sum_{i=1}^{C} DIFF_{td_i',t_i} \tag{14}$$

v. Selecting high DPM terms:
After computing the DPM score for each term $t_i \in td_i'$, select top k terms as the important terms for cluster td_i' and add them to an array A.

Step 4 *Generating the training feature vector*:
Repeat step 3 for all the clusters and merge the important terms to an array A. Remove if any duplicate terms are there in A. Now A is the required reduced training feature vector.

4 Experimental Analysis

4.1 Experimental Setup

Four benchmark datasets are used for experimental work (DMOZ,[2] 20-Newsgroups,[3] Reuters[4] and Classic4[5]). 20-Newsgroups is a standard machine learning dataset and

---

[2]https://www.dmoz.org/.
[3]http://qwone.com/~jason/20Newsgroups/.
[4]http://www.daviddlewis.com/resources/testcollections/reuters21578/.
[5]http://www.dataminingresearch.com/index.php/2010/09/classic3-classic4-datasets/.

Table 3 Parameters used on different datasets

Dataset	Total features	Length of the feature vector	Hidden nodes
20-Newsgroups	20,422	1627	2000
DMOZ	24,320	1988	2400
Classic4	21,299	1597	1800
Reuters	17,582	1353	1600

it has 11,293 training and 7528 test documents classified into 20 classes divided into seven categories. For experimental purpose, three categories are taken into consideration (*alt.atheism*, *soc.religion.christian*, and *misc.forsale*), consisting of total 1663 training and 1107 test documents. The total number of features used is 20,422 and among them, 16,270 are used for training. DMOZ is one of the largest dictionaries on the web. It has 14 categories out of which three categories (*Arts, Homes, and Science*) of 5238 documents are used for experimental purpose. Among them, 3142 number of documents are used for training and rest are used for testing. The total number of features is 24320 out of which 19886 are considered for training. Classic4 is a popular dataset in text mining. It has 4257 training and 2838 test documents classified into four classes—*cacm, cran, cisi, and med*, having 3204, 1460, 1400, and 1033 documents, respectively. All the classes are considered in the evaluation. The total vocabulary contained in all documents is 21,299 and for training documents, it is 15,971. Reuters is a widely used text mining dataset. It has 5485 training and 2189 test documents classified into eight classes, where all class documents are considered for the evaluation. The total number of terms used is 17,582 and among them, 13,531 are used for training. The details of the parameters used on different datasets are shown in Table 3.

4.2 Parameters for Performance Measurement

$$Precision~(p) = \frac{(relevant_{documents}) \cap (retrieved_{documents})}{retrieved_{documents}}$$

$$Recall~(r) = \frac{(relevant_{documents}) \cap (retrieved_{documents})}{relevant_{documents}}$$

$$F\text{-}measure~(f) = 2 * \left(\frac{p * r}{p + r}\right)$$

$$Average_{precision} = \frac{\sum_{i=1}^{n} X_i}{\#documents} \quad (15)$$

$$Average_{recall} = \frac{\sum_{i=1}^{n} Y_i}{\#documents} \quad (16)$$

$$Average_{F\text{-}measure} = \frac{\sum_{i=1}^{n} Z_i}{\#documents} \quad (17)$$

where X_i, Y_i, and Z_i represent $(p_i * d_i)$, $(r_i * d_i)$, and $(f_i * d_i)$, respectively, of ith category. d_i is the number of test documents of the ith category. *#documents* represents the "total number of test documents".

4.3 Discussion

Tables 4 and 5 show the average F-measure of *CCDPM* compared to other feature selection techniques (bold result indicates the highest average F-measure obtained by a classifier using the respective feature selection technique) where the state-of-the-art classifier such as SVM (LinearSVC) and ELM are used for classification. Table 6 shows the average F-measure comparison of ELM with other established classifiers on *CCDPM* (bold result indicates the highest average F-measure obtained by a classifier on a dataset). From the results, it can be seen that the proposed *CCDPM* technique outperforms other feature selection techniques except for Reuters dataset on SVM where Chi-square succeeded others and for Classic4 dataset of ELM where IG has highest average F-measure compared to others. From the results, it is observed that the performance of ELM is better compared to other traditional classifiers.

The reason why ELM performs well compared to other conventional classifiers is due to its ability to map the features from lower to higher dimension space (by making the number of nodes in the hidden layer more than the length of the input feature vector) and hence makes the features linearly separable in the higher dimensional space. The data become more *linear separable* if a large number of nodes are present in the hidden layer of ELM [23]. Also, in ELM implementation, it has been noticed that ELM can give excellent performance for large values of L, and it is not sensitive to the dimension of the hidden layer feature space (L) [24]. No random distortion of inputs in SVM takes place, and this is the hack ELM claims which makes it work so well. The optimization problem one solves in ELM has lesser constraints as compared to SVM. ELMs are optimized very much like SVMs (both are sequential minimal optimization), just that while SVMs are optimized constrained to a hyperplain, ELMs work on an entire hypercube.

Table 4 SVM (LinearSVC)

Technique	20-NG	DMOZ	Classic4	Reuters
Chi-square	87.83	76.72	80.38	**78.72**
BNS	83.41	78.43	77.63	75.32
IG	80.37	76.36	82.45	72.47
Gini	85.76	77.21	78.32	73.67
CCDPM	**89.74**	**80.37**	**83.85**	76.86

Table 5 ELM

Technique	20-NG	DMOZ	Classic4	Reuters
Chi-square	88.10	76.67	81.18	78.76
BNS	83.38	79.17	77.00	74.38
IG	81.45	78.32	**83.44**	71.11
Gini	86.67	77.37	80.17	72.18
CCDPM	**90.33**	**81.14**	82.98	**80.30**

Table 6 F-measure comparisons on CCDPM

Classifier	20-NG	DMOZ	Classic4	Reuters
SVM (LinearSVC)	89.74	80.37	**83.85**	76.86
SVM (Linear kernel)	75.73	72.33	71.28	68.32
ELM	**90.33**	**81.14**	82.98	**80.30**
Multinomial naive Bayes	81.83	75.76	77.82	73.35
Gaussian naive Bayes	74.28	74.23	73.45	70.32
Bernoulli naive Bayes	76.37	73.12	74.55	74.85
Decision trees	67.72	70.18	69.23	63.45
k-NN (k = 5)	66.35	67.42	69.67	61.17
Random forest (10 classifiers)	80.37	77.32	77.23	74.15
Extra trees (10 classifiers)	76.63	78.20	75.16	70.35
Gradient boosting (10 classifiers)	78.62	73.43	71.92	72.40

5   Conclusion

In this paper, a novel feature selection technique is proposed which combined the correlation and discriminative power measure to select the top features from a given corpus. By using the correlation measure, the highly correlated terms are removed from the corpus, and then, the terms are ranked based on their discriminative power measure score. Top k terms are selected to generate the training feature vector. Results on different benchmark datasets have shown that the proposed approach is better than the existing feature selection techniques and ELM is better than other established classifiers. This work can be further extended by combining different conventional classifiers with the feature space of ELM to classify the test documents which will improve the results further. Also, using multiple hidden layers can represent a hierarchical learning and hence boost the performance of the classifier.

References

1. Qi, X., Davison, B.D.: Web page classification: features and algorithms. ACM Comput. Surv. (CSUR) **41**(2), 12 (2009)
2. Aggarwal, C.C., Zhai, C.: A survey of text classification algorithms. In: Mining text Data, pp. 163–222. Springer (2012)
3. Qiu, X., Huang, X., Liu, Z., Zhou, J.: Hierarchical text classification with latent concepts. In: Proceedings of the 49th Annual Meeting of the Association for Computational Linguistics: Human Language Technologies, vol. 2, pp. 598–602. Association for Computational Linguistics (2011)
4. Sebastiani, F.: Machine learning in automated text categorization. ACM Comput. Surv. (CSUR) **34**(1), 1–47 (2002)
5. Roul, R.K., Asthana, S.R., Kumar, G.: Study on suitability and importance of multilayer extreme learning machine for classification of text data. Soft Comput. 1–18 (2016)
6. Roul, R.K., Nanda, A., Patel, V., Sahay, S.K.: Extreme learning machines in the field of text classification. In: 2015 16th IEEE/ACIS International Conference on Software Engineering, Artificial Intelligence, Networking and Parallel/Distributed Computing (SNPD), pp. 1–7. IEEE (2015)
7. Kalaivani, P., Shunmuganathan, K.: Feature selection based on genetic algorithm and hybrid model for sentiment polarity classification. Int. J. Data Min. Modell. Manag. **8**(4), 315–329 (2016)
8. Sarkar, A., Sahoo, G., Sahoo, U.: Feature selection in accident data: an analysis of its application in classification algorithms. Int. J. Data Anal. Tech. Strateg. **8**(2), 108–121 (2016)
9. Lee, J., Kim, D.-W.: Mutual information-based multi-label feature selection using interaction information. Expert Syst. Appl. **42**(4), 2013–2025 (2015)
10. Roul, R.K., Sahay, S.K.: K-means and wordnet based feature selection combined with extreme learning machines for text classification. In: International Conference on Distributed Computing and Internet Technology, pp. 103–112. Springer (2016)
11. Roul, R.K., Bhalla, A., Srivastava, A.: Commonality-rarity score computation: a novel feature selection technique using extended feature space of elm for text classification. In: Proceedings of the 8th Annual Meeting of the Forum on Information Retrieval Evaluation, pp. 37–41. ACM (2016)
12. Hall, M.A.: Correlation-based feature selection for machine learning. Ph.D. dissertation, The University of Waikato, 1999

13. Langley, P., Sage, S.: Induction of selective Bayesian classifiers. In: Proceedings of the Tenth International Conference on Uncertainty in Artificial Intelligence, pp. 399–406. Morgan Kaufmann Publishers Inc. (1994)
14. Kohavi, R., Sommerfield, D.: Feature subset selection using the wrapper method: overfitting and dynamic search space topology, pp. 192–197. In: KDD (1995)
15. Collins, R.T., Liu, Y., Leordeanu, M.: Online selection of discriminative tracking features. IEEE Trans. Pattern Anal. Mach. Intell. 27(10), 1631–1643 (2005)
16. Manning, C., Raghavan, P.: Introduction to Information Retrieval (2008)
17. Forman, G.: An extensive empirical study of feature selection metrics for text classification. J. Mach. Learn. Res. 3, 1289–1305 (2003)
18. Yang, Y., Pedersen, J.O.: A comparative study on feature selection in text categorization. In: ICML, vol. 97, pp. 412–420 (1997)
19. Shang, W., Huang, H., Zhu, H., Lin, Y., Qu, Y., Wang, Z.: A novel feature selection algorithm for text categorization. Expert Syst. Appl. 33(1), 1–5 (2007)
20. Huang, G.-B., Zhu, Q.-Y., Siew, C.-K.: Extreme learning machine: theory and applications. Neurocomputing 70(1), 489–501 (2006)
21. Rao, C.R., Mitra, S.K., et al.: Generalized Inverse of a Matrix and Its Applications, vol. 1, pp. 601–620 (1972)
22. Bezdek, J.C., Ehrlich, R., Full, W.: FCM: the fuzzy c-means clustering algorithm. Comput. Geosci. 10(2), 191–203 (1984)
23. Huang, G.-B., Chen, L.: Convex incremental extreme learning machine. Neurocomputing 70(16), 3056–3062 (2007)
24. Huang, G.-B., Zhou, H., Ding, X., Zhang, R.: Extreme learning machine for regression and multiclass classification. IEEE Trans. Syst. Man Cybern. Part B: Cybern. 42(2), 513–529 (2012)

Survey of Challenges in Sentiment Analysis

Sweety Singhal, Saurabh Maheshwari and Monalisa Meena

Abstract As a result of immense innovation, the measure of information is expanding day by day. This data is used by Internet users who also provide their feedback. They describe the product in detail and evaluate the sentiment of the product. It is essential to explore and analyse their reviews for a better decision-making. For this, we use sentiment analysis process. Sentiment analysis, which is also called as opinion mining is a natural language processing technique which extracts the information and identifies the users' views as positive or negative. This paper provides a review of sentiment analysis, its challenges, issues and also a survey of different approaches and techniques to handle those issues with respective advantages and disadvantages. A comparative study of different approaches has also been included.

Keywords Sentiment analysis · Sentiment analysis challenges
Machine learning · Lexicon-based approach

1 Introduction

Crucial information can be gathered by following a line of investigation over the human contemplation. As the popularity and accessibility of online facilities have enlarged, social networking sites and personal blogs have become the platform for

---

S. Singhal (✉) · S. Maheshwari · M. Meena
Department of Computer Science and Engineering, Govt. Women Engineering College, Ajmer, India
e-mail: sweetysinghal55@gmail.com

S. Maheshwari
e-mail: dr.msaurabh@gmail.com

M. Meena
e-mail: monalisa.meena@gmail.com

© Springer Nature Singapore Pte Ltd. 2018
P. K. Sa et al. (eds.), *Recent Findings in Intelligent Computing Techniques*,
Advances in Intelligent Systems and Computing 709,
https://doi.org/10.1007/978-981-10-8633-5_24

publishing the user's opinions. Sentiment analysis is a procedure to determine the nature of writing. It is also referred as opinion mining. It is a field of research which is used to differentiate the data that sentiment consists of as either positive, negative or neutral with the usage of natural language processing, statistics and machine learning method. Knowledge retrieval and computational linguistics are together combined to form the sentimental analysis. It classifies and extracts subjective information from the source materials.

Consider a review 'Current state of economy is a matter of interest for me'. In this example, there is some emotion regarding the current state of economy. This is an example of sentiment analysis. In sentiment analysis, there are three levels of classification, namely, document level, feature level and sentence level. The sentiment analysis will be more difficult as the level of classification goes higher in the pyramid because it will have a combination of things and combination of opinions [1]. In document-level classification, a whole document is categorized as positive, negative and objective. In sentence level, the individual sentences in the text are classified. This level of sentiment classification is more robust as compared to the previous one. In feature level, analysis of features in a document or sentence is called feature sentiment analysis. In this level of sentiment classification, opinion is determined from the already extracted features.

1.1 Sentiment Classification Techniques

In sentiment analysis, classification techniques can be divided into mainly three categories. They are machine learning, lexicon-based and hybrid approach. Machine learning approach is further divided into three categories, i.e. supervised, semi-supervised and unsupervised approaches. Lexicon-based approach is dependent on sentiment lexicon. It does not need any prior training data for data mining. It uses a predefined list of words associated with specific sentiments. It is divided into two categories, i.e. dictionary and corpus-based approach. It is also called as semantic orientation-based approach. Last one, i.e. hybrid approach, is the combination of both approaches [2].

2 Challenges in Sentiment Analysis

Sentiment analysis can be thought as content categorization task since it classifies text as positive, negative or objective. However, sentiment analysis is challenging as compared with conventional content characterization, although it has only three classes because of the following factors [3].

2.1 Sarcasm and Conditional Sentences

If a user is using a positive sentence about a product but his/her intentions are negative. In other words, we can say that the meaning is just opposite. These kinds of sentences fall in the category of sarcasm. It is very difficult for a system to identify the sarcastic sentence. Different users use sarcasm in a different manner. How this different intention can be understood by the system? It is a challenging task. For example, 'This movie is good enough to waste money' is a sarcastic sentence [3].

To understand the conditional sentences is also a typical task for a system. For example, 'The movie will be perfect if the story is interesting'. Several attempts have been made to detect sarcasm. Twitter has been a crucial source of training such models. Using 'hashtags', Twitter posts as a gold standard, for example, '#sarcasm'. Another major resource is product reviews from www.Amazon.com.

Lexical features are used to classify the sarcastic and non-sarcastic tweets based on the dictionary-based tags. A final classification was performed by SVM and logistic regression technique [4]. Machine learning approach with linguistic features was used to detect sarcasm in Dutch tweets which were hashtagged. A balanced winnow was implemented, and the result was favourable [5]. A rule-based approach with hashtag-based sentiment was also used for Twitter dataset. In this, they not only consider the range of sarcastic modifier to observe meaning of tweets but also sentiment's polarity [6]. Again, a rule-based approach is applied with lexical, implicit incongruity, explicit incongruity and pragmatic features for Twitter data and discussion forum [7].

2.2 Spam Detection

There are so many users who try to post the negative reviews to pamper other's reputation. In today's scenario, it is challenging task to identify the spam among the many reviews. So, it is essential to develop such a system which can identify spam and can remove it [3].

Three approaches, i.e. genre identification text, psycholinguistic deception detection and text categorization, were used for finding misleading opinion. A tool called LIWC—Linguistic Enquiry and Word Count—is used by second approach called Psycholinguistics Deception detection [8]. A hierarchical framework was used for spam detection using singleton review [9]. They find a correlation between volume of single review and rating because as the review increases, the rating decreases or increases dramatically. Different classifications of machine learning algorithms, i.e. support vector machine, decision tree, LogitBoost, Bagging, KNN and AdaBoost, were implemented on two real and large public datasets. By using these approaches, the best overall results were achieved by the bagging of decision tree in this scenario in which they did not combine features. Adaptive Boosting (AdaBoost) attained the highest performance with the combination of features vectors. The evaluated techniques showed the best result for balanced classes [10]. A shallow dependency parser technique is used to compute sentiment score. A relationship between spam reviews and sentiment score was given by them. Spam review detection was combined with sentiment analysis. Furthermore, by using the discriminative rules, the spam reviews can be also identified from the abnormal time window. The case study and the experiment showed the efficacy of these methods [11]. In contrast to earlier work, they noticed deceptive reviews do not express the emotions as strongly as the genuine reviews do. They used rule-based method for deceptive spam dataset. They used deep linguistic features to build a better deceptive spam detection model. The final result of this approach gave an improvement in performance by 1.1% [12].

2.3 Anaphora Resolution

During the sentiment analysis, pronouns are ignored by most of the researchers. It is difficult for a system to identify what a pronoun or noun refers to in the sentence. In many situations, pronouns also play an important role to know about the users' perception. For example, 'The movie is awesome. It contains many good actions as well as emotions'. In this example, the word 'it' referring to movie. We cannot refer 'good' to 'movie' without knowing the reference of 'it' [3]. The problem of source co-reference resolution was proposed by [13]. However, they used partially supervised clustering rather than using simply supervised learning algorithms. Supervised machine learning approach with two semantic features was used to improve the co-reference resolution accuracy [14].

2.4 Negation Handling

Negation handling in sentiment analysis plays an important role in altering the polarity of the associated adjective and hence the polarity of the text. Negation words include not, neither, nor, etc., for example, 'The movie is good' should be classified as positive. 'The movie is not good' should be classified as negative. This type of sentences can be handled by reversing the polarity of the adjective occurring after a negative word. But this solution fails to entertain the cases like 'No wonder the movie is good' and 'Not only the story was interesting, the songs were also entertaining'. Negation has not been tackled completely with the use of mathematical models and language processing techniques.

In French context of sentiment analysis [15], they differentiated different types of negative operators, negative quantifiers and lexical negations by using linguistic features. Tree kernel-based scope detection which uses the parse information which is syntactically structured. Added to this, a way of selecting attributes which are compatible for different PoS, as features have an efficiency which is imbalanced for classifying scope, which is affected by PoS was explored by them [16]. An automatic system was developed to detect negation and speculation cues by using machine learning approach. It is the first system which is trained and tested on the SFU Review corpus annotated with speculative and negative information. The results reported—92.37% in F1 and 89.64% for negation—are encouraging. In scope detection task, the results—F1, 84.07% in negation, 78.88% in speculation, G-mean, 90.42% for negation and 87.14% for speculation, and PCRS, 71.43% in speculation and 80.26% in negation, are very promising [17].

2.5 Word Sense Disambiguation

Word sense disambiguation is identifying which sense of a word is used in a sentence as the single word has multiple meanings. It is controlled by the sense of the word in that context, for example, 'Small'. If we relate small with television, it sounds negative sense. But if talk about a mobile phone, it can be positive. It depends on the user, what he likes or not. So it is difficult to determine this for a system.

Some researchers including [18–20] initiate by creating lexicon dictionaries where words are associated with the prior polarity out of context. The contextual polarity of a word present in a phrase may differ from the word's prior polarity because a word may appear in different senses. Additionally, it is difficult to define the prior polarity for several words such as long, short, think, deeply, entirely, small, feel, practically, etc. because they do not carry specific polarity by themselves. Linguistic features were used to determine the polarity of polar clause [21]. These linguistic features include modifications features, structure features, and sentence features. Instead of disambiguating the word sense, the effect of enhancers, negation and modifiers is determined using the word context. Speech pattern matching method was used to resolve disambiguation of words at the sentence level [22]. In order to determine the polarity of the sentence, parts of speech pattern are extracted and compared with WordNet glossaries in order to identify the appropriate sense in SentiWordNet. However, results achieved through parts of speech pattern matching are not satisfactory because a word used in the same parts of speech pattern may not have the same sense. In order to identify the disambiguate sense of the word, four tasks were proposed [23]. (A) Exact boundaries of the text are determined where opinion about a feature is articulated. (B) Context of word is identified in a sentence using an appropriate method. (C) Context matching mechanism is provided in order to obtain the polarity of the corresponding context from the lexicon. (D) Lexicon dictionary is built which not only contains the senses of words in a particular domain but also supports a context matching mechanism. The results show that these methods considerably improve the overall performance of feature level sentiment analysis. Table 1 explains various methods with its specific features, along with advantages and disadvantages for various challenges.

Table 1 Comparative study of different methods to handle various challenges

Method	Approach	Dataset	Advantage	Disadvantage	Accuracy
SARCASM					
Lexical and pragmatic features [4]	Supervised machine learning approach	Twitter	Compare technique against human performance	Low accuracy, lack of explicit context and features not provide sufficient information to differentiate accurately sarcastic sentences	57.41% by using SMO and LIWC
Linguistic features, unigrams, bigrams, trigrams [5]	Supervised machine learning approach	Dutch Twitter Trained with 78,000 Dutch tweets, test with 3.3 million Dutch tweets	Among the 3.3 million dataset, classifier identify 101 tweets correctly	Difficult to recognize those tweets which were not absolutely apparent with hashtag	AUC—76%
Extraction of common patterns, hashtag-based sentiment, unigrams [6]	Rule-based approach	Twitter corpus of 600 tweets containing hashtag	Deal with sarcastic sentences as well as range of sarcastic modifier on meaning of tweet and polarity of sentiment expressed	Not deal with sarcasm without hashtag	F1—97.25%
Lexical, pragmatic, patterns, explicit and implicit incongruity features [7]	Supervised machine learning approach	Three dataset 1. Twitter (5208 tweets) 2. Twitter (2278 tweets) 3. Discussion forum (1502 discussion forum reviews)	It uses linguistic relationship between context incongruity and sarcasm	In this, following errors occurred: 1. Incongruity due to numbers 2. Subjective polarity 3. No incongruity within text 4. Dataset granularity 5. Politeness	F—88.76/64

(continued)

Table 1 (continued)

Method	Approach	Dataset	Advantage	Disadvantage	Accuracy
SPAM					
POS, LIWC, unigram, bigram, trigram [8]	Genre identification, psycholinguistic deceptive detection, text categorization	Gold-standard deceptive opinion	LIWC with bigram achieved best accuracy	Not give accurate result	89.8% accuracy
Temporal pattern discovery [9]	Time series pattern discovery	Reviews (408,469)	Present a hierarchical framework for spam detection using singleton review	Only effective in detecting singleton review spam	Precision —61.11%
Content-based, link-based, transformed link-based features [10]	Machine learning and data mining techniques	Large public dataset (Web spam-UK)	Best results were achieved by bagging of decision tree within they did not combine features	Classification methods were failed to detect spam for balanced classes	Accuracy — 94.4 ± 0.7
Sentiment lexicon [11]	Shallow dependency parser, time series method	Real-life dataset	First paper who incorporate sentiment analysis into spam review detection	Low accuracy, ignores negation word in reviews	Sentiment score method accuracy —86.3%
Deep level linguistic features [12]	Shallow discourse parsing	Public golden standard dataset (1680 reviews)	Find deceptive reviews tend to accept milder affections than accurate reviews	Syntactic features improve recall but reduces precision	Accuracy —89.5%
Anaphora resolution					
Feature vector [13]	Partially supervised clustering algorithm	MPQA corpus (536 documents)	Used partially supervised clustering algorithm rather than supervised learning algorithm	Do not directly optimize for performance measure	F1— 67.1%

(continued)

Table 1 (continued)

Method	Approach	Dataset	Advantage	Disadvantage	Accuracy
Semantic features, opinion mining features, grammatical, lexical and other [14]	Supervised machine learning algorithm	Forum discussion from three domain-mobile phones, plasma, LCD, TV and cars	Deal with object and attribute co-reference resolution and also proposed new features	Not give accurate result	F—75%
Negation handling					
Linguistic features [15]	Semantic orientation-based approach	French dataset	Showed effect of modality and negation	For French language, so typical to deal this approach for English language	Multiple results
POS-based feature, syntactic-based features [17]	Machine learning approach	Simon Fraser University review corpus	Both negation and speculation information can be identify by their automatic system	Errors in cue detection phase: 1. False negative error 2. False positive error	92.37% in F1 and 89.64% for negation
Word sense disambiguation					
Linguistic features [21]	Coupling niche browsing techniques and affect analysis techniques	Google news	Their application produce score with training and provide score for entities in non-opinion-based text	A number of parameters are fixed	–
Lexicon dictionary [23]	Feature level SA method, heuristic-based feature scope identification method	3000 reviews	Improves overall performance of feature level SA	Not accurately determine the context of word in sentiment	Accuracy —93.2%

3 Conclusion and Future Scope

Sentiment analysis has been widely used in various applications including reviews classification, government policies and many real-time applications. We have studied various techniques used in sentiment analysis to handle the issues. In order to overcome the individual drawbacks of different features extraction and classification algorithms, they have been combined efficiently thereby enhancing the performance of sentiment classification. More research work is required in future to further improve the performance. Sarcasm and spam detection are two most challenging task in the sentiment analysis. Many researchers tried to deal with these but failed to achieve accuracy. The techniques used for sentiment analysis are increasing fast, though many problems in sentiment analysis still remain unresolved which can be targeted for more work in future.

References

1. Pang, B., Lee, L., Vaithyanathan, S.: Thumbs up? sentiment classification using machine learning techniques. In: Proceedings of the 2002 Empirical Methods in Natural Language Processing Conference, ACL (2002)
2. D'Andrea, A., Ferri, F., Grifoni, P., Guzzo, T.: Approaches, tools and applications for sentiment analysis implementation. Int. J. Comp. App. **125** (2015)
3. Cabral, L., Hortaçsu, A.: The dynamics of seller reputation: evidence from eBay (2006)
4. Ibanez, G., Roberto, Muresan, S., Wacholder, N.: Identifying sarcasm in twitter: a closer look. In: Proceedings of the 49th Annual Meeting of the Association for Computational Linguistics: Human Language Technologies, ACL (2011)
5. Liebrecht, C., Kunneman, F., Bosch, A.V.D.: The perfect solution for detection sarcasm in tweets# not. In: Proceedings of the 4th Workshop on Computational Approaches to Subjectivity, Sentiment and Social Media Analysis, ACL (2013)
6. Maynard, D., Greenwood, M.A.: Who cares about sarcastic tweets? investigating the impact of sarcasm on sentiment analysis, LREC (2014)
7. Joshi, A., Sharma, V., Bhattacharyya, P.: Harnessing context incongruity for sarcasm detection. In: Proceedings of the 53rd Annual Meeting of the Association for Computational Linguistics and the 7th International Joint Conference on Natural Language Processing, ACL (2015)
8. Ott, M., Choi, Y., Cardie, C., Hancock, J.T.: Finding deceptive opinion spam by any stretch of the imagination. In: Proceedings of the 49th Annual Meeting of the Association for Computational Linguistic: Human Language Technologies, ACL (2011)
9. Xie, S., Wang, G., Lin, S., S. Yu, P.: Review Spam Detection via Temporal Pattern Discovery. In: Proceedings of the 18th ACM SIGKDD International Conference on Knowledge Discovery and Data Mining, ACM (2012)
10. Silva, R.M., Almeida, T.A., Yamakami, A.: Machine learning methods for spamdexing detection. Int. J. Inf. Security Sci. (2013)
11. Peng, Q., Zhong, M.: Detecting spam review through sentiment analysis. J. Softw. (2014)
12. Chen, C., Zhao, H., Yang, Y.: Deceptive opinion spam detection using deep level linguistic feature. In: National CCF Conference on Natural Language Processing and Chinese Computing, Springer International Publishing (2015)

13. Stoyanov, V., Cardie, C.: Partially supervised co-reference resolution for opinion summarization through structured rule learning. In: Proceedings Conference on Empirical Methods in Natural Language Processing, ACL (2006)
14. Ding, X., Liu, B.: Resolving object and attribute coreference in opinion mining. In: Proceedings of the 23rd International Conference on Computational Linguistics, ACL (2010)
15. Benamara, F., Chardon, B., Mathieu, Y., Popescu, V., Asher, N.: How do negation and modality impact opinions? In: Proceedings of the Workshop on Extra Propositional Aspects of Meaning and Computational Linguistics, ACL (2012)
16. Zou, B., Zhou, G., Zhu, Q.: Tree kernel-based negation and speculation scope detection with structured syntactic parse features. In: Proceedings Conference on Empirical Methods in Natural Language Processing (2013)
17. Cruz, N.P., Taboada, M., Mitkov, R.: A machine-learning approach to negation and speculation detection for sentiment analysis. J. Ass. Inf. Sci. Technol. (2015)
18. Kim, S.M., Hovy, E.: Determining the sentiment of opinions. In: Proceedings of the 20th international conference on Computational Linguistics, ACL (2004)
19. Yu, H., Hatzivassiloglou, V.: Towards answering opinion questions: separating facts from opinion and identifying the polarity of opinion sentences. In: Proceedings Conference on Empirical Method in Natural Language Processing, ACL (2003)
20. Hu, M., Liu, B.: Mining and summarizing customer reviews. In: Proceedings of the 10th ACM SIGKDD International Conference on Knowledge Discovery and Data Mining, ACM (2004)
21. Grefenstette, G., Qu, Y., Shanahan, J.G., Evans, D.A.: Coupling niche browsers and affect analysis for an opinion mining application. In: RIAO (2004)
22. Khan, A., Baharudin, B., Khan, K.: Sentiment classification using sentence-level lexical based semantic orientation of online reviews. In: Trends in Applied Science Research (2011)
23. Farooq, U., Dhamala, T.P., Nongaillard, A., Ouzrou, Y., Qadir, M.A.: A word sense disambiguation method for feature level sentiment analysis. In: 9th International Conference on Software, Knowledge, Information Management and Applications (SKIMA), IEEE (2015)

Data Mining Techniques for Smart Mobility—A Survey

Apeksha Aggarwal and Durga Toshniwal

Abstract With the rapid increase in urbanization, there is essence for making urban areas smart. There are assorted fields of computer science liable for development of a city to make it smart and intelligent. Some of these fields include data mining, sensor networks, Internet of things, web of things, cloud computing techniques and machine learning. Smart city is an umbrella term that encompasses smart mobility, smart governance, smart planning, smart environment and others. Smart mobility is one of the crucial aspects of smart city addressing efficient movement of people and goods from one place to another. This work is an extensive survey of research works related to application of data mining techniques for smart mobility. A comparative study of major works done in the aforementioned field is outlined in this paper.

Keywords Data mining · Smart city · Urban computing · Smart mobility

1 Introduction

According to the report published by United Nations [1], more than half of the world's population lives in urban areas or cities, which was 54% in 2014, and is expected to rise to 66% by 2050. According to this report [1], Delhi is the second most populated city in the world after Tokyo with 25 million inhabitants and is expected to rise to 36 million by 2030. With the rapid progress in urbanization, developing countries are expected to see momentum in urbanization process than developed countries.

A. Aggarwal (✉) · D. Toshniwal
Department of Computer Science and Engineering, Indian Institute of Technology Roorkee,
Roorkee, India
e-mail: aaggarwal@cs.iitr.ac.in

D. Toshniwal
e-mail: durgafec@iitr.ac.in

© Springer Nature Singapore Pte Ltd. 2018
P. K. Sa et al. (eds.), *Recent Findings in Intelligent Computing Techniques*,
Advances in Intelligent Systems and Computing 709,
https://doi.org/10.1007/978-981-10-8633-5_25

239

Urban areas different from rural areas are characterized as human settlements where people have infrastructure facilities, commuting facilities, banking and finance facilities, generally huge population, etc. As the world continues to metropolitanize, there is essence of tackling problems related to urban cities like traffic, environmental degradation, inadequate transportation facilities, lack of proper supply of goods, inadequacy of high-tech equipment and infrastructure, improper waste management, improper land use, lack of health facilities and what not. Hence, crops up the concern for making an urban area a smart city.

Consequently, the smart city aims at decreasing such challenges faced by cities. There are several definitions for any city to be a smart city. One of them defines a smart city as the efficient use of modern techniques to save time and resources but not at the cost of environment. Alternative definition of smart city is the urban area with advanced infrastructure, communication and health facilities. A smart city should be technically advanced, environmentally safe, economically stable and should be able to provide ease of living to its citizens.

1.1 Smart Mobility

Smart city is an umbrella term that encompasses smart environment, smart mobility, smart energy, smart governance and so on, as shown in Fig. 1. The primary focus of this work is on smart mobility. Mobility is the movement of goods and people from one place to another. Smart mobility means efficient movement of goods and people to save time, energy and resources.

Fig. 1 Smart city applications

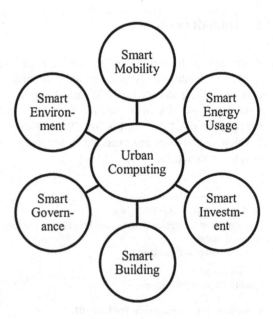

Movement of goods and people such that it reduces fuel wastage, allows smooth and faster commutation lies under smart mobility. Smart mobility deploys intelligent transportation techniques to ensure faster, safer and environment-friendly solution for movement of people and goods in urban space. There are different types of problems addressed in smart mobility such as finding long-term traffic patterns in the city, prediction of driver behaviour, saving time of drivers on roads, better and faster parking schemes, classification of transport, etc. A detailed discussion is provided in Sect. 2.

Besides smart mobility, there are other subdivisions of smart city. One is smart environment directed towards tackling environmental problems effectively. Major environmental problems including waste management, air pollution, noise pollution, water pollution, etc. are required to be handled with modern tools and technologies. Several surveys have shown more wastage than the use of electricity in homes, industries, workplaces, etc. Thus, there is a need for smart lighting system which automatically adjusts itself when not in use. Automatic lightning appliances are required to be installed in urban areas; this may reduce the energy consumption by adjusting lights themselves.

Another one is smart government which includes proper management of huge amount of money issued by government in various schemes so as to give people maximum advantage along with ease of use. To ensure a high level of transparency, smart government is the need of the hour. Furthermore important is involvement of citizens in their process of decision-making and implementation of policies. For fulfilment of smart governance, citizen's feedback, confidentiality and integrity of data obtained, transparency in policy-makers' decision, etc. are monitored.

Buildings have lots of energy consumption even when appliances are not in use. Buildings should be smart and automatically adjustable to the environment. For example, sensors can be deployed in a building to automatically switch off electrical appliances not in use. Smart investment and smart energy are furthermore classifications of smart city.

1.2 Use of Data Mining

Huge amounts of heterogeneous data are generated from various data sources every second in the city such as social networking data, GPS trajectories data, mobile phone data such as call records, nearby points of interests such as restaurants, banks, etc., card swiping data, weather data, stock market records, light metre readings, medical records including past patient historical records of pulse, breathing rate, etc., and so on. Consequently, this data can be utilized for varied applications. This may raise challenges for handling such data as well as efficient use of such data makes a city smart.

Considering smart mobility, there are different sources of data collection in a city. GPS trajectories of vehicles, tweets, card swiping data, etc. Posterior to acquisition of heterogeneous data generated from urban cities, data is then preprocessed in a form relevant for applying data mining methods. Data preprocessing includes removal of missing values, noisy data and other well-known preprocessing methods.

Subsequently, traditional data mining approaches like clustering, prediction, frequent pattern mining, regression, etc. are applied in order to extract knowledge from the data. For instance, finding efficient travel routes in terms of traffic by employing frequent pattern mining techniques. Another example includes clustering of trajectories to find out similar traffic patterns. Knowledge obtained is then visualized in presentable form, so that it is easily understandable. Finally, knowledge acquired from this methodological process is applied to make a city smarter in terms of mobility. Figure 2 shows the basic steps of data mining as knowledge extracting process which includes data mining from urban data and using such knowledge for smart city applications.

This paper is organized as follows. Section 1 introduces the paper about smart mobility along with the motivation for the work. Additionally, the role of data mining techniques towards smart mobility is given in brief. In Sect. 2, the extensive

Fig. 2 Data mining as a knowledge discovery process

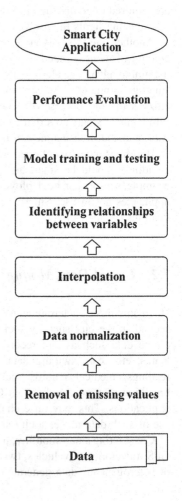

literature work done in the field of smart mobility using data mining techniques as well as other few traditional techniques is discussed. Section 3 concludes the paper with the future scope and some research challenges.

2 Smart Applications of Data Mining Techniques to Smart Mobility

In this section, manifold works on smart mobility are discussed. Primarily, following types of problems are identified from the literature:

- Route identification, i.e. identifying best routes from long-term traffic patterns in the city.
- Driver and passenger recommendation systems, i.e. recommending alternative routes to drivers and passengers moving in the city.
- Smart parking schemes, i.e. suggesting parking places in order to avoid last-minute chaos.
- Other problems include detecting transportation modes in the city, taxi ride-sharing systems, bike sharing, etc.

2.1 Route Identification

One of the oldest shortest route finding algorithms is Dijkstra algorithm. But in real-world scenarios, shortest route may not always be the fastest. Finding fastest routes depends upon the identification of traffic patterns in the city at different time intervals and various other factors. Hence, finding time-dependent shortest paths is the need of the hour, which is worked upon excessively in the literature. Time-dependent routes are those routes whose traversing time changes at different periods of time. For example, traffic time at a road segment is more during morning and evening hence on the same road segment traversing time is more in morning and evening, while less during midday hours. Several time-dependent shortest path algorithms [2–4] are given in early research works. These methods took advantages of various shortest path algorithms like Dijkstra algorithm to find the shortest route at a given time, which depends upon the traffic on that path at that time; hence, the fastest route should be the one with less traffic at that time. These methods have high computational complexity for calculation of best routes.

Later approaches utilized the field of Sensor networks to identify traffic patterns in the city. The researches of Cheung et al. [5] and Gonzalez et al. [4] were based on sensor networks in which sensors were deployed to monitor traffic jams at various places in the city. But there are several disadvantages of these methods. First, it is not feasible to deploy sensor at each and every nook and corner of the city. Second, deploying sensors is quite expensive, and also, it is not affordable to deploy in large volumes. Moreover, sensor networks have limited coverage area, and base stations also need to be installed along with them, which is infeasible.

With the availability of huge platforms on sharing social media data voluntarily, participatory sensing emerged as a very powerful tool to sense the data from social networks such as Instagram, Flicker, Twitter, etc. Participatory sensing is better technique over wireless sensor networks as they have a wider access and scope. Sensing data from all those websites or tools over Internet, accordingly allows the user to share its information voluntarily contributes to participatory sensing. Crowd sensing is another method for data collection, in which instead of installing the sensors over specific road segments to monitor traffic, GPS device installed in vehicles is used as mobile sensors. Yuan et al. [6] proposed T-Drive in which data is collected from GPS points of moving vehicles. T-Drive finds quick routes from traffic flows on the roads and historical GPS trajectories data utilizing a novel concept of landmark graphs. First, GPS trajectory dataset is used to generate a sequence of GPS points. A map matching algorithm [7] is then used to convert GPS points into road segments. Subsequently, landmark graphs are constructed from this sequence. Landmark graphs consist of landmark segments, most frequently used routes in the map, and landmark edges, calculated given the user's departure time based upon traffic conditions of the road at that time. Location variant [6] and time-dependent [6] features of a graph are considered in this method.

Each graph is defined for a set of edges and vertices with a few properties over edges. In case of road networks, location variant graph is a type of dynamic graph where traffic varies with respect to the location of the edge representing roads. For example, roads centrally located in a city are supposed to possess more traffic than roads not in a central location. Henceforth, traversing time of road two road segments of same length varies with respect to locations. Similar is the case with time-dependent graphs. Time-dependent graph depicts edges or road networks with different average traversing times at different time periods of the day on the same road segments.

For each landmark edge at different time partitions, a different distribution needs to be learned. VE clustering [6] method is used to separate similar pattern landmark edges into one cluster. VE clustering algorithm is a two-phase clustering algorithm. In first phase, landmark edges with similar travel times are clustered together. In second phase, for a single landmark edge, different travel times are considered and clusters depending upon travel times are formed. Henceforth, fastest routes are computed with the segments resulted from final clusters.

Landmark graph concept is newly introduced but there are some limitations of this graph which are overlooked in previous works. First, spatial features of these landmark graphs are not considered. For example, traffic of next real-time interval is dependent only on the traffic history trajectory on that road segment as well as real-time traffic of previous time interval, but the traffic dependency of other landmark segments on these segments are completely ignored. Second, the region where the landmark segment lies should be considered. For example, some landmark segments which are even full of traffic and do not lie in the middle of the city have less impact on navigating inside the city than the landmark segments which lie in the middle area of the city.

Later, time variant dynamic routing algorithms [3, 5] are applied to find out the practically the fastest driving routes given a user departure time. Factors like optimism index, i.e. driver's intelligence, are taken into account in this method for the first time, unlike previous methods which only considered GPS trajectories to mine the traffic patterns. This method outperformed previous traditional data mining approaches which used probabilistic models to predict best routes directed towards predicting practically fastest driving route.

Pan et al. [8] worked on two crowd sensing-based techniques: one is using GPS trajectories data and other using social networking data such as tweets. The main idea behind was the identification of anomalies based on driver's driving routines. Anomaly detection is one of the important technique of data mining which aims at separating anomalous data from normal data points. Anomalous data points are those which behave differently than other normal data points. Unlike few traditional works which were based on detecting anomalies in the traffic flow, caused by some events which include accidents, natural calamity, protest march or maybe some celebrations, this work identified anomalies based on driver's driving routines.

2.2 Recommendation Systems

Yuan et al. [9] proposed an extension of T-Drive [10] whereby along with recommendation of fastest path, future traffic conditions on a road segment is also predicted using hidden Markov model [11] along with prediction of best routes. In this work, the problem of sparse trajectories is addressed using GPS trajectory dataset based upon driver's intelligence which is calculated by his previous historical trends of driving. For example, even in the case when there is no traffic on the road, some drivers drive slowly with their own speed. This factor is learned by keeping a watch over historical data of that driver. Results showed that this method outperforms other models. Furthermore, Yuan et al. [9] also proposed a cloud-based system in which user sends his real-time queries giving its current location and destination. The cloud-based system calculates the fastest path using historical data as well as driver's intelligence, and this best route is suggested in reply to the user query.

Ge et al. [12] proposed recommender systems for taxi drivers, to provide better and fastest routes to them. These systems are not designed for passenger's advantage and basically target at providing more profit to taxi drivers and reduced fuel wastage. Such systems do not take into account traffic conditions in the city while predicting faster routes from one location to another.

In Yuan et al. [13], a recommender system for both taxi drivers and the passengers is proposed. Data are collected [13] for 12,000 taxi trajectories, continuously for 110 days for Beijing city in 2010. Parking candidates are then identified using bagging classifier [11] model. OPTICS clustering [11] technique is used to detect the parking places in the city. Bagging classifier performs the prediction of unknown class labels on consecutive portions of dataset and then performs accurate prediction by taking cumulative average of all these predictions. Bagging classifier helps avoiding overfitting of data points.

In addition to this, recommender system for taxi drivers and passengers is suggested. First is taxi recommender system for taxi drivers which suggested a good parking place where there is a high probability of finding a passenger, instead of cruising over roads which lead to high traffic [13] as well as fuel wastage. A good parking place is suggested out of those identified, on basis of four criteria, i.e. high probability of driver to get a passenger, short waiting time at the taxi stand, long distance of the next trip after finding a passenger and the short queuing time at the parking place. Second is passenger recommender system for passengers which are suggested to each passenger while he is searching for it based upon his current location. Passenger recommender system suggests the nearby parking places where there is probability of finding unoccupied taxi at the walking distance from the current location of the passenger and if there is no such place or availability of taxi, this system suggests nearby road segments which are having more possibility for finding a taxi with the minimum waiting time.

Data Sparseness problem is addressed by clustering similar road segments (segments with similar features) using a density-based clustering algorithm. OPTICS [11] a density-based clustering algorithm is used to detect same parking places. OPTICS is among the most popular data mining technique which gives good results in case of sparse data by clustering regions of similar density. OPTICS continues to add data points in its cluster until there are no points in the reachability of current cluster's neighbourhood. In this work, recommender system for both taxi drivers and passengers on real time was given for the first time.

2.3 Smart Parking

As the number of car traffic has increased in the city, 30% of traffic is caused by the driver's contention for appropriate places to park the vehicles. This leads to fuel wastage as well as loss of productive hours. To alleviate such problems, few of the researches have been done in this area. Parking guidance and Information systems have developed previously in which automatic detection of parking space is done by installing sensors at different parking locations in the city. While this approach [14] did not produce much effect after implementation in reduction of traffic congestion in the area. It allowed earlier booking of slots in the parking area. Later few works identified dynamic pricing policies for smart parking in order to generate better parking schemes.

Later another smart parking system was developed in Geng et al. [15], which aimed at maximizing the use of resource and minimizing the driver's cost function this depends upon driver's proximity and parking cost. This approach makes use of Mixed Integer Linear Programming Problem (MILP), which is calculated on every time a decision for a new vehicle is to be placed, such that resource should be allocated without any conflicts and driver should not be assigned the resource whose cost function is greater than the driver's current cost function. But the problem in this approach is that the driver is allowed to park for a limited period. Moreover, during rush hours, these approaches did not yield good results.

Work presented in Kotb et al. [16] is a combination of smart resource scheduling, dynamic resource allocation as well as dynamic pricing policy named iParker. Unlike the previous approaches where the driver is allowed to park for a limited time period, in iParker, the driver is allowed to reserve its parking space for any time in the future. Hence, there is a lot of scope in this area to make a smart parking system which minimized the user waiting time as well as increased the probability of successful parking.

2.4 Other Works

Other valuable contribution to reduce traffic and pollution emissions and a very popular method these days are taxi ride sharing or carpooling. In Shuo et al. [17], a real-time taxi ride-sharing service with efficient updating of route schedules and hence reduction in total average travel distance is introduced. Whenever user raised a query specifying his current location and destination, the taxi ride-sharing schedule is updated according to the space available in taxis which are currently traversing the same path and are nearby. The whole city is divided into a set of regions, and site is formed at the centre of each region. The distance between sites is used to approximate the travelling distance and time of the taxi so that it can identify that user requests can be served or not. This type of work [17] is proposed first time which reduced overall travelled distance with 40% efficient routes than the total distance travelled when each taxi is dedicated to set of people as in case of advance booking. Bike sharing systems were also suggested with the similar concept in later a few works.

Other works included traffic prediction and route estimation for public transport such as buses, use of transfer learning to estimate the traffic in cities where there is lack of infrastructure, use of smart card data to find the passenger routes in smart railway transportation systems [18].

Van et al. [18] proposed this work in the Netherlands by using real-life dataset of railway transportation systems of the Netherlands. Previous approaches took into consideration that passengers always follow the shortest paths. This work justified the results stating that passenger may not always follow the shortest path. For example, if a passenger is willing to go from source to destination and there is no direct commuting service available on the shortest route, then passenger may decide to go via a little longer route which provides direct commutation from source to destination based upon travelling comforts. Unlike previous approaches, in this method, data is collected from automated ticket collection system. Consequently, this approach generated better results with 90% accuracy from a four-step methodology to compute the route a passenger is willing to choose.

Shang et al. [19] utilized the GPS trajectories dataset of China, to infer the gas consumption in the city. Based on a number of vehicles travelling on a road and estimated travel time, the total gas consumption and pollution emissions on that road are attempted to be identified with the use of basic popular data mining methodology for classification, i.e. Bayesian networks. Bayesian networks [11] are the graphical representation of conditional dependencies between random variables, utilized to generate probability distributions of these random variables.

3   Conclusion and Future Work

Smart city is an upcoming concept, and a few of the cities around the world are recognized as smart cities. There are various aspects of smart cities which include smart mobility, smart governance, smart environment, smart energy and so on. In this work, our main focus is on smart mobility. Few smart mobility problems like finding traffic patterns, route identification and recommendation systems are discussed in detail in this work. In contrast to finding shortest and quick routes, dynamic routing problem is addressed. In real time, traffic on a particular route changes; hence, travel time of that route changes. So to monitor similar road segments with similar traffic patterns, clustering, one of the most important data mining techniques is used in these works. VE clustering algorithm is discussed in this work.

Further, recommendation systems for taxi drivers as well as passengers are suggested in related works so that instead of wasting useful time, drivers can move towards the roads where there are high probabilities of finding passengers and passengers can move towards the road where there is a high probability of finding drivers, respectively, thus saving precious time and fuel consumption of taxis. Such systems mainly target traffic conditions in the city in order to suggest best routes to drivers as well as passengers.

Smart parking is yet another category that lies under smart mobility, and there are a few researches in this field. Some of the research works in smart parking are also reviewed in this work. There is a lot of future scope for improvement of existing techniques in order to avoid rush at parking places and thus increasing the chances of vehicle to grab the parking space without wastage of time.

As far as smart mobility concept is concerned, more focus is done on researches in developed countries, in future and there is a lot of scope for applying these methods to developing countries. Additionally, a huge amount of heterogeneous data is generated in cities like commuting data, card swipe data, transactions data and so on. All of these data can be utilized towards smart mobility.

References

1. U. Nations: World Urbanization Prospects: The 2014 Revision, Highlights, United Nations (2014). https://esa.un.org/unpd/wup/Publications/Files/WUP2014-Highlights.pdf. Accessed 23 Nov 2016
2. Cooke, K.L., Halsey, E.: The shortest route through a network with time-dependent internodal transit times. J. Math. Anal. Appl. **14**(3), 493–498 (1996)
3. Dean, B.C.: Continuous-Time Dynamic Shortest Path Algorithms, Doctoral Dissertation, Massachusetts Institute of Technology (1999)
4. Gonzalez, H., Han, J., Li, X., Myslinska, M., Sondag, J.P.: Adaptive fastest path computation on a road network: a traffic mining approach. In: Proceedings of the 33rd International Conference on Very Large Data Bases, Vienna, Austria (2007)
5. Cheung, R., Sing-Yiu, C., Varaiya, P.P.: Traffic surveillance by wireless sensor network: final report, University of California, Berkeley, CA, USA (2007)
6. Yuan, J., Zheng, Y., Xie, X., Sun, G.: T-Drive: enhancing driving directions with taxi drivers' intelligence. IEEE Trans. Knowl. Data Eng. **25**(1), 220–233 (2010)
7. Yuan, J., Zheng, Y., Zhang, C., Xie, X., Sun, G.-Z.: An interactive-voting based map matching algorithm. In: Proceedings of the 2010 Eleventh International Conference on Mobile Data Management, Washington, DC, USA (2010)
8. Pan, B., Zheng, Y., Wilkie, D., Shahabi, C.: Crowd sensing of traffic anomalies based on human mobility and social media. In: Proceedings of the 21st ACM SIGSPATIAL International Conference on Advances in Geographic Information Systems, Orlando, Florida (2013)
9. Yuan, J., Zheng, Y., Xie, X., Sun, G.: Driving with knowledge from the physical world. In: Proceedings of the 17th ACM SIGKDD international conference on Knowledge discovery and data mining, NY, USA (2011)
10. Yuan, J., Zheng, Y., Xie, X., Sun, G.: T-Drive: enhancing driving directions with taxi drivers' intelligence. IEEE Trans. Knowl. Data Eng. **25**(2), 220–233 (2013)
11. Han, J., Kamber, M., Pei, J.: Data Mining: Concepts and Techniques, 3rd edn. Morgan Kaufmann Publishers, USA (2011)
12. Ge, Y., Xiong, H., Tuzhilin, A., Xiao, K., Ruteser, M., Pazzani, M.: An energy-efficient mobile recommender system. In: Proceedings of the 16th ACM SIGKDD international conference on Knowledge discovery and data mining, Washington, DC, USA (2010)
13. Yuan, N.J., Zheng, Y., Zhang, L., Xie, X.: T-Finder: recommender system for finding passengers and vacant taxis. IEEE Trans. Knowl. Data Eng. **25**(10), 2390–2404 (2013)
14. Vera-Gómez JA., Quesada-Arencibia A., García CR., Suárez Moreno R., Guerra Hernández F.: An intelligent parking management system for urban areas. Sensors. **16**(6), 931 (2016)
15. Geng, Y., Cassandras, C.G.: New "smart parking" system based on resource allocation and reservations. IEEE Trans. Intell. Transp. Syst. **14**(3), 1129–1139 (2013)
16. Kotb, A.O., Shen, Y.-C., Zhu, X., Huang, Y.: iParker—a new smart car-parking system based on dynamic resource allocation and pricing. IEEE Trans. Intell. Transp. Syst. **17**(9), 2637–2648 (2016)
17. Shuo, M., Zheng, Y., Wolfson, O.: T-share: a large-scale dynamic taxi ridesharing service. In: IEEE 29th International Conference on Data Engineering (ICDE), Brisbane, Australia (2013)
18. Van, E., Hurk, D., Kroon, L., Maróti, G., Vervest, P.: Deduction of passengers' route choices from smart card data. IEEE Trans. Intell. Transp. Syst. **16**(1), 430–440 (2015)
19. Shang, J., Zheng, Y., Tong, W., Chang, E., Yu, Y.: Inferring gas consumption and pollution emission of vehicles throughout a city. In: Proceedings of the 20th SIGKDD Conference on Knowledge Discovery and Data Mining, New York, USA (2014)

EDARC: Collaborative Frequent Pattern and Analytical Mining Tool for Exploration of Educational Information

Karan Sukhija, Naveen Aggarwal and Manish Jindal

Abstract Educational data mining is an emerging trend in the field of data mining concerned with developing methods to explore the distinctive data obtained from educational settings. It helps in the study of student's behavior and the learning environment. This paper demonstrates novel methodology in form of EDM analytical tool that comforts educational system using various kinds of expertises. The recent 6 years data from Punjab School Education Board, India was collected containing around 2 million students' records by structuring whole data into incremental datasets. Collaborative customization of various frequent pattern-mining algorithms was designed and implemented that explicitly handle the educational dataset. The online analytical processing feature is incorporated within the system to illustrate the graphs of result rule repository. This system also integrates with visual analytics that empower better exploration of educational system.

Keywords Education data mining · Collaborative frequent pattern mining Visual analytics · Dataset generation · OLAP

1 Introduction

The EDM process takes input from various educational settings and transforms into valuable facts that could possibly have a massive influence on educational research and training [1]. The education scenario in developing countries has changed

K. Sukhija (✉)
DCSA, Panjab University, Chandigarh, India
e-mail: rs.karansukhija@gmail.com

N. Aggarwal
UIET, Panjab University, Chandigarh, India

M. Jindal
RC, Muktsar, Panjab University, Chandigarh, India

© Springer Nature Singapore Pte Ltd. 2018
P. K. Sa et al. (eds.), *Recent Findings in Intelligent Computing Techniques*,
Advances in Intelligent Systems and Computing 709,
https://doi.org/10.1007/978-981-10-8633-5_26

251

rapidly in terms of size and diversity over last few decades. Nowadays, expansion of education system has increased the outreach and availability, but the quality of the system in terms of success ratio and retention rate is still unsatisfactory. The education survey done by Ministry of Human Resource Development, India in 2011 emphasized that the dropout rate is alarmingly in schools education. After the primary level, the dropout rate is 28.9%, which increases to 42.4% after middle school and further increases to 52.8% after high school. Even though significant work has been done in this field, most of the popular studies concentrated only on simplistic [2] metric analysis and led toward solitary institutions and on single course. Many issues need to be identified and addressed in order to improve the success rate of school education system. Figure 1 portraits the components of educational data mining system [3].

Therefore, we introduce a hybrid methodology in the form of EDM analytical tool to analyze learner behavior and find out the multi-hidden association among attributes [4]. Student data of six academic sessions' have been collected that contains around 2 million students records with various personal and academic details of the school students of different courses. The improvisation of mining algorithms has been implemented to perform the constraint-based multidimensional frequent pattern mining for association rule-based classification system on educational dataset. In addition, online analytical processing system (OLAP) feature has been incorporated within methodology to identify the interesting patterns those emerge from the mined data. This EDM analytical tool facilitates to expand the eminence of education by analyzing the student's performance [5] and determine those elements that affect the academic result.

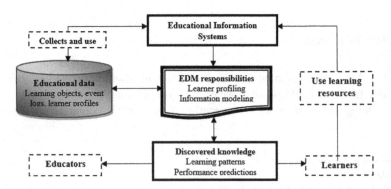

Fig. 1 Components of educational data mining in a nutshell

2   Background Work

Frequent pattern mining is one of the most established methods to discover hidden knowledge and to describe a close correlation among attributes. Table 1 describes some prominent work done in this area with different mining tools and method corresponding to diverse educational dataset.

Even though significant work has been done in this field, most of the popular studies concentrated only on simplistic metric analysis and led toward solitary institutions and on single course.

Table 1 Educational data mining research trends

Tools	Methodology	Database	References
WUM	Pattern-mining techniques to extract the useful patterns to evaluate online courses	Web Log of Technical University, British Columbia	Zaiane and Luo [9]
Multistar	Association and classification technique leads to knowledge discovery that helps in the assessment of distance learning	Students enrolled for distant learning courses	Silva et al. [10]
SPSS data analyzer	Log analysis techniques assist in identification of student learning behavior	Wire website interactive dataset	Sheard et al. [11]
EPRules	Discover the prediction rules to provide feedback for courseware authors	Logs of web-based education system	Romero et al. [12]
TADA-ED	Association and pattern-mining techniques assist in discerning relevant patterns	Student's online exercises	Merceron and Yacef [13]
TRAC system	Generalized sequential pattern algorithms improve the working and monitoring of team operation	Team data of software development	Kay J. et al. [14]
MINEL	Clustering techniques used to analyze the navigational behavior of the learner	Web log data	Bellaachia et al. [15]
OLAP, DELPHI	Multidimensional model is applied for construction and evaluation of online scenarios	Relationship data	Mansmann and Marc [16]
Excalibur	Novel method for learning classifier from both study-related and linked data	Social dependencies data gathered from, e-mail, discussion, boards conversation	Bayer et al. [17]
E-learning data analysis	Prediction rules applied to analyze the learner behavior in learning management systems	Student data of LMS	Psaromiligkos et al. [18]
WEKA, SQL server	Cross validation is compared with all the results	Real data of 670 middle school students	C. Marquez et al. [19]

3   System Architecture

We adopted the multilayer architecture having three layers, namely, presentation layer, application layer, and data layer. Figure 2 shows the integrated multilayer architecture built for the education system.

1. The presentation layer is the uppermost level of EDM tool. It is responsible for handling the visualization of EDM analytical tool. Visual analytics is used to view the data statistics. The administrators and management can easily use this menu-based GUI system to build the strategy for educational framework.
2. The middle-level layer of this architecture is application layer. It covers the functionality of this tool and sets up the constraints or rules for data processing. Three constituents of this layer are extract transform and load, frequent pattern-mining algorithms, and expert rule repository.
3. The last layer in this architecture is the data layer, constructed using 6 years (2011–2016) student dataset of matriculation and senior secondary courses conducted by Punjab School Education Board, India. The dataset contains around 2 million student records regarding examination, personal, and infrastructure details.

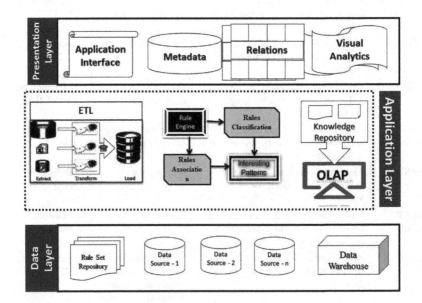

Fig. 2 Multilayer architecture of EDM analytical tool

4 System Methodology and Implementation

This section elaborates the methodology and implementation of the proposed research work for education setting. The building process is carried out in a number of phases. The initial work started with acquisition of relevant data from different federated educational sources, setting up of working attributes [6], and the transformation of data into requisite format. The next phase is data extraction; here generated dataset is loaded into EDM analytical tool, which is further used as input to mining algorithms. Subsequent phase is the ARCS; here hybridization of constraint-based frequent pattern mining is executed that specifically handle the educational dataset [4]. The last but not the least one is OLAP; online analytical processing is performed on result rule set to illustrate the graphical view of mined dependencies. The visual analytics [7] feature also been incorporated to graphical visualizing the course outcomes for better understanding of educational scenario.

4.1 Data Extraction

Exploration and interactive visualization of multivariate data remains a challenge. Several analytical tools [5] assist the analytical progression using static visualizations and support animations. The API of various standalone applications facilitate analysts to cartel their functionality with conformist data mining. The integration of visual analytics and JFreeChart library is incorporated in EDM analytical tool to

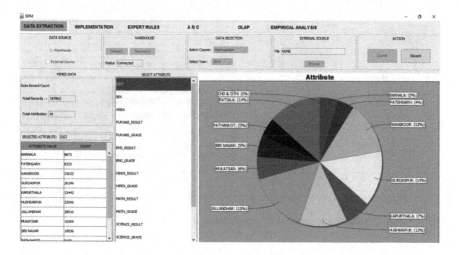

Fig. 3 Data extraction and loading

better exploration of the educational dataset. Figure 3 demonstrates multidimensional data extraction process incorporated in our EDM analytical tool to demonstrate the metadata of different attributes.

4.2 Collaborative Frequent Pattern Mining

This section describes the hybridization of mining algorithms for better harnessing of the features of different approaches to perform the constraint-based multidimensional frequent pattern mining on association rules-based classification system. This improvisation of EDM analytical tool [8] allows the educators and management to find out the multi-hidden dependencies among the attributes of education [3] framework. This module also enables the involvement of domain expert with customized frequent pattern-mining process. The rule selection module introduced here to accept the rules returned by association algorithm along with those rules that are fed as input by domain expert. The combination of both these rules is followed in association rule-based classification operation. The classification of these combined rule sets is based on desired target attribute. The result rule repository is used to illustrate the classification tree. Figure 4 demonstrates the result window of constraint-based association rules classification system.

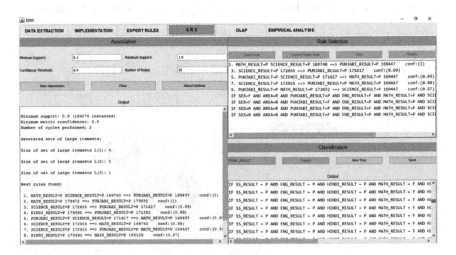

Fig. 4 Constraint-based association rules classification system

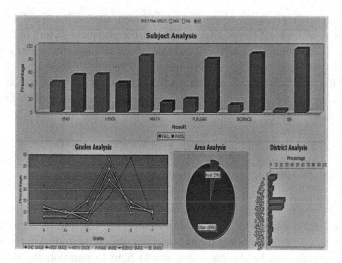

Fig. 5 Online analytical process window

4.3 OLAP

The result set of collaborative association rule-based classification is input to the online analytical process (OLAP) to illustrate the hidden picture of input dataset. The various types of charts are incorporated within this module to determine the interesting patterns those emerge from mined data. The entire graphs and charts returned by OLAP are used to analyze the behavior and relationship among attributes. Figure 5 demonstrates the online analytical process of this EDM analytical tool.

5 Conclusion

This research paper demonstrates novel methodology in form of EDM analytical tool that offers the benefit using school education dataset. The improvisation of collaborative frequent pattern-mining algorithms may weigh the academic management system to find out the multi-hidden dependency among different attributes. The new feature introduced to accept the rules set that has been input by domain expert also heighten the functionality of mining process. The combination of both these rules is followed in association rule-based classification operation to discover the hidden relationship among association rules. The final association rule-based classification result rule set has been input to the online analytical process (OLAP) to illustrate the hidden picture of the educational dataset. The outcome of the proposed research predicted to be a better-formed approach toward EDM. The collaboration with visual analytics demonstrates the graphical picture of discovered pattern sets. The solutions proposed will contribute to shape the system in a way that teachers, students, and the administration will be able to get the best out of the system. Mutual dependencies among various entities will be identified and the administration will be able to get a better insight into the causes of high failure rate of student academics.

6 Future Scope

As literature made known that major analysis technique functional to diverse datasets and did not consider the complex factors that may contribute to the more interesting outcomes. Hereafter, subject-oriented multidimensional framework for education domain can be designed that would help to partition the dataset into different classes based on hidden relationships among entities and their attributes. The proposed methodology may effectively engage to bear out the optional hypothesis for school education system. In addition, for validation of result rule set, the empirical analysis of multidimensional entities will be executed to build the confidence in EDM. The novel experiment will be accomplished to generalize the outcomes that will help users to apply them to their respective educational domain.

References

1. Sukhija, K., Jindal, M., Aggarwal, N.: Educational data mining towards knowledge engineering: a review state. Int. J. Manag. Educ. **10**(1), 65–76 (2016)
2. Chandra, E., Nandhini, K.: Predicting student performance using classification techniques. In: Proceedings of SPIT-IEEE Colloquium and International Conference, pp. 83–87, Mumbai, India (2005)
3. García, E., Romero, C., Ventura, S., De Castro, C.: A collaborative educational association rule mining tool. Internet High. Educ. **14**(2), 77–88 (2011)
4. Sukhija, K., Jindal, M., Aggarwal, N.: The recent state of educational data mining: a survey and future visions. In: IEEE 3rd International Conference on MOOCs, Innovation and Technology in Education (MITE), pp. 354–359, Amritsar, India (2015)
5. Jiménez-Gómez, M.Á., Luna, J.M., Romero, C., Ventura, S.: Discovering clues to avoid middle school failure at early stages. In: Proceedings of the 5th International Conference on Learning Analytics And Knowledge, pp. 300–304. Springer, New York (2015)
6. Dara, R., Satyanarayana, C., Govardhan, A.: A novel approach for data cleaning by selecting the optimal data to fill the missing values for maintaining reliable data warehouse. Int. J. Modern Educ. Comput. Sci. **8**(5), 64–70 (2016)
7. Geryk, J., Popelinsky, L.: Analysis of student retention and drop-out using visual analytics. In: Proceedings of the 7th International Conference on Educational Data Mining, pp. 331–332, London, UK (2014)

8. Kumar, A.S.: Edifice an educational framework using educational data mining and visual analytics. Int. J. Educ. Manag. Eng. **2**, 24–30 (2016)

9. Zaiane, O.R., Luo, J.: Towards evaluating learners' behaviour in a web-based distance learning environment. In: Proceedings of IEEE International Conference on Advanced Learning Technologies, pp. 357–360 (2001)

10. Silva, D.R.: MTP: using data warehouse and data mining resources for ongoing assessment of distance learning. In: Proceedings of IEEE International Conference on Advanced Learning Technologies, pp. 40–45, Kazan, Tatarstan, Russia (2002)

11. Sheard, J., Ceddia, J., Hurst, J., Tuovinen, J.: Inferring student learning behaviour from website interactions: a usage analysis. Educ. Inf. Technol. **8**(3), 245–266 (2003)

12. Romero, C., Ventura, S., De Bra, P.: Knowledge discovery with genetic programming for providing feedback to courseware authors. User Model. User Adap. Interact. **14**(5), 425–464 (2004)

13. Merceron, A., Yacef, K.: Educational data mining: a case study. AIED, 467–474 (2005)

14. Kay, J., Maisonneuve, N., Yacef, K., Zaïane, O.: Mining patterns of events in students' teamwork data. In: Proceedings of the Workshop on Educational Data Mining at the 8th International Conference on Intelligent Tutoring Systems, Jhongli, pp. 45–52, Taiwan (2006)

15. Bellaachia, A., Vommina, E., Berrada, B.: Minel: a framework for mining e-learning logs. In: Proceedings of the 5th IASTED International Conference on Web-based Education, pp. 259–263, Puerto Vallarta, Mexico (2006)

16. Mansmann, S., Scholl, M.H.: Exploring OLAP aggregates with hierarchical visualization techniques. In: Proceedings of the 2007 ACM Symposium on Applied Computing, pp. 1067–1073, Center Seoul, Korea (2007)

17. Bayer, J., Bydzovská, H., Géryk, J., Obšıvac, T., Popelınský, L.: Improving the classification of study related data through social network analysis. In: Proceedings of 7th Doctoral Workshop on Mathematical and Engineering Methods in Computer Science, pp. 3–10, Lednice, Czech Republic (2011)

18. Psaromiligkos, Y., Orfanidou, M., Kytagias, C., Zafiri, E.: Mining log data for the analysis of learners' behaviour in web-based learning management systems. Oper. Res. **11**(2), 187–200 (2011)

19. Marquez-Vera, C., Romero, C., Ventura, S.: Predicting school failure using data mining. In: Proceedings of the 4th International Conference on Educational Data Mining, pp. 271–276, Eindhoven, Netherlands (2011)

Rule-Based Method for Automatic Medical Concept Extraction from Unstructured Clinical Text

Ruchi Sahu

Abstract Medical concept extraction was the part of i2b2 challenge 2010 in which three concepts like problem, treatment, and test were targeted. This paper presents a rule-based method for automatic concept extraction from clinical notes. The method is compound of two modules that are text preprocessing and automatic rules creation. Rules creation module generates rules to recognize single and composite words for concept identification and mapping these concepts with their semantic types using medical dictionary UMLS. The method is applied to two different training datasets by Beth Medical Center with 73 annotated clinical notes and by Partners Healthcare with 97 annotated clinical notes, and then evaluated its performance using a test dataset with 256 annotated notes. The method achieved an average precision of 70% and average recall of 60%.

Keywords Unified medical language system · Clinical notes · Medical concept extraction · Semantic type

1 Introduction

Medical concept extraction is divided into two sequential subtasks: first one is identification of medical entities, and other is classification of the semantic category for each detected medical entity. I2b2 had organized an NLP challenge for clinical data in 2010. Extracting clinical concepts from natural language text, to include medical problems, tests, and treatments was one of the three tasks of that challenge [1]. In this paper, a rule-based method is proposed for the automatic medical concept extraction. The method contains two modules text preprocessing and rule template creation. Rule template creation module generates some rules for single and composite word identification and rules for concept mapping with their

R. Sahu (✉)
Shri Ramdeobaba College of Engineering and Management, Nagpur 440013,
Maharashtra, India
e-mail: sahur1@rknec.edu

© Springer Nature Singapore Pte Ltd. 2018
P. K. Sa et al. (eds.), *Recent Findings in Intelligent Computing Techniques*,
Advances in Intelligent Systems and Computing 709,
https://doi.org/10.1007/978-981-10-8633-5_27

semantic types using UMLS. Performance of the method is evaluated using precision and recall and comparison with MetaMap results.

2 Background

Many NLP challenges such as the i2b2 Challenge Shared Tasks [2] and the ShARe/ CLEF eHealth Shared Task [3, 4] had focused on medical concept extraction. In numerous previous works, rule-based approaches were used for natural language processing research for clinical notes. MetaMap was developed to recognize Metathesaurus concepts from biomedical texts by utilizing the UMLS [5]. Experimentation of MetaMap 2013v2 on i2b2 2010 clinical data with the 2013AB NLM relaxed database is performed in [6] and gives low score of precision (47.3%) and recall (36%). Other than rule-based, some machine learning approaches, ensemble-based approach [7], and hybrid approaches [8] have been used for concept extraction. We still identified some issues related to classification and entity boundary identification which has some scope to improve recall of the system. For text preprocessing, many natural language tools have been used like openNLP, tree tagger, Stanford parser [9], Lingpipe, Splitta, SPECIALIST, c-TAKES, and Stanford CoreNLP. Evaluation of sentence boundary detection using these tools is performed in [10]. For semantic types mapping, many systems have used UMLS [11]. Unsupervised biomedical named entity recognition is performed on GENIA corpus and i2b2 2010 dataset [12]. Various issues are still present in the identification and classification of clinical concepts because of unstructured nature of clinical notes. Common challenges like boundary identification of single and multi-adjacent words for designing and developing clinical decision support systems had focused in [13].

3 Proposed Method and Dataset

The proposed method has used clinical records provided by I2b2 National Center in 2010 NLP challenge. The dataset consisted of discharge summaries from Partners Healthcare and Beth Israel Deaconess Medical Center. All these records had been manually annotated for three types of concepts (medical problems, tests, and treatments), according to guidelines provided by the i2b2/VA challenge organizers [2]. Gold dataset of Beth Center contains 73 annotated notes, Partners Healthcare contains 97 annotated notes, and for system evaluation, they have provided test dataset which contains 256 annotated notes. For the experiment, the method has used both training and test annotated notes and for evaluation, gold dataset is used. The proposed rule-based method is divided into two sub-modules: text preprocessing and rules template creation.

3.1   Text Preprocessing

I2b2 clinical notes require text preprocessing because of theirs unstructured and semi-structured nature of the text. These notes contain some sections such as discharge date, admission date, allergies, history of present illness, past medical history, etc. Every section contains some information related to every patient with some special characters, colons, semicolons, punctuations, hyphens, etc. In this method, Natural Language Toolkit (NLTK) is used for line tokenization, word tokenization, and POS tagging. Special characters identification is performed by some regular expressions, which is used for word tokenization. After text preprocessing, numerous single words with their POS tags have been identified, which is applied as input for concepts generation.

3.2   Rules Template Creation

After getting words with POS tags, concepts have been identified. Concepts can be single word or composite word. For concepts identification, some rule templates are created, which are some common patterns and takes different values for different conditions. Rule template creation is divided into two subparts: (1) rules for composite words identification, and (2) then rules for map these words as concepts with UMLS using their semantic types. For multiword or composite word identification, some features like word, previous word, next word, previous word POS, and next word POS are used. For POS feature, same rules as word feature are used for rule template generation, only few POS tag combinations such as noun, pronoun, adjective, and determiner have considered for concept extraction. Rules using word feature are defined below: Suppose w = "single word" and cw = "composite word":

Rule 1: if w is middle word, then $y2 = w$, $cw = y1 + y2 + y3$, where $y1$ = previous one word and $y3$ = next one word. **Rule 2**: if w is middle word, then $y3 = w$, $cw = y1 + y2 + y3 + y4 + y5$, where $y1$ = previous two words from $y3$, $y2$ = previous one word, $y4$ = next one word, $y5$ = next two words from $y3$. **Rule 3**: if w is the first word, then $y1 = w$, $cw = y1 + y2 + y3$, where $y2$ = next one word and $y3$ = next two words from $y1$. **Rule 4**: if w is the last word, then $y3 = w$, $cw = y1 + y2 + y3$, where $y2$ = previous one word and $y1$ = previous two words from $y3$. **Rule 5**: if w is the first word, then $y1 = w$, $cw = y1 + y2$, where $y2$ = next one word. **Rule 6**: if w is the last word, then $y2 = w$, $cw = y1 + y2$, where $y1$ = previous one word.

Rule template for concept mapping with UMLS

Some rule templates have defined below in which Semantic type is one attribute of medical information which is defined in UMLS Metathesaurus database. Table 1 show some categories of semantic type which is used for three medical concepts.

Table 1 Semantic type categories of medical concepts

Medical concept	Semantic types
Problem	Disease or syndrome, sign or symptom, finding, pathologic function
Test	Tissue, cell, laboratory procedure, laboratory or test result, clinical attribute
Treatment	Antibiotic, organic chemical, therapeutic or preventive procedure, pharmacologic substance, diagnostic procedure

Rule 7: (Semantic type = x1) ∨ (Semantic type = x2) ∨ (Semantic type = x3) ∨ (Semantic type = x4) ∨ (Semantic type = x5) → Class = X, where X = 1 to n are semantic types corresponding to their concept category which is used for classification such as *if* x1 = "Disease or Syndrome", x2 = "Sign or Symptom", x3 = "Finding", x4 = "Pathologic Function", then X = Problem; *if* x1 = "Tissue", x2 = "Cell", x3 = "Laboratory Procedure", x4 = "Laboratory or Test Result", x5 = "Clinical Attribute" then X = Test; and *if* x1 = "Antibiotic", x2 = "Organic Chemical", x3 = "Therapeutic or Preventive Procedure", x4 = "Pharmacologic substance", x5 = "Diagnostic Procedure", then X = Treatment. If word is matched with any of semantic type of categories (given in Table 1), then it can be correctly mapped with appropriate concept class. In the proposed method, exact matching is performed.

4 Results and Discussions

The proposed method has performed experiments on I2b2 2010 clinical notes. It contains 73 annotated clinical notes provided by Beth Medical Center and 97 annotated clinical notes provided by Partners Healthcare and test dataset with 256 annotated notes. Rules created for these clinical notes and evaluated using gold dataset. The system has achieved an average precision of 70% and average recall of 60%, average of all concepts problem, test, and treatment. The system has performed better than MetaMap 2013v2. MetaMap gave a low score of precision (47.3%) and recall (36%). Figure 1a shows a comparison of performance of proposed method with MetaMap 2013v2 for all concepts based on precision and recall. Performance is measured for every concept individually also and compare Beth and Partners results concept wise (see Fig. 1b). The precision of Beth data for problem and treatment concept is more than Partners data, but for the test is equal. Recall of Partners data is more than Beth data for every concept.

After error analysis, it has been found that recall still has some scope for improvement. Few concepts have been missed because of incorrect boundary identification of composite words. Boundary identification issue is more observed in problem concept because gold standard contains some composite words of problem like "burst of atrial fibrillation" and found in treatment concept also like

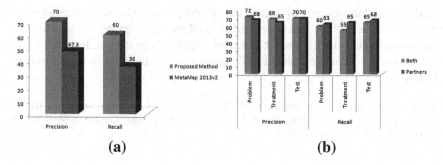

Fig. 1 Comparison of performance of proposed method based on precision and recall **a** with MetaMap 2013v2 for all concepts **b** on Beth data and Partners data for problem, treatment, and test concepts

"saphaneous vein graft -> posterior descending artery". The method has performed strict matching but not used partial matching; large composite words are not correctly or matched in UMLS database as medical concept. This issue can be resolved using relaxed or partial matching. Rules for composite words creation are designed in such a way which considers word features for maximum five words. Some concepts are combination of 4–5 or more words such "mild postoperative widening of the cardiomediastinal silhouette", which are not identified in the proposed method. Text preprocessing included some regular expressions for special character identification, which are used as word splitter, but few words identified in gold standard which contains these characters in between composite words such as "severe 3 vessel disease, "heel/shin", "leg, emg &apos", etc., which are incorrectly recognized as concept using proposed method. POS feature has also used for composite word identification; few combinations of POS patterns have been defined in rules like "NN, NNP, NST", "DT, NNP", "DT, JJ, NNP". These errors can be resolved in future work by designing some other rules or using hybrid approaches.

5 Conclusions

In the proposed rule-based method, medical concepts have been recognized using rules template generation for multiword identification and concept mapping with UMLS. It has been found that the performance of the system is better than Meta-Map 2013v2 in terms of precision and recall. Still recall can be improved by more accurate multiword boundary identification; for this, rules with more features like stemming, prefix, and suffix can be added. Rules are generalized not domain-dependent and regardless of the semantics of the sentences. In future work, this method can be applied to other corpuses for entity extraction with different regular expressions for text preprocessing, it will give better results. The method has not used any machine learning approach, that is why performance is not dependent on the size of dataset. Rules can be applied to small and large dataset in a similar way, but system processing time will be increased for the large dataset. In future work, this method can be implemented in a distributed environment for fast processing of large dataset.

Acknowledgements I would like to thank the 2010 i2b2/VA challenge organizers for the development of training and test corpora. I also thank U.S. National Library of Medicine for providing UMLS for the research work.

References

1. Jiang, M., Chen, Y., Liu, M., Rosenbloom, S.T., Mani, S., Denny, J.C., Xu, H.: A study of machine-learning-based approaches to extract clinical entities and their assertions from discharge summaries. J. Am. Med. Inform. Assoc. JAMIA **18**, 601–606 (2011)
2. Uzuner, Ö., South, B.R., Shen, S., DuVall, S.L.: 2010 i2b2/VA challenge on concepts, assertions, and relations in clinical text. J. Am. Med. Inform. Assoc. JAMIA **18**, 552–556 (2011)
3. Suominen, H., Salanterä, S., Velupillai, S., Chapman, W.W., Savova, G., Elhadad, N., Pradhan, S., South, B.R., Mowery, D.L., Jones, G.J.F., Leveling, J., Kelly, L., Goeuriot, L., Martinez, D., Zuccon, G.: Overview of the ShARe/CLEF eHealth Evaluation Lab 2013. In: Forner, P., Müller, H., Paredes, R., Rosso, P., Stein, B. (eds.) Information Access Evaluation. Multilinguality, Multimodality, and Visualization. Proceedings of 4th International Conference of the CLEF Initiative, CLEF 2013, Valencia, Spain, pp. 212–231. Springer Berlin Heidelberg, Berlin, Heidelberg, 23–26 Sept 2013
4. Kelly, L., Goeuriot, L., Suominen, H., Schreck, T., Leroy, G., Mowery, D.L., Velupillai, S., Chapman, W.W., Martinez, D., Zuccon, G., Palotti, J.: Overview of the ShARe/CLEF eHealth evaluation lab 2014. In: Kanoulas, E., Lupu, M., Clough, P., Sanderson, M., Hall, M., Hanbury, A., Toms, E. (eds.) Information Access Evaluation. Multilinguality, Multimodality, and Interaction. Proceedings of 5th International Conference of the CLEF Initiative, CLEF 2014, Sheffield, UK, pp. 172–191. Springer International Publishing, Cham, 15–18 Sept 2014
5. Aronson, A.R., Lang, F.-M.: An overview of MetaMap: historical perspective and recent advances. J. Am. Med. Inform. Assoc. JAMIA **17**, 229–236 (2010)
6. Kim, Y., Riloff, E., Hurdle, J.F.: A study of concept extraction across different types of clinical notes. In: AMIA Annual Symposium Proceedings 2015, pp. 737–746 (2015)
7. Kang, N., Afzal, Z., Singh, B., van Mulligen, E.M., Kors, J.A.: Using an ensemble system to improve concept extraction from clinical records. J. Biomed. Inform. **45**, 423–428 (2012)
8. Minard, A.-L., Ligozat, A.-L., Ben Abacha, A., Bernhard, D., Cartoni, B., Deléger, L., Grau, B., Rosset, S., Zweigenbaum, P., Grouin, C.: Hybrid methods for improving information access in clinical documents: concept, assertion, and relation identification. J. Am. Med. Inform. Assoc. **18**, 588 (2011)
9. Xu, H., AbdelRahman, S., Jiang, M., Fan, J.W., Huang, Y.: An initial study of full parsing of clinical text using the Stanford Parser. In: 2011 IEEE International Conference on Bioinformatics and Biomedicine Workshops (BIBMW), pp. 607–614 (2011)
10. Griffis, D., Shivade, C., Fosler-Lussier, E., Lai, A.M.: A quantitative and qualitative evaluation of sentence boundary detection for the clinical domain. AMIA Summits Transl. Sci. Proc. **2016**, 88–97 (2016)

11. Bodenreider, O.: The Unified Medical Language System (UMLS): integrating biomedical terminology. Nucleic Acids Res. **32**, D267–D270 (2004)
12. Zhang, S., Elhadad, N.: Unsupervised biomedical named entity recognition: experiments with clinical and biological texts. J. Biomed. Inform. **46**, 1088–1098 (2013)
13. Dehghan, A., Keane, J.A., Nenadic, G.: Challenges in clinical named entity recognition for decision support. In: 2013 IEEE International Conference on Systems, Man, and Cybernetics, pp. 947–951 (2013)

Part II
Operating System, Databases, and Software Analysis

DevOps with Continuous Testing Architecture and Its Metrics Model

Jayasri Angara, Sridevi Gutta and Srinivas Prasad

Abstract The advent of DevOps is to take full advantage of iterative model of development, bring agility in software development life cycle and achieve time to market goal. However, testing becomes roadblock and reduces the rate of speed. Hence, there is a critical need to strategize testing process and align it to continuous planning, continuous integration, continuous deployment and continuous monitoring, and feedback goals of DevOps practice. There is a vital difference between test automation and continuous testing. The former is subset of latter. Continuous testing identifies integration issues much earlier in the life cycle; makes defect resolution cheaper, faster; and frees tester's precious time for exploratory testing and value-added test activities. This paper conducts literature survey on various strategies applied for continuous testing and proposes a continuous testing architecture for better implementation. It also presents the conceptual design of few important testing metrics for successful implementation of continuous testing function in the context of DevOps.

Keywords DevOps · Continuous testing · Continuous integration
Behavioral-driven development · Agile testing

1 Introduction

The ultimate goal of software development is to solve customer problems and provide solution to the true behavior of user needs. Anything more or less leads to variation in the cost, productivity loss, non-user acceptance, and non-compliance to

J. Angara (✉) · S. Gutta
K.L. University, Vijayawada, AP, India
e-mail: anagara.jayasri@gmail.com

S. Gutta
e-mail: sridevi.gutta2012@gmail.com

S. Prasad
GMR Institution of Technology, Rajam, AP, India
e-mail: srinivas.prasad@hotmail.com

© Springer Nature Singapore Pte Ltd. 2018
P. K. Sa et al. (eds.), *Recent Findings in Intelligent Computing Techniques*,
Advances in Intelligent Systems and Computing 709,
https://doi.org/10.1007/978-981-10-8633-5_28

total quality. The ideal software tester is the customer or user. However, it is an ideal scenario. The role of testing is to map and verify business requirements with technology requirements and assess verifiable business value. DevOps is an emerging cross-disciplinary practice, which enhances communication and collaboration between business, development, QA, and IT operations teams. It brings radical change in the way of traditional software development and testing takes place. Shift left is the key theme in the DevOps. Testing and deployment take place early in the life cycle, detect defects early, and reduce the cost of fixing. DevOps is all about continuous testing. Continuous testing does a quantitative assessment of all risks, and corresponding mitigation plans before project move to next phase of SDLC [1]. It helps developers code faster and write better code [2]. The success of DevOps lies in how test design, test case development, and test automation take place, and how they are executed on continuous platform. Continuous testing introduces a new set of tools, technologies, processes, and methodologies with an ultimate goal of capturing the market opportunities; reduces the feedback loop; and improves the quality of software code and deployment and overall organization performance.

The objective of this paper is to discuss various nuances of continuous testing practices and develop systematic continuous testing architecture in DevOps context. This paper has been organized as follows. Section 2 presents related work. Section 3 proposes continuous testing architecture and conceptual design of testing metrics models. Section 4 presents the conclusion.

2 Related Work

Business demands uninterrupted service with seamless continuous integration of service upgrades. This model results in shorter, frequent, and efficient releases. This is possible through continuous testing function of DevOps. Domain understanding and grasp on application behavior through modeling seamlessly supports software development, testing, and maintenance [3]. It is critical in the case of continuous testing as testing function fulfills high coverage, early detection of defects, better utilization of resources, and seamless communication between business users, domain experts, testers, and developers. Continuous testing brings three major business benefits—decision to go or no-go in SDLC, new features to market faster, and trade-off between time, quality, and functionality [4]. Communication and collaboration are critical in continuous testing process. Metrics and dashboards provide confidence and action among all stakeholders. Typical metrics which can be considered for implementation are percentage of test cases automated, availability of infrastructure, release frequency, test efficiency, requirement traceability, test optimization, etc. Continuous testing requires systematic stitching between people, processes, and technology.

Few best practices like automatic test scripts (part of version control system), automation suit (integrated with build deployment tools) [1, 5], making build self-testing, self-communicating, and collaborating with all stake holders [6, 7], regression testing in integration with IDE [8–10], optimization and prioritization of test cases [11, 12], BDD techniques [13–15], model-based approach/emulators [16, 17], unified test framework (separation of test cases (behavior + test case configuration layer) [18], test quarantine (reverses the builds back) [19], test automation framework [20, 21], real-time monitoring [22], GUI test automation [23], and dynamic test wares [24] are essential for continuous testing success. Testing function is most successful when it follows systematic hierarchical test strategy (Level 1, Level 2, Level 3) [25].

3 Proposed Continuous Testing Architecture and Testing Metrics Models

3.1 Proposed Continuous Testing Architecture

The success of continuous testing lies in systematic implementation of following four strategies: (**a**) tools and technology strategy (build and test automation, VM hosting for tests, BDD/TDD/models, robustness, performance, less technical debt, open source for low cost, infrastructure as code, and micro-services). (**b**) Skills deployment strategy (intelligent test prioritization, automation scripting, story articulation, multi-skilled resources (DevTest), and communication management). (**c**) Organization culture strategy (thinking shift (automation to continuous testing), shift form test coverage to risk coverage, collaboration between users (BA-Dev-Test-Ops), and less management debt). (**d**) Process and metrics strategy (story-based estimation techniques, real-time metrics dashboard, automated feedback, and tractability of stories to test scenarios to test cases to code files).

Figure 1 depicts proposed continuous testing architecture which allows high test coverage, early feedback (shift left testing), manages test defects, analytics, test data, service virtualization, etc., and brings down the cost of fixing drastically.

3.2 Proposed Conceptual Design of Continuous Testing Metrics

Metrics and measures represent good information flow. Information flow happens between development environment, integration environment, production environment, document management system, bug tracking system, version control system, project management tools, and other organization-specific systems. DevOps needs design of alternative metrics/measures, which gives real-time status of the project. DevOps metrics may not be sacrosanct numbers but should also measure soft aspects like trust, confidence, culture strength, etc. They are difficult to measure; however, they are critical and important to design. DevOps demands measurement

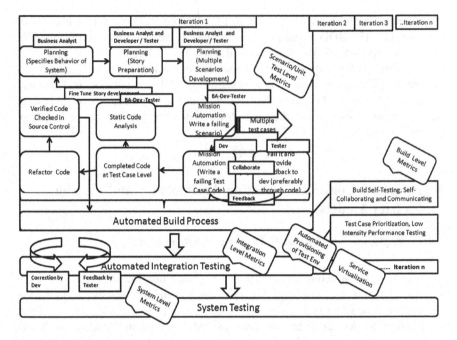

Fig. 1 Proposed continuous testing architecture

of other application-specific or domain-specific metrics like data access, knowledge dissemination level, application release automation, test automation, performance monitoring, overall customer experience level, etc. One of the important outcomes of the study is to develop few important metrics which helps in seamless data communication between all stakeholders.

In this section, few test case-dependent models/metrics are proposed.

1. Test Case Complexity Model: Planning, resource allocation, and test case prioritization.
2. Test Case Business Priority: It determines business priority for each test case.
3. Test Resource (Human Resource) Maturity: It determines tester's maturity.
4. Test Case Complexity versus Test Human Resource Maturity Matrix: It is a 2-D matrix which is useful for human resource allocation process.
5. Test Case Execution Effort Estimation: It helps in the estimation of testing effort.
6. Test Case Prioritization and Allocation: It improves test execution productivity.

These models can be used in test case management process. The operational sequence goes as follows: 1. develop test cases; 2. determine test case complexity; 3. determine test case business priority; 4. assess tester's maturity; 5. develop test case complexity versus tester maturity matrix; 6. determine test case execution effort; 7. test case prioritization; 8. allocation for execution; and 9. repeat this as necessary.

3.2.1 Test Case Complexity Model

Test case complexity is a measure of how difficult and complex to run the test case during the testing process [26]. Test case complexity is determined by following five factors: (a) Product/Application Criticality (AC), (b) Product/Application Stability (AS), (c) Product/application Technical Complexity (TC), (d) Product/application Domain Complexity (DC), and (e) Project Management/Process Maturity (PM). These calculations are explained below. The following notations are used: Long (LO), Medium (MD), Low (LW), Frequent (FR), Moderate (MD), New Development (ND), Significant Enhancements (SE), Bug Fixes (BF), Online (OL), Online and Batch (OB), Batch (BT), Rare Tools (RT), Common Tools (CT), and No Tools (NT). Each of these sub-factors has different ratings and corresponding weightages which are supplied by Test Manager/Project Manager to calculate weightage of particular parameter. Weightage (Wt) value lies between 2 (less complexity) and 10 (more complexity).

(a) Product/Application Criticality (AC)

Application Criticality (AC) [27, 28] is determined by sub-factors like Turnaround Time (*TART*) (LO (2), MD (5), LW (10)), Business Users Usage (BUUS) (FR (10), MD (5), LW (2)), Data Sensitivity (DATS) (HI (10), MD (5), LW (2)), Revenue Generating Features (REVG) (HI (10), MD (5), LW (2)), and Application Security (ASEC)) (HI (10), MD (5), LW (2)). Total Application Criticality (AC$_i$) score pertained to that particular test case is calculated as follows:

$$AC_i = TART\,Wt_i + BUUS\,Wt_i + DATS\,Wt_i + REVG\,Wt_i + ASEC\,Wt_i \quad (1)$$

The value range AC$_i$ is $10 \leq AC_i \leq 50$, $i = 1, ..., n$.

(b) Product/Application Stability (AS)

Application Stability (AS) is determined by sub-factors like Application Development Stage (ADES) (ND (10), SE (5), BF (2)), Availability of Documentation (ADOC) (HI (2), MD (5), LW (10)), Interdependency on other Systems (ISYS) (HI (10), MD (5), LW (2)), Code Stability (CSTB) (HI (2), MD (5), LW (10)), and Previous Iterations/Sprint Accuracy (PIAC) (HI (2), MD (5), LW (10)). Total Application Stability (AS$_i$) score pertained to that particular test case is calculated as follows:

$$AS_i = ADES\,Wt_i + ADOC\,Wt_i + ISYS\,Wt_i + CSTB\,Wt_i + PIAC\,Wt_i \quad (2)$$

The value range AS$_i$ is $10 \leq AS_i \leq 50$, $i = 1, ..., n$.

(c) Product/Application Technical Complexity (TC)

Application Technical Complexity (TC) is determined by sub-factors like Product/Application Environment (AENV) (OL (2), OB (5), BT (10)), Technical Skills Required (TECS) (HI (10), MD (5), LW (2)), Tools Used (TOOL) (RT (10), CT

(5), NT (2)), Technical Standards Required (TSTD) (HI (10), MD (5), LW (2)), and Test Data Required (TDAT) (HI (10), MD (5), LW (2)). Total Technical Complexity (TC$_i$) score pertained to that particular test case is calculated as follows:

$$TC_i = AENV\,Wt_i + TECS\,Wt_i + TOOL\,Wti + TSTD\,Wt_i + TDAT\,Wt_i \qquad (3)$$

The value range TC$_i$ is $10 \leq TC_i \leq 50$, i = 1, ..., n.

(d) Product/Application Domain Complexity (DC)

Application Domain Complexity (DC) is determined by sub-factors like Business Complexity (BUSC) (HI (10), MD (5), LW (2)), Legal Compliance/Business Risk (LEGC) (HI (10), MD (5), LW (2)), Business Users Usage (BUUS) (HI (10), MD (5), LW (2)), Business Documentation (BDOC) (HI (2), MD (5), LW (10)), and Availability of Business Domain Experts(BDEX) (HI (2), MD (5), LW (10)). Total Domain Complexity (DCi) score pertained to that particular test case is calculated as follows:

$$DCi = BUSC\,Wti + LEGC\,Wti + BUUS\,Wti + BDOC\,Wti + BDEX\,Wti \qquad (4)$$

The value range DCi is $10 \leq DCi \leq 50$, i = 1, ..., n.

(e) Project Management/Process Maturity (PM)

Project Management/Process Maturity (PM) is determined by sub-factors like Project-Specific Process Artifacts/Manuals/Documentation Availability (PDOC) (HI (2), MD (5), LW (10)), Organization-Specific Standard Quality System Availability (STDQ) (HI (2), MD (5), LW (10)), Tools Usage (Bug Tracking/ Quality Management Tools) (TOlU) (HI (2), MD (5), LW (10)), Acceptance toward Opportunity for Improvements (AOIM) (HI (2), MD (5), LW (10)), Process Oriented Resources Availability (PORA) (HI (2), MD (5), LW (10)). Total Project Management/Process Maturity (PMi) score pertained to that particular test case is calculated as follows:

$$PMi = PDOC\,Wt_i + STDQ\,Wt_i + TOlU\,Wt_i + AOIM\,Wt_i + PORA\,Wt_i \qquad (5)$$

The value range PMi is $10 \leq PMi \leq 50$, i = 1, ..., n.
Finally, the total test case complexity TTC per test case is calculated as follows:

$$\text{Total Test Case Complexity } (TTC_i) = AC_i + AS_i + TCi + DC_i + PM_i \qquad (6)$$

The value range TTCi is $50 \leq TTC_i \leq 250$, i = 1, ..., n.
Normalized value of one particular test case is calculated as follows:

$$TTC_i = (TTC_i - 50)/(250 - 50) \qquad (7)$$

3.2.2 Test Case Business Priority

Once test case complexity in normalized form has been calculated, the next step is to determine business priority. Business Priority is calculated using sub-factors like Release Priority Feature (REPF) (HI (10), MD (5), LW (2)), Multiple Approvals Required to implement in Business Function (MARB) (HI (10), MD (5), LW (2)), Shared Business Resources (Customer/Partners/Vendors) (SBRE) (HI (10), MD (5), LW (2)), Interdependent Business Feature (INBF) (HI (10), MD (5), LW (2)), and Test Data Preparation Complexity (TDPC) (HI (10), MD (5), LW (2)). Total Business Priority (BPi) score pertained to that particular test case is calculated as follows:

$$\text{BPi} = \text{REPF Wti} + \text{MARB Wti} + \text{SBRE Wti} + \text{INBF Wti} + \text{TDPC Wti} \qquad (8)$$

The value range BPi is $10 \leq \text{BPi} \leq 50, i = 1, \ldots, n$.
Normalized value of one particular test case business priority Normalized BPi =

$$(\text{BPi} - 10)/(50 - 10) \qquad (9)$$

3.2.3 Test Resource (Human Resource) Maturity

Tester's maturity is calculated using following sub-factors like Tester's Overall Testing Experience of the Resource (EXPR) (HI (10), MD (5), LW (2)), Tester's Experience in Previous Iterations (EXPI) (HI (10), MD (5), LW (2)), Tester's Experience in Development/Operations area (TEDO) (HI (10), MD (5), LW (2)), Tester's Domain/Functional Experience (TDFE) (HI (10), MD (5), LW (2)), and Tester's Collaboration Attitude (TCDO) (HI (10), MD (5), LW (2)). Total Tester Maturity (TM_i) score pertained to that particular test case is calculated as follows:

$$TM_i = \text{EXPR Wt}_i + \text{EXPI Wt}_i + \text{TEDO Wti} + \text{TDFE Wt}_i + \text{TCDO Wt}_i \qquad (10)$$

The value range TM_i is $10 \leq TM_i \leq 50, i = 1, \ldots, n$.
Normalized value of one particular tester

$$\text{Normalized TMi} = (\text{TMi} - 10)/(50 - 10) \qquad (11)$$

3.2.4 Test Case Complexity Versus Test Human Resource Maturity Matrix

The Total Test Case Complexity (TTCi) and the Test Resource Maturity (TMi) are assessed based on previous sections. The Total Test Case Complexity (TTCi) is measured in terms of a normalized score, varying from 0 to 1. The Test Resource Maturity (TMi) is measured in terms of a normalized score, varying from 0 to 1.

Based on the scores, they are mapped on to one of the four quadrants. They are broadly classified in terms of the quadrants in which the test complexity vis-à-vis the test resource maturity are mapped using four-quadrant classification model [29]. As per the model, the first quadrant ($0 < TTCi < 0.5$, $0 < TMi < 0.5$) testers are categorized as budding/seekers. Testers and test cases mapped on the second quadrant ($0.5 < TTCi < 1$, $0 < TMi < 0.5$) are categorized as adoptive/challengers. Testers and test cases mapped on the third quadrant ($0 < TTCi < 0.5$, $0.5 < TMi < 1$) are categorized as progressive/loungers, while testers and test cases mapped on the fourth quadrant ($0.5 < TTCi < 1$, $0.5 < TMi < 1$) are categorized as matured/performers. Figure 2 depicts these four quadrants. Once this matrix is defined, it is easy to lock test resources for specific test cases. This 2-D matrix is useful for resource allocation based on test case complexity and resource maturity. This step is useful in locking resources against complexity, prioritization, and allocation of test cases.

3.2.5 Test Case Execution Effort Estimation

Total test case complexity, business priority, resource maturity values are used in determining the test case estimation. The team can use test case as the basis for estimation instead of user story for effort estimation. Informal testing or uncertified test personnel dent the quality [30]. Hence, it is important to do this activity diligently. Alternatively, effort estimation can be derived from User story. Each User story may have one or more acceptance tests. Each acceptance test consists of one or more test cases. Hence, if user story is estimated using planning poker estimation method, then a mechanism can be developed to derive test case level effort (top-down estimation method). Another way is bottom-up estimation method. Few dependent test cases together can build User story. Effort estimation can be derived using test cases. This process also helps in creating a platform for test-driven development and acceptance test-driven development.

3.2.6 Test Case Prioritization and Allocation

Test case prioritization and allocation is a most vital process in software testing. This is important to minimize the number of faults in future releases and minimization of testing resources while maximizing the software reliability. A lot of studies have been conducted using genetic algorithm, fuzzy logic, neuro-fuzzy, and soft computing techniques for run-time allocation of testing resources [31–33]. Test case prioritization aims to maximize the early fault detection [34]. The control variables listed in previous sections are critical to steer this process. Test case complexity, test case business priority, total test case complexity versus test resource maturity matrix classification, software environment (shared resources), test case execution effort estimation, and test case dependency matrix are vital in determining prioritization and allocation.

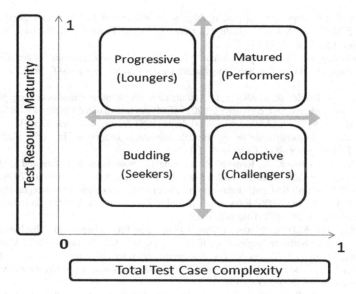

Fig. 2 Total test case complexity versus test resource maturity matrix classification model

4 Conclusion

Continuous testing, a critical function of DevOps practice, brings two major business benefits—go or no-go decision, and manages trade-off between time, quality, and functionality. Automation, communication, and collaboration are critical parameters in continuous testing process. Metrics drive continuous testing and show the health of the project. DevOps needs design of alternative metrics/ measures with good benchmark base. It elevates the culture of an organization and improves the collaboration between stakeholders. The traditional isolated QA/tester teams may find challenge and limited role in DevOps. They have to scale beyond regular testing function and integrate with development and operations teams. Continuous testing is the tone of an organization's mindset and strength for the effective testing process in DevOps setting.

References

1. Continuous Testing for IT Leaders. https://alm.parasoft.com/continuoustestingbook
2. Saff, D., Ernst, M.: An experimental evaluation of continuous testing during development. In: Proceedings of the 2004 International Symposium on Software Testing and Analysis, pp. 76–85, Boston, MA, USA (2004)
3. Schur, M., Roth, A., Zeller, A.: Mining workflow models from web applications. IEEE Trans. Softw. Eng. **41**(12), 1–1 (2015)

4. Ariola, W.: DevOps: are you pushing bugs to your clients faster? In: Thirty-Third Annual Pacific Northwest Software Quality Conference, World Trade Center Portland, Portland, Oregon, 12–14 Oct 2015
5. Jumpstarting DevOps with Continuous Testing. https://www.cognizant.com/content/dam/Cognizant_Dotcom/whitepapers/Jumpstarting-DevOps-with-Continuous-Testing-codex1719.pdf
6. Gmeiner, J., Ramler, R., Haslinger, J.: Automated testing in the continuous delivery pipeline: a case study of an online company. In: 2015 IEEE 8th International Conference on Software Testing, Verification and Validation Workshops (ICSTW), pp. 1–6, Graz (2015)
7. Stolberg, S.: Enabling agile testing through continuous integration. In: Agile Conference, AGILE '09, pp. 369–374, Chicago, IL (2009)
8. Cannizzo, F., Clutton, R., Ramesh, R.: Pushing the boundaries of testing and continuous integration. In: Conference on AGILE '08 Conference, pp. 501–505, Toronto (2008)
9. Reich, C., Scharpf, B.: Continuous software test distributed execution and integrated into the globus toolkit. In: 2006 Fifth International Symposium on Parallel and Distributed Computing, pp. 185–190, Timisoara (2006)
10. Saff, D., Ernst, M.D.: Continuous testing in Eclipse. In: Proceedings of the 27th International Conference on Software Engineering, ICSE '05, pp. 668–669, St. Louis, 18–20 May 2005
11. Marijan, D., Gotlieb, A., Sen, S.: Test case prioritization for continuous regression testing: an industrial case study. In: Proceedings of 29th IEEE International Conference on Software Maintenance, pp. 540–543 (2013)
12. Tricentis 2015: Risk coverage optimization. http://www.tricentis.com/wp-content/uploads/2015/04/20150421-Risk-Coverage-Optimization-Factsheet-A4_PRINT.pdf
13. Chelimsky, D., Astels, D., Dennis, Z., Hellesoy, A., Helmkamp, B., North, D.: The RSpec Book: Behaviour Driven Development with RSpec, Cucumber, and Friends. The Pragmatic Programmers, United States (2010)
14. Evans, E.: Domain-Driven Design: Tackling Complexity in the Heart of Software, 1st edn. Addison-Wesley Longman, Amsterdam (2003)
15. Hatko, R., Mersmann, S., Puppe, F.: Behaviour-driven development for computer-interpretable clinical guidelines. In: Proceedings of ECAI (2014)
16. Brajnik, G., Baruzzo, A., Fabbro, S.: Model-based continuous integration testing of responsiveness of web applications. In: 2015 IEEE 8th International Conference on Software Testing, Verification and Validation (ICST), pp. 1–2, Graz (2015)
17. Hill, J.H.: CUTS: a system execution modeling tool for realizing continuous system integration testing. In: 2010 ACM/IEEE 32nd International Conference on Software Engineering, pp. 309–310, Cape Town (2010)
18. Liu, H., Li, Z., Zhu, J., Tan, H., Huang, H.: A unified test framework for continuous integration testing of SOA solutions. In: ICWS 2009 IEEE International Conference on Web Services, 2009, Los Angeles, pp. 880–887, CA (2009)
19. Experiences from Continuous Testing at Siemens Healthcare. https://www.infoq.com/news/2015/02/continuous-testing-siemens
20. Kim, E.H., Na, J.C., Ryoo, S.M.: Test automation framework for implementing continuous integration. In: ITNG '09. Sixth International Conference on Information Technology: New Generations, 2009, pp. 784–789, Las Vegas, NV (2009)
21. Rathod, N., Surve, A.: Test orchestration a framework for continuous integration and continuous deployment. In: 2015 International Conference on Pervasive Computing (ICPC), pp. 1–5, Pune (2015)
22. InfoStretch. http://www.qmetry.com/casestudy-stanford/
23. RBCS. http://rbcs-us.com/site/assets/files/1274/case-studies-of-free-test-tool-success.pdf
24. Vos, T., Tonella, P., Prasetya, W., Kruse, P.M., Bagnato, A., Harman, M., Shehory, O.: FITTEST: a new continuous and automated testing process for future internet applications, 3–6 Feb 2014
25. Spirent: A solution blueprint for DevOps. http://www.spirent.com/Assets/WP/WP_A_Solution_Blueprint_for_DevOps

26. Thillaikarasi, M., et al.: A test case prioritization method with weight factors in regression testing based on measurement metrics. Int. J. Adv. Res. Comput. Sci. Softw. Eng. 3(12), 390–396 (2013)
27. Muthusamy, T., Seetharaman, K.: A new effective test case prioritization for regression testing based on prioritization algorithm. Int. J. Appl. Inf. Syst. (IJAIS) 6(7), 21–26 (2014). Foundation of Computer Science FCS, New York, USA. ISSN 2249-0868
28. neoIT: The offshore assessment: building your offshore road map, offshore insights. http://www.neogroup.com/PDFs/Whitepapers/neoIT-Aug03-OffshoreAssessment.pdf
29. Saripalle, R., Kumar, P., Tatavarti, R.: Individual Innovation Index (I^3): assessment and enhancement. Int. J. Innov. Technol. Manag. 11(5) (2014)
30. Software Quality in 2012: A survey of the state of the art, Namcook, Analytics LLC. http://sqgne.org/presentations/2012–13/Jones-Sep-2012.pdf
31. Huang, C.Y., et.al.: Optimal allocation of testing-resource considering cost, reliability, and testing-effort. In: Proceedings of 10th IEEE Pacific Rim International Symposium Dependable Computing, pp. 103–112 (2004)
32. Huang, C.Y., Lyu, M.R.: Optimal testing resource allocation, and sensitivity analysis in software development. IEEE Trans. Reliab. 54(4) (2005)
33. Nasar, M., et al.: Software testing resource allocation and release time problem: a review. Int. J. Modern Educ. Comput. Sci, 248–255 (2014)
34. Yoo, S., Harman, M.: Regression testing minimization, selection and prioritization: a survey. Softw. Test. Verif. Reliab. 22, 67–120 (2012)

Optimized MBT-Test Case Generation for Embedded System Controller Using LabVIEW and Sequence Graphs

Parampreet Kaur and Rajeev Sobti

Abstract System testing using model-based approach is gaining momentous acceptance worldwide not only in academics but also in industrial context. Rapid technological advancements in the software industries drive the software engineers to produce highly expedited applications in minimum time without spending more efforts on testing and maintenance activities. The design artifacts are modeled using LabVIEW software and validated to produce optimized test paths. A rising inclination is observed in the Model-Based Testing (MBT) of software systems embedded on various devices. This article primarily focusses upon the model-based testing of embedded application of elevator controller using state machines and transforming these to graphical notations to generate test paths. An algorithm is designed to traverse all nodes and provide full coverage of system. A taxonomy of test case generation is described, which can serve as a fundamental basis to develop strategies to derive potential testing techniques.

Keywords Model-based testing (MBT) · SUT · Test path · Test case
Embedded applications

1 Introduction

Quality is one of the most crucial and indispensable aspects of today's software systems [1]. It assists in establishing enduring relationships with customers and helps achieve customer satisfaction. Quality assurance plays a vital role to attain these targets. Hence, testing is considered as one of the most dominant areas of quality assurance [1]. To execute testing, there are basically two major forms of

P. Kaur (✉) · R. Sobti
School of Computer Science & Engineering, Lovely Professional University,
Phagwara, Punjab, India
e-mail: paramnagpal16@gmail.com

R. Sobti
e-mail: sobtirajeev@gmail.com

© Springer Nature Singapore Pte Ltd. 2018
P. K. Sa et al. (eds.), *Recent Findings in Intelligent Computing Techniques*,
Advances in Intelligent Systems and Computing 709,
https://doi.org/10.1007/978-981-10-8633-5_29

tests broadly classified as black box and white box tests. In the subsequent sections, we will be focusing on one of the former testing processes of black box tests or functional testing method commonly known as model-based testing.

1.1 Model-Based Testing

Model-Based Testing (MBT) is a testing paradigm where the system under test is demonstrated with a formal modeling method such as finite state machines or UML and can be used for automatic generation of tests [2]. MBT considers software specifications as a fundamental base for formal modeling and test cases are extracted from these models [3]. Testing case studies report 20–85% effort reduction on executing model-based testing approach [1]. The escalating complexity and demand for highly qualitative applications, having small production times, led organizations to invest more on automated test suites generation, and fascinated the tester's interest on model-oriented testing methodologies and tools. MBT is associated with several advantages with respect to other approaches like manual testing or code-based testing [4]. Abstract tests are produced using MBT approach which must be transformed into executable tests by the test engineers. Comparisons between manual model-based evaluation and automated model-based evaluations claim systematic and consistent discovery of defects covering more functional issues. This fact was illustrated by a severity score summary which was 60% higher for MBT than manual approaches [5]. Since requirements are transformed into models and used for validating the System Under Test (SUT), therefore, models are excellent source of communication with the stakeholders proving to be effective in managing the complexity of the system. Model-based testing provides inexpensive mode of validating the iterations in software leading to easier and faster production of an application. Finally, the greatest advantage is that testing starts earlier in the software development life cycle and detection of irregularities or faults can be rectified at earlier phases.

1.2 Why MBT is More Successful for Embedded Software Domain?

In our paper, we have primarily focused upon embedded domain of software. Embedded systems are getting more and more complex due to a higher rate of integration and shared usage of sensor signals [6]. For instance, systems embedded in the automobiles have become increasingly intricate as they are supposed to make vehicles more secure, fuel-efficient, and luxurious [7].

Errors in embedded software systems can lead to catastrophic failures if not eliminated timely [8]. Along with the increased number of sensors involved in the systems, the development is getting increasingly complex. Hence, in order to ensure a reliable and safe operation, a thorough test is required for validating the expected behavior by the system [9]. According to a recent survey conducted in the year 2014 and 2016 by Binder et al. [10, 11], key industrial areas adopting the usage of MBT in their services were identified. It was found that maximum contribution for adoption of MBT is practised by embedded domain sector. Embedded domain (both real time and not real time) covers approximately 37% of the respondents as compared to various other domains broadly categorized as Enterprise IT which accounts for another 30%, web applications accounting to around 20%. For ensuring effectiveness in safety-critical systems it becomes crucial to test them using more advanced techniques [7]. Hence, the technique of model-based testing proves to be effective for conducting validation and verification of such systems in an efficient manner [1]. The survey mentioned above serves as a motivational factor for this paper to study various techniques and implementation strategies for executing model-based testing in the field of embedded system domain [8].

The following paper is structured into literature review in Sect. 2 followed by a discussion on broad testing strategies in Sect. 3. It provides an overview of techniques which can be combined for generating tests. Section 4 presents the actual implementation carried out and then test paths are generated. Section 5 discusses the conclusion and future work.

2 Literature Review

Weissleder et al. [1] in their research reported a pilot project to apply MBT in embedded software domain. They have concluded that if model-based testing is initiated well in the project's starting phases, and then there is an instant advantage as time for designing the models is usually lesser than manual development of test cases. The technique used is manually designing the UML state machines in a software enterprise architect and then importing this model in conformiq designer tool using XMI 2.1 interchange format. Later generating test sequences using coverage criteria all-states and all-transitions is done. Figure 1 depicts the efforts needed for manual and automatic test creation. If there is no test suite for the project under consideration, it may result in some initial effort by both manual and MBT approaches but the application of MBT pays off early during second iteration only.

From Fig. 1, it can be deduced that although some initial effort is incurred to apply MBT; afterwards, there is a constant graph for all subsequent iterations. Hence, MBT proves to be efficient in incorporating the change easily and produce the results early. The graph in Fig. 2 depicts the effort for different scenarios which clearly indicates that if MBT is applied to embedded system domains least effort is applied when a model needs alteration.

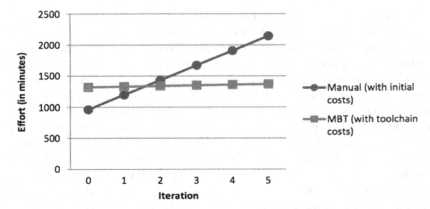

Fig. 1 MBT versus manual test effort [1]

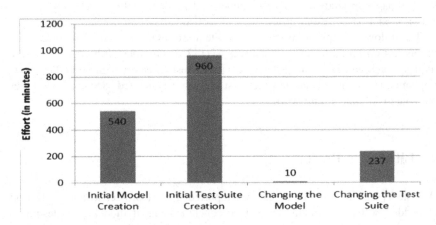

Fig. 2 Effort applied for various processes of MBT [1]

Research work [4, 12] introduces the technique of obtaining colored UML models based on state machine charts. The technique is based on coverage criteria methodology in which a novel Model Coverage Analysis Tool MoCAT is implemented, which analyzes the coverage of a state machine achieved by the test suite. In this paper, the authors have presented their experiences of introducing model-based testing in an industrial context. The approach used in this research claims to expedite the model-based testing of embedded systems and assures a better quality of test suites achieved by supporting coverage analysis and coloring of complex guard conditions [4]. Research work conducted by [13] proposed MilEST (Model in the Loop of Embedded System Test) a test development method. The methodology assumes that SUT is already available and input–output interfaces are clearly defined. The test design phase consists of a transformation function, a test stimulus, and a test algorithm applied to the SUT.

A study [14] described the steps toward model-based testing framework for testing printer controllers at Oce (a printer company). The approach proved to be promising for boosting the testing process and quality of test cases. The research highlights the advantages of MBT in terms of enhanced performance and maintenance, measurable coverage as well as better understanding of test outcomes. The study presented by [15] demonstrates an experimental MBT in Hardware-in-the-Loop prototype (HIL) which provides an insight into the online and offline MBT benefits. They have implemented a prototype platform which was evaluated with SUT control algorithm along with mutation testing which proves to be successful in identifying the discrepancies in embedded software. The research work conducted by [16] introduces multilevel test models to derive multilevel test cases. The researchers worked upon the headlights functionality of modern vehicles. In their paper, they have considered four different test levels for testing of embedded applications of software component testing, software integration testing, software–hardware integration testing, and system integration testing [16].

The paper [17] discusses in detail the criteria to test real-time embedded software which basically makes three major contributions: (i) development of a new architectural model for component-based embedded system, (ii) generating tests from this model, and (iii) application of new model and tests generated to industrial embedded software. The major contribution of this paper is the introduction of three intermediate graphs which model the integration and interaction aspects of the embedded software such as connectivity and dependency relationship among states of system. And at the end, tests are generated from the above diagrams [17].

3 Classes, Categories, and Options for Executing Test Generation in Embedded System Domain

In the embedded area domain, a broad taxonomy for model-based testing is explained taken from [18, 19] as shown in Fig. 3. In this taxonomy, four classes: model, test generation, test execution, and test evaluation are identified. Each class is then further classified into different categories. The model class consists of type of model (system or test) which will be considered. The test generation class consists of test selection criteria, technology supported, and result categories [8, 18]. The test execution class contains execution options. The last class is test evaluation which classifies specifications and MBT mode such as offline or online and manual or automatic type. The chart presented in Fig. 3 clearly classifies various approaches that can be combined to generate test cases for validating the embedded software designs.

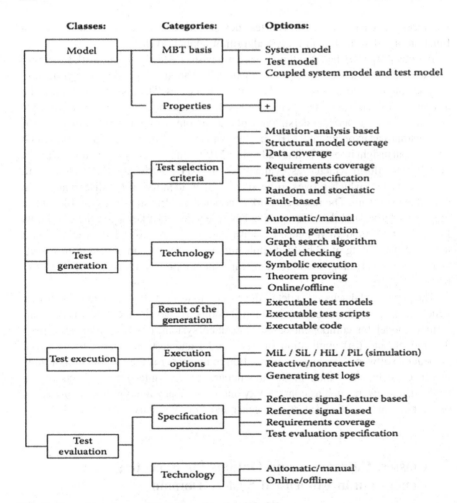

Fig. 3 General taxonomy of model-based testing [19, 20]

4 Approach Used in the Research

For testing using model-based approach, we are using a system model such as a UML state machine diagram of a cargo elevator where test selection criteria are the structural model coverages by using random generation technique. We have used LabVIEW software to construct the functional aspect of the freight/cargo elevator system. It is designed in the front panel of the software. Initially, the button is in the idle state and corresponding LED is OFF. When some weight is placed and the button is pressed, the elevator LED turns to ON state. A floor indicator is used to indicate the floor level where the weight is supposed to be delivered. The system is designed in this manner so as to improve the customer interaction with the overall behavior in advance. Any changes proposed can then be easily accommodated to avoid later defect detection. Introducing LabVIEW in software engineering design phase is very valuable both for the coder and the tester. It improves the overall process of visualizing the system under development in a more realistic manner and increases the effective test generation. We have tried to derive simpler models from the complicated systems to efficiently generate test paths. Our approach basically focusses upon modeling a system using LabVIEW and then following the traditional UML methodology to design and produce test cases. Consider the following state machine in Fig. 4 for the cargo elevator (an embedded software) for transferring the freight, which moves when the weight placed is lesser or equal to the permissible limit of weight otherwise it does not move.

Moreover, the speed of elevator increases if the weight of actual cargo is equal to minimum weight set whereas it slows down if weight is more than minimum weight placed. The system can then be modelled as shown in Fig. 5. This provides a detailed version of a state machine with transitions describing the conditions to move from one state to another. There are five states in this elevator (idle, button pressed, start moving, move fast, and move slow). First, the system is in idle state. When the button is pressed, the following conditions must be satisfied to start the movement in the elevator.

(a) The floor number is less than or greater than the current floor number of elevator.
(b) The designated floor should be greater than the basement (assume a single basement).
(c) Or the designated floor should be greater than minimum floor present. Actual weight placed should not be more than maximum permissible weight. The

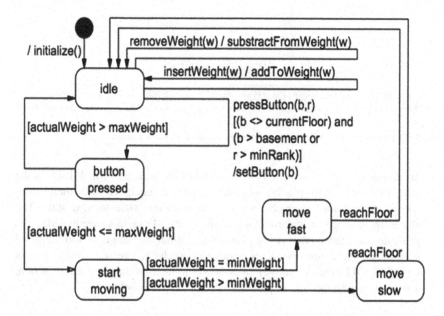

Fig. 4 State machine design of cargo elevator [21]

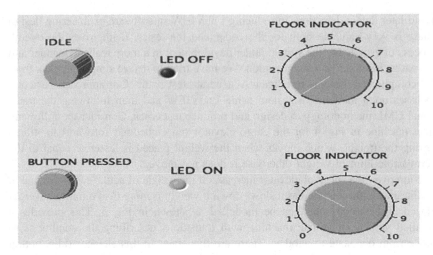

Fig. 5 Abstract view of system using LabVIEW

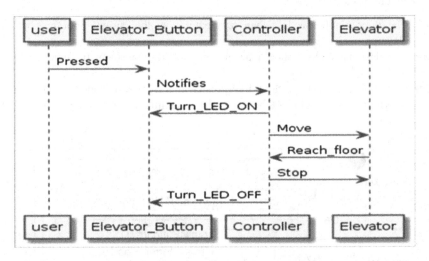

Fig. 6 Sequence chart showing message passing between elevator phases

mentioned conditions are mentioned on one of the transition which leads to the start_moving state. Similarly, other transitions are equipped with their corresponding conditions to enable a transition from one state to next state. The above figure shows only partial view of the system. From this state machine, we generate a sequence chart using plant-unified modeling language tool depicted in Fig. 6. There are messages sequences such as pressed, notifies, turn LED ON, OFF, etc., to convey information to next module to take an action. The code to generate the sequence chart is mentioned below.

```
@startuml
user -> Elevator_Button : Pressed
Elevator_Button -> Controller : Notifies
Controller -> Elevator_Button : Turn_LED_ON
Controller -> Elevator : Move
Elevator -> Controller : Reach_floor
Controller -> Elevator : Stop
Controller -> Elevator_Button : Turn_LED_OFF
@enduml
```

There are several coverage criterions specified by various studies [17, 18, 22] such as node coverage, edge coverage, and edge-pair coverage. In the following case, edge-pair coverage is utilized which generates more number of shorter test paths using algorithm for test path generation. For execution options and simplicity sake, we chose test logs specifying the requirements coverage. A state intermediate graph is obtained from the complex state machine to derive test paths shown in Fig. 7.

Graph edges of above figure are fed as input in the Graph Coverage Web Application developed by Paul Ammann and Jeff Offut [9]. It is an online module to generate test paths on specifying the type of coverage needed. The tester needs to specify the test requirements for which testing is sought. Test requirements describe test path properties. In edge-pair coverage, a pair of edges is required. Test requirements accommodate each reachable path from initial node to other nodes. Test criteria determine the rules which define test requirements. The graph in Fig. 7 is the simplified version of the state machine in Fig. 5.

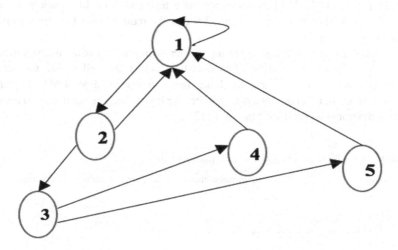

Fig. 7 State intermediate graph of cargo elevator

Table 1 Test paths using edge-pair coverage (direct tour)

Test paths	Test requirements that are toured by test paths directly
[1, 1, 2, 1]	[1, 2, 1], [1, 1, 2]
[1, 1, 1]	[1, 1, 1]
[1, 2, 3, 4, 1, 2, 1]	[1, 2, 1], [1, 2, 3], [2, 3, 4], [3, 4, 1], [4, 1, 2]
[1, 2, 3, 4, 1, 1]	[1, 2, 3], [2, 3, 4], [3, 4, 1], [4, 1, 1]
[1, 2, 3, 5, 1, 2, 1]	[1, 2, 1], [1, 2, 3], [2, 3, 5], [3, 5, 1], [5, 1, 2]
[1, 2, 3, 5, 1, 1]	[1, 2, 3], [2, 3, 5], [3, 5, 1], [5, 1, 1]

Algorithm for test path generation

Input: A directed SIG (state intermediate graph) having a set of initial nodes (i_1, i_2...) and other nodes n (n_1, n_2...)
Output: A set of test paths TP that cover all the nodes in SIG.
1→for every starting node i_a do:
2→ execute Breath-First-Search (BFS) algorithm to cover the nodes in SIG ;
3→ create a BFS-tree with the root node i_a;
4→ add the path from i_a to the node n;
5→ end for loop
6→return P

From Table 1, it is found that in order to traverse the state graph directly in Fig. 7, six test paths are needed for edge-pair coverage. In Table 1, we obtained test paths to check the validity of whether the states are traversed successfully or not. The Test Path [1, 1, 2, 1] indicates the state traversal from idle state to button pressed state as shown in Fig. 5. Table 2 shows the requirements that are traversed by the test paths indirectly.

Thus, the above process of triggering the generation of test paths is a significant method for effective utilization of models to design tests well before the actual coding initiates. Not only state machines but also several other UML diagrams which exhibit functional behavior can be deployed such a sequence diagram, activity diagrams as used by Mall et al. [23].

Table 2 Test paths with side-trips coverage (indirect tour)

Test paths	Test requirements that are toured by test paths with side-trips
[1, 1, 2, 1]	None
[1, 1, 1]	None
[1, 2, 3, 4, 1, 2, 1]	[3, 4, 1]
[1, 2, 3, 4, 1, 1]	None
[1, 2, 3, 5, 1, 2, 1]	[3, 5, 1]
[1, 2, 3, 5, 1, 1]	None

5 Conclusion and Future Work

This article has focused primarily on the inclination toward a novel model-based testing methodology for different applications. Various aspects of testing embedded applications have been discussed which ultimately provide evidences that MBT is a constructive technique to test software with a dedicated functionality. MBT also has a good hold in the automotive industry for the design of modern high-end vehicles. The next work is focused on embedded software such as advanced driver assistant and safety control systems such as seat belt system, automatic cruise control, headlight systems, and remote self-parking systems.

References

1. Weissleder S., Schlingloff, H.: An evaluation of model-based testing in embedded applications. In: Proceedings—IEEE 7th International Conference on Software Testing, Verification and Validation, ICST 2014, art. no. 6823884, pp. 223–232 (2014)
2. Takala, T., Katara, M., Harty, J.: Experiences of system-level model-based GUI testing of an android application. In: Proceedings—4th IEEE International Conference on Software Testing, Verification, and Validation, ICST 2011, pp. 377–386 (2011)
3. Muniz, L.L., Netto, U.S.C., Maia, P.H.M.: TCG: a model-based testing tool for functional and statistical testing. In: ICEIS 2015—17th International Conference on Enterprise Information Systems (2015)
4. Ferreira, R.D.F., Faria, J.P., Paiva, A.C.R.: Test coverage analysis of UML state machines. In: 2010 Third International Conference on Software Testing, Verification, and Validation Workshops (ICSTW), Paris, pp. 284–289 (2010)
5. Schulze, C., Ganesan, D., Lindvall, M., Cleaveland, R., Goldman, D.: Assessing model-based testing: an empirical study conducted in industry. In: Companion Proceedings of the 36th International Conference on Software Engineering (ICSE Companion 2014). ACM, New York, NY, USA, 135–144 (2014)
6. Caliebe, P., Herpel, T., German, R.: Dependency-based test case selection and prioritization in embedded systems. In: 2012 IEEE Fifth International Conference on Software Testing, Verification and Validation, Montreal, QC, pp. 731–735 (2012)
7. Siegl, S., Caliebe, P.: Improving model-based verification of embedded systems by analyzing component dependences. In: 6th IEEE International Symposium on Industrial and Embedded Systems, Vasteras, pp. 51–54 (2011). https://doi.org/10.1109/sies.2011.5953678
8. Iyenghar, P., Pulvermueller, E., Westerkamp, C., Wuebbelmann, J.: Infrastructure support to convey test data from state diagrams for executing MBT in embedded systems. In: EUROCON 2013 IEEE, Zagreb, pp. 651–659 (2013)
9. Liu, S.: Evaluation of model-based testing for embedded systems based on the example of the safety-critical vehicle functions. Ph.D. thesis. Oct 2012
10. Binder, R.V., Legeard, B., Kramer, A.: Model-based testing: where does it stand? Qual. Assur. 13(1), 1–9 (2015)
11. Binder, R.V., Legeard, B., Kramer, A.: In: 2016/2017 Model-based Testing: User Survey Results of Quality Assurance (2017)
12. Aggarwal, M., Sabharwal, S.: Test case generation from UML state machine diagram: a survey. In: 2012 Third International Conference on Computer and Communication Technology (2012)

13. Zander-nowicka, J.: Model-based testing of real-time embedded systems, Ph.D. thesis, In: Technical University Berlin (2009)
14. Olsen, P., Foederer, J., Tretmans, J.: Model-based testing of industrial transformational systems. In: Lecture Notes in Computer Science (including subseries Lecture Notes in Artificial Intelligence and Lecture Notes in Bioinformatics), vol. 7019, pp. 131–145. LNCS (2011)
15. Keränen, J.S., Räty, T.D.: Model-based testing of embedded systems in hardware in the loop environment. IET Softw. **6**(4), 364 (2012)
16. Pérez, A.M., Kaiser, S.: Multi-level test models for embedded systems. Softw. Eng. Conf. Proc. **2010**, 213–224 (2010)
17. Guan, J., Offutt, J.: A model-based testing technique for component-based real-time embedded systems. In: 2015 IEEE Eighth International Conference on Software Testing, Verification and Validation Workshops (ICSTW), pp. 1–10 (2015)
18. Utting, M., Legeard, B.: Practical Model-Based Testing: A Tools Approach. Morgan Kaufmann Publishers Inc., San Francisco, CA, USA (2007)
19. Zander, J., Schieferdecker, I., Mosterman, P.: A taxonomy of model-based testing for embedded systems from multiple industry domains. In: Model-Based Testing for Embedded Systems, pp. 1–22 (2011)
20. Utting, M., Legeard, B., Bouquet, F., Fourneret, E., Peureux, F., Vernotte, A.: In: Recent Advances in Model-Based Testing, 1st edn. Elsevier Inc. (2016)
21. Weissleder, S., Schlingloff, B.H.: Quality of automatically generated test cases based on OCL expressions. In: Proceedings of the 1st International Conference on Software Testing, Verification and Validation, ICST 2008, no. May, pp. 517–520 (2008)
22. Ammann, P., Offutt, J.: Introduction to Software Testing (2008)
23. Sarma, M., Kundu, D., Mall, R.: Automatic test case generation from UML sequence diagram. In: 15th International Conference on Advanced Computing and Communications (ADCOM 2007), pp. 60–65 (2007)

JSON as ORM Mapping Database Layer for the SaaS-Based Multi-tenant Application

Rajalingam Raghu and N. Sandeep Varma

Abstract SaaS-based multi-tenant application deployment model is becoming more popular as it is a more cost-effective way of deployment, quickly go online in less time. There are numerous technical or deployment issues/challenges needs to be addressed when applications are deployed in multi-tenant SaaS model, so that all tenants co-exist together and continue their operations smoothly without impacting their business operations. One of the complex and challenging design issues in multi-tenant SaaS application model is database design. When single database (shared schema with shared table) is shared among all the tenants it leads to complex database design. If all the tenants require different columns along with the core columns within the same table, it will become extremely difficult to design and maintain the database by itself. In this paper, we propose a JSON-based solution to address the complex database design for the multitenant SaaS model.

Keywords Multi-tenant · JSON · SaaS · ORM mapping

1 Introduction

As a new approach of software deployment, Software as a Service (SaaS) is getting more popular with advantages of cost-effectiveness, no need to invest and maintain the infrastructure as the service provider takes care of it. This model operates as pay per service, metered payment, etc. Over a period of time, SaaS model has been evolved and matured paving way to multitenant operating model. When the application is hosted as multi-tenant environment it may or may not require tenant-specific business need-based customization. For example, enterprise appli-

R. Raghu (✉) · N. S. Varma
Department of ISE, BMSCE, Bengaluru, India
e-mail: hairags@gmail.com

N. S. Varma
e-mail: sandeepvarma.ise@bmsce.ac.in

© Springer Nature Singapore Pte Ltd. 2018
P. K. Sa et al. (eds.), *Recent Findings in Intelligent Computing Techniques*,
Advances in Intelligent Systems and Computing 709,
https://doi.org/10.1007/978-981-10-8633-5_30

cations such as HRMS, CRMS and so on requires a considerable amount of customization in multi-tenant mode to meet the business requirements of each tenant.

2 Background

A common solution or practice to address the database customization is through either by the configuration of extended data column via relationships defined in the databases. This approach allows to extend the columns seamlessly during runtime. There are three approaches exist, viz., custom field, pre-allocation field and name-value pairs [1]. A comprehensive analysis of these approaches is discussed below.

2.1 Custom Field

One of the common solutions to address the tenant-specific business requirement is to add custom fields with the specific data type. For example, consider the B2C e-commerce application; tenants want to persist product information with the customized column as well as data as shown in Table 1 Product Description.

In Table 1, TenantID identifies specific tenants and core three columns which are common (manufacturer, location and discount) are added for column customization requirement of tenant 1003, 1004 and 1006. For tenant 1001 and 1002, these 3 columns are meaningless and filled with null values. Custom field solution is a simpler database design and implementation. In this way, we are ensuring a very simple model without having complex extended data. However, in future, if the same data set is used for data analysis or for the data warehouse it requires a considerable amount of data cleaning and data filtering. Also, more requirement for adding more custom fields to the table may have an impact on the table structure as well as for the most of the tenants these custom fields become meaningless.

2.2 Pre-allocation Field

In pre-allocation field method, extension columns are defined in the table where column extensions are required. These extension columns are used wherever extension columns are required. For example, consider B2C e-commerce system, capturing and storing the product information differs from tenants to tenants. In such scenario, extension pre-allocated columns can be used to capture and store such data as shown in Table 2.

Table 1 Main table for the product

TenantID	SKUID	ProdID	ProdName	Price	Warranty	Manfacturer	Location	Discount
1001	SKU001	001	Asus laptop	250.00	2 years	Null	Null	0
1002	SKU009	009	Lenovo laptop	300.00	3 years	Null	Null	0
1003	SKU110	110	Hard disk	50.00	Null	WD	Japan	0
1004	SKU201	201	Keyboard	10.00	Null	Logitech	Null	0
1006	SKU301	301	Mouse	13.00	Null	Zebronics	Null	10

Table 2 Column extension table where extension column such as Ext1, Ext2 and Ext3 defined

TenantID	SKUID	ProdID	ProdName	Price	Warranty	Ext1	Ext2	Ext3
1001	SKU001	001	Asus laptop	250.00	2 years	Null	Null	0
1002	SKU009	009	Lenovo laptop	300.00	3 years	Null	Null	0
1003	SKU110	110	Hard disk	50.00	Null	WD	Japan	0
1004	SKU201	201	Keyboard	10.00	Null	Logitech	Null	0
1006	SKU301	301	Mouse	13.00	Null	Zebronics	Null	10

Table 3 Relation table which defines the relation ship between Main table and Extension table for the extended column

ConfigID	TenantID	Table	Column	Content	DataType
11	1001	Product	Ext1	Warranty	String
12	1002	Product	Ext2	Location	String
13	1003	Product	Ext3	Discount	String

In Table 2, TenantID is used to identify different tenants in product information table. Along with core fields which are common to all tenants, additional three extension columns Ext1, Ext2 and Ext3 are introduced for each tenant based on their business requirement. When compared with customized field model, both of them cater to the extension column by adding an additional column to the table. However, the columns in pre-allocation strategy do not have meaning and may vary from tenant to tenant as per their business requirement. For example, Ext1 column serves to tenant 1001 to persist warranty period, tenant 1002 uses the same Ext1 column to persist location information. Having different requirements of tenants both columns wise and data wise, it is suggested in pre-allocation columns should have a generic datatype of certain fixed length which makes design simple and efficient use of extension column at runtime. In this case, a metadata configuration table is required to describe the semantics of the extension columns (Table 3).

As shown metadata persisted in Table 3, which confirms the mapping of extension data column and data type for all tenants for the table 'Product'. The extension data column is a simple way of achieving data extension, easily mapped to the corresponding data type to achieve better configurability and scalability. The limitation of pre-allocation field is that number of pre-allocated columns should be determined during the initial database design, otherwise database design will lead to complexity and difficult to maintain as the number of tenants is high or keep increasing. For efficient database design for pre-allocation field determine the number of pre-allocation fields during the initial design phase. Existence of many pre-allocated field which were not used by the tenant is meaningless as well as leading to data access space wastage. Pre-allocation fields will have a performance impact on CRUD operation as it involves primary and foreign key relationships. For better performance, complex foreign key relationships should be avoided.

2.3   Name-Value Pairs

In this model, the original and extension tables are separated to archive the data extension. Here separate extension table and configuration table are used to achieve the data extension, while the original table is used to persist the data associated with a configuration(metadata table) table to create an extension data.

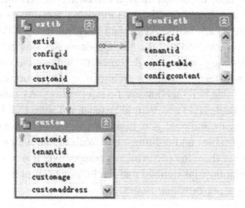

Create Custom table (tenant details), configtb (meta data information), exttb (extended data information).Identify and setup relationship between three tables for the name-value pairs model as shown in the above figure. This model on paper accommodates unlimited number of custom field extensions which lead to better scalability and flexibility for the extension data. As per the business needs of a tenant can decrease or increase the custom column. This model has better flexibility with increased data operation complexity. Data addition, modification, and deleting are complex tasks as it requires an user to follow a proper mapping between actual data and user-defined data to persist or remove from database. The complex primary and foreign key relationship will always make database query complex as well as it will decrease the database performance.

3   Related Work

There have been numerous models of multi-tenant database schema with varying columns depending upon the business requirement of the tenant has been analysed and implemented to overcome the complexities or issues posed by the tenant. Some of the studies such as private tables, extension tables, universal table, chunk folding and XML [2–6]. However the techniques discussed, designed or implemented are not mature enough, still has the relationship model which is performance wise bad and makes the design more complex. With Extensible Markup Language (XML) and relational database [7] approach, tenant's business requirement for the additional data extension can be archived without impacting the actual relational data structure. Parsing the XML documents are easier as they are well-formatted documents, easy to read as they are self-explanatory documents. As the XML documents follow the hierarchical tree data structure which easily facilitates for data extension. Many relational databases support the XML data type, to query the XML data use XPath within the SQL statements. Using XML data type, we can perform various operations seamlessly as database supports it. XML data type provides flexibility for the data extension then database-based relation model. However, XPath is the only way to retrieve the data imposing pressure on developers to learn and be fluent with XPath skills. Also, data retrieval using XPath leads to performance issues. As of now, most of the NoSQL databases support JSON data format.

Multi-tenant database schema is designed to address the common problem or issues faced in multi-tenant database architecture. The proposal included such as Elastic Extension Table (EET), contains the information on Common Tenant Tables (CTT), Extension Table (ET) and Virtual Extension Tables (VET) [8, 9]. As per this proposal, tenants create and configure their own virtual database schema including a required number of tables and columns, etc. With the above proposal, it is a clean RDBMS solution but from the performance perspective, it is not a good design as it involves lots of relationships defined and fetching data from these relations is quite complex and causes performance issues.

4 Proposed JSON Data Exchange Format Between Application and Database

The proposed solution is to use the JSON as the data exchange between the application and database. As of now most of the database supports the JSON format data persisting. We can use either fully JSON-supported database such as NoSQL DB or relational database which has the data type as JSON or partial support for the JSON data which stores the JSON as a string and has the functions to check the JSON format. With this approach, we can address the dynamic column inclusion without compromising the performance. Presentation tiers are completely implemented with a JavaScript framework such as AngularJS which uses JSON as the data object to render the presentation which simplifies the integration between presentation tier and business tier as the data exchange format will be a JSON object

Databa	se JSON data type	JSON checking function	JSON data persist
Oracle	No	Yes	No
MySql	Yes	No	Yes
Mongo	DBNo	No	Yes
couchD	B No	No	Yes
MSS QL	No	No	No

4.1 Analysis on JSON-Supported Databases

1. MySQL relationship RDBMS supports JSON column as data type. With this mechanism, database supports both relational and non-relational data.
2. MongoDB is document-based NoSQL database, which stores data in the JSON format in the database.
3. CouchDB is a JSON format database where data is stored as JSON object.
4. Oracle has the JSON-type check functions to determine as data stored in the column is JSON data or normal data.
5. MSS QL Server supports the JSON data types.

4.2 JSON as the Data for Dynamic Columns with MongoDB

To meet the requirement of the dynamically varying column, we can use MongoDB which persists the data in the form of JSON object. We need to define the bean such as below

```
public class Product {
        String      productId    ;      String
productName ; int price ; }
```

With additional column 'skuid' for the tenant1

```
public class ProductTenant1 extends Product{
        String skuid ; }
```

With additional different column 'warranty' for the tenant2

```
public class ProductTenant2 extends Product {
        String warranty ; }
```

Persist each bean in the MongoDB using the MongoDB client as below

```
MongoClient mongo = new MongoClient("localhost" , 27017);
DB db = mongo.getDB("journaldev" );
DBCollection productCollection = db. getCollection ("product" );
WriteResult result = productCollection . insert (createProductDBObject ());
WriteResult result1 = productCollection . insert (createProductDBObjectTenant1 ());
WriteResult result2 = productCollection . insert (createProductDBObjectTenant2 ());
```

Here, we can see the additional columns are added dynamically to the core table without impacting the table structure.

The JSON object format for the product is as follows

```
{
    " – id" : ObjectId ("5860023ca5781ee685a373ea") ,
    "productId" : "100" ,
    "name" : "Hard– Disk" ,
    "price" : 120
}
```

The JSON object format for the Tenant1 is as follows.

```
{
    " – id": ObjectId ("5860023ca5781ee685a373eb") ,
    "productId": "100" ,
    "name": "Hard– Disk" ,
    "price": 120,
    "skuid": "SKU100"
}
```

The JSON object format for the Tenant2 is as follows.

```
{
    " –id": ObjectId ("5860023ca5781ee685a373ec") ,
    "productId": "100",
    "name": "HardDisk", "price": 120,
    "warranty": "1year"
}
```

Update or deleting the data is just simple. instead of insert we have to use update () and remove() API.

```
DBObject query = BasicDBObjectBuilder . start (). add(" –id" ,–14).get ();
DBCursor cursor = col . find (query );
DBObject doc = createDBObject(product ); WriteResult result = col . update(query ,
doc ); result = col . remove(query );
```

Reading the data from the MongoDB is as follows.

```
DBObject query = BasicDBObjectBuilder . start (). add(" –id" ,–14).get ();
DBCursor cursor = col . find (query );
```

4.3   JSON Data Format with CouchDB

In case of CouchDB, the data accepted is in the form of JSON and stored in the database as JSON form only. Below is the snapshot of the code

```
Product product = new Product("100" ,"Hard– Disk" ,120);
ProductTenant1 product1 = new ProductTenant1("100" ,"HardDisk" ,120 ,"SKU100" );
ProductTenant2 product2 = new ProductTenant2("100" ,"Hard– Disk" ,120 ,"1– year" );
Session dbSession = new Session ("localhost" ,5984);
Database db = dbSession . getDatabase("productinformation" ); Document doc = new Document( obj . fromObject(product ));

db. saveDocument(new Document(new JSONObject()
                . fromObject(new Product("100" ,"Hard– Disk" ,120))));

db. saveDocument(new Document(new JSONObject()
                . fromObject(new ProductTenant1("100" , "HardDisk" , 120, "SKU100" ))));

db. saveDocument(new Document(new JSONObject()
                . fromObject(new ProductTenant2("100" ,"Hard Disk" ,120 ,"1– year" ))));
```

Here the true OR mapping of the table details maintained with different column names according to the business need. Product JSON data representation

```
{
    "id": "12e0be0e068625ad83a5e59fde0332b3" ,
    " –rev": "1–1acb787d237d68f1d00350aadeb6c6b4" ,
    "price": 120,
    "productId": "100" ,
    "productName": "Hard– Disk"
}
```

ProductTenant1 JSON data representation with additional column skuId

```
{
    "id": "12e0be0e068625ad83a5e59fde034181" ,
    " –rev": "1–3395f178aef4eec00c2a411f6d94c12b" ,
    "price": 120,
    "productId": "100" ,
    "productName": "Hard– Disk" ,
    "skuId": "SKU100"
}
```

ProductTenant2 JSON data representation with additional column warranty

```
{
    "id": "12e0be0e068625ad83a5e59fde034bd1",
    "-rev": "1-214b78b106d6e57b3bbb8b539cfe03ab",
    "price": 120,
    "productId": "100",
    "productName": "Hard Disk",      --
    "warranty": "1 year"      --
}
```

4.4   JSON as Column Data Type in MySQL

Create a table product as per below SQL statement

```
CREATE TABLE 'product' (
        'productId' varchar(50) DEFAULT NULL,
        'productName' varchar(45) DEFAULT NULL,
        'price' int (11) DEFAULT NULL,
        'productAdditionalInfo' json DEFAULT NULL
) ENGINE=InnoDB DEFAULT CHARSET=utf8 ; SELECT * FROM sampledatabase .

product;
```

Here, the ProductAdditionalInfo column is declared as the data type JSON. Now with the column data type declared as JSON, we can add additional varying columns in JSON object and persist in the database. Here is the screenshot of the table how data persisted

ProductId	ProductName	Price	ProductAdditionalInfo
100	Hard disk	100	Null
101	Hard disk	100	{"skuId":"SKU101"}
102	Hard disk	100	{"warranty":"1 year"}

We can create another flavour of the table to avoid the null values for the tenant who does not have any additional column requirement. Maintain the ProductId in the table column and rest as a JSON object as shown below

```
CREATE TABLE ' productInfo ' (
' productId ' varchar(50) DEFAULT NULL,
        ' productAdditionalInfo ' json DEFAULT NULL ) ENGINE=InnoDB DEFAULT
CHARSET=utf8 ;
```

With this table setup, we can avoid the null values as some time null values causes performance issues while retrieving the data from the table.

ProductId	ProductAdditionalinfo
100	{"price":120,"ProductName":"Hard Disk"}
101	{"price":120,"skuId":"SKU100","productName":"Hard Disk"}
102	{"price":120,"warranty":"1 year","productName":"Hard Disk"}

This is a combination of both relational as well as non-relational data. The data conversion from bean to JSON format as well as JSON to Java object format can be done simply by using the conversion framework. Here design and implementation are clean.

4.5 JSON as Column Data Type in Oracle

Oracle with 12.x version onwards started supporting the JSON object in the database.

We can create a table for the JSON support by adding constraint as shown below

```
CREATE TABLE "SYSTEM"."PRODUCT"                    (
    "PRODUCTID" VARCHAR2(20 BYTE),
    "PRODUCTNAME" VARCHAR2(20 BYTE),
    "PRICE" NUMBER,
    "PRODUCTADDITIONALINFO" CLOB
        CONSTRAINT ensure – json CHECK ("PRODUCTADDITIONALINFO" IS JSON)                    );
```

Oracle provides lots of inbuilt functionality, as well as conditions, check to handle the JSON data while retrieving the value or persisting the values in the database.

Below are the few functions as well as the conditions listed.

1. Functions—json value, json query and json table
2. Conditions—json exists, is json, is not json, and json textcontains.

5 Caution on Using JSON Data as the Data Layer

Using the JSON for the data layer may not be applicable for all scenarios. It may not be applicable in some cases. It should be used after thorough analysis whether it fits the business needs and improve the performance as well simplifies database design. At present most of the databases supports the JSON data as the database persisting format.

6 Merits of JSON Data as the Database Layer

There are numerous advantages of using the JSON data as the database layer.

1. By using the JSON data as the database layer will simplify the database design. There will not be any complex inner, outer or left joins query required to retrieve the data from the database.
2. When there is no complex relation tables there will be good performance while storing, update or reading the data from the database.

3. As the JSON data will be stored as a key, value documents in the MongoDB or CouchDB as a collection. We can index the document collection for the better performance for report generation or search functionality.
4. As the JSON data will be stored as a key, value documents in the MongoDB or CouchDB as a collection. We can index the document collection for the better performance for report generation or search functionality.
5. NoSQL database has evolved over a period of time and has improved a lot from the performance. NoSQL database supports transaction as well as ACID properties as supported by the relational database.

7 Conclusion

In this paper, we addressed data layer by using JSON data format for persisting the data with the support for extension columns of different data types as per the tenant requirement. User has to either go for pure document way of handling the data without any relation among the extended columns using JSON data format supported by pure documentation persistence databases like MongoDB, CouchDB, etc. If the user wants to maintain relationships with non-relation data then go with a few named databases such as MySQL, Oracle databases, etc. There is an advantage of using pure JSON format database design as the user can maintain the ORM mapping with retaining OOPS concepts of the hierarchical structure while storing or retrieving the data from the database. Our approach has advantages: simplifies the database design, performance is improved and provides flexibility and scalability to address the need of dynamic columns not compromising on performance. Our proof of concepts clearly demonstrates the said advantages.

References

1. Chong, F., Carraro, G., Wolter, R.: Microsoft Corporation, "Multi-Tenant Data Architecture". https://msdn.microsoft.com/enus/library/aa479086.aspx (2006). Accessed June 2006
2. Hudli, A.V., Shivaradhya, B., Hudli, R.V.: Level-4 SaaS applications for healthcareindustry. In: Proceedings of the 2nd Bangalore Annual Compute Conference, Bangalore, India, p. 19 (2009)
3. Gao, B., An, W.H., Sun, X., Wang, Z.H., Fan, L., Guo, C.J., Sun, W.: A non-intrusivemulti-tenant database software for large scale SaaS application. In: e-Business Engineering, Beijing, China, pp. 324–328 (2011)
4. Xia, C., Yu, G., Tang, M.: Efficient implement of ORM (object/relational mapping) usein J2EE framework: hibernate. In: Computational Intelligence and Software Engineering, pp. 1–3 (2009)
5. Liu, G.: Research on independent SaaS platform. In: Information Management and Engineering, Chengdu, China, pp. 110–113 (2010)
6. Chen, W., Shen, B., Qi, Z.: Template-based business logic customization for SaaS applications. In: Progress in Informatics and Computing, vol. 1, Shanghai, China, pp. 584–588 (2010)

7. Du, J., Wen, H.Y., Yang, Z.J.: Research on data layer structure of multi-tenant ecommerce system. In: 2010 IEEE 17th International Conference on Industrial Engineering and Engineering Management, Xiamen, pp. 362–365 (2010). https://doi.org/10.1109/icieem.2010.5646593
8. Yaish, H., Goyal, M.: A multi-tenant database architecture design for software applications. In: 2013 IEEE 16th International Conference on Computational Science and Engineering, Sydney, NSW, pp. 933–940 (2013). https://doi.org/10.1109/cse.2013.139
9. Yaish, H., Goyal, M., Feuerlicht, G.: An elastic multi-tenant database schema for software as a service. In: 2011 IEEE Ninth International Conference on Dependable, Autonomic and Secure Computing, Sydney, NSW, pp. 737–743 (2011). https://doi.org/10.1109/dasc.2011.127

Secure Query Processing Over Encrypted Database Through CryptDB

Ashutosh Kumar and Muzzammil Hussain

Abstract Cloud computing has changed the world of technology business by providing more flexible services over the Internet. DaaS is a service, where the organization can store their data at the heavy server at remote sites of the service provider and do their transaction seamlessly and without much effort. But this has raised new security issues from both client and server side. In this paper, we have addressed both the issues separately to secure the data at server side by encrypting data using CryptDB and secure it from client-side queries and data are encrypted using cipher technique. We use asymmetric key encryption mechanism for confidentiality and a digital signature for authentication.

Keywords CryptDB · Asymmetric key · Digital signature · Confidentiality
Integrity Authentication

1 Introduction

Today, in the era of modernization there is a race between organizations to provide online service and applications to users as cheaper as possible. In these races, new organizations are coming every day and it has become difficult for them to store large amount of data. So they have found an intermediate way. Rather than purchasing new hardware, they are hiring databases, big server, and network administrators. This idea is to not only cut the maintenance cost but also increases the efficiency. As our work is on databases, we are concerned with the database. Database as a service (DaaS) provided by database service provider (DSP) to store, retrieve and modify data anywhere in the world with the help of Internet.

A. Kumar · M. Hussain (✉)
Department of Computer Science and Engineering, Central University
of Rajasthsan, Kishangarh 305817, Rajasthan, India
e-mail: mhussain@curaj.ac.in

A. Kumar
e-mail: Akashutosh09@gmail.com

© Springer Nature Singapore Pte Ltd. 2018
P. K. Sa et al. (eds.), *Recent Findings in Intelligent Computing Techniques*,
Advances in Intelligent Systems and Computing 709,
https://doi.org/10.1007/978-981-10-8633-5_31

Database service provider provides these services on nominal rental packages. This model generates a solution to manage a large amount of database [1].

When it comes to database as a service (DAAS) model, then security and privacy are major issues. Every organization has some confidential data and when it goes to a third-party database, curious database administrators try to access these confidential information from the database. The DSP's main concern is to protect data from such type of database administrators. The vulnerable point is not only this curious database administrator, but an attacker also. Attackers attack on database server and extract all sophisticated data of the users or any organization and can do forgery with this information.

Now DSP has two challenges; first is to protect data from attackers and second is to maintain secrecy and privacy to protect data from databases as well as network administrators. One of the solutions to this problem is to store data in an encrypted format and send an encrypted query to the database server. Database server processes this encrypted query by fetching results from the encrypted database and these encrypted results are sent to the user or client side, client or user decrypts the data and get actual plain text result. This is a good solution but not all SQL queries run over the encrypted database. Till now, there is no complete system which can process all encrypted queries.

Second challenge is our motivation and improving the efficiency of query processing over encrypted database in our proposed algorithm. In our proposed work, we mainly focus on the second challenge of DaaS model.

In this proposed mechanism, all data are being stored at database service provider. Client sends encrypted query to database server. It processes encrypted query over the encrypted database and sends the retrieved result to the client side. Client processes encrypted data, decrypts it, and after decryption, he or she gets actual results. For this work, client as well as the server performs different encryption and decryption techniques which support different types of encrypted database [2, 3].

2 Related Work

Reluka Ada Popa developed CryptDB [4, 5], it mainly deals with encrypted query and encrypted database. In CryptDB, proxy server is kept between the users and the database server, which takes normal plain text query from users and converts it into encrypted format and these encrypted queries are sent to database server. Database server, which is in encrypted format processes query and fetches the encrypted result from database and sends it to the proxy server. It decrypts the query result and sends it to the user. Column encryption techniques are used in the CryptDB, which means column wise encryption technique is used in CryptDB or we can also say that different columns encrypt with different column encryption techniques. The benefit of doing this, is intentionally or unintentionally when DBA will be trying to access the information from database server he or she will not be able to do it [6, 7]. Onion layer encryption techniques are used to encrypt a column. It means multiple layers of encryptions are used to encrypt data in a column. When any request comes for query processing first, the column is decrypted in such a way that data privacy does not break, and then query processing is done on that column [8–11].

3 Proposed Work

In our mechanism, two things are proposed, first to provide data integrity during query processing on the encrypted database and second to provide partitioning data storage so as to increase the efficiency of data processing. Previous works mostly focus on data confidentiality, in the proposed algorithm we try to address authentication, integrity, availability, confidentiality.

Figure 1 shows data communication between the client and server. Following steps are followed during this client, server communication

Step1: Client shares public key with the server, for the verification of digital signature
Sent by client.
Step2: Server shares public key with the client, for the verification of digital signature
Sent by server.
Step3: Encrypted queries are transferred to database server, for further processing of
Data and encrypted data are stored in the database.
Step4: Encrypted results are transferred to client, for further processing of query.
Step5: Client processes encrypted results and converts them into plain text.

The proposed mechanism consists of two entities, first entity is trusted client and second section entity is untrusted database service provider. Architecture and

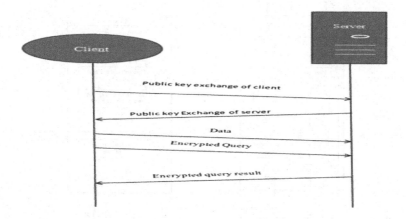

Fig. 1 Data flow diagram

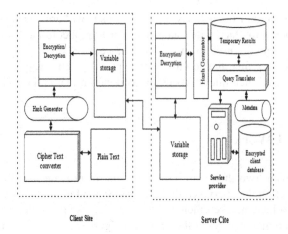

Fig. 2 Proposed system

workflow are shown in Fig. 2 client performs encryption, decryption.DSP performs storage of encrypted database and processes query over encrypted database. Client performs encryption of data, plain text query into cipher text hash generator, encryption and decryption, etc. User poses plain text query it goes to ciphertext converter, ciphertext converter converts plain text to cipher text, hash generator generates hash of ciphertext and then generates signature and encrypts cipher query and signature and send it to server. Server consists of encryptor/decryptor, hash generator, query translator, and database server. Server receives encrypted data, it decrypts data with its own private key and extract cipher query and calculate hash if both hashes match then proceed for further otherwise discard. Query translator translates encrypted ciphertext into cryptDB format and fetch results and send back to client followed by reverse process Fig. 3.

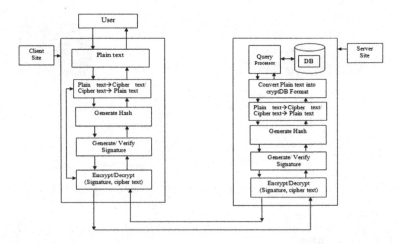

Fig. 3 Flow diagram

Proposed Algorithm

The proposed algorithm is divided into sections. First section is algorithm for database creation and second is query processing.

Algorithm for database creation

Algorithm 3.1 Database Creation
---

1. For Every relation $R(x_1, x_2, x_3, \ldots\ldots\ldots\ldots\ldots\ldots\ldots\ldots\ldots\ldots x_n)$.
2. Convert R into cipher text $R^c, R^c: R \rightarrow R^c$.
3. Calculate Hash of R^c.
4. Generate signature: signature = (Hash $(R^c))_{KRC.}$
5. Then encrypt signature and cipher text with public key of server: X= (signature, $R^c)_{KUS}$.
6. Send X to server.
7. Server receives X.
8. Decrypt X with private key of server.
9. Generate hash and verify signature.
10. Convert cipher text R^c into plain text R, R: $R^c \rightarrow R$.
11. Convert plain text query into encrypted format, RS: R $\rightarrow R^s$.
    $R^s = <$etuple,c_1-rnd,c_1-det,c_1-ope$\ldots\ldots\ldots\ldots\ldots\ldots\ldots\ldots c_1$-rnd,$c_1$-det,$c_1$-ope $>$
    Where etuple = $E_k(x_1, x_2, x_3, \ldots\ldots\ldots\ldots\ldots\ldots\ldots\ldots\ldots x_n)$.
    c_1-rnd = E_k rnd(x_1).
    c_1-det = E_k det(x_1).
    c_1-ope = E_k ope(x_1).
    .
    .
    .
    C_n-rnd = E_k rnd(x_n).
    C_n-det = E_k det(x_n).
    C_n-ope = E_k ope(x_n).
12. Store R^s on database.

Here Step 10 to 11 process in Secure query Translator
Where: R→Database Relation.
        R^c→Cipher form of Relation.
        KRC→Private Key of client.
        KUS→Public key of server.
        Rnd→Random.
        Det→ Deterministic.
        Ope→Order-Preserving Encryption.

Algorithm for Query processing

Algorithm 3.2 Query Execution

---

1. Query generated by user: Q.
2. Convert Q into cipher text Q^c,Q^c:Q$\rightarrow$ Q^c.
3. Calculate Hash of Q^c.
4. Generate signature: signature = (Hash (Q^c))$_{KRC.}$
5. Then encrypt signature and cipher text with public key of server: Y= (signature, Q^c)$_{KUS}$.
6. Send Y to server.
7. Server receives Y.
8. Decrypt Y with private key of server.
9. Generate hash and verify signature.
10. Convert cipher text Q^c into plain text Q, Q: $Q^c$$\rightarrow$Q.
11. Convert plain text query into CryptDB format ,Q^s: Q $\rightarrow$ Q^s.
12. Database server Process the query and send back result
    Q^p: Q^s $\rightarrow$ Q^p.
13. Result is generated by server in cryptDB format. Res^s: Q^p $\rightarrow$ Res^s.
14. Reg^s again convert into plain text Reg.
    Res: Res^s $\rightarrow$ Res .
15. Convert Res into cipher text Res^c, Res^c: Res $\rightarrow$ Res^c.
16. Calculate Hash of Res^c.
17. Generate signature: signature = (Hash (Res^c))$_{KRS.}$
19. Then encrypt signature and cipher text with public key of client:
    Y= (signature, Res^c)$_{KUC}$.
20. Send Y to client.
21. Client received Y.
22. Decrypt Y with private key of client.
23. Generate hash and verify signature.
24. Convert cipher text Res^c into plain text Res, Res: Res^c$\rightarrow$ Res.
25. User gets query result.

Steps 11 to 15 execute on Secure Query Translator
Where: Q $\rightarrow$Query request by user.
        Q^c $\rightarrow$Cipher form of query.
        $Q^s$$\rightarrow$Convert into cryptDB format.
        $Q^p$$\rightarrow$Process Query.
Res$\rightarrow$Result.
        Res^s$\rightarrow$Query Result in CryptDB format.
        Res^c$\rightarrow$Result in cipher text.
        KRC$\rightarrow$private key of client.
        KUS$\rightarrow$public key of server.
        KRS$\rightarrow$private key of server.
        KUC$\rightarrow$public key of client.

Query Execution with Example

To understand query execution, we use user_info table of users' database in which data is stored in plain text. This table is stored in the database server but before storing into the database, it is transformed into encrypted form so that any curious database administrator as well as any intermediary intruder will not be able to read or tamper with queries Tables 1, 2, 3, 4, 5, and 6.

Table 1 User_info table encrypt in the form of CryptDB

User_id	Name	Email	Password
001	Aishwarya	Aishwarya@gmail.com	Aishwarya123
002	Ashutosh	Ashutosh@gmail.com	Ashutosh786
003	Chand	Chand@gmail.com	Chand007
004	Bharti	Bharti@gmail.com	Bharti467
005	Deepak	Deepak@gmail.com	Deepak124

ET = etuple = EK(001, Aishwarya, Aishwarya@gmail.com, Aishwarya123)
ET = etuple = EK(002, Ashutosh, Ashutosh@gmail.com, Ashutosh786)
ET = etuple = EK(003, Chand, Chand@gmail.com, Chand007)
ET = etuple = EK(004, Bharti, Bharti@gmail.com, Bharti467)
ET = etuple = EK(005, Deepak, Deepak@gmail.com, Deepal124)

Table 2 User_id attribute are encrypted as

RND1 = EKrnd(001) DET1 = EKdet(001) OPE1 = EKope(001)	RND4 = EKrnd(004) DET4 = EKdet(004) OPE4 = EKope(004)
RND2 = EKrnd(002) DET2 = EKdet(002) OPE2 = EKope(002)	RND5 = EKrnd(005) DET5 = EKdet(005) OPE5 = EKope(005)
RND3 = EKrnd(003) DET3 = EKdet(003) OPE3 = EKope(003)	

Table 3 Name attributes are encrypted

Search01 = EKSearch(Aishwarya)
Search02 = EKSearch(Ashutosh)
Search03 = EKSearch(Chand)
Search04 = EKSearch(Bharti)
Search05 = EKSearch(Deepak)

Table 4 Email encrypted as

Search11 = EKsearch(Aishwarya@gmail.com)
Search12 = EKsearch(Ashutosh@gmail.com)
Search13 = EKsearch(Chand@gmail.com)
Search14 = EKsearch(Bharti@gmail.com)
Search15 = EKsearch(Deepak@gmail.com)

Table 5 Password encrypted as

Search21 = EKSearch(Aishwarya123)
Search22 = EKSearch(Ashutosh786)
Search23 = EKSearch(Chand007)
Search24 = EKSearch(Bharti467)
Search25 = EKSearch(Deepak124)

Table 6 After encryption user_info table encrypt like this

etuple	C1_rnd	C1_det	C1_ope	X_Search	Y_search	Z_search
ET1	RND1	DET1	OPE1	SEARCH01	SEARCH11	SEARCH21
ET2	RND2	DET2	OPE2	SEARCH02	SEARCH12	SEARCH22
ET3	RND3	DET3	OPE3	SEARCH03	SEARCH13	SEARCH23
ET4	RND4	DET4	OPE4	SEARCH04	SEARCH14	SEARCH24
ET5	RND5	DET5	OPE5	SEARCH05	SEARCH15	SEARCH25

Step wise execution of query followed by the proposed algorithm.

Step: 1 "SELECT * FROM users WHERE name LIKE='Bharti'";

Step: 2 Query= (SELECT * FROM users WHERE name LIKE='Bharti')$_{Cipher}$

Step: 3 Hash= H(Query)

Step: 4 Signature=$E_{kClient\_Secret}$(Hash)

Step: 5 Y= $E_{kserver\_public}$(Signature, Query)
   Y sends to step 6

Step: 6 Decrypting-----> $E_{kserver\_private}$(Signature, Query)

Step: 7 Hash= H(Query)

Step:8 $E_{kClient\_Public}$(Hash)

Step: 9 Hash matches SELECT * FROM users WHERE nameLIKE='Bharti'

Step: 10 SELECT etuple FROM users^s WHERE X_Search='SEARCH04'

Step: 11 (004, Bharti, Bharti@gmail.com, Bharti467)

Secure Query Translator

   Step: 9, step: 10 and Step: 11 processes on secure query translator on server side and these three steps process at one phase

Step: 12 Hash= H(Query)

Step: 13 Signature=$E_{kserver\_prive}$(Hash)

Step: 14 Y= E_{client_public}(Signature, Query)
Y sends to the server....

Step: 15 Decrypting----->$E_{kclient\_private}$(Signature, Query)

Step: 16 Hash= H(Query)

Step: 17 $E_{kserver\_Public}$(Hash)

Step: 18 Query-> Plain text(004, Bharti, Bharti@gmail.com, Bharti467)
   And that is the final query result
   004, Bharti, Bharti@gmail.com, Bharti467

4 Security Analysis

X.1 Environment
The proposed protocol is analyzed for security by modeling it in AVISPA under various back ends

```
sonal@ubuntu:~$ avispa relation.hlpsl --ofmc
% OFMC
% Version of 2006/02/13
SUMMARY
  SAFE
DETAILS
  BOUNDED_NUMBER_OF_SESSIONS
PROTOCOL
  /home/sonal/avispa-1.1/testsuite/results/relation.if
GOAL
  as_specified
BACKEND
  OFMC
COMMENTS
STATISTICS
  parseTime: 0.00s
  searchTime: 0.02s
  visitedNodes: 3 nodes
  depth: 2 plies
```

Fig. 4 For first protocol (Relation) relation.hlpsl Adversary model: Dolev–Yao model Goals: Secrecy of relation table, Authentication of client at server

Avispa Version	: Avispa-1.1
Type of tool	: Protocol Checker modelling tool
Modeling Language	: High Level Protocol Specification Language (HLPSL)
Back-end	: On–the–fly Model–Checker(OFMC)
Interface	: Shell Prompt
Roles	: Client (Initiator), Server (Responder)
Adversary model: Dolev-Yao model	**Adversary model: Dolev-Yao model**
Goals: Secrecy of relation table,	**Goals: Secrecy of data,**
Authentication of client at server	**Authentication of client at server**

The result generated by AVISPA in Figs. 4 and 5 shows that the proposed mechanism is safe and immune to many potential security attacks.

```
sonal@ubuntu:~$ avispa cryptDB.hlpsl --ofmc
% OFMC
% Version of 2006/02/13
SUMMARY
  SAFE
DETAILS
  BOUNDED_NUMBER_OF_SESSIONS
PROTOCOL
  /home/sonal/avispa-1.1/testsuite/results/cryptDB.if
GOAL
  as_specified
BACKEND
  OFMC
COMMENTS
STATISTICS
  parseTime: 0.00s
  searchTime: 0.03s
  visitedNodes: 4 nodes
  depth: 3 plies
```

Fig. 5 For second protocol (CryptDB) cryptDB.hlpsl Goals: Secrecy of data, Authentication of client at server

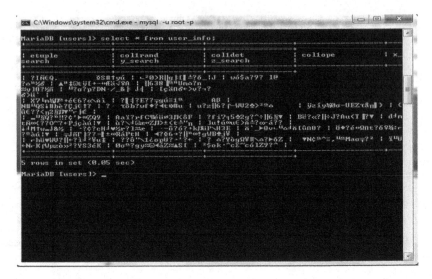

Fig. 6 Encrypted query result

5 Simulation Results

The proposed protocol is simulated using the following parameters given below to
verify its performance

Experimental Setup
Database Use:	Mysql 5.5.54
Front/Backend:	Php 5.3.10
Server:	Xampp 5.5.38.3

If a curious database administrator (DBA) tries to access data from database
server it gets nothing. If DBA at back end tries to access user_info table, then the
result is like as shown in Fig. 6.

6 Time and Storage Chart

The following bar chart shows [12] time and storage taken during normal query
execution and query execution using CryptDB. This is the comparison between
normal query execution in MySql and query execution using Cryptdb Figs. 7 and 8.

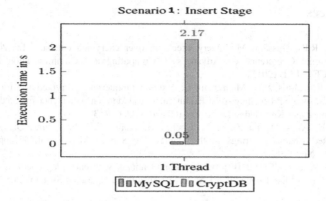

Fig. 7 Time comparison between simple query execution and CryptDB query execution [12]

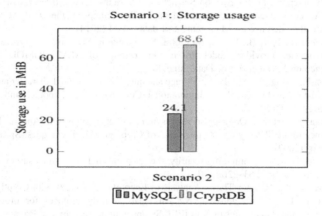

Fig. 8 Storage comparison between simple query execution and CryptDB query execution [12]

7 Conclusions

Nowadays security of data is one of the biggest challenges for enterprises and users. Data as a service (DaaS) provides the facility to users to access the data anywhere in the world. But these data needs to be exchanged securely through the Internet. In our proposed algorithm, client encrypts and signs the data using public key encryption method and digital signature, and then sends it to DB Server. The proposed protocol verify in AVISPA and it has been found that the proposed protocol is more efficient than CryptDB. The proposed mechanism also ensures authenticity, integrity, and confidentiality also reducing communication overhead.

References

1. Kumar, R.R., Hussain, M.: Query execution over encrypted database. In: 2015 Second International Conference on Advances in Computing and Communication Engineering (ICACCE). IEEE (2015)
2. Ravan, R.R., Idris, N.B., Mehrabani, Z.: A survey on querying encrypted data for database as a service. In: the Proceedings of IEEE International Conference on Cyber-Enabled Distributed Computing and Knowledge Discovery (CyberC), Oct 2013
3. Wu, Z.D., Xu, G.D., Yu, Z., Yi, X., Chen, E.H., Zhang, Y.C.: Executing SQL Queries Over Encrypted Character Strings in the Database-As-Service Model, vol. 35, pp. 332–348. Elsevier, Nov 2012
4. Popa, R.A. et al.: CryptDB: protecting confidentiality with encrypted query processing. In: Proceedings of the Twenty-Third ACM Symposium on Operating Systems Principles. ACM (2011)
5. Popa, R.A., Zeldovich, N., Balakrishnan, H.: CryptDB: a Practical Encrypted Relational DBMS (2011)
6. Xiao, L., Yen, I.-L.: Security analysis for order preserving encryption schemes. In: 2012 46th Annual Conference on Information Sciences and Systems (CISS). IEEE (2012)
7. Arasu, A., et al.: Querying encrypted data. In: Proceedings of the 2014 ACM SIGMOD international conference on Management of data. ACM (2014)
8. Hacigumus, H., Iyer, B., Li, C., Mehrotra, S.: Executing SQL over encrypted data in the database service provider model. In: the Proceedings of ACM SIGMOD International Conference on Management of Data, June 2002
9. Wang, Z.-F., Wang, W., Shi, B.-L.: Storage and query over encrypted character and numerical data in database. In: The Fifth International Conference on Computer and Information Technology (CIT'05). IEEE (2005)
10. Boldyreva, A., et al.: Order-preserving symmetric encryption. In: Annual International Conference on the Theory and Applications of Cryptographic Techniques. Springer, Berlin, Heidelberg (2009)
11. Ryan, Mark D.: Cloud computing security: the scientific challenge, and a survey of solutions. J Syst Softw **86**(9), 2263–2268 (2013)
12. Skiba, M., et al.: Bachelor Thesis Analysis of Encrypted Databases with CryptDB
13. Popa, R.A., Li, F.H., Zeldovich, N.: An ideal- security protocol for order-preserving encoding. In: the Proceedings of 34th IEEE Symposium on Security and Privacy (IEEE S&P/Oakland), May 2013
14. Liu, L., Gai, J.: A method of query over encrypted data in database. In: the Proceedings of 09th IEEE International Conference on Computer Engineering and Technology, Jan 2009
15. Arasu, A., Eguro, K., Kaushik, R., Ramamurthy, R.: Querying encrypted data. In: the Proceedings of 29th IEEE International conference on Data Engineering (ICDE), Apr 2013
16. Liu, L., Gai, J.: A method of query over encrypted data in database. In: International Conference on Computer Engineering and Technology, 2009. ICCET'09, vol. 1. IEEE (2009)
17. Hacıgümüş, H., Iyer, B., Mehrotra, S.: Query optimization in encrypted database systems. In: International Conference on Database Systems for Advanced Applications. Springer, Berlin, Heidelberg (2005)
18. Hacigumus, V.H., Iyer, B.R., Mehrotra, S.: Query optimization in encrypted database systems. US Patent No. 7,685,437. 23 Mar 2010

19. Smart, N.P., Vercauteren, F.: Fully Homomorphic Encryption with Relatively Small Key and Ciphertext Sizes. International Workshop on Public Key Cryptography. Springer, Berlin, Heidelberg (2010)
20. Ge, T., Zdonik, S.: Fast, secure encryption for indexing in a column-oriented DBMS. In: 2007 IEEE 23rd International Conference on Data Engineering. IEEE (2007)
21. Bao, F., et al.: Private query on encrypted data in multi-user settings. In: International Conference on Information Security Practice and Experience. Springer, Berlin, Heidelberg (2008)

Sustainable Software Development with a Shared Repository

Tribid Debbarma and K. Chandrasekaran

Abstract Cloud-based computing services trend has significantly increased the building of new cloud data centres in the recent times. These data centres consume a large amount of electrical energy to run the Information Communication Technology (ICT) devices and for its cooling, as a result of that emission of CO_2 and Greenhouse Gas (GHG) has increased largely. This paper proposes a Sustainable Software Development Life Cycle (SDLC) Model. This model would contribute towards making the cloud computing systems more energy efficient and the ICT a green and sustainable development ecosystem. The Green Cloud Computing will be considered as an implementation platform for the proposed Sustainable SDLC. It is proposed that reusable, modular and cloud-shared development approach be used to make the software development and the software more energy efficient. The SDLC will use a cloud repository which will be accessible as per software development organizations cloud deployment model.

Keywords SDLC · Green cloud · Green software · Sustainable software Repository

1 Introduction and Motivation

Cloud computing systems has started evolving since the last decade of the twenty-first century. Initially, technologists were concerned only toward computing power and availability of the computing resources, without giving much concern about its energy consumption issues in the cloud computing data centres. Many studies by

T. Debbarma (✉)
Computer Science and Engineering Department, National Institute
of Technology Agartala, Agartala, Tripura, India
e-mail: dtribid@gmail.com

K. Chandrasekaran
Computer Science and Engineering Department, National Institute
of Technology Karnataka, Surathkal, Mangalore, Karnataka, India

© Springer Nature Singapore Pte Ltd. 2018
P. K. Sa et al. (eds.), *Recent Findings in Intelligent Computing Techniques*,
Advances in Intelligent Systems and Computing 709,
https://doi.org/10.1007/978-981-10-8633-5_32

321

different organizations and researchers give us an idea about the alarming situation of the worldwide environment pollution and the contribution of data centres and ICT at present and in the future. U.S. datacenters consumed an estimated 91 billion KWh of electricity in 2013 and it is estimated that by year 2020 it would increase to approximately 140 billion KWh. This would cause an estimated 150 million metric tons of carbon emission on the basis of coal-fired power plants emission [1]. According to [2] analysis based on US EPA's report 113 ICT companies in the US consumed 59 billion KWh of electricity in 2014 which was 1.5% of the total electricity consumption for the country. Out of the 59 billion KWh, only 8.3 billion KWh (14%) was sourced from renewable electricity. It was predicted that by the year 2020 the group of 113 companies could use 18.5 billion KWh to above 37 billion kWh of renewable electricity, which is 31 and 48% renewable electricity use on industry average basis.

GeSI SMARTer [3] in 2011 data centers emission was only 0.16 $GtCO_2e$ which was 17% of all ICT emissions, it is expected to reach 0.29 $GtCO_2e$ in 2020 assuming 7.1% growth rate.

Corcoran and Andrae [4] for ICT electricity consumption, it is estimated that by year 2017 datacenters power consumption will grow from 15% to an estimated 21% and Networks based consumption will increase upto 39% due to high data traffic requirements.

In one of the studies, it is found that Small and Medium organizations are responsible for approximately half of all U.S. server electricity consumption but almost 50% of that is wasted due to lack of awareness and less efficient management as highlighted in [5].

1.1 Motivation

The requirements for computing power is increasing day by day which increases the electricity requirement. To meet this computing requirement, it is necessary the cloud computing systems be made energy efficient with proper measures and steps, so that the negative effects of ICT is at the minimal. Many studies were done to minimise the energy consumption of cloud computing systems using the hardware-based approaches. Limited numbers of studies have been done so far to resolve these issues by means of sustainable software development in cloud computing platforms.

1.2 Contributions

A sustainable software development lifecycle model. A repository-based support system for software development in the cloud environment is also proposed as part of the proposed SDLC. Some important procedures and guidelines for sustainable ICT development ecosystem is also suggested.

The rest of the paper has been organised as follows- Sect. 2 discusses about different works in the field of green sustainable software development engineering, efficiency, and awareness on green engineering in ICT. Section 3 proposes model for green and sustainable software development. Section 4 discusses about the proposals to achieve sustainability in ICT. Section 5 conclusion of the paper.

2 Background and Related Works

In [6] the green engineering processes adopted at the earlier stages of software and hardware development engineering would greatly help us in attaining green environment, sustainable software and green cloud computing.

Tom Mochal and Andrea Krasnoff of Tenstep [7] has coined some concepts to manage the software development process, they suggested organizations to incorporate environmental policies into the project management processes.

Cloud computing was designed and developed to overcome the rising needs of computing powers by the organizations and individuals. Various efforts and studies were done to make ICT systems computationally efficient and 24/7 available through different networks. With the increase in computing power, consumption of electrical energy increases and produces heat as by-product and energy is wasted because of inefficient management of the cloud computing data centres. To control this heat, data centres use cooling systems which in turn uses large amount of electrical energy and most of the electrical energy is wasted in this process, making the cloud computing technology a major contributor of the worldwide CO_2 and GHG emissions. Many research works have been done and many works are going on to reduce the energy consumption of ICT systems by means of making the cloud computing more energy efficient and sustainable one. In this regard, different methods have been considered to address the energy consumption issues, e.g., adopting different hardware control approaches such as power-aware CPU scheduling, virtual machine management, network and cooling systems management, storage devices management, etc. Research is also being carried out with the use of different software engineering approaches to make ICT system more energy efficient and sustainable.

"GreenSoft" [8] suggested different green software development steps and guidelines to make the software development sustainable. The authors discussed a cradle-to-grave inspired life cycle to show the three different order effects of a software in its life cycle.

In [9] the proposed Green Software Development life cycle with seven phases each with guidelines to manage the green qualities of software and also two supporting processes namely Quality and Infrastructure. Chauhan and Saxena [10] in their paper proposed an SDLC model with supporting processes to meet the green qualities in the SDLC.

In [11], green software life cycle includes four green phases in different activities of the macro-spiral SDLC to meet the green constraints. The authors provided the green efficiency with sub-characteristics as green computation efficiency, green data management, green data communication, energy consumption awareness in the green external/internal quality model.

3   Proposal Towards a Sustainable and Energy Efficient Software Development Life Cycle Model

This paper proposes a Green and Sustainable Software Development Model. It will be deployed in the cloud platform and output of the software development will be compatible for cloud and standalone ICT devices. It emphasises on different steps to reduce the energy consumption by means of effective use of software techniques and adaptation of green and sustainable utilization of the ICT ecosystem. Software development processes can have a great impact on the power consumption of ICT systems and it also must be ensured that deployed software should be developed with all the sustainability metrics and quality metrics to make the target system energy efficient. In [12], it is asserted that for achieving computational efficiencies through software by considering different techniques like efficient algorithms, multithreading, vectorization. As efficient software takes less time to complete a work, it will consume less energy for the same work compared to the system using an inefficient software.

In [13], creation of a performance knowledge base which could act as repository for later use and sharing among designers for Software Performance Engineering (SPE) issues is suggested. Current software development scenarios differ from the earlier days software development which used to be independent or local datacenters/servers are used in independent systems.

The proposed Software Development Life Cycle model consists of seven generic development phase, three supporting phase to meet the green and sustainability requirements, three sub-phase for verification and a repository service to support the green and sustainability requirements of SDLC. The proposed Sustainable Software Development Life Cycle model is shown in the Fig. 1. The various phases, its functionalities and requirements are discussed below.

3.1   Requirements Analysis with Green and Sustainable Criteria

It is an important phase of the sustainable SDLC. It must be ensured to collect all the requirements not only to meet the computational efficiency requirements but also the sustainability requirements and energy efficiency for each and every possible components and modules. Different tools and guidelines of sustainability metrics and standards must be followed. Based on the software where it is intended to be deployed the criteria must be finalized.

Fig. 1 Proposed sustainable software development life cycle model

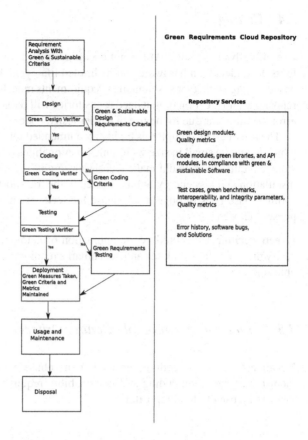

3.2 Design

The proper design of a software can greatly reduce the later burden of maintenance and it must consider all the integrity, interoperability issues, security and sustainability requirements also must be considered. A sub-phase is added with this phase which verifies the green and sustainability requirements and next phase is selected based on the verification results.

3.3 Green and Sustainable Design Requirements Criteria

This phase becomes active when the green verification sub-phase gives a negative result. Based on the target deployment platform different design and models must be finalized. This phase with the help of repository access makes additions to the design, then it is proceeded for the next phase. A software designed specific to a hardware very much depends on lifecycle of hardware, it is suggested to avoid this type of approach [9].

3.4 Coding

Use of hardware specific API is not encouraged [9] for software sustainability due to its dependency on hardware and its limited life-cycle. Re-use of codes, modular-based coding, and other sustainability requirements must be met, use of existing code repository and creation of managed repositories will reduce the coding time and less error would occur due to the use of earlier tested codes.

The use of green compiler is highly recommended for creating a software product which would be small in size with same functionalities of larger software. The small storage requirement would make it easier during its deployment by means of online installation and less energy will be wasted in the transportation process. At the end, coding phase goes through a sub-phase for verification and based on that the next phase is chosen.

Green Verifier This could be a combination of automated and manual process to verify the codes for standards and benchmarks requirements as in green and sustainable software.

3.5 Green and Sustainable Coding Criteria

Negative result at the coding verification sub-phase would activate this phase for inclusion of the green coding and sustainability requirements. This process would involve the use of cloud repositories.

3.6 Testing

A proper planned testing is important to maintain the sustainability of SDLC and it also makes the software sustainable. Test for compliance with different green-engineering metrics and standards must be complied. Re-use of test cases available from the repository and creating new test cases and histories to make it available in the repository for future uses will help in creating the sustainable ecosystem.

Green Verifier—This sub-phase must check for green testing requirements and would make the necessary recommendation to meet the criteria.

3.7 Green Requirements Testing

This phase ensures that the software meet the various standard green and sustainability requirements of the software being tested. All the standards of green software tests are ensured and it will also ensure that the developed software be energy efficient in all the target platforms where it is to be deployed.

3.8 Deployment

Minimal use of physical media such as DVD/CD/Flash drives- based deployment can have a positive effect on software development and can reduce the CO_2, GHG emissions due to less manufacturing process and e-waste management requirements. Instructions, manuals and other documentation also could be distributed in soft form which will highly reduce the use of papers in the development process.

Small footprint software will be easier to be transported through the Internet and that way it would reduce energy consumption by the networks.

3.9 Usage and Maintenance

Automated bugs, error reporting systems make the early detection of faults possible, this makes the maintenance time short and easier without onsite deployment of personnel in the user organizations. The cloud repositories of software bugs and errors could simplify and reduce the maintenance time and make the life cycle sustainable. Software must be used at its optimal hardware and software platform and configurations to make the system most efficient and sustainable one. Software not meeting the efficiency, sustainability may be maintained and requirements may be checked, the cycle may be repeated internally with all the quality and efficiency metrics criteria.

3.10 Disposal

When the software reached its end of life or it is no more required it may be disposed off, archives can be made and the modules may be used in other software considering the license and copyright issues.

3.11 Green Verifier

A sub-phase has been added to design, coding and testing phase to confirm the development process with all the green and sustainability requirements. This sub-phase has different requirement specifications according to the main phase it is associated with.

The additional green phase is activated when the verifier's result is negative (not conforming to its specified specifications). And it provisions the green criteria by both automated and manual intervention process and once it is done it goes to the general development phase. These phases are linked to the cloud/local repository (private/public/community/hybrid model) which can be both automated, manual with readily available modules and it can also provide standard guidelines for green and sustainable SDLC.

3.12 Repository Service Access Model

For taking advantage of the cloud platform, cloud repository has been proposed in four categories to achieve the efficiency and sustainability qualities in SDLC. The cloud repositories are made available to meet the software efficiency and sustainability requirements of Design, Coding, Testing and Maintenance. These supporting services are used to achieve the non-functional qualities in the software development. The repositories are made available in the cloud parallelly to all the developers and module-based development works can be carried out in parallel without breaking the requirements objective.

The repository services of the green sustainable SDLC model has three main components (Fig. 2):

- Repository Service Management Module
- Deployment Module Types
- Repository Database Module.

Repository Service Management Module (RSMM): The services are accessed by different phases of the development life cycle through the cloud services for sustainable development of software in the cloud platform. The RSMM module is interface between the repository and the different phases of software life cycle. It contains the Cloud services Manager and connects to the Database Module for different deployment types of software. First module is the interface between SDLC and the repository services. Through this, the developers and users of a software can have different level of access schemes of the repository database module. This module selects the deployment platform of the software which connects to the RDM for both the access and addition or creation of the repository. The cloud repository service provides access to readily available designs, codes, tests and maintenance support with the specified requirements, standards, benchmarks, metrics, guidelines suitable for a particular access or contributory request from the SDLC phases. This request could be of automated and manual options.

Deployment Module Types: It has four types of deployment—Private, Public, Community and Hybrid, [14]. Based on the Software development it is designed for, it will be selected and will connect to the repository database as per licensing and copyright criteria of the software.

Fig. 2 Proposed repository
service access model

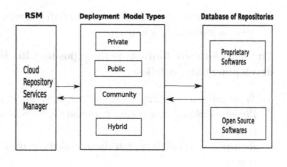

RSM   Deployment Model Types   Database of Repositories

Cloud
Repository
Services
Manager

Private

Public

Community

Hybrid

Proprietary
Softwares

Open Source
Softwares

Proprietary and Open Source
Repository Resources:

Green design modules,
Quality metrics

Code modules,
green libraries, API modules,
compliance with green and sustainable
Software

Test cases, green benchmarks,
Interoperability, and Integrity parameters,
Quality metrics

Error history, software bugs,
and Solutions

Repository Database Module (RDM): It is responsible for storage and maintenance of the repositories. It is categorised into two parts based on the environment of the software development. Proprietary for proprietary licensed software and open source for free and open-source software license-based software.

Based on the deployment module manager, the databases must be maintained in the RDM with all criteria such as the integrity, security adaptability and integration characteristics towards the green, efficient and sustainable aspects of SDLC. The RDM is responsible for management of the four repositories (Design, Coding, Testing, Maintenance) store in proprietary and open-source repository database.

The steps for accessing the repository in different phases of the software development are as follows:

1. *Repository service selects one of the deployment types.*
2. *Based on cloud deployment type and software licensing the repository database is selected.*
3. *Repository database selects the requested repository and looks for the available resources.*
4. *Available resources are pulled into the software development lifecycle and applied.*

5. *If not available, the guidelines, green criteria, metrics and standards are provided for development of a new one.*

The steps for contributions to the repository in different phases of the software development are as follows:

1. *Repository service selects one of the deployment types*
2. *Based on cloud deployment type and software licensing the repository database is selected*
3. *Repository database selects the intended repository and compares for any redundancy*
4. *New resources are added into the repository.*

In both the proprietary and open-source repository, it stores the green and sustainable requirements standards, metrics, guidelines as well as previous Design, Coding, Testing and Maintenance records of software life cycles.

4 Discussions and Suggestions

Woodside et al.[13] suggested for creation of a performance knowledge base which could act as repository for later use and sharing among designers for Software Performance Engineering (SPE) issues.

In this paper, a repository separately for Proprietary and FOSS is proposed (Repository Service Access Model) which will store green design, codes, test and maintenance data.

Data centers could be located in different geophysical locations, the servers could create a network of data centers (hardware) without using the cloud services (which are made available to customers), [15] algorithms are designed to make the server communications effective (efficient) in that geodispersedly located datacenters.

In [16], the authors proposed two improvement cycles, where one focuses on effects and impact of the development process itself and second one focuses on the impacts due to its distribution and use of that software.

5 Conclusion and Future Works

With the proposed model, this paper suggests to build the energy efficient, green and sustainable software development life cycle and ICT as a whole. To reduce CO_2 and GHG emissions by the datacenters, which deploys the Cloud Computing systems and ICT at large it would require to make the whole ecosystem of ICT be energy efficient and sustainable including both software and ICT hardware. As ICT hardware are the components which consume most electrical energies and to produce electrical energy GHGs and CO_2 emissions occur it must take necessary steps to reduce the power consumption by the ICT devices as many ways and as effectively possible. It is important that the hardware management and software-based reduction of power consumption techniques should be utilised for achieving green ICT and a green environment without sacrificing its computational efficiency.

In our future work, we will implement different green standards and guidelines explicitly in design and coding phases. After implementation of the proposed works, the systems energy efficiency will be extensively studied for finding its benefits.

References

1. Natural Resources Defence Council. https://www.nrdc.org/energy/files/data-center-efficiency-assessment-IP.pdf
2. https://www.nrel.gov/publications
3. GeSI SMARTer: The Role of ICT in Driving a Sustainable Future (2020). https://www.gesi.org
4. Corcoran, P., Andrae, S.G.: Emerging Trends in Electricity Consumption for Consumer ICT (2012)
5. Cloud Computing. https://www.nrdc.org/energy/files/cloud-computing-efficiency-IB.pdf
6. Definition of Green Engineering. https://www2.epa.gov/green-engineering/about-green-engineering#promote
7. Green Project Management: A Project Management Focus on the Environment. https://www.green-pm.com
8. Naumann, S., Dick, M., Kern, E., Johann, T.: The GreenSoft model: a reference model for green and sustainable software and its engineering sustain. Comput. Inf. Syst. 1, 294304 (2011)
9. Shenoy, S.S., Eeratta, R.: Green software development model: an approach towards sustainable software development. In: 2011 Annual IEEE, India Conference (INDICON), p. 16 (2011)
10. Chauhan, N., Saxena, A.: A green software development life cycle for cloud computing. In: IT Pro, pp. 28–34 (2013)
11. Beghoura, M.A., Boubetra, A., Boukerram, A.: Green software requirements and measurement: random decision forests-based software energy consumption profiling. In: Requirements Engineering. Springer (2015)
12. Developing Green Software. https://www.software.intel.com/sites/default/files/developing_green_software.pdf
13. Woodside, M., Franks, G., Petriu, D.C.: The future of software performance Engineering. In: Proceedings of the Future of Software Engineering, pp. 171–187 (2007)
14. The NIST Definition of Cloud Computing. https://csrc.nist.gov/publications/nistpubs/800-145/SP800-145.pdf
15. Toosi, A.N., Buyya, R.: A Fuzzy logic-based controller for cost and energy efficient load balancing in geo-distributed data centers. In: IEEE International Conference on Cloud Computing Technology and Science (CloudCom'13), Bristol, UK (2013)
16. Dick, M., Drangmeister, J., Kern, E., Naumann, S.: The software engineering with agile methods. In: GREENS 2013, Sanfrancisco, CA, USA (2013)

Automated Methodology to Streamline Business Information Flow Embedded in SRS

Shivanand M. Handigund, B. N. Arunakumari
and Ajeet Chikkamannur

Abstract Information technology software tools are used in all walks of life. These software are developed using appropriate process models that are built on software development life cycle (SDLC) stratiform stage activities in linear or nonlinear order. These stratiform stages transform the requirements of ensuing software from general to human and machine understanding. The requirements gathered in the form of software requirements specification (SRS) of ensuing software is always bidirectional between each pair of end users and is non-sequenceable. This information flow is captured in the form of graph. The end software executed by machine is in a control flow-ordered segments of sequel of statements with one start and many ends depending on values of branching attributes analogical to hierarchical tree structure. Thus, the streamlining process should be in the direction from graph to tree structure. This paper, attempts to transform the quintessential part of information flow of the requirements present in SRS linguistically, reaching directionally and with respect to business aspect towards the stratiform analysis stage. The correctness and completeness of the transformation is authenticated through the maintenance of the scope.

Keywords Software requirements specification (SRS)
Software development life cycle (SDLC) stages · Control flow

S. M. Handigund (✉)
Department of Information Science & Engineering, Vemana Institute of Technology,
Bengaluru 560 034, India
e-mail: smhandigund@gmail.com

B. N. Arunakumari
Department of Computer Science & Engineering, Research Resource Center,
Visvesvaraya Technological University, Belagavi 590 018, India
e-mail: arunakbn@gmail.com

A. Chikkamannur
Department of Computer Science & Engineering, R. L. Jalappa Institute of Technology,
Bengaluru 561 203, India
e-mail: ajeetac@rediffmail.com

© Springer Nature Singapore Pte Ltd. 2018   333
P. K. Sa et al. (eds.), *Recent Findings in Intelligent Computing Techniques*,
Advances in Intelligent Systems and Computing 709,
https://doi.org/10.1007/978-981-10-8633-5_33

Data flow · Usecase · Work process · Work · Business process
Asynchronous/synchronous information flow

1 Introduction

The goal of this paper is represented in the state-of-the-art form of vision mission and objectives as follows.

Vision: To resuscitate the software requirements specification (SRS) to be amenable for next stratiform SDLC stages.

Mission: To develop an automated methodology to transform dexterous SRS towards human and machine understandable concise document to be amenable for subsequent stratiform SDLC stages.

Objectives:

- To stratify conflation of semiotics of both language and business into congenial flow and transform information flow to be compatible for software.

- To mould the embedded business semiotics of SRS to be amenable for next stratiform stage of SDLC.

- To inculcate logical positivism to authenticate the developed tools and techniques.

1.1 Motivation

The software development starts with requirements gathering, performs all stratiform SDLC stage activities and ends up with the requisite software. The information flow in the SRS is turbulent as is bidirectional. But the ensuing software is to be executed in the control flow-ordered segments of sequel of statements. This indicates that the information flow moves from turbulence to hierarchical tree structure from root to all leaves. This is realized if semiotics of both embedded business and embedding language are interleavingly lustrated to achieve the control flow order. The requirements specification is at the human understanding level and the ensuing software need to be at the machine understanding level. Therefore, the requirements stage needs to lustrate the SRS in the direction from human to machine understanding. Thus, the end document of the requirements stage should reach the milestone at the junction of human to machine understanding, turbulent to unidirectional information flow with implicit containment of business semiotics. The current methodology has suffered due to manual processes which is even not carried for achieving that milestone.

1.2 Literature Survey

The lustration from asynchronous SRS into synchronous software is to be carried out in two phases at each stage. In the first phase of each stage, language semiotics is lustrated to facilitate the second phase, the lustration of business semiotics. The lustration of language semiotics of SRS is to be carried in the following sextuplet framework of activities, viz. formation of SRS statements into single statement sentences, transformation of first, second and third person nouns into attribute/ attributes values present in the immediate prior referential and definitional attributes, conversion of passive voice to active voice statements, elimination GOTOs, resolution of synonyms and heteronyms and reorganization of statements to enable the definitional attributes (objects) precede their references (subjects). All these activities except the last are fully or partial carried out through semi-automated methods by [1–3].

In the second phase, the extant paradigms are accounted from nouns and noun phrases, the role played them, visibility, the actors and their interface attributes, the object methods and the link between different object methods [4, 5]. This facilitates the design of referential and definitional table. The SRS statements and the ensuing software lines of code (LOC) both are voluminous, the nature of information flow of these can be identified by the control flow graph (CFG) [2], where an entry/LOC represents a node and the flow (logical connection to next entry/LOC) should be represented by an edge. The nature of flow is clear if the sequential entries/LOCs which are in logical control flow order, are collapsed into single vertex. The CFG of SRS and the ensuing software may contain cycles indicating iterative recursive nature of segments of statements. Since, the ensuing software contains hierarchical tree structure form of the graph. The realization of each statement/LOC takes different sets of values to the attributes in referential defintional table. The information flow of both SRS and ensuing software are organized in tree structure by reorganizing referential defintional entries to form data flow table (DFT) [2, 6]. The DFT not only serve for analysis stage but also helps to slice [7] the requirements into object methods satisfying good software engineering principles [8].

2 Proposed Methodology

In this paper, an attempt is made to establish stratiform transformation of SRS to be amenable to the 'analysing' through transformation of asynchronous SRS to synchronous one. In the requirement gathering stage, the lustration of SRS is carried out in two phases, viz. lustration of semiotics of representative language and perpetual software business. The lustration of language semiotics is discussed by number of researchers [1, 2, 9]. This paper proposes the methodology for lustration of both language and business semiotics interleavingly to form synchronous document from asynchronous SRS. The SRS contains turbulent information flow.

This is streamlined by design of control flow graph (CFG) for SRS and is designed in two phase. In the first phase, we design CFG for SRS document such that the definitional attributes precede their references iteratively till their closure. Then this CFG is used to form data flow table (DFT) comprising entries comprising statement number, referential and definitional attributes of each statement in the CFG order. This process is briefed in Sect. 2.1. This DFT in the control flow order is quintessential to derive activities to be expected of the ensuing software. In the second phase, we design CFG through the reticulation of the activities in a reticular flow either in the form of works each embedded between pair of deliverables indicating progressive stratiforms and conjoining these successive works or in the form of work processes of each user and forming reticular network with information flows between the work processes or with reticular sequence of activities bounded between information flows of actors of the client organization. This process is briefed in Sect. 2.2. Each software development is a Sui Generis project and hence desiderates the schedule, scope, resources, risks and cost. The clear understanding and use of the scope serves the base for the computation of other factors and other project ingredients [10]. Hence, the SRS needs to be modified to be amenable for the abstraction of scope in solidity and the lacunae (hiatuses) are bridged through referential and definitional attributes of the activities used in analogizing the scalar triple product and is discussed in Sect. 2.3.

2.1   First Cut CFG for SRS Document

Input: Refined SRS as per language semiotics

i. Read each statement of SRS in their physical sequence, categorize the nouns/ noun phrases of subject and object into referential and definitional attributes, respectively, and store them in appropriate entries of referential-definitional table as shown in Table 1.

ii. Take projection on Table 1 based on statement number and referential attributes to form referential table and statement number and definitional attributes to form definitional table.

iii. Consider referential table. Organize the attributes of each entry in the canonical form as shown in Table 2.

iv. Sort the entries using merge sort technique [11] in lexicographic order of referential attribute as primary sort key and statement number as secondary sort key. The structure of the sorted table is shown in Table 3.

Table 1 Referential-definitional

Stmt. No.	Referential attributes	Definitional attributes
–	–	–

v. Collapse statement numbers representing the same attribute into a single entry with list of statement numbers in the ascending order. The structure of the table is shown in Table 4.

vi. Repeat the Steps iii–v for definitional table with appropriate relevant names stored it in Table 5.

vii. Now, we have collapsed sorted referential and collapsed sorted definitional tables.

viii. Take natural join of collapsed sorted referential and definitional tables based on attribute name and store it in Table 6.

Table 2 Canonical referential

Referential attribute name	Statement number
<attribute e>	26
<attribute k>	52
<attribute d>	45
<attribute i>	33
<attribute d>	11

Table 3 Sorted canonical referential

Referential attribute name	Statement number
<attribute d>	11
<attribute d>	45
<attribute e>	26
<attribute i>	33
<attribute k>	52

Table 4 Collapsed sorted referential

Referential attribute name	Statement numbers	
<attribute d>	11	45

Table 5 Collapsed sorted definitional

Definitional attribute name	Statement numbers		
<Attribute>	07	25	36

Table 6 Merged sorted referential definitional

Attribute name	Definitional statement numbers			Referential statement numbers	
<Attribute>	07	25	36	11	45

ix. Consider Table 6, compare referential and definitional statement numbers. If referential is less than definitional statement number then mark that referential statement number and attribute name and proceed to next statement number.

x. If referential and definitional statement numbers are equal then search the entries in the original referential-definitional Table 1 and grasp other definitional attributes. Store it in first cut class table.

xi. Consider the marked statement numbers and their referential attribute of Table 6. For each marked statement number, identify other attributes in Table 1 and for each such attribute search their definitional statement numbers in Table 3 using binary search technique [11]. Replace the entry of the least referential statement number immediately after the highest definitional statement number in the Table 1.

xii. Repeat the Step xi for all marked attributes in Table 6. The refined Table 1 forms the DFT organized in the CFG for SRS document.

2.2 Lustration of Business Semiotics from SRS to DFT

Input: DFT (Concise version of SRS), First cut class table

i. Many researchers [1, 2, 9] have developed methodology first cut object method from Table 1 after slicing table with slicing criteria <last statement number of DFT, class attributes>. We presume that cluster of object methods are available and the sliced portion of the SRS as shown in Table 7. In the next procedure these object methods are interconnected to satiate usecase.

ii. Consider each object method say (A_{im1}), take intersection of referential attributes of (A_{im1}) with definitional attributes of other object methods slice say (A_{lm3}, A_{km2},... A_{um2}). If intersection is not null then draw edge from definitional cluster (A_{lm3}) to referential cluster (A_{im1}), then take attributes of A_{lm3}, repeat this process till referential attributes are defined anywhere.

iii. Next, consider each object method say (A_{im1}), take intersection of definitional attributes of (A_{im1}) with referential attributes of other method slices. If

Table 7 Sliced portion of SRS document

Slice name	Activity names			
Slice 1	A_{im1}	A_{im2}	A_{im3}	A_{im4}
Slice 2	A_{jm1}	A_{jm2}	A_{jm3}	A_{jm4}
Slice 3	A_{km1}	A_{km2}	A_{km3}	
Slice 4	A_{lm1}	A_{lm2}	A_{lm3}	A_{m4}
–	–	–	–	–
Slice n	A_{um1}	A_{um2}		

    intersection is not null then draw edges from (A_{im1}) to other method slices. Repeat this process till defined attributes are not referenced in other activities.

iv. Each path from object method containing referential attributes to definitional attributes form a usecase as shown in Fig. 1.

Similarly DFT can be designed based on either fasciculate tasks of the stakeholders (work processes), or on the sequence of activities embedded between one start and one end documents (work) [12].

2.3 Solidification of Scope of SRS Eliminating Hiatus

Consider the scalar triple product (u ● v × w) [13] where u is the vector containing usecase slices perpendicular to the vector. Similarly v, w are the vectors containing respective slices of work processes and works. The expression (u ● v × w) represents the parallelepiped satisfying the cyclic changed order of the components. The slices take the input from the left, bottom and back faces for appropriate slices. These are indicated from the referential attributes of three perspective views. And the right, front and top faces represent the output from appropriate slices represented through definitional attributes. Thus, parallelepiped represents the scope as shown in Fig. 2. The lacuna in slice of one of the

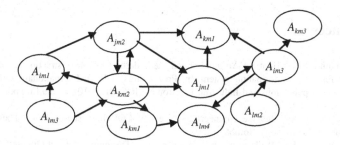

Fig. 1 Reticular network containing usecases

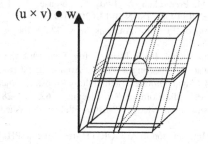

Fig. 2 Scope of the business process

perspective views ramifies into smaller lacunae in different slices of other two perspective views. This means the lapse in an attribute value is identified through the convolution of lapses in other two perspective views attribute values using divide and conquer algorithm [14]. The lapse can be regained without the need of domain knowledge and information provider. Thus, through the analogy of scalar triple product applies to perspective views, we can procure solidified scope of SRS which paves the way to optimize the cost and schedule.

3   Conclusion

This paper discusses a semi-automated lustration technique that moulds the requirements, projecting the SRS statements vertically through the needy part required by the paradigm. The turbulent information flow is transformed to hierarchical tree structure flow from root to leaves. Further, the resulted DFT entries are reorganized without violating the reaching definition into one of the three perspective views.

Acknowledgements Words are insufficient to express our deep sense of gratitude to Dr. D. B. Phatak, Chair Professor of Department of Computer Science and Engineering, IIT Bombay, for his inspiration.

References

1. Handigund, S.M., Bhat S.: An ameliorated methodology for the abstraction of object class structures for an information system. In: Proceedings of IEEE International Conference on Computer Applications and Industrial Electronics (ICCAIE-2010), Kuala Lumpur, Malaysia, pp. 362–367 (2010)
2. Handigund, S.M.: Reverse engineering of legacy COBOL systems. Ph.D. Thesis, Indian Institute of Technology Bombay (2001)
3. Wirfs-Brock, R., Wilkerson, B., Wiener, L.: Designing Object-Oriented Software. Prentice-Hall of India Private Limited, New Delhi (2000)
4. Booch, G., Rumbaugh, J., Jacobson, I.: Unified Modeling Language User Guide, 2nd edn, p. 496. Addison-Wesley Professional, published May 19, 2005
5. Rumbaugh, J., Blaha, M., Premerlani, W., Eddy, F., Lorensen, W.: Object-Oriented Modeling and Design. Prentice Hall (1996). ISBN 0–13-629841-9
6. Frank Tips: Generation of Program Analysis tools. Ph.D. thesis, Amsterdam University (1995)
7. Weiser, Mark: Program Slicing. IEEE Trans. Softw. Eng. **10**(4), 352–357 (1984)
8. Jalote's, Pankaj: Software Engineering A Precise Approach, 1st edn. WILEY, India (2010)
9. Handigund, S.M., et al.: An ameliorated methodology to abstract object oriented features from programming system. Procedia Comput. Sci. Elsevier J. **62**, 274–281 (2015)
10. Project management body of knowledge. PMBOK 5th edn. Published by project management institute (PMI) 2012
11. Coremen, T.H., et al.: Introduction to ALGORITHMS, 2nd edn. PHI Learning Private limited (2009)

12. Handigund, S.M.: Presented a tutorial on 'Unraveling the hidden treasures of research avenues in object oriented technology'. In: The 2011 World Congress in Computer Science, Computer Engineering and Applied Computing, Las Vegas, Nevada, USA, July 18–21 (2011)
13. Kreyszig, E.: Advanced Engineering Mathematics, 10th edn. Wiley Publisher, Dec 2010
14. Gregorg Heileman, L.: Data Structures, Algorithms and Object Oriented Programming. McGra-Hill Book Co. International Editions, Singapore (1996)

An Approach for Improving Classification Accuracy Using Discretized Software Defect Data

Pooja Kapoor, Deepak Arora and Ashwani Kumar

Abstract Predicting software defects in software systems at early stages of its development has always been a very crucial and desirable aspect of software development industry. Today the machine learning algorithms are playing a massive role in classifying and predicting the possible bugs during the design phase. In this research work, the authors have proposed a discretization method based on metrics threshold values in order to gain better classification accuracy on a given data set. For the experimentation purpose, the authors have chosen the defect data sets from NASA repositories. In this Jedit, Lucene, Tomcat, Velocity, Xalan, Xerces software systems have been considered for experimentation using WEKA. The authors have also considered object-oriented CK metrics specifically for the study. Two very common and popular classifiers namely Naive Bayes and voted perceptron for the classification purpose. In the proposed work, various performance measures like ROC, RMSE values have been considered and analyzed. The results show that classification accuracy improvements can be made while using proposed discretization method with both classifiers.

Keywords Discretization · Software defect prediction · Classification
CK metrics

P. Kapoor (✉) · D. Arora
Department of Computer Science & Engineering, Amity University, Lucknow, India
e-mail: pkhanna@lko.amity.edu

D. Arora
e-mail: deepakarorainbox@gmail.com

A. Kumar
Area of IT & Systems, IIM Lucknow, Lucknow, India
e-mail: ashwani@iiml.ac.in

© Springer Nature Singapore Pte Ltd. 2018   343
P. K. Sa et al. (eds.), *Recent Findings in Intelligent Computing Techniques*,
Advances in Intelligent Systems and Computing 709,
https://doi.org/10.1007/978-981-10-8633-5_34

1 Introduction

Software defect prediction is the process of identifying probable defective modules in software, prior to the testing phase. Software defect prediction plays important role especially in the case where development time is a crucial issue or when developing system is too large to perform exhaustive testing. Various software metrics are mentioned in the literature that include static code attributes like size of code, complexity of code, coupling, cohesion, number of children, depth of inheritance, previous releases of software with the logs of defects encountered. Different researches show the potential of software metrics in software defect prediction and estimation of the final software quality [1]. Machine learning introduces four types of learning techniques like supervised, unsupervised, semi-supervised, and reinforcement learning. Inductive/supervised learning systems, learn a general rule from a set of observed instances. Inductive learning systems can be effectively, used for classification of input data to a particular output class. Inductive learning algorithms can be applied in various domains with continuous attributes in order to map discrete output class. CK defined the six metrics considering specific software quality attributes in mind. CK suggested that value of these six metrics in suits: wmc, noc, dit, rfc, lcom, cbo should be maintained to low, medium, high range using software designs so that it meets the software quality requirement specifications. But previous literature does not define values for the linguistic terms used here: low, medium, high. For example, classes with large number of methods are likely to be more application specific, restricting RE-USE of code, so low WMC (Weighted Methods per Class) is desirable if re-usability is the desired quality attribute of the quality requirement specifications. The method for discretization adopted for this study is so chosen keeping the concept, that in future we can set a range for each metrics to convert these linguistic terms to fuzzy variables. In this paper, a threshold-based discretization method is proposed and analyzed during experimentation, it is found that this proposed threshold based discretization method has significantly improved the classification performance of two existing learning algorithms like Naive Bayes and Voted Perceptron.

2 Background

A large amount of the total development cost is kept aside for testing phase. But sometimes due to size complexity, exhaustive testing becomes a hard problem with time and cost constraints. Software Defect Prediction (SDP) is an important activity before testing phase, which identifies the probable defective module. Software Defect prediction (SDP) faces various issues to come up with an accurate Defect Prediction model [2]. Arora et al., has identified some of the major issues of SDP. The study represents an exhaustive research on how to address those issues like Relationship between Attributes and Fault, No Standard Measures for Performance Assessment, Issues with Cross-Project Defect Prediction, No General Framework Available, Economics of Software Defect Prediction, Class Imbalance Problem [3]. Various researches are working to build a prediction model. A study presented a model that can predict the defective module on the basis of complexity metrics like size complexity [4]. The study proposed a direct relation between complexity of code and defects. Though the study was unable to present an accurate model it raised many research objectives that say there is need to identify better metrics for defect perdition model and how that can integrate with the development process. Industry and Academics both are concerned to solve the problem of SDP resulting several software defect repositories. These software repositories are a collection of data, for the software metrics and the defective state of each module. The study, presents a data mining approach that can predict the defective state of software modules. The study shows better prediction capabilities when all the algorithms are combined using weighted votes. In the study, the author introduces kernel-based asymmetric learning for software defect prediction. To eliminate the negative effect of class imbalance problem, the author proposed two algorithms called the asymmetric kernel partial least squares classifier and the asymmetric kernel principal component analysis classifier [5]. In a study, the authors presented the use of the CK software metrics suite in predicting the level of maintainability of software using threshold [6]. The study identified maintainability can be predicted on the basis of most effective metrics out of the dozens of available metrics.

3 Proposed Discretization Transformation

Discretization is the process of transferring continuous functions, models, and equations into equivalent discrete values. Discretization of continuous features/ attributes plays an important role in Machine learning data preprocessing phase. Various discretization methods are present in literature [7–10]. Deciding a CUT-POINT for discretization is the most crucial issue for good classification accuracy and achieving better efficiency of any machine learning algorithm. Better classification accuracy can be achieved on the basis of discretized data [7]. The discretization method proposed for this study is a supervised dichotomization method. The error-based discretization is a supervised method, given in literature study that considers the performance of the learning algorithm itself. It tries to find number of intervals on the basis of error observed. The discretization function $f(x)$ used for the current study is defined as 1 if value of x lies within the boundary range of the function (CUT-POINT). The boundary of the function is defined using MMV (Mean Metric value) that is computed using Naive Bayes classifier in a study [11]. Like in error-based discretization, the current discretization uses Naive Bayes supervised method for calculation of CUT-POINT. The study suggested that if central tendency of CK metrics for the system is according to MMV so defined, will ensure less occurrence of fault in the final system and using MMV we can predict the probability of fault occurrence at the design level.

$$f(x) = \begin{cases} 1, & T1 < x < T2 \\ 0, & OTHERWISE \end{cases} \qquad (1)$$

In Eq. (1), T1 and T2 are the boundary values, i.e., computed using MMV suggested in a by Kapoor et al. [11].

4   Experimental Design

The software defect data set for the study is collected from NASA repository (promise repository) [12]. The primary objective of the current study is to discretize CK metrics using proposed discretization method. For the purpose of study, we are concentrating on CK metrics namely cbo, rfc, wmc, dit, and noc. In study, the authors have established a co-relation between these five metrics and occurrence of fault. lcom is not considered because of its unpredictable response to fault [11]. In current work, the authors used this MMV as CUT-POINT for discretizing the data. The value of f(x) is set 1 if x lies between T1 and T2. Value of T1 and T2 is (MMV ± deviation). WEKA 3.6 is used for the empirical study of current work, using the Naive Bayes and Voted Perceptron classifiers. The primary objective of the study is to analyze the effect of proposed discretization functions on classification accuracy of two classifiers using Naive Bayes and Voted Perceptron classifier.

5   Result and Discussion

Looking at the continuous data set it is visible that in some cases the number of distinct value for a feature is too high, and for classifiers like Naive Bayes and neural network it increases the computational complexity [13]. Jedit 3.2 has a total number of instance 196 and RFC feature contain 97 different values, i.e., 32% different values of total instances, similarly in case of velocity 1.4 RFC feature contains 33% distinct values. The two classifiers Naive Bayes and Voted Perceptron ran twice, one for each discretized and continuous data set. This is being done for 18 different defect sets and different statistical measures are recorded and represented using Figs. 1, 2, 3 and 4 for Naive Bayes and voted perceptron. All these statistical measures have been considered for comparing the classifier performance [14].

Figures 1 and 2 represent the accuracy performance of Naive Bayes and voted perceptron. Figure 1 represents the accurate comparison of discrete and continuous data obtained using Naive Bayes classifier.

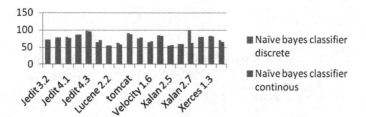

Fig. 1 Represents accuracy value comparison for discrete and continuous data for Naive Bayes classifier. X-axis represents 18 different sets under consideration and y-axis represents the accuracy values

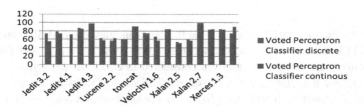

Fig. 2 Represents accuracy value comparison for discrete and continuous data for voted perceptron classifier. X-axis represents 18 different sets under consideration and y-axis represents the accuracy values

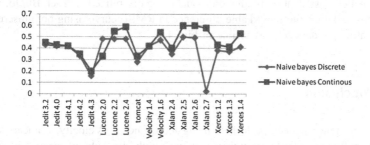

Fig. 3 Represents RMSE value comparison for discrete and continuous data for Naive Bayes classifier. X-axis represent 18 different sets under consideration and y-axis represents the RMSE values

The graph presents better accuracy in case of discretized data for both the classifiers. Though from the figure, it is quite obvious that in some cases like for Lucene 2.0 Naive Bayes classifier accuracy for continuous data is better than the accuracy obtained through discredited data set. Similarly, results obtained in Fig. 2 for voted perceptron classifier indicates, an overall accuracy improvement, except few cases like in Xerces 1.4, graph shows better results for continuous data set in this case. But sometimes accuracy is faced with accuracy paradox. Consider a situation when TP = 0, FP = 0, TN = 115, FN = 15, this says that classifier is

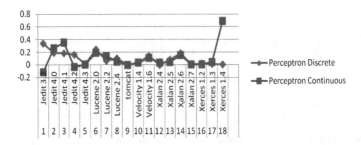

Fig. 4 Represents RMSE value comparison for discrete and continuous data for voted perceptron classifier. X-axis represent 18 different sets under consideration and y-axis represents the RMSE values

unable to predict TRUE class (TP = 0), but can classify FALSE class well (TN = 115). In such a case, accuracy comes out to be 88%. To deal with this kind of situation many other measures are considered to check the performance of the classifier. For the study authors considered RMSE. RMSE is a measure of correctness of classifier. It tells how distinct is classification model from actual prediction. Figures 3 and 4 represent the RMSE value comparison for discrete and continuous data of two classifiers used in the study.

Figure 3, the curve generated using Naive Bayes classifier, indicates better RMSE curve for discretized data. Generally, for all systems under the study, it shows better classification/prediction model using discretized data set. But for voted perceptron, there is no remarkable difference in RMSE curve using both discretized and continuous data sets, as visible in Fig. 4.

6 Conclusion

The experiment has been performed using Jedit, Lucene, Velocity, Tomcat, Xalan, Xerces software systems and their versions with the help of open-source tool WEKA. This research work is intended to analyze the effect of, proposed discretization method, on improving the classification accuracy. During experimentation, it is found that the proposed discretization function, based on threshold values can be used to improve the classification accuracy of classifiers like Naive Bayes and voted perceptron. Naive Bayes classifier works on frequency count for each value from the feature set, it performs well in case of categorical input variables compared to numerical variable [13, 15, 16]. Similarly in case of voted perceptron as we reduced the number input signals to the perceptron it showed mostly better efficiency in classification accuracy. The study suggests that using proposed discretization function the efficiency of CK metrics in the prediction of software quality in terms of defect prediction at early stages of development, can be improved. The boundary values of the proposed descritization function would be dependent on, various factors like software development environment, and software quality requirement of the system [17].

References

1. Arora, D., Khanna, P., Tripathi, A., Sharma, S., Shukla, S.: Software quality estimation through object oriented design metrics. IJCSNS Int. J. Comput. Sci. Netw. Secur. **11**(4) (2011)
2. A systematic literature review of software defect prediction: research trends, datasets, methods and frameworks. J. Softw. Eng. (2015)
3. Arora I., Tetarwala, V., Sahaa, A.: Open issues in software defect prediction. Elsevier (2015)
4. Zimmermann, T., Premraj, R., Zeller, A.: Predicting defects for eclipse. In: Proceedings of the Third International Workshop on Predictor Models in Software Engineering, p. 9. IEEE Computer Society (2007)
5. Ren, J., Qin, K., Ma, Y., Luo, G.: On software defect prediction using machine learning. J. Appl. Math. (2014)
6. Bakar, A.D., Sultan, A., Zulzalil, H., Din, J.: Predicting maintainability of object-oriented software using metric threshold. Inf. Technol. J. **13**, 1540–1547 (2014)
7. Salvador, G., et al.: A survey of discretization techniques: taxonomy and empirical analysis in supervised learning. IEEE Trans. Knowl. Data Eng. **25**(4), 734–750 (2013)
8. Kotsiantis, S., Kanellopoulos, D.: Discretization techniques: a recent survey. GESTS Int. Trans. Computer Sci. Eng. **32**, 47–58 (2006)
9. Kaya, F.: Discretizing continuous features for Naive Bayes and C4. 5 classifiers. University of Maryland publications, College Park, MD, USA (2008)
10. Kohavi, R., Sahami, M.: Error-based and entropy-based discretization of continuous features. KDD (1996)
11. Kapoor, P., Arora, D., Kumar A.: Effects of mean metric value over CK metrics distribution towards improved software fault predictions. In: Proceedings of the Springer's International Conference IC4S (2016)
12. http://promisedata.googlecode.com
13. Rish, I.: An empirical study of the Naive Bayes classifier. In: IJCAI Workshop on Empirical Methods in Artificial Intelligence, vol. 3, no. 22, p. 111. IBM New York (2001)
14. Metz, C.E.: Basic principles of ROC analysis (PDF). Semin. Nucl. Med. **8**(4), 283–298 (1978)
15. Tomar, Divya, Agarwal, Sonal: Twin support vector machine for multiple instance learning based on bag dissimilarities. Adv. Artif. Intell. **2016**, 1–18 (2016)
16. Rish, I.: An empirical study of the Naive Bayes classifier. In: IJCAI 2001 Workshop on Empirical Methods in Artificial Intelligence, vol. 3, no. 22. IBM New York (2001)
17. Yousef, A.H.: Extracting software static defect models using data minig. Elsevier (2015)

Deadlock Recovery for NOC Using Modified Encoding Scheme

H. R. Shashidhara and T. R. Prateek kumar

Abstract Current technology trends stress upon scaling down size and area in processors and chip designs. There is a paradigm shift toward highly complex circuits, such as System on Chip (SoC), Complex chip Multi-Processors (CMP) and so on. Effective Communication between many such components is achieved through Network on Chip (NoC), which has a trade off correctness assurance of transfer data. This is observed as a deadlock, where data loss is inherent. Previous methods for deadlock recoveries are multi-path topology and transitive closer method. This leads to more traffic, higher complexity and increased transmission time. A methodology is proposed to achieve deadlock recovery through encoding and decoding technique. This technique reduces the network traffic by avoiding re-request.

Keywords NoC · Deadlock detection · Deadlock recovery · Arithmetic encoder · ASIC

1 Introduction

The steadily increasing System on Chip (SoC) complexity leads to increase in data transfer over a network between communication partners. Due to the support of high-speed networks, the data transfer rate and data handling requirement are increasing rapidly. The data handling is done by the networks, which includes the NoC (Network on Chip) to maintain reliable communication between the nodes from transmitter end to receiver end. The user node connection makes the network more and more complicated, as each one node has to connect to every other node

H. R. Shashidhara (✉) · T. R. Prateek kumar
Department of Electronics and Communication Engineering, JSS Academy of Technical
Education, Visvesvaraya Technological University, Bengaluru, Karnataka, India
e-mail: shashidharahr@gmail.com

T. R. Prateek kumar
e-mail: trprateek@gmail.com

© Springer Nature Singapore Pte Ltd. 2018
P. K. Sa et al. (eds.), *Recent Findings in Intelligent Computing Techniques*,
Advances in Intelligent Systems and Computing 709,
https://doi.org/10.1007/978-981-10-8633-5_35

with high-end security and reliability, due to this the bandwidth of the channel is increasing. The network traffic becomes hectic due to increased data packet transfer. On an average, 60% of the packets are due to re-request from the receiver endpoint to the transmitter end.

Deadlock is a condition in NoC's, in which two or more packets are waiting for one another for channel liberate and unable to move forward. As a result, the network gets struck due to a chain of waiting packets leading to an unending cycle. A simple illustration for deadlock shows the dependency cycle of four packets in Fig. 1. Every packet corresponds to holding some channel. The current holding packet cannot proceed further until and unless the previous packet releases the channel. Therefore, in such a situation, packets are deadlocked and will stay in this state. This will cause halting of all the occupied channels until some external interference is applied to bring them back to life by breaking the dependency cycle. Deadlock can paralyze the communication by halting the channels occupied by the deadlocked packets which will in turn block other packets which are dependent on the same channel [1]. Hence, it is essential to take out the deadlocks. There are two strategies to compact with deadlocks: they are, deadlock avoidance and deadlock recovery [2, 3]. The deadlock avoidance contains separate fixed paths for transmission of packets with no overlapping of channels. This is done based on a turn-model, where the routing algorithm is prohibited from making certain turns in the network [4, 5] or based on the firm ordering of the implicit channels [2]. In common, the deadlock avoidance technique requires constrained routing functions in the network or additional channels to prevent deadlocks [6]. The turn-model is popular in NoCs because of its simplicity, even though it constrains the routing alternatives, as it has restricted routing functions and diminishes fault tolerance capabilities.

For arbitrary network topology, the turn-model is not applicable [7]. Alternatively, in deadlock recovery the packets are provided grant to all the resources of the network without any routing restrictions. This has the capability of potentially outperforming the deadlock avoidance [1, 2, 8–11]. Deadlock may occur in the networks, hence an efficient deadlock detection and recovery mechanisms are required to intervene the cycle. However, the main challenge is detecting the deadlock. It should be accurate since dummy deadlock may occur due to their distributed nature. The deadlock conditions on the network of the chip increase the

Fig. 1 Block diagram

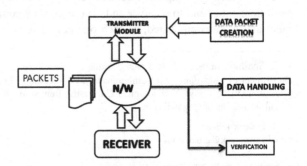

heavy packet traffic which in turn causes a delay in switching. This leads to loss of packets. One of the reasons may be that the receiver end is not efficient enough to receive the data and request for re-transmit of the packet again. This results in an increase in network traffic.

An efficient and promising technique to deal with deadlock for network communication is data recovery for NoC. Here, the data is recovered at the receiver end by decoding the packet. This holds good for the packet that gets corrupted in a deadlock situation. The framing of the packet at the transmitter end was proposed by Kenichi Mori, Abderazek Ben Abdullah, and Kenichi Kuroda [12]. The main aim of the technique is switching and routing at a faster rate and reliable communication by designing the packet. To understand the base connection in a network, the network unit is placed in between the transmitter and the receiver. The general methods of IPv4 checksum have been adopted in encoding techniques; another encoding method uses the binary compression techniques.

2 Methodology

The proposed technique, deadlock recovery involves packet design at the transmitter side, deadlock detection and data recovery at the receiver side. The recovery of the data involves decoding the packet received through the transmitter. The efficient data recovery requires the best encoding techniques. The encoding technique consists of checking for repeating of the data and storing the repeated values using zeros and ones and storing the repeating value. The detailed analysis of these techniques is shown below.

2.1 Data Recovery

An efficient data encoding algorithm is designed to recover the data. The encoding algorithm depends on the number of data bits a packet contains. Using this, data is segmented into equal no of bits. The similarity is considered in these segmented data bits. These similar data bits are positions are considered and created an encoded data values. The below algorithm states the proposed encoding technique
     The proposed encoding algorithm is as follows.

Step1. Consider the data bits and divide them into an equal number of divisions using following equation.

$$\text{No. of Division} = \text{Total No. of data bits}/\text{Reference bits}$$

Step2. Consider one division as a reference value and compare with other remaining divisions.

Step3. The repetitions of divisions are stored in registers and this operation is performed on all the available data.
Step4. Find number of zeros in each recorded register and store it separately.
Step5. Find the maximum number of zeros and select respected position register and reference register as encoded data.

2.2 Packet Design

The packet is constructed with a new encoding technique. The packet is designed with some minimum criteria at the receiver end. They are

1. Should contain start and stop bits.
2. The receiver should receive a minimum of starting and encoding bits to data recovery (Minimum data reception).
3. If the stop bit is not received and the all the data bits are received then data recovery should not be considered (Maximum data reception).

Using the above minimum criteria, the packet is designed and used for data recovery. The packet contains start bit followed by 24 bits of the encoded field and 128 bits of data, finally a stop bit. As the data bits go on varies the encoded data bits also varies. The minimum 16% data repetition is considered for the encoding, if it is below that the encoding algorithm fails to encode and packet designs also fail to generate the encoded packet, the whole data itself will be considered for a packet that is Encoded field and Data field. The encoding structure of the frame is as shown Fig. 2, Table 1.

Fig. 2 Packet design

Field	Description	Number of bits
St	Start bit	1
Ed	Encoded bits	24
Data	Data bits	128
Sp	Stop bit	1

Table 1 Signal description

2.3 Decoder

A new encoding technique is proposed. Here, consider the data bits and divide them into an equal number of divisions. Now check for repetition of divisions, by considering one of the divisions as reference. Record the data according to the repetition of the divisions. Repeat this by considering all divisions one by one as reference division. The design implementation of the encoder consists of 128-bit data input, which has to be copied into internal memory and divide it according to the reference number as given by the following equation. Here, consider 128 bits as data bits for illustration purpose and reference bits as 8, hence the number of division that is, number of bits of the position register is equal to 16 bits. Now let us divide the data bits according to the reference bits and save it in two-dimensional arrays of 'eight cross sixteen', that is 16 row bits and 8 is a column.

Let $a[8:16]$ be two-dimensional array, $a[1] = data[7:0]$, $a[2] = data[15:7]$ and so on. Then check for repetition of 8-bit registers, by considering one of the 8 bits as a reference check for repetition of 8-bit registers with reference register and store the values of the repetition into the position register according to following equations.

$$\text{reference byte} = a[1]; \text{Check}$$
$$a[i] = \text{reference byte},$$
$$\text{pos\_reg}[i] = 0, (\text{put zero in the ith position of the position register})$$
$$\text{or}$$
$$a[i] = \sim \text{reference byte},$$
$$\text{pos\_reg}[i] = 1, (\text{put one in the ith position of the position register})$$

Repeat this checking for all the 16, 8-bit registers of the array as reference register and store the values of the position registers in sixteen different registers. Now, we have to count number of zeros in each of the position register, as switching power for is low and store it aside. For this following equations are used.

$$\text{Check} \qquad \text{pos\_reg}[i] = 0,$$
$$\text{count} = \text{count} + 1,$$
$$\text{or}$$
$$\text{count} = \text{count}$$

Now find the maximum value of the count among all the 16 counts, so that we can fix the value of the reference register value. For this, we have to assign a count value to a temporary register and start comparing with other values, if any value find greater than assigned value then replace the temporary value, otherwise leave it as it is.

$$\text{assign temp} = \text{count}[1],$$
$$\text{temp} < \text{count}[i],$$
$$\text{temp} = \text{count}[i],$$
$$\text{temp} = \text{temp}$$

Now to find the reference register, find the position of the maximum count value and then assign the reference register as well as position register of the same position as encoded values of the registers.

$$\text{Check} \quad \text{temp} = \text{count}[i], \text{then}$$
$$\text{pos\_register} = \text{pos\_reg}[i],$$
$$\text{ref\_reg} = \text{reference\_register}[i]$$

2.4 Deadlock Detection

Deadlock detection is done by checking the stop bit is received, if the stop bit is received then there is no loop formation which is termed as deadlock. If the stop bit is not received then there is a deadlock, then the deadlock decoding technique are applied to recover the data. Before applying decoding technique, the minimum packet information of the data is checked for availability at the received end. The decoding technique fails for nonavailability of packet information. Minimum of 17% of the data is required for decoding the packet else the data cannot be recovered.

3 Results

The designed packet using the encoded process is verified functionally using ModelSim[R] and implemented using Cadence[R] Tool. The simulation results shown in the figure verifies the functionality of the encoding of the data bits. For every positive edge of the 'clk' and positive logic of 'rst', the data is enabled for encoding and frame the packet (Fig. 3).

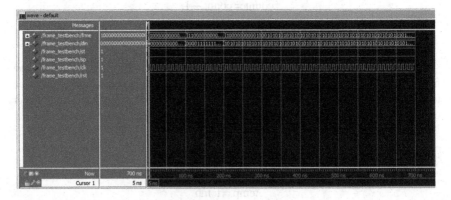

Fig. 3 Simulation results of encoder

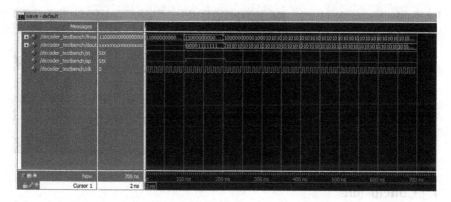

Fig. 4 Simulation results of decoder

In the simulation results, 128 bits of data are considered. The data bits are reduced to 24 bits in the encoded format. Hence, more number of bits can be sent by encoding the data in the proposed technique. The decoder shows the validation of the proposed encoding scheme and also shows the recovery of data. The decoder recovers the data for the positive edge of 'clk' and positive logic of 'rst'. Figure 4 shows the decoding process.

In the above simulation result of the decoder, with three different conditions.

1. Under deadlock with no minimum information of the packet. Here data cannot be recovered by decoding the encoded data, as the encoded data is incomplete. Hence, the output is not generated and the packet is discarded.
2. Under deadlock with minimum information of the packet. Here data can be decoded as the encoded data is completely received and the output is recovered.
3. Under no deadlock condition, where data can be taken the packet directly (Figs. 5 and 6).

Fig. 5 Synthesis results of encoder

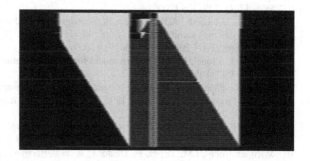

Fig. 6 Synthesis results of
decoder

4 Conclusions

This work implements the deadlock recovery for NoC using encoding scheme. The
proposed encoding scheme shows higher performance with respect to deadlock
appearance. The encoding scheme shows an area of 24861 sq-μm, the power of
17322 nw and the speed as 89632 ps. The deadlock recovery is done for the
minimum packet appearance of 16–20%.

References

1. Warnakulasuriya, S., Pinkston, T.M.: Characterization of deadlocks in interconnection
   networks. In: Parallel Processing Symposium, 1997. Proceedings, 11th International, pp. 80–
   86. IEEE (1997)
2. Duato, J., Yalamanchili, S., Ni, L.M.: Interconnection Networks: An Engineering Approach.
   Morgan Kaufmann (2003)
3. Dally, W.J., Towles, B.P.: Principles and Practices of Interconnection Networks. Elsevier
   (2004)
4. Glass, C.J., Ni, L.M.: The turn model for adaptive routing. J. ACM (JACM) **41**(5), 874–902
   (1994)
5. Chiu, G.-M.: The odd-even turn model for adaptive routing. IEEE Trans. Parallel Distrib.
   Syst. **11**(7), 729–738 (2000)
6. Mori, K., Abdallah, A.B., Kuroda, K.: Design and evaluation of a complexity effective
   network-on-chip architecture on fpga. In: Proceedings of the 19th Intelligent System
   Symposium (FAN 2009), pp. 318–321 (2009)
7. Glass, C.J., Lionel, M.N.: The turn model for adaptive routing. J. Assoc. Comput. Mach. **41**
   (5), 874–902 (1994)
8. Anjan, K., Pinkston, T.M.: Disha: a deadlock recovery scheme for fully adaptive routing. In:
   Parallel Processing Symposium, 1995. Proceedings. 9th International, pp. 537–543. IEEE
   (1995)
9. Kim, J.H., Liu, Z., Chien, A.A.: Compressionless routing: a framework for adaptive and
   fault-tolerant routing. IEEE Trans. Parallel Distrib. Syst. **8**(3), 229–244 (1997)
10. Martinez-Rubio, J.M., Lopez, P., Duato, J.: A cost-effective approach to deadlock handling in
    wormhole networks. IEEE Trans. Parallel Distrib. Syst. **12**(7), 716–729 (2001)

11. Lankes, A., Wild, T., Herkersdorf, A., Sonntag, S., Reinig, H.: Comparison of deadlock recovery and avoidance mechanisms to approach message dependent deadlocks in on-chip networks. In: 2010 Fourth ACM/IEEE International Symposium on Networks-on-Chip (NOCS), pp. 17–24. IEEE (2010)
12. Mori, K., Esch, A., Abderazek, B.A., Kuroda, K.: Advanced design issues for oasis networkon-chip architecture. In: International Conference on Broadband, Wireless Computing, Communication and Applications, pp. 74–79 (2010)

PNSDroid: A Hybrid Approach for Detection of Android Malware

Satish Kandukuru and R. M. Sharma

Abstract Android devices are more vulnerable to malware due to its open architecture. As the rapid growth in Android malware, traditional static approaches are ineffective to detect malware against encryption, code transformation, and polymorphic techniques. Dynamic approaches are run time but higher resource consumption. In this paper, we proposed a hybrid model to detect the android malware by analyzing permission bit vector, network traffic, and system call invocations. A decision tree classifier to detect the Android malware was constructed. The results show that combination of permission bit vector, network traffic, and system calls analysis is highly efficient by achieving 97.5% of detection accuracy.

Keywords Android malware · Permission bit vector · Network traffic
System calls · Detection

1 Introduction

Android is the most popular operating system for the mobile platform. It has the highest 87.6% share of the worldwide market in second quarter of 2016. Although Android has become world popular mobile operating system, at the same time it is more vulnerable to security attacks. The authors who are interested in malware attacks are attracted to the Android platform because of millions of Android users and openness of the internal design. The growth of Android malware is drastically increased from 350 K to 10.6 M between 2012 and 2015. In 2015, the rate of

S. Kandukuru (✉) · R. M. Sharma
Department of Computer Engineering, National Institute of Technology Kurukshetra,
Kurukshetra, India
e-mail: cybersatish12@gmail.com

R. M. Sharma
e-mail: rmsharma123@rediff.com

© Springer Nature Singapore Pte Ltd. 2018
P. K. Sa et al. (eds.), *Recent Findings in Intelligent Computing Techniques*,
Advances in Intelligent Systems and Computing 709,
https://doi.org/10.1007/978-981-10-8633-5_36

growth is doubled in every quarter as compared to 2014. The Kaspersky's mobile security analysis lab report Q1-2016 shows 2,045,323 malware packages, 4,146 banking Trojans, and 2,896 ransomware has been installed in the first quarter of the 2016. The detected malware packages are 11 times higher than the fourth quarter of 2015 and 1.2 times greater than the third quarter of 2015. Symantec internet security report stated that, they detected 3,944 new unique variants of Android malware from 18 new android malware families. The objective is to develop a cloud-based hybrid solution which is highly efficient to detect Android malware by consuming the limited resources of the mobile platform. The rest of this paper is organized as follows: Sect. 2 gives a brief survey of related work on detection android malware. Section 3 explains about the proposed framework. Section 4 presents experiment and results and finally Sect. 5 clearly describes the conclusion.

2 Related Work

Researchers proposed many static approaches. Android Application Analyzer [1] is using K-mean clustering algorithm and it fails to detect new malware variants. Kirin [2] is a permission-based static approach to detect the malicious repackaged applications. It has predefined rules based on the permission set policy. DroidMOSS [3] is a static approach, but it fails to detect repackaged application if it is successfully uploaded into official appstore. AndroSimilar [4] is another static approach to detect android malwares. It divides the .apk file into fixed size segments and calculates the entropy values for each segment. ScanDroid [5] is a static approach to detect repackaged applications. But it detects only specific repackaged applications. APK-DNA [6] is a static approach to detect malicious applications. This approach generates the DNA fingerprint for Android applications using fuzzy technique. It has a drawback to detect unknown malware families. SCSdroid [7] is the dynamic approach that aims to detect malicious repackaged applications. But it fails to detect unknown malware families. Crowdroid [8] is a cloud-based solution to detect the Android malware. The remote server is analyzing behavioral data by applying 2-mean clustering algorithm. Anshul Arora [9] proposed a dynamic approach to detect the android malware based on the network traffic analysis. It has a drawback that, it only detects the malware that are need to connect some malicious remote server. Andromaly [10] is a distributed dynamic tool to detect the malware based on anomalies but it consumes a lot of resources on the device.

3 The Proposed Framework

The proposed framework has two components: client application and remote server. Next two subsections describe the complete details about client application and remote server.

3.1 Client Application

It is a signature-based static analysis to detect the known malware as much as faster. The training and evaluation phases are explained below.

Training Phase: Initially, we collected a number of malicious applications a1, a2, a3 ... an, which are belonging to the same malware family MFi. Every application is disassembled using apk tool. Androidmanifest.xml file is converted into readable format then extract all the permissions requested by the malicious application. From all requested permission sets, we extracted most common and dangerous permissions. In order to generate a signature for androidmanifest.xml file, we are using permission bit vector which is having a predefined order for all Android permissions. Initially, all bits are 0s in the permission bit vector. Only the extracted common permission bits are flagged as 1s in the bit vector. Figure 1 shows that malware family signature generation process. The size of the permission bit vector is 256 bits, in fact, the total number of android permissions are does not exceed this number.

Evaluation Phase: In this phase, extract all the dangerous permissions requested by the new application. Then generate the signature by flagging the corresponding bits as 1's in the permission bit vector. Here, we are using bitwise Jaccard similarity technique to calculate a similarity score between two permission bit vectors. Jaccard bitwise similarity technique has been explained below.

$$Jaccard - bitwise(X, Y) = \frac{Ones(X \cdot Y)}{Ones(X + Y)}$$
$$0 \leq Jaccard - bitwise(X, Y) \geq 1 \tag{1}$$

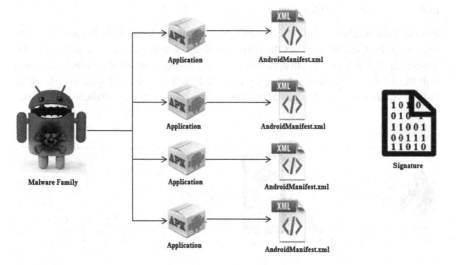

Fig. 1 Android malware signature generation process

Now, we compute the similarity score between new application signature and existing malware family signatures using Eq. (1). Out of all, choose the highest similarity score, if the score is greater than the threshold value then we tagged the new application is belonging to that malware family. The client application will declare that testing app is malicious without proceeding to the next level of analysis. If the highest similarity score is less than the threshold value, then new application is allowed to install on the device for the dynamic analysis in the next level.

3.2 Remote Server

The remote server receives the text file and .pcap file periodically. To understand run-time behavior of the application, our model analyzes network traffic and system call patterns. Figure 2 shows the architecture of the PNSDroid.

Network traffic analysis: In our research, we find that most of the android malware wants to communicate with some remote server which is maintained by the attacker. During communication, the application is generating some network traffic. Hence, we are interested to analyze both TCP conversation and HTTP packets. The received network traffic (.pcap) file is opened with the Wireshark for further analysis. Table 1 shows the most distinguishable traffic features from TCP conversations. Android malware sends the confidential data like IMEI number, SMS conversations, bank OTP's, contacts, OS details, etc., to the remote server. Hence, this model performs the content-based analysis on HTTP packets. Every keyword is having some confidence score, the value depends on the how much confidential data it is and a number of times it was detected in trained data. The malicious score from http analysis will depend on the number of keywords were detected in the analysis and their confidence scores.

System calls analysis: Android operating system is built on Linux kernel. So every activity in the android, invokes the corresponding system calls. Hence analyzing the system call sequences is very effective to understand the application interaction with the kernel. In training phase, we captured the thousands of system calls from android malware and benign applications. To perform any malicious activity,

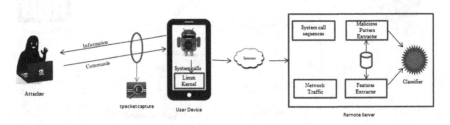

Fig. 2 Architecture of the PNSDroid

Table 1 Network traffic features

Notation	Network traffic features	Malicious traffic	Benign traffic
Aps	Average packet size	89–360	570–900
Riob	Ratio of incoming to outgoing bytes	0.1–4.6	6–8
Abrs	Average number of bytes received per second	16–1800	9000–16000
Aprf	Average number of packets received per flow	1–18	25–58
Apsf	Average number of packets sent per flow	6–13	20–45
Abrf	Average number of bytes received per flow	50–5000	30000–88000
Absf	Average number of bytes sent per flow	490–2100	2700–6000

malware needs to invoke the corresponding system calls in a sequence. For effective analysis, we captured thread-level system calls instead of process level to understand better about the malicious activity. We analyzed thousands of system calls generated by Android malware and extracted malicious system call patterns. Official Linux kernel documentation is used for the analysis of each system call with its arguments and return value. Based on the extracted patterns we make the rules to detect malicious system call patterns in the evaluation phase.

Classifier: For classification of the evaluated application, we need an efficient classifier. Although many machine learning algorithms are used by many researchers, for our selected features decision tree classifier is the most suitable one. In fact decision trees having many advantages like easy to construct, easy for trained data, robust, fast prediction based on nodes and it can easily handle the irrelevant parameters. Also, decision trees exhibit the high detection accuracy.

4 Experiment and Results

For our simulation work, we selected randomly 180 android malware samples from Malgenome dataset [11] and also 100 benign applications from Google Play Store. The selected malware samples belong to 20 android malware families. In training phase, we analyzed 97 malicious applications belongs to different families and 65 benign applications from Google Play Store. In the static analysis phase, our model successfully detected 136 malicious applications as malware out of 180 malicious applications based on trained data. Hence, our model saves the lot of resources like battery consumption, RAM usage, CPU usage, etc. To understand the run-time behavior of the application, we analyzed http packets, TCP conversation from captured network traffic and also searched for the malicious patterns in the recorded system call. Here, we need a classifier which should exhibit high detection accuracy, hence we are using decision tree as a classifier to classify the evaluated sample as either malicious or benign. Figure 3 shows that the Plankton malware leaks the confidential data through http request and response packets.

POST /v2/api.php HTTP/1.1..
Content-Length: 652..
Content-Type: application/x-www-form-urlencoded..
Host: api.airpush.com..
Connection: Keep-Alive..
Accept-Encoding:gzip..
apikey=1323159754872974151&appId=508966&imei=d08ad418a89659db1a0c9c8da40e1ca96token=d1e843fe4b5a1541ecf379ffa7197977&request_timestamp=Wed+Dec+28+23%
3A00%3A05+GMT%2B05%3A30+2016&packageName=com.macte.JigSaw.Abstract&version=22&carrier=Jio+4G&networkOperator=Jio+4G&phoneModel=LS-4505&manufacturer
=LYF&longitude=0&latitude=0&sdkversion=4.02&wifi=0&useragent=Mozilla%2F5.0+%28Linux%3B+Android+5.1.1%3B+LS-4505+Build%2FLMY47V%3B+wv%29+AppleWebKit%
2F537.36+%28KHTML%2C+like+Gecko%29+Version%2F4.0+Chrome%2F46.0.2490.76+Mobile+Safari%2F537.36&android_id=dad67559ce80a6d &longitude=0&latitude =0&
model=user&action=setuserinfo&APIKEY=1323159754872974151&type=app

Fig. 3 Leakage of confidential data through HTTP packets by Plankton malware

$$\text{Detection Accuracy} = \frac{N_{mal} * \text{TP} + N_{ben} * \text{TN}}{N_m + N_b} \times 100 \qquad (2)$$

The detection accuracy is calculated using True Positive and True Negative. The total numbers of malicious and benign applications are denoted by N_{mal} and N_{ben} respectively. Detection accuracy (DA) is calculated using the Eq. (2). We conducted the experiment with 180 malicious applications and 100 benign applications. Our model detected 176 malicious applications as malware, hence the TP is (176/180) = 0.97 and also detected 97 benign applications as benign, hence TN is (97/100) = 0.97. Finally, the detection accuracy is **97.5%** calculated using the Eq. (2).

5 Conclusion

In this paper, we proposed an efficient hybrid model to detect the android malware based on permission bit vector, network traffic patterns, and system call sequences. It leverages the advantages of both static and dynamic approaches to overcome the drawbacks of existing models. We are using decision tree classifier to detect android malware based on trained data. Our results show that combination of permission bit vector, system call sequences, and network traffic analysis is highly efficient by achieved **97.5%** of detection accuracy. PNSDroid is faster like static approach and efficient to detect the unknown malware like dynamic approach. Hence, our hybrid model is very efficient to detect the malware on Android device.

References

1. Bushra, S., Chatterjee, M.: A novel approach to detect android malware. In: ICACTA-2015, Procedia Computer Science, pp. 407–417 (2015)
2. Enck, W., Ongtang, M., McDaniel, P.: On lightweight mobile phone application certification. In Proceedings of the 16th ACM Conference on Computer and Communications Security, pp. 235–245 (2009)

3. Zhou, W., Zhou, Y., Jiang, X., Ning, P.: Detecting repackaged smartphone applications in third-party android marketplaces. In: Proceeedings of the 2nd ACM CODASPY, New York, NY, USA, pp. 317–326 (2012)
4. Faruki, P., Ganmoor, V., Laxmi, V., Gaur, M.S., Bharmal, A.: AndroSimilar: robust statistical feature signature for android malware detection. In: Eli, A., Gaur, M.S., Orgun, M.A., Makarevich, O.B. (eds.) Proceedings of the SIN, pp. 152–159 (2013)
5. Fuchs, A.P., Chaudhuri, A., Foster, J.S.: SCanDroid: automated security certification of android applications. Manuscript
6. Karbab, E.B., Debbabi, M., Mouheb, D.: Fingerprinting android packaging: generating DNAs for malware detection. Dig. Investig. **18**, S33–S45 (2016)
7. Lin A, Y.-D., Lai, Y.-C.: Identifying android malicious repackaged applications by thread-grained system call Sequences. Comput. Secur. **39**, 340–350 (2013). Elsevier
8. Burguera, I., Zurutuza, U., Nadjm-Tehrani, S.: Crowdroid: behavior-based malware detection system for android. In: Proceedings of the 1st Workshop on Security and Privacy in Smartphones and Mobile Devices, CCSSPSM'11 (2011)
9. Anshul, A., Shree, G., Sateesh, K.P.: Malware detection using network traffic analysis in android based mobile devices. In: Eight International Conference on Next Generation
10. Shabtai, A., Kanonov, U., Elovici, Y., Glezer, C., Weiss, Y.: Andromaly: a behavioral malware detection framework for android devices. J. Intell. Inf. Syst. **38**(1), 161–190 (2012)
11. Yajin, Z., Jiang, X.: Dissecting android malware: characterization and evolution. In: IEEE Symposium on Security and Privacy, pp. 95–109 (2012)

Part III
Advanced Image Processing Methodologies

Detection and Analysis of Oil Spill in Ocean for Reduced Complexity in Extraction Using Image Processing

T. R. V. Anandharajan, R. Jijendiran, K. K. Vignajeth and G. Abhishek Hariharan

Abstract Oil spills occurring in oceans are difficult to detect and require sophisticated measures to obtain and analyze the images. In this chapter, both color image using high-resolution cameras and Synthetic Aperture Radar (SAR) images are analyzed and certain useful results are obtained to reduce the complexity in extracting the oil spills. The recognition and examination of the oil spill images are done using image processing technique. Furthermore, if the oil spill is scattered as patches, the algorithm classifies the patches into smaller patches and larger ones by using k-means clustering. Hence, the patches depending on the size or intensity can be extracted on a simpler basis.

Keywords Synthetic aperture radar (SAR) images · Image processing
K-means clustering · Machine learning

1 Introduction

Due to the transportation of oil through pipelines beneath the ocean floors, the cost of transportation has reduced to a large scale. But there are risks involved in the pipeline transportation of crude oil or any form of fuel under seabed. Most importantly the large oil tanker ships are employed in exporting oil from one region to another. In both the cases, a minor leak may cause major disasters to ocean life

T. R. V. Anandharajan (✉) · R. Jijendiran · K. K. Vignajeth · G. Abhishek Hariharan
Velammal Institute of Technology, Chennai, Tamil Nadu, India
e-mail: trvanandharajan@gmail.com

R. Jijendiran
e-mail: jijendiranravichandran@gmail.com

K. K. Vignajeth
e-mail: vignajeth@gmail.com

G. Abhishek Hariharan
e-mail: abhishekhariharan1995@gmail.com

© Springer Nature Singapore Pte Ltd. 2018
P. K. Sa et al. (eds.), *Recent Findings in Intelligent Computing Techniques*,
Advances in Intelligent Systems and Computing 709,
https://doi.org/10.1007/978-981-10-8633-5_37

and nearby coast. The spills occur due to many reasons. In pipelines, it may be due to the wearing and tearing of pipelines, negligence in continuous monitoring, equipment breakdown or some internal cracks. In the ships, leakage occurs due to internal and external factors. The internal factors are unclamped oil barrels, old storage tanks (rusted) or improper construction of ship base. The external ones are due to storms (climatic conditions), collision with other materials on sea (other ships or marine mammals).

After the leakage, slowly a layer of oil forms over the sea. Mostly in all the cases the oil floats over the ocean and disperses. If it is the case of crude oil, the entire process takes place slowly depending on climatic conditions. In times of heavy currents, the dispersion of oil takes place randomly and instantly. Due to the continuous spreading of oil, the layer becomes thinner and thinner and resembles sleek like structure. There comes a condition when the entire oil has spread and becomes as lighter as in can. In this condition, it is harder to extract as it mixes with water and becomes dilute and distributed. The major effects of oil spill are marine pollution, salt water deposition and coral reef degradation. As the majority of life species exist in ocean, the harm caused due to marine pollution is significant. Initially, the sunlight is trapped as the oil surface engulfs the sea, causing the aquatic plants to die. Oxygen content is also affected leading to the death of small organisms instantly. Later, the large mammals and big fishes slowly begin to starve if the oil spill extent is large. The fur of the mammals gradually deteriorates and vanishes. Thus, the entire aquatic life results in a chaos. The tedious process to extract the oil spill involves various methods like skimming, adding sorbents, chemical dispersants, etc. Skimming involves the scooping of oil from sea using boats. Sorbents involve sponges to suck the oil from sea. Washing off the oil patches using high pressure tubes and separating them is also an efficient method.

2 Formulae

2.1 Distance Formula

The distance formula is used in K-means algorithm to give the distance between the centroids and its neighbors. It is given by:

$$c^{(i)} = \left|\left|x^{(i)} - \mu_j\right|\right|^2 \tag{1}$$

$x^{(i)}$ represents the data points or neighbors, μ_j represents the position of the centroids. $c^{(i)}$ is the index of the least nearest neighbor to the centroids.

2.2 Mean

The mean is used to compute the average of the values. It is represented as

$$\mu_k = \frac{1}{|c_k|} \sum_{i \in c_k} x^{(i)} \tag{2}$$

Here, $x^{(i)}$ represented the value of neighbors present near the centroid and the total number of neighbors is represented as $|c_k|$.

2.3 Gaussian Distribution

The Gaussian distribution is used as a filter to smoothen the image. The Gaussian distribution in terms of image is given as

$$H_{ij} = \frac{1}{2\pi\sigma^2} \exp\left(-\frac{(i - (k+1))^2 + (j - (k+1))^2}{2\sigma^2} \right) \tag{3}$$

$$1 \le j, j \le (2k+1)$$

And σ is the value of the variance.

2.4 Gradient

The gradient is the square of the sum of G_x and G_y. It is represented as

$$G = \sqrt{G_x^2 + G_y^2} \tag{4}$$

Angle

$$\theta = \text{atan}\left(\frac{G_y}{G_x} \right) \tag{5}$$

where

$$G_x = \begin{bmatrix} -1 & 0 & 1 \\ -2 & 0 & 2 \\ 1 & 0 & 1 \end{bmatrix} \tag{6}$$

$$G_y = \begin{bmatrix} -1 & -2 & -1 \\ 0 & 0 & 0 \\ 1 & 2 & 1 \end{bmatrix} \hspace{3cm} (7)$$

The above patches are applied to the images and the resulting image is taken for finding the gradient and angle θ.

3 Methodology

3.1 Image Processing

The obtained SAR images [1] from the satellite should first undergo image processing to acquire the oil-spilled region from the sea. The oil-spilled region is found using edge detection technique. The edge detection works based on the variation in pixel intensity. In this chapter, canny edge detection method is used rather than Sobel edge detection method as Canny edge detection method gives better edges and less noise because of its properties. In order to understand the working of canny edge detection method, it is required to know how Sobel edge detection method works.

The Sobel edge detection works by using image derivatives where the patch G_x and G_y made using the coefficient of image derivatives is applied to the image that needs to be processed. As a result, the gradient of the image is obtained by using (4) and the angle of the edges are found using (5). The gradient of the image is the edge detected by the algorithm. The disadvantage of using Sobel edge detection method is the occurrence of noise is very high as a small patch (6 and 7) is applied to the entire image and, moreover, the edges have varying thickness because of using gradient. To avoid these problems, canny edge detection is used and its procedure is represented in flow graph as shown in Fig. 1.

In the Canny edge detection method, the SAR image undergoes Gaussian Filter (3) to smoothen the image; this is done to remove noise in the image that can give false edges. The image is then processed to find the gradient (4) by applying the patches G_x and G_y to the image. The obtained image has varying edge thickness. This Varying edge thickness is removed using non-maximum suppression where the neighboring pixel values of the edges is selected using (5) and (4). The maximum edge intensity value is chosen and the rest are suppressed. The suppressed image still has minute noises hence it is removed by using a threshold. The Hysteresis edge tracking is an important feature in Canny edge detection, where the patches of the edges formed are joined together to form boundaries and this becomes very useful in finding the area and other features for implementing machine learning algorithm.

The area of the patches (oil spill) detected by edge detection is found by counting the total number of pixel present inside the boundaries. The perimeter is found using the position of oil patch boundaries. The coordinates of the oil patch boundaries are taken and Euclidean distance is applied to its neighboring coordinates, this is done to all the boundaries coordinates.

Fig. 1 Flowchart of Canny
edge algorithm

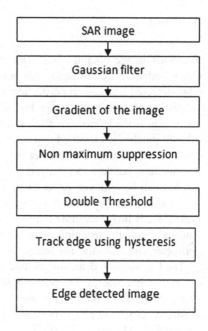

3.2 Machine Learning

K-means clustering [2] is purely an unsupervised learning technique. That means that there are no initial labels for clustering the data. It is an iterative procedure which has to update its values based on certain looping structures. Initially based on our training set the data to be clustered is the area, perimeter, intensity values of the oil patches. First, any two of the available parameters is taken for obtaining a 2D plot, showing the clustered data. But any number of features can also be taken. The algorithm works like this: The cluster centroid is assigned randomly. It may be from the data set itself. After that, the distance between each of the data set value with the cluster index is calculated, squared and minimized, so that the data set value is assigned to the nearest centroid as in (1). This is not yet over as the centroids have to converge to a minimum value although the k-means is not a converging function. The mean value of the cluster of data set belonging to a particular cluster is found and the new centroid is calculated as in (2). And finally, the above steps from assigning the data set values to a cluster and the computation of mean so that the new centroid value found, is repeated till the centroids are minimized to a certain value beyond which the minimisation is useless. In this application, we have made use of three clusters so the number of initial centroid value is given as three sets of the point. We can give any color to the labels of the clusters distinguishing from others [3].

4 Plots

The SAR image obtained is shown in Fig. 2 and the obtained image undergoes Canny edge detection algorithm and result in the output as in Fig. 3. The edge-detected image's properties like area, intensity, perimeter are obtained and its values are plotted in the image Fig. 4 the clustering done with x-axis as the Area and y-axis as the Intensity is shown in Fig. 5. The final image Fig. 6 shows how the oil spill patches is classified depending on that the ways to extract oil from the sea can be implemented. The smaller patches can be removed by adding bacteria to it; the medium patches can be removed using oil Coagulating agents. The larger patches can be removed either by absorbents or by suction pumps.

Another scenario is that the color image taken by drones. Here, as the image is a normal RGB image as shown in (Fig. 7) the same technique is applied to detect the patch and suitable techniques are employed to classify the patches based on area, perimeter, and intensity. The image is resized and is converted into a grayscale image.

The Canny edge detection method is employed here with a suitable threshold to outline the boundaries. The points in the Fig. 8 represent the centroid of each patch. The parameters of the patches are found and based on the available data set the clustered (k-means) plot is obtained. In Fig. 9, the plot on area versus intensity is determined. Majorly it is clustered based on the area [4]. When we look at the intensity, both the clusters somewhat have equal maximum and minimum values.

Fig. 2 SAR image

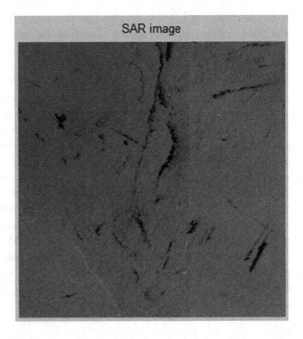

Fig. 3 SAR image after edge
detection

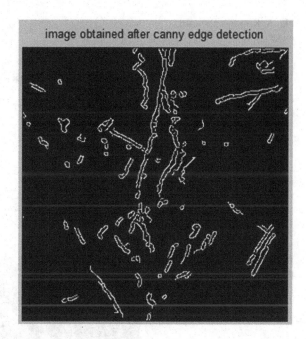

Fig. 4 Image with area and
perimeter

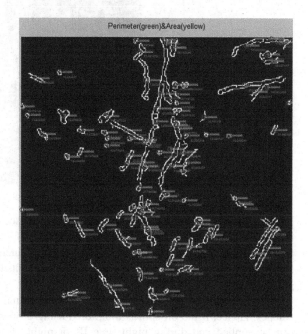

Fig. 5 K-means clustered image

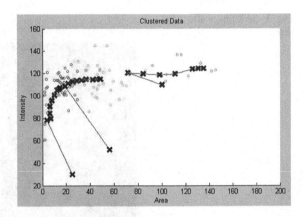

Fig. 6 Image marked with the classification

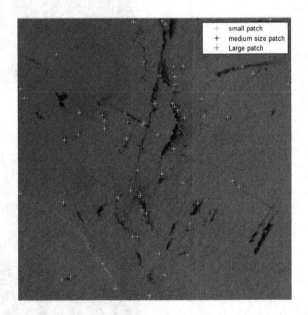

Figure 10 helps in labeling the patches over the grayscale image. Figure 11 shows the area versus perimeter which indicates at an area of 75 units the perimeter of the oil spill is 90 units. The red points denote the small patch clearly and the ones with large patches are denoted by green. Figure 12 represents the labeled cluster over the grayscale plot.

The IR image of the oil spills [3] can also be analyzed for the clustering of patches. Hence, if the oil spill is monitored using drones, then an 8 h surveillance can be applied and during night time IR or night vision lenses can capture the images and is processed in a similar way.

Fig. 7 Color image of oil spill patch

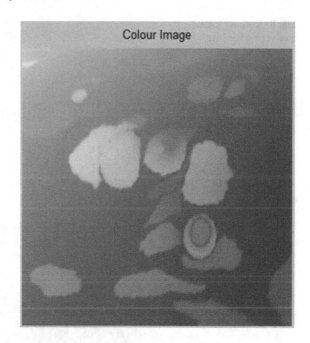

Fig. 8 Canny edge-detected image

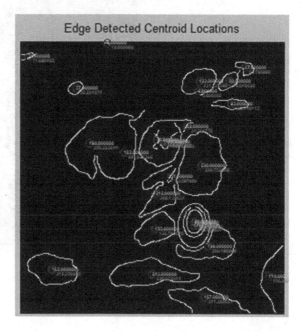

Fig. 9 K-means clustering
area versus intensity

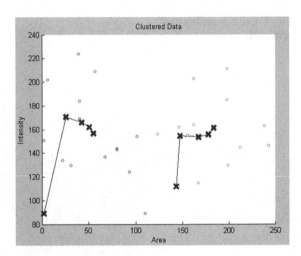

Fig. 10 Image marked with
clustered data points

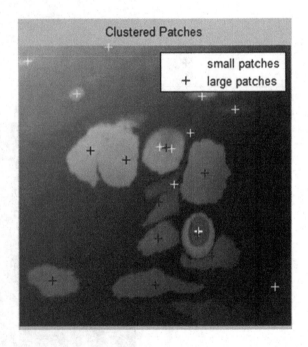

Fig. 11 K-means clustering —area versus perimeter

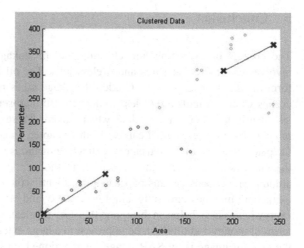

Fig. 12 Image with a clustered index

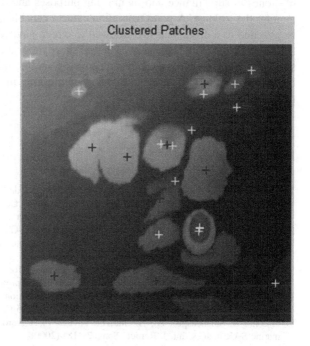

5 Conclusion

The process of extracting and cleaning the oil leakage from the ocean surface involves expensive measures and tireless jobs. The oil leak may be distributed in form of sleeks or as patches. Crude oil leakage takes more time to spread in dry oceans where currents are of less frequency. But kerosene and compressed gas are less dense and they tend to float when leaked. There are traditional methods to remove the oil as explained earlier. In this chapter, since the algorithm can cluster the patch areas clearly, the usage of extracting methods needs not to be employed in the entire spill area. As the extraction methods involved in larger patches are arduous and expensive, this paper helps to identify the larger patches such that the extraction methods are only employed there and are not employed randomly everywhere. This helps in bringing the cost of cleaning down and efficiency in terms of money and time can be improved to a large extent. The best way to analyze the oil spill image is by SAR images as mentioned earlier. The algorithm can also be applied to high-resolution color cameras. Nowadays there is the adequate usage of drones in surveillance and monitoring purposes and the same can be applied to monitor the leakage. A good resolution camera capable of 360° image capture can be attached to a drone and it is allowed to circumnavigate a predefined perimeter. So that if there is any leakage near the off shore rig or the container ship, the drone sends the image to control centre immediately and the images are analyzed to take the above steps to reduce the complexity in cleaning up the oil spill and can save marine life.

References

1. Topouzelis, K.N.: Oil spill detection by SAR images: dark formation detection, feature extraction and classification algorithms. In: Joint Research Centre (JRC), European Commission, Via Fermi 2749, 21027, Ispra (VA), Italy, 23 October 2008
2. Del Frate, F., Petrocchi, A., Lichtenegger, J., Calabresi, G.: Neural networks for oil spill detection using ERS-SAR data. IEEE Trans. Geosci. Remote Sens. 38(5), 2282–2287 (2000). ISSN: 0196-2892
3. Kubat, M., Holte, R.C., Matwin, S.: Machine Learning for the Detection of Oil Spills in Satellite Radar Images, vol. 30, Issue 2–3, pp. 195–215. Springer Link. ISSN: 1573-0565
4. Solberg, A.H.S., Brekke, C., Husoy, P.O.: Oil spill detection in Radarsat and Envisat SAR Images. IEEE Trans. Geosci. Remote Sens. 45(3), 746–755 (2007). ISSN: 0196-2892
5. Fiscella, B., Giancaspro, A., Nirchio, F., Pavese, P., Trivero, P.: Oil spill detection using marine SAR images. Int. J. Remote Sens. 21(18) (2000)

SOMES: An Efficient SOM Technique for Event Summarization in Multi-view Surveillance Videos

Krishan Kumar, Deepti D. Shrimankar and Navjot Singh

Abstract In this work, we propose an unsupervised learning network model based on Self-Organizing Map (SOM) to detect and summarize the events in multi-view surveillance videos. Action recognition in multi-view surveillance videos can be a primary element of a multimedia surveillance systems. The basic elements: feature extraction, structuring, mapping, and visualization are counted by our proposed model while video skimming as action recognition of the activities are subjected as the event detection and summarization. The model can work better for real-time applications, for e.g., surveillance, security systems, etc. The qualitative as well as quantitative assessment is done in order to compare the performances of our proposed SOM-based summarization technique and state-of-the-art models. Experimental results on three benchmark data-sets with various types of videos indicate that the proposed method outperforms the state-of-the-art models with the best *F-measure*.

Keywords Event summarization · Multi-view · Surveillance videos · SOM

1 Introduction

During this digital era, an expeditious growth in the amount of digital audiovisual data around the clock, captured by multiple cameras in the last few years. Moreover, due to advancement in the computing as well as network infrastructure and the recurrent use of the digital video technology, a large number of multimedia applications

K. Kumar (✉) · D. D. Shrimankar
Department of Computer Science & Engineering, VNIT, Nagpur, India
e-mail: kkberwal@nituk.ac.in

D. D. Shrimankar
e-mail: dshrimankar@cse.vnit.ac.in

K. Kumar · N. Singh
Department of Computer Science & Engineering, NIT Uttarakhand, Srinagar, India
e-mail: navjot.singh.09@gmail.com

© Springer Nature Singapore Pte Ltd. 2018   383
P. K. Sa et al. (eds.), *Recent Findings in Intelligent Computing Techniques*,
Advances in Intelligent Systems and Computing 709,
https://doi.org/10.1007/978-981-10-8633-5_38

are forthwith required. So, a big volume of audiovisual data is rapidly generated for such numerous applications such as content-based navigation and search, video browsing and retrieval, and semantic annotation, saliency [1], etc. Video summarization (*VS*) is one of the main fundamental keys for the success of these applications in order to abate the size of audiovisual content adequately and comprehensively, including the user-friendly access from a huge amount of video content. *VS* can be either *key-frames* or *skimming* [2]. Cong et al. [3] mapped the *key-frames-based VS* task as a sparse dictionary selection problem using the selected *key-frames* via reconstruction of optimal videos. Various local as well as global representations based semi-supervised learning approaches [4, 5] have been suggested to summarize the video content through structural analysis with multiple features. Summarizing the videos by preserving the most interesting activities/highlights with their semantics often referred as *dynamic-based VS*. The highlights usually known as *Event Summarization* (*ES*). The primary issue with the existing *ES* models is not detecting the appropriate event boundaries, to state the effectiveness of the summarized video by keeping the adequate event content from the video. Moreover, *multi-view* video comprises huge content which is captured by the surveillance cameras for even a few hours. In addition to this, most of the video content captured by fixed camera systems are not much interesting which creates the problem for extracting the useful information from such recorded videos. *multi-view VS* comprises some distinct (from *mono-view*) issues. Therefore, *mono-view VS* approaches cannot be helpful to resolve the *multi-view* issues addressed in [6–9]. Hence, an integrated system for multi-view videos is urgent needed, for presenting the effectiveness and efficiency of the comprehensive video content, with the inter-view and inter-relationship between different views of the video; for e.g., various events are simultaneously recorded as a set of videos in monitoring/surveillance systems. Thus, conventional *VS* techniques become inadequate for the applications of *multi-view* videos including inspection of post-accidental situation, fast event recognition in case of suspect as theft and burglary at public places. To resolve the above challenges, Fu et al. [6] proposed the graph- based summarization solution for *multi-view* videos by using random walks, while the correlations between different views are represented by a hyper-graph, to obtain the summaries of the interesting information/events in minimum time. However, all the previous existing methods are capable only for off-line processing which requires the huge computational power and memory resource requirements; so it becomes more difficult to apply on the online or wireless video sensors. Moreover, a low-complexity online processing *multi-view VS* approach presented by Ou et al. [9] for wireless video sensors; it also saves compression and transmission power together by preserving the critical information from the videos. Recently, a bipartite matching graph-based *multi-view VS* technique was presented to generate the video summaries through optimum path forest clustering constraints [8]. Our purposed model is not only detecting the events, but also locating the events boundary. The main contribution of our work is mentioned below.

- We formulate the multi-view *VS* problem with an unsupervised learning model based on a Self-Organization Map (SOM). Specifically, intra-view redundancy eliminated at the stage-I for the fast summarization at stage-II where the input *key-frames* are very less as the resultant of the stage-I.

- Our proposed SOM-based system captures the dependencies among multiple views with more accuracy and more efficient in comparison with the related existing methods.
- From the point of view of action recognition using *SOM* in multi-view surveillance videos to produce highlights/important events has not been employed to the best of our knowledge. Our proposed model still has good summarization capability in comparison to previously online as well as off-line-based models.

2 Problem Formulation

We first extract N frames in total with size $W \times H$ from a video **V**, where W and H indicate the width and height of a frame respectively. The variation in the visual content of the successive frames is not much but still computation in processing these frames needs processing all the three planes of a frame. In order to save the computation time, all *N-colored* (RGB) frames of a video are converted into grayscale frames where each frame is resized into 1-dimensional vector as an input for the neurons in unsupervised learning network using SOM.

Algorithm 1 *key–frames* extraction from a view using SOM

1: **procedure** $SOM(\{f_1, f_2, , f_N\}$ INPUT FROM VIEW $V_p)$
2:     Initialize the weights $W_{i1}, W_{i2}, ..., W_{iM}$, where $i = 1, 2, ..., N$, and
        Set learning rate $\alpha = 0.9$
3:     **Do**
4:     **for** each input vector f **do**
5:         **for** each neuron i **do**
6:             Compute the Euclidean distance $D_i = \sqrt{\sum_{j=1}^{N}(f_i - W_{ji})}$
7:         **end for**
8:         $I = min(D_i)$
9:         **for** each neuron $i \in$ neighborhood (I) **do**
10:            **for** each map unit j **do**
11:                $W_{ji}[t + 1] = W_{ji}[t] + \alpha(f_j - W_{ji}[t])$ weight update at time $t + 1$.
12:            **end for**
13:        **end for**
14:    **end for**
15:    Update learning rate α - decreasing function with number of epochs.
16:    Reduce radius of topological_neighborhood[t].
17:    **doWhile** $W[t + 1] \neq W[t]$ & $\alpha \approx 0.01$
18:    All N frames of view V_p are grouped into clusters. Select the frames close to
        the centroids of the clusters and announce them as *key–frames*
19:    **return** E <– final *key–frames* from view V_p
20: **end procedure**

Multi-view videos correlation with SOM: Numerous supervised or semi- super-vised training approaches [4, 5] have been investigated to produce the summaries of videos. Now, we move to unsupervised learning which is possible without human intervention during the training phase. Competitive learning is one of popular exam-ples of unsupervised system, where output neurons participate among themselves to become active. In result, there is only one activated at one time, such neuron are referred as winner. This type of competition can employed with having lateral hin-dering connections between these neurons in order to organize themselves. Such network is referred as a Self-Organizing Map (SOM) [10]. In the SOM, the points which are available near to each other in the input space is mapped to nearby map units only. Neurons in the network are well connected by neighborhood relation to adjacent neurons. Therefore, SOM approach is employed as a clustering analysis tool for high-dimensional data such as multi-view videos. SOMES model are divided into two stages as shown in Fig. 1. At stage-I, a mono-view summarization is performed using SOM-based *key-frames* extraction algorithm (see Algorithm 1) to filter out unimportant frames from the individual view of a dataset. At stage-II (for inter-view summarization), where again algorithm 1 is employed to extract the final *key-frames* to achieve the event summarization from different views of a particular dataset.

Event Summarization: video skimming using key-frames: In order to achieve a good event summarization as video skimming through *key-frames*, deciding the events boundary is very difficult task. It rarely happens that an event has only one frame. Thus, we estimate a parameter $[b_{min}, b_{max}]$ boundary frame number to locate the boundary of an event which belong to a particular view. The value of the param-eter $[b_{min}, b_{max}]$ is determined based on the similarity (*Euclidean distance*) between all frames of a cluster and *key-frame* of the cluster. The parameter Event Boundary Threshold (EBT [%] $= \frac{10*W}{N*H} * 100$) is experimentally estimated based on H (height of a frame), W (width of a frame) and W (total frame). If similarity between a frame and the *key-frame* of the cluster is greater than or equal to threshold EBT value, then only the current frame is consider for the current event, otherwise discard. We start comparison from the lowest frame number to the largest frame number with *key-frame* of a cluster. The b_{min} value is fixed as the first frame number which is similar

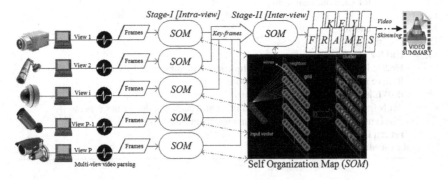

Fig. 1 Two-stage cascading SOM model for summarizing the video

with *key-frame* of the cluster and b_{max} value is fix with the previous frame number where current frame is dissimilar from *key-frame* of the cluster and then stop the comparison process and declare an event from the frame number b_{min} to b_{max}. Hence, event summarization can be achieved with video skimming process employing the extracted *key-frames* based on the boundary frames $[b_{min}, b_{max}]$ in multi-view surveillance video systems.

3 Experiments and Results

We elected totally *26* videos from three *multi-view* video data-sets: *Lobby* (*3 views*), *Office* (*4 views*), *BL-7F* (*19 views*) with collectively *26* videos in total, along with their user summaries/ground truth summaries from [6–9].

Qualitative Analysis: Figure 2a, b indicate *key-frames* of the detected summarized events are arranged in temporal order by bipartite graph [8] (18 events) and by our proposed model (22 events) receptively. In all, our proposed model quantitatively works better than existing models.

Quantitative Analysis: To expanse quality of each summarization model, summaries of video produced from multiple methods are equated to ground summaries rely on the three evaluation metrics, which consists often $Precision = \frac{TruePositive}{TruePositive + FalsePositive}$, $Recall = \frac{TruePositive}{TruePositive + TrueNegative}$, and $F\text{-}measure\,(F_\beta) = \frac{(1+\beta^2) \times Precision \times Recall}{\beta^2 \cdot Precision + Recall}$. For $\beta = 1$, i.e., F_1 is the Harmonic mean of *Precision* and

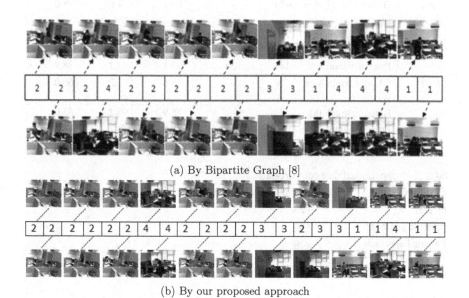

(a) By Bipartite Graph [8]

(b) By our proposed approach

Fig. 2 *key-frames* arranged in temporal order on *Office* data-set

Recall. Hence, maximum value of F_1-*measure* represents more accurate model. From Table 1, the following observations are made where the best results are shown in bold.

- *Summary length* is just considered to indicate that the length of summarized video is always smaller than the actual video in *multi-view* summarization.
- *Extracted Events* by our approach for *multi-view* summarization on all three datasets is maximum among all other models. It indicates that our proposed model is robust to keep the important information as events.
- *F-measure* of our proposed approach for *multi-view* summarization on all three benchmark data-sets is also maximum among all the other existing models, which represents the better performance of our proposed model.

Computational Complexity: For *Office* data-set which comprises four views, our proposed model requires processing time altogether less than 480 s to complete as compared to the reported processing time of 600 s in bipartite graph [8]. The average computation time per video is equated between the proposed model and the existing models in Table 2.

Table 1 *Multi-view* comparison of our proposed model with with state-of-art

Dataset	Algorithm	Summary length (s)	Events	Precision (%)	Recall (%)	F-measure (%)
Office	Bipartite graph [8]	59	18	**100**	69.2	81.8
	Metric learning [7]	–	20	**100**	76.9	86.9
	Proposed approach	75	**22**	81.7	**94.6**	**87.6**
Lobby	GMM + Multi-view [9]	484	–	60.2	77.0	68.0
	Bipartite graph [8]	176	33	**100**	76.7	86.8
	Fu et al. [6]	158	**34**	**100**	79.0	88.3
	Proposed approach	149	**34**	93.1	**85.2**	**88.9**
BL-7F	GMM + Multi-view [9]	516	–	58.0	61.2	60.0
	Bipartite graph [8]	633	–	75.9	98.2	85.0
	Proposed approach	314	**45**	**98.8**	78.2	**87.3**

Table 2 Computational time comparison

Algorithm	Sampling rate [frame/s]	Average time per video (s)
Fu et al. [6]	25–30	225
Bipartite graph [8]	30	150
Proposed approach	30	**120**

4 Conclusion

In this work, we proposed SOMES technique to recognize the action in terms of events; with the minimum storage as well as accessing time of the multi-view videos. The experimental results on all three benchmark data-sets with various types of videos demonstrate that our model outperforms the state-of-the-art models with the best *F-measure*. In future, we will work with the large duration *multi-view* surveillance videos to certify the scalability and integrity of our approach to implement more dedicated summarization techniques for different applications.

References

1. Singh, N., et al.: A novel position prior using fusion of rule of thirds and image center for salient object detection. MTAP (2016). https://doi.org/10.1007/s11042-016-3676-8
2. Vermaak, J., et al.: Rapid summarization and browsing of video sequences. In: BMVC, pp. 1–10 (2002)
3. Cong, Y., Yuan, J., Luo, J.: Towards scalable summarization of consumer videos via sparse dictionary selection. IEEE TMM **14**(1), 66–75 (2012)
4. Krishan, K., et al.: Equal partition based clustering approach for event summarization in videos. IEEE SITIS (2016). https://doi.org/10.1109/SITIS.2016.27
5. Kumar, K., et al.: Multimed. Tools Appl. (2017). https://doi.org/10.1007/s11042-017-4642-9
6. Fu, Y., et al.: Multi view video summmarization. TMM **12**(7), 717–729 (2010)
7. Fu, Y., et al.: Multi-view metric learning for multi-video video summarization (2014)
8. Kuanar, S., et al.: Multi-view video summarization using bipartite matching constrained optimum-path forest clustering. IEEE TMM **17**(8), 1166–1173 (2015)
9. Ou, S., et al.: On-line multi-view video summarization for wireless video sensor network. IEEE J. Sel. Topi. Signal Process. **9**(1), 165–179 (2015)
10. Teuvo, K.: Self-Organization and Associative Memory. Springer Science+Business Media (2012)

A Two-Stage Hybrid Operator for Tone-Mapping HDR Images

N. Neelima, Y. Ravi Kumar and K. Ravindra

Abstract The devices used for displaying the images need to have a mapping between luminance values of the image to be displayed and characteristics of the device. It has been more common to acquire High-Dynamic Range (HDR) images using the available capturers. In order to match the luminance characteristics between captured image and display device, tone mapping operators will be used. Tone mapping operators convert HDR images to Low-Dynamic Range (LDR) images. A new TMO has been proposed in this paper. Two different HDR images having different dynamic ranges and different sizes have been used for applying the proposed model. To validate the proposed model, the images have also been processed with existing TMOs such as Raman and Drago. The parameters derived for comparison are mean, MSE, PSNR, mPSNR, and Luminance. The results presented in this paper prove that the performance of the proposed TMO is better in producing LDR image with enhanced quality to show on conventional LDR display device.

Keywords HDR · LDR · TMO · Luminance · Mean · mPSNR

1 Introduction

Tone mapping is the process of converting luminance from the real world to display. The quality of acquiring real-world scenes depends on the operator's control on maintaining proper lighting and camera settings [1]. Tone Mapping Operators

N. Neelima (✉)
CMR Institute of Technology, Hyderabad 501401, India
e-mail: nnrani2729@gmail.com

Y. Ravi Kumar
Defence Electronics and Research Laboratories, Chandrayangutta, Hyderabad, India
e-mail: dr.ravikumaryeda@gmail.com

K. Ravindra
Malla Reddy Institute of Technology and Science, Hyderabad 500100, India
e-mail: kasa_ravi@yahoo.com

© Springer Nature Singapore Pte Ltd. 2018
P. K. Sa et al. (eds.), *Recent Findings in Intelligent Computing Techniques*,
Advances in Intelligent Systems and Computing 709,
https://doi.org/10.1007/978-981-10-8633-5_39

391

(TMOs) have been useful for developing direct numerical control on display values rather than depending on the physical limitations of the capturers. An attempt has been made in this work to evolve a Tone Mapping Operator which can effectively convert High-Dynamic Range (HDR) images into Low-Dynamic Range (LDR) images. Two HDR images with different sizes and different dynamic ranges have been considered as sources images in this work. These images have been processed using existing Raman and Drago TMOs in addition to the application of proposed TMO. To assess the conversion efficiency, Objective Assessment Parameters (OAPs) have been tabulated and compared. The OAPs such as Mean, Luminance, MSE, PSNR, and mPSNR are compared for quality assessment. The approach followed for the proposed TMO has been described in Sect. 2. The objective assessment parameters have been defined in Sect. 3. The results obtained have been presented in Sect. 4.

2 Tone Mapping Operators

As mentioned in the above section, Tone Mapping Operators have been used to convert HDR images into LDR images. This conversion has been necessitated to map the luminance values in the HDR images with the luminance values of pixels in LDR display devices. The process of conversion of HDR images into LDR images is shown in Fig. 1. An image capturer such as a high definition camera is used to render the natural scene. The result is the generation of HDR image. As depicted in Fig. 1, the HDR image is processed using a TMO to remodel the luminance values to make them suitable for displaying on an LDR display device. Several TMOs have been developed to meet this requirement [2–5]. However, each of the TMOs has their own merits and demerits. Global TMOs will apply a non-linear function to the entire image irrespective of variation in luminance of each pixel in the image. The result is poor quality in the LDR image. Local TMOs have been developed using several approaches to map the luminance values pixel wise. The objective assessment parameters evaluated for different Local TMOs have shown vide variation with the poor quality of image.

To overcome these difficulties a new TMO has been proposed in this work. The block diagram for the proposed TMO is shown in Fig. 2. The new TMO has been

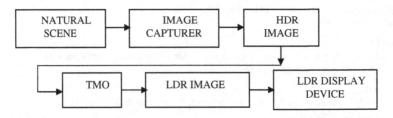

Fig. 1 General block diagram for generation of LDR image from HDR using TMO

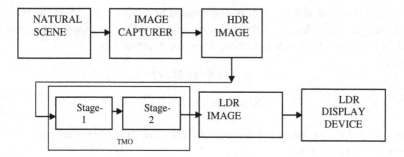

Fig. 2 Block diagram for proposed TMO

developed with an approach of two-stage hybridized processing of the source image. In the first stage, the luminance of the image have been processed globally using Eq. 1. In the second stage, the image is further processed using a logarithmic mapping function given in Eq. 2 to modify the luminance values of the image locally to produce LDR image with better quality.

2.1 Proposed TMO

The proposed Two-Stage Hybrid Tone Mapping Operator can preserve the local and visual contrast. A Tone Mapping function that can preserve local contrast is

$$C_d(x) = \frac{C_w(x)f(C_w, a(x))}{L_w, a(x)} \tag{1}$$

where f is tone mapping function, C_w, $a(x)$ are local luminance adaptation and $C_w(x)$ is the luminance for the pixel location x.

To preserve visual contrast, the tone mapping function is

$$C_d(x) = f(C_w, a(x)) + \frac{VCIf(C_w, a(x))}{VCI(L_w, a(x))} \tag{2}$$

where VCI is Visual Contrast versus Intensity function. The VCI function is given in Eq. 3.

$$C_d(x) = \begin{cases} \frac{x}{0.014} & if\ x \le 0.0034 \\ 2.4483 + \log(\frac{x}{0.034})/0.4027 & if\ 0.0034 \le x \le 1.0 \\ 16.5630 + \frac{x-1.0}{0.4027} & if\ 1.0 \le x \le 7.2444 \\ 32.0693 + \log(\frac{x}{7.2444})/0.0556 & otherwise \end{cases} \tag{3}$$

After performing the corrections for local and visual contrasts, the images have been subjected to bilateral filtering for further improvement in local particulars.

Let the image of size M × N be represented by function f(x, y). In the bilateral filter two standard deviation functions σ_S and σ_r which are used and represent special and range Gaussian functions. They are expressed as

$$\sigma_s = P_1 X min(M, N) \tag{4}$$

$$\sigma_r = P_2 X(\max(f(x, y) - \min(f(x, y)))) \tag{5}$$

where P_1 and P_2 are positive real constants. They are varied to obtain smoothing at edges in the range 1–0.1.

3 Objective Assessment Parameters

The objective quality assessment on tone mapped images [6] can be rated based on the following parameters:

(a) mPSNR (Modified peak signal-to-noise ratio):
(b) MSE (Mean square error):
(c) Mean
(d) Luminance

3.1 Modified PSNR

Peak signal-to-noise ratio (PSNR) is one of the simplest and commonly used parameters for objective assessment of image quality. Peak signal-to-noise Ratio is the ratio between the reference signal and the distortion in the signal of an image. It is expressed in decibels. The higher the PSNR, the lesser the distortion. PSNR is a measure of the quality of a TMO image.

Modified PSNR metric is based on Human Visual System (HVS) characteristics and correlates well with the perceived image quality. It is expressed as

$$mPSNR = 10 \log_{10} \frac{(col * 255^2)}{MSE} \tag{6}$$

3.2 MSE (Mean Square Error)

Mean Squared Error is the average squared differences of average between a reference image and a distorted image. It is computed pixel-by-pixel by taking the addition of the squared differences of all the pixels and dividing by the total pixel count.

If M denotes the number of pixels in an image then we write $A = \{a1, ..., a_M\}$ and $B = \{b1, ..., b_M\}$. The MSE for images A and B is calculated using Eq. 7

$$MSE(A,B) = \frac{1}{M} \sum_{i=1}^{M} (a_i - b_i)^2 \qquad (7)$$

The squaring of the deviations dampens small differences between the 2 pixels but increases large ones.

3.3 Mean

Let $f(x, y)$ be the input image with the intensity of A of size $M \times N$, then the mean of the image is given by

$$\mu = \frac{1}{MN} \sum_{i=1}^{M} \sum_{j-1}^{N} A_{ij} \qquad (8)$$

3.4 Luminance

The luminance of different images is calculated by taking log-average luminance of an image. The log-average luminance is determined by calculating the geometric mean of the luminance values of all pixels. In a grayscale image, the value of luminance is the pixel value. In a color image, the value of luminance is found by a weighted sum.

$$luminance = 0.213\,red + 0.715\,green + 0.072\,blue \qquad (9)$$

4 Results and Discussion

For better comparison and to know the versatility of the proposed model, two HDR images have been considered for evaluation. They are named as image1 and image2. The dynamic range of image1 is 3.5061 and the dynamic range of image2 is 4.5306. The size of image1 is 401×535 and the size of image2 is 713×535. Images with different dynamic ranges and different sizes have been used to validate the application of proposed TMO for any HDR image. The HDR images are subjected three TMOs. Out of three, two are existing TMOs. They are Raman TMO and Drago TMO. The third TMO is the proposed two-stage hybrid TMO. The HDR image1 along with its LDR images obtained as result of application Raman, Drago and proposed TMO are shown in Fig. 3a–d.

Fig. 3 **a** HDR image1. **b** Raman image1. **c** Drago image1. **d** Proposed image1

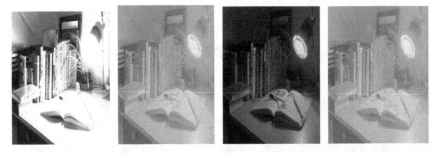

Fig. 4 **a** HDR image2. **b** Raman image2. **c** Drago image2. **d** Proposed image2

The HDR image2 along with its LDR images obtained as result of the application of Raman, Drago and proposed TMO are shown in Fig. 4a–d.

The objective assessment parameters derived for above two images are given in Table 1. The parameters derived are mean, MSE, PSNR, mPSNR, and Luminance.

The profiles of mean and MSE are plotted and shown in Fig. 5 simultaneously for both the images and for all three TMOs. The mean value is measure of smoothening of the image. The higher the mean value the more smoother the image.

MSE indicates the suitability of the converted image to use with conventional LDR display devices.

The lesser the MSE the better the conversion. It is clear from Fig. 5 that optimum values for these two parameters are obtained for the proposed TMO. Figure 6 depicts the variation in the values of PSNR and mPSNR for both the images with all three TMOs. These parameters indicate the proportion of noise content in the LDR images. The magnitude of these parameters also determines the merit of a TMO over the other TMOs. The more the PSNR and mPSNR parametric values, the less

Table 1 Objective assessment parameters obtained for TMOs under study

OAP	HDR to Raman LDR		HDR to Drago LDR		HDR to Proposed LDR	
	Image 1	Image 2	Image 1	Image 2	Image 1	Image 2
MSE	0.2415	0.125	0.0946	0.2821	0.0133	0.0183
Luminance	0.4849	0.4194	0.2296	0.2043	0.7724	0.7057
mPSNR	2.8535	12.1075	3.9318	9.9701	25.8923	23.7951
Mean	0.4852	0.4108	0.2398	0.1903	0.7725	0.6972
PSNR	54.3025	57.1631	58.4416	53.6267	66.8981	65.5022

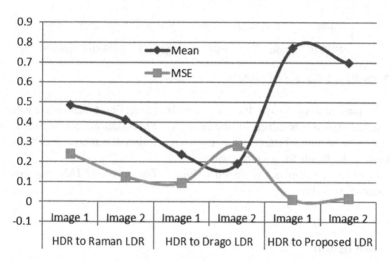

Fig. 5 Profiles of Mean and MSE for the TMOs under study

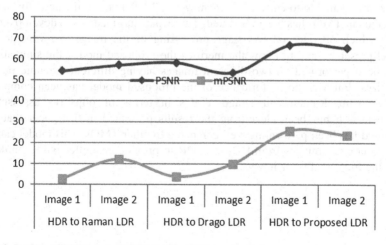

Fig. 6 Profiles of PSNR and mPSNR for the TMOs under study

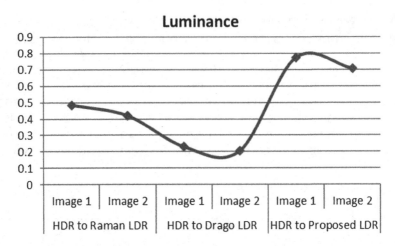

Fig. 7 Profile of luminance for the TMOs under study

the noise present in the image. It can be visualized from Fig. 6 that higher values for these parameters are obtained for the proposed TMO. Luminance is also a major parameter to show a better image on a conventional display. The higher the luminance, the better the picture quality.

The profile of luminance is shown in Fig. 7. The proposed TMO has given the best luminance value.

5 Conclusion

The significance of Tone Mapping Operators has been presented. Though several TMOs are available to convert HDR images to LDR images, the need for newer and efficient TMOs has been identified. This aspect has made us to develop a new TMO which operates on two-stage processing of the image. To validate the model presented and to know the merit of the proposed model, the simulations have been performed on two different images having different dynamic ranges and sizes. Further, the performance of the proposed model has been compared with existing Raman and Drago TMOs in terms of objective assessment parameters. It has been clear from the results presented in this paper that the proposed has been performing well compared to other TMOs. This model can be used to convert images from HDR to LDR to provide cost-effective technology with the high quality of images.

References

1. Larson, G.W., Rushmeier, H., Piatko, C.: A visibility matching tone reproduction operator for high dynamic range scenes. IEEE Trans. Visual Comput. Graph. **3**(4), 291–306 (1997)
2. Drago, F., Myszkowski, K., Annen, T., chiba, N.: Adaptive logarithmic mapping for displaying high contrast scenes. Comput. Graph. Forum **22**(3), 419–426 (2003)
3. Raman, S., Chaudhuri, S.: Bilateral filter based compositing for variable exposure photography. In: Eurographics (2009)
4. Tumblin, J., Rushmeier, H.E.: Tone reproduction for realistic images. IEEE Comput. Graph. Appl. **13**(6), 42–48 (1993)
5. Ma, K., Zeng, K., Wang, Z.: Perceptual quality assessment for multi-exposure image fusion. IEEE Trans. Image Process. **24**(11), 3345–3356 (2015)
6. Yeganeh, H., Wang, Z.: Objective quality assessment of tone-mapped images. IEEE Trans. Image Process. **22**(2), 657–667 (2013)

Use of Textural and Structural Facial Features in Generating Efficient Age Classifiers

Sreejit Panicker, Smita Selot and Manisha Sharma

Abstract This paper presents a computational and feasible solution to age classification in humans using facial feature parameters. Computer vision and machine learning have its relevance in research and industry because of its potential applicability and ease of implementation with hidden convolution. Human age classification has gained significant importance in recent times with technological advancement and widespread use of Internet services. In this paper, we propose a methodology that implements chronological growth parameter as well as the textural factor. The facial structural parameter has its significance in the early phase of growth (0–20), whereas texture has its importance in the later span of life (25 onwards), when fine lines start forming in selective facial regions. The proposed method is implemented using statistical techniques (Euclidean distance) and Local Binary Pattern (LBP) for structural and textural feature extraction. As compared to previous work, the parameters used for structural and textural are more in number that contributes to better prediction and classification in these age groups. The results obtained are noteworthy and significant.

Keywords Aging · Texture · Facial features · Structural · LBP

1 Introduction

With the technological uprising in all spheres of mankind, proper usage of these resources can lead to efficient utilization of these advancements. Age classification using facial features can be applied in applications where access permission needs be to granted based on age. Many applications that provide these services lack in efficiency, due to the external transformation that does affect the performance of

S. Panicker (✉) · S. Selot
Department of Computer Application, SSTC, Bhilai, Chhattisgarh, India
e-mail: sreejit.bhilai@gmail.com

M. Sharma
Department of Electronics and Telecommunication, BIT, Durg, Chhattisgarh, India

© Springer Nature Singapore Pte Ltd. 2018   401
P. K. Sa et al. (eds.), *Recent Findings in Intelligent Computing Techniques*,
Advances in Intelligent Systems and Computing 709,
https://doi.org/10.1007/978-981-10-8633-5_40

the system. These include performing plastic surgery, extensive makeup, usage of anti-aging drugs, and cosmetics. In this paper, both facial structural and textural features are used because it is not possible to manipulate the facial structure and hide the wrinkles that are prominent in some areas in the facial region, thus overcome those challenges. With this apprehension, it increases the complexity of age classification thus giving scope for improvement by applying new algorithms and methodologies to improve the overall performance of the proposed system.

Human face exhibits a lot of information that needs to be processed in such a manner that useful information can be retrieved upon subsequent processing of facial images. The aging begins at the age of 25, when the fine lines get to be distinctly obvious on the facial skin. In the earliest reference point, a barely recognizable difference shows up which progressively swing to wrinkles, and after some time it loses its volume and thickness that makes it more observable. Understanding the life structures of facial aging, the surface and subsurface are known as epidermis, dermis, and subcutaneous layer changes its auxiliary frame in numerous layers which incorporates fat and muscle. During the facial aging process, skin changes its surface, for example, more slender skin, drier skin, less flexible skin, wrinkle, and a decrease in collagen.

A human being is unique as far as aging is considered, which cannot be known by his gene, however different elements additionally contribute to aging such as, the wellness, living methodology, working style, and sociality. Classifying human age using facial features have gained significant importance in the field of computer vision and machine learning [1–3].

The paper is organized as follows: Sect. 2 discusses the role of researchers in terms of the classification problem. Section 3 is about the approach used for feature extraction by using various approaches. Section 4 shows the experimental results achieved by using various approaches. Section 5 provides the conclusion of the proposed system.

2 Related Work

Duong et al. [4] proposed a blend of both global and local features using Active Appearance Model (AAM) and LBP approach, in which LBP for the images in FGNET was used for the binary classifier to identify youth from adults using Support Vector Machine (SVM).

Ramanathan et al. [5] proposed an image gradient-based texture transformation function that distinguishes facial wrinkles often seen during aging. The rate of wrinkles visible for individuals varies from person to person. Jana et al. [6] proposed a technique, which provides a robust method that validates the age group of individuals from a set of different aged face images. The vital features such as distances between various parts of the face, analysis of wrinkle characteristics, and computation of face position are observed.

Yen et al. [7] proposed a methodology based on the edge density distribution of the image. In the preprocessing stage, a face is estimated to an ellipse, and genetic algorithm is applied to look for the finest ellipse region to match. In the feature extraction stage, a genetic algorithm is applied to find out the facial features, which include the eyes, nose, and mouth, in the earlier-defined subregions.

Jana et al. [8] provide a method to calculate the age of human by analyzing wrinkle area of facial images. Wrinkle characteristics are detected and features are extracted from facial images. Depending on wrinkle features, the facial image is grouped using fuzzy C-means clustering algorithm.

Lanitis et al. [9] generated a model of facial appearance that uses statistical methods. It was further used as the source for generating a set of parametric depiction of face images. Based on the model, classifiers were generated that accepted the form of representation given for the image and computed an approximation of the age for the face image. With the given training set, based on different clusters of images, classifiers for every age group were used to estimate age. Thus as given requirement in terms of age range, the most appropriate classifier was selected so as to compute accurate age estimation.

Ramesha et al. [10] proposed age classification algorithm with extracted features using small training sets which gives improved results even if one image per person is available. It is a three-stage process which includes preprocessing, feature extraction, and classification. The facial features are identified using canny edge operator for detecting facial parts for extraction of features, and are subjected to classification using Artificial Neural Network.

Gu et al. [11] proposed automatic extraction of feature points from faces. A possible approach to find the eyeballs, close to and distant corners of eyes, the center of the nostril, and corners of mouth was adopted.

3 Methodology

3.1 Local Binary Pattern

The transformation of an image into an array depicting small-scale appearance of the given image is carried out using Local Binary Pattern operator. The threshold values with weights of the corresponding pixels are multiplied and then added up to get the result, known as LBP code for a given neighborhood.

This formulation of texture is based on a model that depicts texture as a sample of a two-dimensional stochastic process that can be described by its statistical parameters. The changes in values that mathematically correspond to a derivative are referred as texture. Thus, by subtracting the values of neighbors with the center value and then dividing by distance, we find the first derivative in each direction.

$$\text{LBP}_{P,R}(x_c, y_c) = \sum_{P=0}^{P-1} s(g_p - g_c)2^P \tag{1}$$

where

g_c is the gray value of the center pixel.
P is number of gray value neighbors.
R is radius.
S is the step function.
2^P binomial weight.

3.2 Euclidean Distance

The Euclidean distance is the straight line distance between two points in the Euclidean space. The Euclidean distance between points p and q is the length of line segment connecting them in the plane with coordinates (x, y) and (a, b) is given by

$$\text{dist}((x, y), (a, b)) = \sqrt{(x-a)^2 + (y-b)^2} \tag{2}$$

3.3 Our Approach

In the proposed system, the input image is cropped to extract the region of interest which includes the facial part that needs to be processed for feature extraction. The cropped image is then subjected to preprocessing so as to achieve uniformity that enables the system to perform well. The preprocessed image is then applied for feature extraction involving both structural and textural aspects of facial images. While applying the structural feature extraction mechanism with the Euclidean distance that is more powerful as compared to others in finding exact values for a given coordinate. Simultaneously to the same image textural feature extraction is applied to find out the wrinkles within the region of interest. The detailed process flow is depicted in Fig. 1.

After extracting the features from the given images, classification is performed based on different groups classified for the purpose. The proposed approach is implemented using four groups classified as Child (0–15), Young (16–30), Adult (31–45), and Senior (46 above).

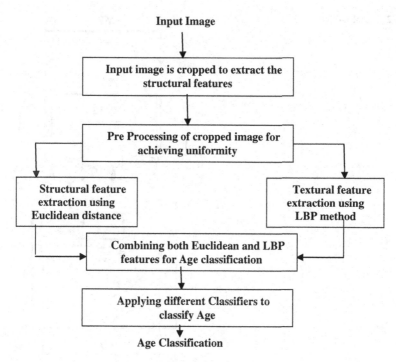

Fig. 1 Block diagram depicting the overall process

3.4 *Structural Feature Extraction*

The cropped image is subjected to preprocessing to have uniformity in size and shape for extraction of features. After these preprocessing steps are done to the input image, we compute the mean value within the cropped image area. The cropped image is then applied for feature extraction by using facial parameters. The proposed method in our approach is Geometric Facial Measurement Model (GFMM) which considers various landmark points which comprise the feature set for further analysis using various classifiers. The facial model with parameters is revealed in (see Fig. 2) [12, 13]. The details of each feature ID is elaborated in Table 1.

The approach used is implemented in FGNET aging database. The GFMM is a graphical-based implementation for feature extraction from the input image. The normalized image is subjected to feature extraction here the distance between the points for the given feature ID is computed. After plotting all the facial feature parameters, the values generated are visualized for age classification problem. In finding the features usually landmark points are used, whereas our approach deals with Euclidean distance for extracting features. Our approach uses ten facial parameters which are more in number as compared to others. The working model of our approach is shown in Fig. 3.

Fig. 2 Facial model

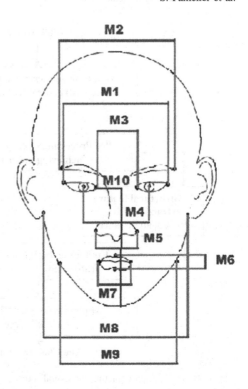

Table 1 Facial features used in model

Feature Id	Feature parameter
M1	Endpoints of left and right eye
M2	Endpoints of left and right eyebrow
M3	Left and right eye points between nose
M4	Between left and right iris
M5	Nose endpoints
M6	Lips vertical measurement
M7	Lips horizontal measurement
M8	Ear points left and right
M9	Cheek points left to right
M10	Vertical measurement from nose

These facial features are used to compute the distance between the given points for different persons in our FGNET facial aging database. The computed values are then structured in different groups for age classification.

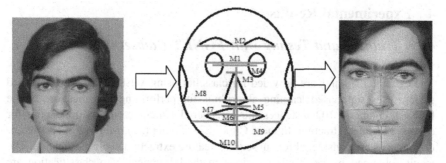

Fig. 3 Cropped image subjected to distance vector feature extraction

3.5 Textural Feature Extraction

In order to extract the local features from facial images, we applied the Local Binary Patterns (LBP) for feature extraction. It is applied to the preprocessed image and further provided to integral filters; it is implemented for removing the noise that further enhances the result. After performing these steps feature extraction is done at the specified areas where the wrinkles are prominent. Our approach considers four areas where the potential difference is noticeable in facial images. The areas considered for local feature extraction is shown in (see Fig. 4).

The wrinkle details extracted using LBP are grouped into different age groups as classified earlier. Each image in different age groups from FGNET database is considered, that gives four different LBP values per image. The computed values from both the approach are used to train the network.

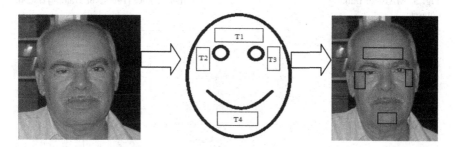

Fig. 4 Input image subjected to texture feature extraction

4 Experimental Results

4.1 Training and Testing with FGNET Dataset

The extracted features are provided for classifier using WEKA tool. The proposed system uses supervised learning algorithms for performing classification. In the FGNET database, 102 instances are considered which resembles good pose for effective feature extraction. In these Child (27), Young (37), Adult (26), and Senior (12) images are considered for subsequent feature extraction and classification. The results obtained by performing various methods applicable to classification are listed in Table 2.

Table 2 Results of training and testing datasets applied to various classifiers

Algorithm	Training dataset			Testing dataset		
	MAE	Correctly	Incorrectly	MAE	Correctly	Incorrectly
Multilayer perceptron	0.12	89.4	10.7	0.11	92	8
Logistics	0.15	77.6	22.3	0.18	68	32
AdaboostM1	0.31	48.6	51.3	0.3	52	48
Logitboost	0.15	94.7	5.3	0.1	100	0
Multiclass classifier	0.19	73.6	28.3	0.21	68	32
Iterative classifier optimizer	0.19	88.1	11.8	0.17	88	12
Randomizable filtered classifier	0.01	100	0	0.01	100	0
PART	0.09	88.1	11.7	0.1	88	12
J48	0.1	85.5	14.7	0.08	88	12
Random forest	0.1	100	0	0.1	100	0

Fig. 5 MAE for training dataset

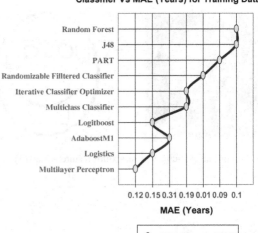

Classifier Vs MAE (Years) for Training Dataset

The above data can be visualized by plotting graph to understand the results obtained. Figures 5 and 6 show classifier and MAE results compared for training dataset which clearly shows that better MAE results are achieved as compared to others.

The comparison of correctly classified instances and incorrectly classified for the training and testing dataset is shown in Figs. 7 and 8. It represents the percentage of correctly and incorrectly classified instances, it is evident from the above table that to some methods its performance is to the mark.

Fig. 6 MAE for testing dataset

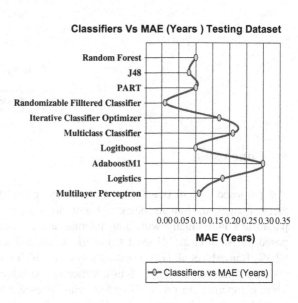

Fig. 7 Training dataset (correctly/incorrectly)

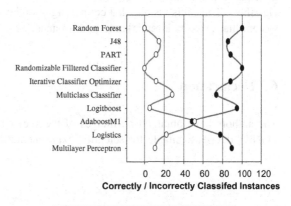

Fig. 8 Testing dataset
(correctly/incorrectly)

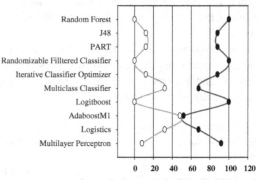

5 Conclusion

The proposed system is a fusion of local and global facial features that exist in facial images of human beings. There are factors that influence both these parameters individually with both internal and external factors. The system proposed by Nabil et al. [4] used landmark points and achieved the performance of 82.3%. Ramesha et al. [10] proposed a system which implemented both texture and distance parameter with only four distance ratios, achieved a performance of 90%. Thus, comparing our proposed system with distance parameter and texture features to be different with others makes it distinct. The number and feature for distance parameter are different and generates good result compared to others. Our system generates a performance range of 52–100%, with varied performances for different classifiers. Thus, it is visible, that combining both the feature parameters generates a system that exhibits a remarkable performance.

6 Declaration

The authors have obtained all images of the subjects involved in this study from FGNET dataset available for scientific experimentation.

References

1. Hewahi, N., Olwan, A., Tubeel, N., Asar, S.E., Sultan, Z.A.: Age estimation based on neural networks using face features. J. Emerg. Trends Comput. Inf. Sci. **1**(2), 61–67 (2010)
2. Othman, Z.A., Adnan, D.A.: Age classification from facial images system. IJCSMC **3**(10), 291–303 (2014)
3. Panicker, S., Selot, S., Sharma, M.: Human age estimation through face synthesis: a survey. i-manager's J. Pattern Recogn. **1**(2), 1–6 (2014)
4. Duong, C.N., Quach, K.G., Luu, K., Le, H.B., Ricanek, K.: Fine tuning age-estimation with global and local facial features. IEEE Int. Conf. ICASSP 2032–2035 (2011)
5. Ramanathan, N., Chellappa, R.: Modeling shape and textural variations in aging faces. In: IEEE International Conference on Automatic Face and Gesture Recognition, Amsterdam, pp. 1–8 (2008)
6. Jana, R., Datta, D., Saha, R.: Age estimation from face image using wrinkle features (ICICT). Elsevier Procedia Comput. Sci. **46**, 1754–1761 (2015)
7. Yen, G.G., Nithianandan, N.: Facial feature extraction using genetic algorithm. In: Congress on Evolutionary Computation, Honolulu, pp. 1895–1900 (2002)
8. Jana, R., Datta, D., Saha, R.: Age group estimation using face features. Int. J. Eng. Innov. Technol. (IJEIT) **3**(2), 130–134 (2013)
9. Lanitis, A., Draganova, C., Christodoulou, C.: Comparing different classifiers for automatic age estimation. IEEE Trans. Syst. Man Cybern.—Part B: Cybern. **34**(1), 621–628 (2004)
10. Ramesha, K., Raja, K.B., Venugopal, K.R., Patnaik, L.M.: Feature extraction based face recognition, gender and age classification. Int. J. Comput. Sci. Eng. **02**(01), 14–23 (2010)
11. Gu, H., Su, G., Du, C.: Feature Points Extraction from Faces. In: Image and Vision Computing, Palmerston North, pp. 154–158 (2003)
12. Panicker, S., Selot, S., Sharma, M.: Measuring the craniofacial growth for determination of human age through classifiers. BVICAM, BIJIT (2016)
13. Panicker, S., Selot, S., Sharma, M.: Fusion of structural and textural facial features for generating efficient age classifiers. In: ACM International Conference Proceedings Series, p. 95 (2016)

Evidence-Based Framework for Multi-image Super-Resolution

Ujwala Patil, Ramesh Ashok Tabib, Channabasappa M. Konin and Uma Mudenagudi

Abstract In this paper, we propose to address spatial Super Resolution (SR) from multiple registered Low-Resolution (LR) observations. We formulate a scheme to estimate a High-Resolution (HR) image from weighted combination of LRs. We propose an evidence- based technique to compute confidence factor, assigned as weights to each LR observation. We generate evidence parameters using variations in intensity and distance of registered LR images and combine them using Dempster–Shafer Combination Rule (DSCR) to generate confidence factor. We show that spatial super-resolution is obtained while retaining the high-frequency information in the image and demonstrate the same. We perform quality analysis of the super- resolved image using qualitative and quantitative approaches.

Keywords High-resolution (HR) · Low-resolution (LR) · Dempster– Shafer combination rule (DSCR) · Evidence parameter · Confidence factor (CF)

1 Introduction

The thirst for HR images is increasing due to demand for better digital imaging applications. Super-resolution is the process of reconstruction of a high-resolution image from a single or multiple low-resolution observations. We address the problem of spatial super resolution from a set of registered multiple low-resolution images

U. Patil · R. A. Tabib (✉) · C. M. Konin · U. Mudenagudi
School of Electronics Engineering, KLE Technological University, Vidyanagar,
Hubballi 580031, Karnataka, India
e-mail: ramesh_t@bvb.edu

U. Patil
e-mail: ujwalapatil@bvb.edu

C. M. Konin
e-mail: channukonin@gmail.com

U. Mudenagudi
e-mail: uma@bvb.edu

© Springer Nature Singapore Pte Ltd. 2018
P. K. Sa et al. (eds.), *Recent Findings in Intelligent Computing Techniques*,
Advances in Intelligent Systems and Computing 709,
https://doi.org/10.1007/978-981-10-8633-5_41

defined by affine motion model. We model evidence-based framework for multiple image super-resolutions. Super-resolved images find applications in entertainment, medical imaging, remote sensing, etc.

Different methods of multiple image spatial super-resolution are proposed in the literature [1, 3, 7, 8, 10–12, 15], where registration is the preprocessing step for multiple image spatial super-resolution [1]. Reconstruction-based multiple image spatial super-resolution maps multiple LR observations to the defined HR space. The above relation is used to model high- resolution image as an inverse problem [7] which is typically ill conditioned due to necessity of smoothness and priori information [12]. Reasonable solutions for well conditioning this problem are proposed in [7, 8]. The author in [7] has addressed the ill-posed problem by setting up a zone of influence to consider the effect of multiple LRs contributing to the HR reconstruction. Multiple LRs contributing to a HR pixel are averaged and used for reconstruction of HR pixel in [7]. This results in smoothening of the HR image. To address this problem, we propose to assign weights to each LR pixel based on confidence factor.

Towards this:

1. We provide an evidence-based technique to estimate the contribution of each LR pixel for the reconstruction of HR pixel by considering confidence factors as weights for LR pixels.
2. We propose to generate evidence parameters based on *Euclidean distance* and deviation in intensity of LR observations.
3. We generate confidence factor by combining evidence parameters using DSCR.

In Sect. 2, we discuss the proposed framework, we demonstrate the results in Sect. 3, and conclude in Sect. 4.

2 Multi-image Spatial Super-Resolution

In this section, we discuss the proposed framework for evidence- based multi-image spatial super-resolution. The image observation model discussed in [4, 8] is considered to model the proposed framework.

2.1 Image Observation Model [4, 8]

The imaging system generates observation 'g' influenced by different transformations. In image observation model for N multiple images, the actual image f undergoes affine transformation M_k and is influenced by space invariant PSF of camera B_k, decimation factor D with observation noise η_k to obtain kth LR observation g_k.

$$g_k = DB_kM_kf + \eta_k \qquad 1 \leq k \leq N \qquad (1)$$

Given the magnification factor X and low-resolution observations, the D is fixed. We estimate M_k for registration of LRs using hierarchical model-based registration by [9]. We model the HR image as an evidence-based weighted combination of LRs using Dempster–Shafer Combinational Rule (DSCR). We maximize probability of HR pixel given LR observations [6] using MLE framework. From [6], we model HR as $\hat{f}_{ML}$ using estimate $\hat{f}$ and is given by

$$\hat{f}_{ML} = arg\ \min_{\hat{f}}||g_k - Hf||^2 \qquad (2)$$

where $H = DB_k M_k$.

2.2 Evidence-Based Framework

The author in [7] discusses about well conditioning of SR deconvolution. Towards well conditioning of the inverse problem, we define a region of influence (ROI) with radius R for every LR pixel. We reconstruct the HR pixel by considering only those LR pixels whose ROI overlaps the HR pixel region. Consider the HR pixel $f^{(m,n)}$ at location (m, n) maps to LR pixel $g_k^{(x,y)}$ of the kth LR observation where, $x = \lfloor \frac{m}{X} \rfloor$ and $y = \lfloor \frac{n}{X} \rfloor$ for a given magnification factor X. In Fig. 1, it is observed, $g_1^{(x,y)}$ and $g_2^{(x,y)}$ are the pixels of low-resolution images related with affine motion model. The HR pixel $f_1^{(m_1,n_1)}$ is reconstructed from the weighted combination of LR pixels $g_1^{(x,y)}$ and $g_2^{(x,y)}$ as ROI of $g_1^{(x,y)}$ and $g_2^{(x,y)}$ are overlapping with HR pixel $f_1^{(m_1,n_1)}$. Similarly, the HR pixel $f_2^{(m_2,n_2)}$ is reconstructed by the contribution of LR pixel $g_2^{(x,y)}$. Neither LR pixel $g_1^{(x,y)}$ nor $g_2^{(x,y)}$ contribute to the reconstruction of HR pixel $f_3^{(m_3,n_3)}$. It is generalized as

Fig. 1 HR pixel influenced by region of influence of LR pixel

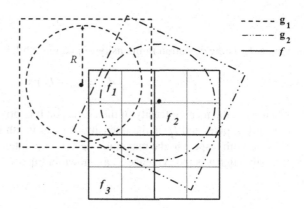

$$f^{(m,n)} = \frac{\sum_{k=0}^{N} \phi_{g_k}^{(x,y)} * I_{g_k}^{(x,y)}}{\sum_{k=0}^{N} \phi_{g_k}^{(x,y)}} \qquad (3)$$

where $\phi_{g_k}^{(x,y)}$ and $I_{g_k}^{(x,y)}$ are the weight and intensity value of the LR pixel at (x, y) of kth LR image respectively. The ϕ_{g_k} is the confidence factor calculated by combining the evidences using DSCR as explained in Sect. 2.3.

2.3 Generation of Evidence Parameters

In this section, we discuss the computation of evidence parameters $\mathcal{E}_1$ and $\mathcal{E}_2$. $\mathcal{E}_1$ is generated using *euclidean distance* between corresponding HR pixel and LR pixel, $\mathcal{E}_2$ is generated considering the deviation in intensity of corresponding LR pixel in ROI. $\mathcal{E}_1$ and $\mathcal{E}_2$ are calculated as shown in Eqs. 4 and 6 respectively.

$$\mathcal{E}_1 = \begin{cases} 1 - (ed_{g_k})_n & \text{if } ed_{g_k} < R, \\ 0 & \text{otherwise.} \end{cases} \qquad (4)$$

where $(ed_{g_k})_n$ is the normalized *euclidean distance* between pixel of kth LR observation $g_k^{(x,y)}$ and corresponding HR pixel $f^{(m,n)}$, normalized with respect to R and is given by

$$ed_{g_k} = euclidean(f^{(m,n)}, g_k^{(x,y)}) \qquad (5)$$

Lesser the *euclidean distance* between corresponding HR pixel and LR pixel, higher is the evidence $\mathcal{E}_1$ as shown in Eq. 4.

Normalized deviation of the intensity $(d_{g_k})_n$ of the LRs influencing the HR is used to calculate evidence parameter $\mathcal{E}_2$ given by

$$\mathcal{E}_2 = 1 - (d_{g_k})_n \qquad (6)$$

where d_{g_k} is the deviation in intensity of pixel in kth LR and is given by

$$d_{g_k} = (I_\mu - I_{g_k}) \qquad (7)$$

where I_μ is the mean intensity of the corresponding pixel in LR observations influencing HR pixel and I_{g_k} is the intensity of pixel in kth LR observation. d_{g_k} is normalized with respect to the maximum deviation in intensity. Lesser the deviation in intensity, higher is the evidence $\mathcal{E}_2$ as shown in Eq. 6.

Table 1 The combination table [5, 13]

∩	$m(\mathcal{E}_1^{belief})$	$m(\mathcal{E}_1^{disbelief})$	$m(\mathcal{E}_1^{ambiguity})$
$m(\mathcal{E}_2^{belief})$	$\psi_1 \leftarrow m(J_1)$	$\emptyset \leftarrow m(J_8)$	$\psi_1 \leftarrow m(J_2)$
$m(\mathcal{E}_2^{disbelief})$	$\emptyset \leftarrow m(J_9)$	$\psi_2 \leftarrow m(J_4)$	$\psi_2 \leftarrow m(J_5)$
$m(\mathcal{E}_2^{ambiguity})$	$\psi_1 \leftarrow m(J,3)$	$\psi_2 \leftarrow m(J_6)$	$\Omega \leftarrow m(J_7)$

2.4 Generation of Confidence Factor

We combine evidence parameters using DSCR to generate confidence factor. The
evidence parameters $\mathcal{E}_1$ and $\mathcal{E}_2$ calculated in Sect. 2.3 are considered as the mass of
belief function for DSCR [5]. We propose to use confidence factor as weights of LR
pixel from multiple LR images contributing to the reconstruction of HR pixel. Let
ψ_1 be the hypotheses supporting the belief and ψ_2 be the hypotheses supporting the
disbelief towards the contribution of LR for reconstruction of HR. We denote the
hypothesis in a set as,

$$2^{\Omega} = \{\emptyset, \{\psi_1\}, \{\psi_2\}, \Omega\} \tag{8}$$

where $\emptyset$ denotes the conflict between the set hypothesis and $\Omega = \{\psi_1, \psi_2\}$ denotes
ambiguity. We combine the evidence parameters as masses $m(\mathcal{E}_1)$ and $m(\mathcal{E}_2)$ using
the combination rule discussed in [2, 5, 14] to generate the mass of combined
hypothesis $m(\mathcal{H})$ and is shown in Eqs. 9 and 10. The accumulated evidence from
$\mathcal{E}_1$ and $\mathcal{E}_2$ is shown in numerator of Eqs. 9 and 10, which supports the hypothesis
$\mathcal{H}$. The mass of conflict towards the combined hypothesis is the summation term
in the denominator of Eq. 9. The total mass due to belief, disbelief and ambiguity
towards the set hypothesis is the denominator of Eq. 10 used as the normalization
factor towards the set hypothesis (Table 1).

The combined hypothesis $\mathcal{H}$ is,

$$m(\mathcal{H}) = \frac{\sum_{\mathcal{E}_1 \cap \mathcal{E}_2 = \mathcal{H} \neq \emptyset} m(\mathcal{E}_1) \cdot m(\mathcal{E}_2)}{1 - \sum_{\mathcal{E}_1 \cap \mathcal{E}_2 = \emptyset} m(\mathcal{E}_1) \cdot m(\mathcal{E}_2)} \tag{9}$$

which can be written as,

$$m(\mathcal{H}) = \frac{\sum_{\mathcal{E}_1 \cap \mathcal{E}_2 = \mathcal{H} \neq \emptyset} m(\mathcal{E}_1) \cdot m(\mathcal{E}_2)}{\sum_{\mathcal{E}_1 \cap \mathcal{E}_2 \neq \emptyset} m(\mathcal{E}_1) \cdot m(\mathcal{E}_2)} \tag{10}$$

The subset contributing towards belief of set hypothesis is $\{\psi_1\} = \{m(J_1), m(J_2),$
$m(J_3)\}$. Similarly, the subset contributing towards disbelief of set hypothesis is,
$\{\psi_2\} = \{m(J_4), m(J_5), m(J_6)\}$. The subset contributing towards ambiguity is, $\{\Omega\}$
$= \{m(J_7)\}$. The subset contributing towards conflict is null set, $\{\emptyset\} = \{m(J_8), m(J_9)\}$.
The masses of hypothesis $\{\psi_1\}$ and $\{\psi_2\}$ as per the Eq. 10 are given by

(a)

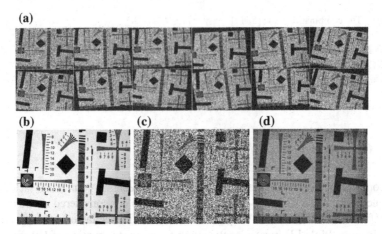

Fig. 2 **a** Input images of size 160 × 128. **b** Ground iruth image. **c** Bilinear interpolated image (*rms* = 105.228). **d** SR Image with 40 LR observations (*rms* = 67.461)

$$m(\psi_1) = \frac{\sum_{k=1}^{3} m(J_k)}{\sum_{k=1}^{7} m(J_k)}, m(\psi_2) = \frac{\sum_{k=4}^{6} m(J_k)}{\sum_{k=1}^{7} m(J_k)} \qquad (11)$$

The masses $m(\psi_1)$ and $m(\psi_2)$ are confidence factors in favor and against the contribution of LR respectively. We assign the mass of hypothesis ψ_1 as the weight ϕ_{g_k} for pixel of kth LR image. The intensity assigned to the HR pixel is calculated using Eq. 3. The mass for disbelief function is considered to be zero because, we assume that all LR observations contribute to the reconstruction of HR pixel with varying weights. The weights of LR pixels that lie outside the Region of Influence are set to zero. If HR pixel is not influenced by any of the corresponding LR pixel, then the weights for corresponding LR pixel for all N observations is considered to be 1 (Fig. 2).

3   Results and Discussions

In this section, we demonstrate the results obtained for the proposed method for spatial super-resolution from multiple registered [9] low-resolution observations on both synthetic data set and real data set. The attempted magnification is 4x (in each dimension) of the input LR observation. We compare our results with interpolated image and method proposed in [7]. It is observed, the proposed method eliminates the holes formed in the reconstructed images due to lack of LR pixels in contributing to HR image as described in [7] and also shows the contribution of each LR image towards the reconstruction of HR image. We perform the qualitative and quantitative quality analysis of the super-resolved image obtained from the proposed framework.

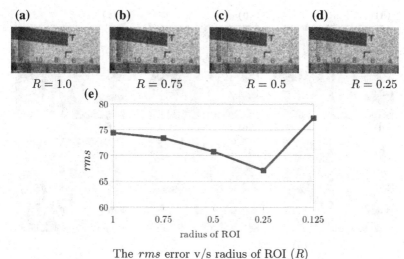

(a)   **(b)**   **(c)**   **(d)**

$R = 1.0$ $R = 0.75$ $R = 0.5$ $R = 0.25$

(e)

The rms error v/s radius of ROI (R)

Fig. 3 Effect of radius (R) on the super-resolution reconstruction using 20 input images each of size 64 × 32 and Magnification factor of 4x

The qualitative analysis is performed with a sample set of 100 and 98% opine, the visibility of super-resolved image using evidence-based framework is better than single-image super-resolved and interpolation LR image. For quantitative analysis, we calculate root mean square (rms) error of super-resolved image with ground truth and observe the reduction in the rms for our results as shown in Figs. 3e and 4g.

3.1 Super-Resolution of Synthetic Data Set

In this section, we discuss the effect of ROI and the number of LR observations on reconstruction of HR image using synthetic data set. We randomly generate input images, related by affine motion model with uniform additive noise of SNR = 2.

Effect of ROI R: We set the radius of region of influence R as 0.25, 0.5, 0.75 and 1.0 pixel width of LR pixel to observe the effect of ROI on HR reconstruction. Figure 3a–d shows the reconstructed HR images with different values of R. We observe nonlinear reduction in rms error with decrease in value of R, shown in Fig. 3e. The reduction in the rms error is because of more number of LR observations having high confidence considered for reconstruction of HR. After certain reduction in R, contribution of LR observations is not seen in many regions of the HR image resulting in an increase in the rms error as shown in Fig. 3e (for $R = 0.125$).

Number of Input LR Observations: We demonstrate the effect of number of LR images on the reconstruction of HR image. We show the reconstruction of HR image using 8, 12, 16, 20, 32 and 40 LR images in Fig. 4a–f. The radius of region of influence R is set to 0.25 as reduction in rms is optimal for the case. We calculate rms

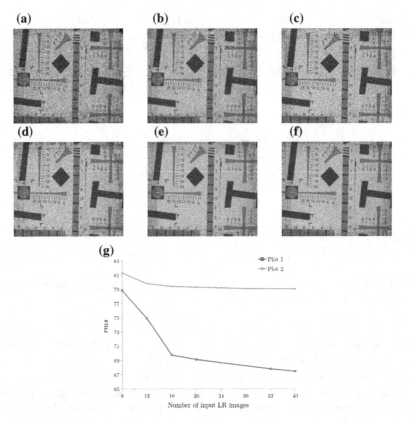

Fig. 4 HR image reconstruction of size 640 × 512 using **a** 8 input images. **b** 12 input images. **c** 16 input images. **d** 20 input images. **e** 32 input images. **f** 40 input images. **g** Variation in *rms* error with number of input images used for HR image reconstruction. Plot 1 shows the *rms* error of HR image for different number of images using the proposed method. Plot 2 shows the *rms* error as shown in [7]

error for the results obtained using proposed method and compare with the results of reconstruction of HR explained in [7]. We observe reduction in *rms* with increase in number of input observations, as the contribution of confident LR observations increase with the number of LR input images as shown in Fig. 4g. The reduction in *rms* error is nonlinear. These results are compared with the *rms* error values given in [7] and obsrved them to be lesser, shown in Fig. 4g.

3.2 Super-Resolution on Real Data Set

In this section, we demonstrate the results of proposed SR framework on real dataset. As discussed in Sect. 3.1 it is observed, the quality of HR image is better for $R=0.25$

Fig. 5 **a** Input images of
size 266 × 300 [Real data
set-1]. **b** Bicubic interpolated
image [Real data set-1]. **c** 4x
HR Image with 40 LR
observations [Real data
set-1]. **d** Input images of size
240 × 288 [Real data set-2].
e Bicubic interpolated image
[Real data set-2]. **f** 4x HR
Image with 40 LR
observations [Real data set-2]

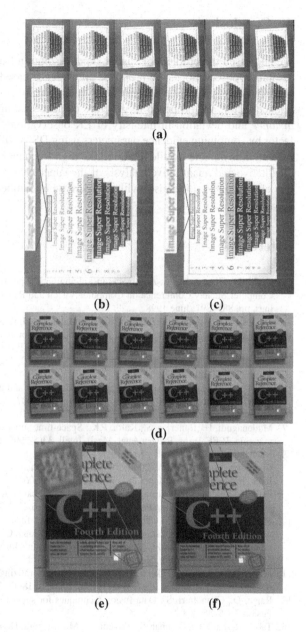

and 40 LR observations (N). We demonstrate super-resolution on real dataset R to
0.25 and number of real LR observations N to 40. Figure 5a, d show the LR observations of the real data set-1 and set-2 generated using a 2 MP smartphone camera
respectively. Figure 5 (b) and (e) shows the bicubic interpolation of the reference LR
observation and (c) and (f) shows the HR image of data set-1 and set-2 respectively.

4 Conclusions

We addressed the problem of spatial super-resolution from multiple registered low resolution observations. We proposed an evidence based technique to estimate the contribution of each LR pixel for the reconstruction of HR pixel by considering confidence factor. We proposed to generate evidence parameters based on *Euclidean distance* and deviation in intensity of LR observations. We generated confidence factor by combining evidence parameters using DSCR. It is observed, spatial super-resolution is obtained while retaining the high frequency information in the image. Qualitative and quantitative analysis of the super-resolved image is performed and it is observed, *rms* idecreases with the increase in number of LR observations.

References

1. Capel, D., Zisserman, A.: Automated mosaicing with super-resolution zoom. In: 1998 IEEE Computer Society Conference on Computer Vision and Pattern Recognition, 1998. Proceedings, pp. 885–891, June 1998
2. Dempster, A.P.: A generalization of Bayesian inference. J. R. Stat. Soc. **30**, 205–247 (1968)
3. Elad, M., Feuer, A.: Restoration of a single superresolution image from several blurred, noisy, and undersampled measured images. IEEE Trans. Image Process. **6**(12), 1646–1658 (1997)
4. Gonzalez, R.C., Woods, R.E.: Digital Image Processing, 2nd edn. Prentice Hall, Upper Saddle River, N.J. (2002)
5. Kay, R.U.: Fundamentals of the Dempster-Shafer theory and its applications to system safety and reliability modeling. RTA 3-4 Special Issue, Dec 2007
6. Milanfar, P.: Super-Resolution Imaging. CRC Press (2010)
7. Mudenagudi, U., Banerjee, S., Kalra, P.K.: Space-time super-resolution using graph-cut optimization. IEEE Trans. Pattern Anal. Mach. Intell. **33**(5), 995–1008 (2011)
8. Mudenagudi, U., Singla, R., Kalra, P.K., Banerjee, S.: Super resolution using graph-cut. In: Computer Vision—ACCV 2006, 7th Asian Conference on Computer Vision, Hyderabad, India, Proceedings, Part II, pp. 385–394, 13–16 Jan 2006
9. Patil, U., Mudengudi, U.: Image fusion using hierarchical PCA. In: 2011 International Conference on Image Information Processing (ICIIP), pp. 1–6, Nov 2011
10. Patil, U., Mudengudi, U., Ganesh, K., Patil, R.: Image fusion framework. In: Computer Networks and Information Technologies: Second International Conference on Advances in Communication, Network, and Computing, CNC 2011, Bangalore, India, 2011. Proceedings, pp. 653–657. Springer, Berlin, Heidelberg, 10–11 Mar 2011
11. Protter, M., Elad, M., Takeda, H., Milanfar, P.: Generalizing the nonlocal-means to super-resolution reconstruction. IEEE Trans. Image Process. **18**(1), 36–51 (2009)
12. Rajan, D., Chaudhuri, S.: Data fusion techniques for super resolution imaging. Elsevier Inf. Fusion **3**, 25–38 (2002)
13. Tabib, R., Patil, U., Ganihar, S., Trivedi, N., Mudenagudi, U.: Decision fusion for robust horizon estimation using dempster shafer combination rule. In: 2013 Fourth National Conference on Computer Vision, Pattern Recognition, Image Processing and Graphics (NCVPRIPG), pp. 1–4, Dec 2013

14. Thomas, C., Balakrishnan, N.: Modified evidence theory for performance enhancement of intrusion detection systems. In: 11th International Conference on Information Fusion, pp. 1–8, July 2008
15. Zomet, A., Rav-Acha, A., Peleg, S.: Robust super-resolution. In: Proceedings of the 2001 IEEE Computer Society Conference on Computer Vision and Pattern Recognition, 2001. CVPR 2001, vol. 1, pp. I–645–I–650 (2001)

Review and Comparative Evaluation of Compressive Sensing for Digital Video

Rohit Thanki, Komal Borisagar and Vedvyas Dwivedi

Abstract In this paper, the application of compressive sensing theory for digital video has been reviewed and analyzed. The compressive sensing (CS) is a new signal processing theory which overcomes the limitation of Shannon–Nyquist sampling theorem. The compressive sensing exploits the redundancy within the signal to get samples of the signal at sub-Nyquist rates. The signal can be reconstructed from these samples when it is fed to CS recovery algorithm. Here challenges and various approaches to CS theory for digital video are discussed, which motivate the future research. The comparative evaluation of CS theory for color digital video using different transform basis is also discussed and analyzed in this paper.

Keywords Compressive sensing · Digital video · Greedy algorithms Measurements

1 Introduction

In digital signal processing, the Shannon–Nyquist sampling theorem is very important for any discrete-time signals. The numbers of samples must be double or greater than double for the size of the signal being sampled and recovered. With increasing number of various applications of signal processing, the Shannon–Nyquist theorem creates problem sometimes. For example, nowadays mobile phone comes with high definition and resolution which require a large amount of data for

R. Thanki (✉) · V. Dwivedi
C. U. Shah University, Wadhwan City, Gujarat, India
e-mail: rohitthanki9@gmail.com

V. Dwivedi
e-mail: vedvyasdwivediphd@gmail.com

K. Borisagar
Atmiya Institute of Technology & Science, Rajkot, Gujarat, India
e-mail: krborisagar@aits.edu.in

© Springer Nature Singapore Pte Ltd. 2018
P. K. Sa et al. (eds.), *Recent Findings in Intelligent Computing Techniques*,
Advances in Intelligent Systems and Computing 709,
https://doi.org/10.1007/978-981-10-8633-5_42

capturing images. As a motivation large amount of data is required for representing an image, researchers introduced the new signal theory and it is proposed for acquiring images [1–3]. This theory is known as compressive sensing or sampling (CS). The CS theory overcomes the limitation of Shannon–Nyquist sampling theorem and exploits redundancy within the acquired image or signal to get samples at less than Nyquist rates. The image or signal must be sparse when associated with CS theory. Most of the images or signals is sparse when it is converted into its transform basis [1–3]. In past 10 years, various CS theory approaches and algorithms are designed for still image camera by researchers [4]. In these approaches, CS camera takes a small amount of CS measurements of the scene and using these measurements, the image is reconstructed. The CS camera is used measurements for reconstruction of the image instead of pixels of the image.

Recently, researchers have described various research challenges and opportunities for digital video. R. Baraniuk and his research team are described and reviewed the various algorithms and approaches for compressive video sensing [4]. They are given various video models, structure, and video sensing systems with a combination of CS theory. They are given video recovery techniques using CS theory. They point out challenges related to compressive video sensing such as real-time implementation of CS video recovery algorithm, reconstruction of video from fewer bits instead of fewer measurements. They are also given a comparative analysis of different video models for CS video recovery algorithm for grayscale digital video. M. Wakin and his research team are compressive video sensing approaches using 2D wavelet transform and Lifting based Invertible Motion Adaptive Transform (LIMAT) [5, 6]. They have described the application of CS theory for grayscale video using above two transforms. Stankovic and his research team are CS theory using block-based processing and discrete cosine transform (DCT) for grayscale video sensing [7]. They are described reconstruct video frame from CS measurements using a greedy algorithm. The CS measurements of the video frame are generated using DCT coefficients.

J. Flower described block-based compressed sensing for digital image and video [8]. In this approach, image or video frame is divided into the block. Then a block of image or video frame is converted into CS measurements using CS theory and then the image or video frame is reconstructed. This approach is easy to implement and fast. J. Yang and his research team described Gaussian mixture model-based algorithm for video frame reconstruction from CS measurements [9]. They are GMM based algorithm compared with KSVD algorithm and KSVD-OMP algorithm for grayscale video. After reviewing papers, it is clearly seen that CS theory is applied to the individual frame of digital video. In above algorithms, the first digital video is divided into the frame. Then CS measurements are taken for an individual video frame and then reconstruct frame from incomplete CS measurements. These above algorithms are applied for grayscale video. The fewer approaches are available for color video acquisition using CS theory.

So in this paper, the application of CS theory for color digital video is proposed and analyzed. Here first the color digital video is divided into frames. Then CS measurements of each frame are generated using various signal processing transforms. The signal processing transforms such as fast Fourier transform (FFT), discrete Cosine transform (DCT), and discrete wavelet transform (DWT) are used for CS measurements generation. The different greedy algorithms such as orthogonal matching pursuit (OMP) [10], compressive sampling matching pursuit (CoSaMP) [11], and subspace pursuit (SP) [12] are used the reconstruction of the frame from CS measurements.

The rest of paper is organized as follows: information and mathematical steps for the proposed system for color video are given in Sect. 2. The result and discussion of a proposed system for different CS recovery algorithms are given in Sect. 3. Finally, the conclusion is given in Sect. 4.

2 Proposed System for Color Video Using CS Theory Framework

The digital video contains various frames where some objects are dynamic and some objects are stationary in frames. The color video has three different channel frame such as R channel frame, G channel frame, and B channel frame. In this paper, CS theory framework is proposed for color video. Here for easy processing and understanding, the first color video stream is divided into frames. Then CS theory is applied on R channel, G channel, and B channel of the individual frame. This process is performed for every frame of video to get reconstructed video frames. Finally, all reconstructed video frames are combined to get reconstructed video stream. Figure 1 is shown frame for the proposed system using CS theory for color video. The mathematical steps for the proposed system are given below.

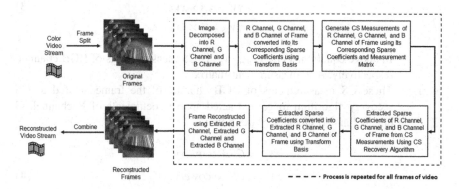

Fig. 1 Framework for the proposed system

Step 1: Take a color video stream V and then video split into frame V_f. Then below process is performed for every individual frame.

$$V_f(n) = \sum_{i=1}^{n} V \tag{1}$$

In above equation, n is no. of frames.

Step 2: The frame is decomposed into R channel, G channel, and B channel of the frame.

Step 3: Generate transform basis matrix with the equal size of R channel, G channel and B channel of the frame.

Step 4: Then, the value of R channel, G channel and B channel are converted into its sparse coefficients x using transform basis matrix and its inversion version.

$$xV_{fR} = \Psi \times V_{fR} \times \Psi'$$
$$xV_{fG} = \Psi \times V_{fG} \times \Psi' \tag{2}$$
$$xV_{fB} = \Psi \times V_{fB} \times \Psi'$$

In above equation, xV_{fR}, xV_{fG}, xV_{fB} is sparse coefficients of RGB channel of the frame, respectively, Ψ is transform basis matrix.

Step 5: Generate measurement matrix A using Gaussian distribution with mean $= 0$ and standard deviation $= 1$.

Step 6: The CS measurements of RGB channel of the frame are generated by multiplication of its sparse coefficients and measurement matrix.

$$yV_{fR} = A \times xV_{fR}$$
$$yV_{fG} = A \times xV_{fG} \tag{3}$$
$$yV_{fB} = A \times xV_{fB}$$

In above equation, yV_{fR}, yV_{fG}, yV_{fB} is CS measurements of RGB channel, respectively, A is measurement matrix.

Step 7: These CS measurements of RGB channel of the frame are fed to CS recovery algorithm to get extracted sparse coefficients of R channel, G channel and B channel of the frame.

$$xV_{fR}' = CS\_Recov\,ery(yV_{fR}, A)$$
$$xV_{fG}' = CS\_Recov\,ery(yV_{fG}, A) \tag{4}$$
$$xV_{fB}' = CS\_Recov\,ery(yV_{fB}, A)$$

In the above equation, xV_{fR}', xV_{fG}', xV_{fB}' is extracted sparse coefficients of RGB channel of the frame respectively.

Step 8: Then get the actual value of R channel, G channel and B channel from its extracted sparse coefficients using inverse transform basis matrix and its original version.

$$V'_{fR} = \Psi' \times x V'_{fR} \times \Psi$$
$$V'_{fG} = \Psi' \times x V'_{fG} \times \Psi \qquad (5)$$
$$V'_{fB} = \Psi' \times x V'_{fB} \times \Psi$$

In the above equation, $V'_{fR}, V'_{fG}, V'_{fB}$ is reconstructed RGB channel of the frame respectively.

Step 9: Then combine the reconstructed RGB channel of the frame to get the reconstructed frame of the color video stream. Repeat the steps 2–8 for every frame of the color video stream to get reconstructed frames of the color video stream. Then combine all reconstructed frames to get reconstructed color video stream.

$$V'(n) = \sum_{i=1}^{n} V'_f \qquad (6)$$

In above equation, V' is a reconstructed color video stream.

3   Experimental Results

The testing of the proposed system using different transform basis and greedy algorithms with quality measures are discussed in this section. The 190×190 pixel color video having 16 frames are used for testing of the proposed system. The first four frame of video is shown in Fig. 2. A steam of 25270 CS measurements was acquired from tested color video using Gaussian measurement matrix.

Then the color video is reconstructed from these CS measurements using different CS recovery algorithms. The CS measurements of color video are generated using various transform basis matrices such as FFT, DCT, and DWT. These matrices are 2D in nature. The color video is reconstructed separately using various

Fig. 2 Original video frames

greedy algorithms such as OMP, SP, and CoSaMP that were implemented in MATLAB. These three algorithms are chosen because it performs better and faster than L minimization and TV minimization. The quality of color video reconstruction is measured by peak signal-to-noise ratio (PSNR) [13] and structural similarity measure index (SSIM) [14]. The PSNR is the measured picture quality of reconstructed video frame and original video frame. The SSIM is measure similarity between the reconstructed video frame and original video frame. The processing time is measured for comparative analysis of various approaches. Here first calculate PSNR and SSIM value for every individual reconstructed video frame and original video frame. Then calculate PSNR and SSIM value for entire video stream find the average value of PSNR and SSIM value of all frames. The processing time is a summation of time required for all frame reconstruction.

Figure 3 shows the reconstructed first four frames of color video using FFT basis matrix and different greedy algorithms. The FFT basis matrix is generated using MATLAB command dftmtx with equal size of the video frame. Table 1 shown quality measures value using FFT basis matrix and different greedy algorithms.

Figure 4 shows the reconstructed first four frames of color video using DCT basis matrix and different greedy algorithms. The DCT basis matrix is generated using standard DCT equation [15] with equal size of the video frame. Table 2 shown quality measures value using DCT basis matrix and different greedy algorithms.

Figure 5 shows the reconstructed first four frames of color video using DWT basis matrix and different greedy algorithms. The DWT basis matrix is generated

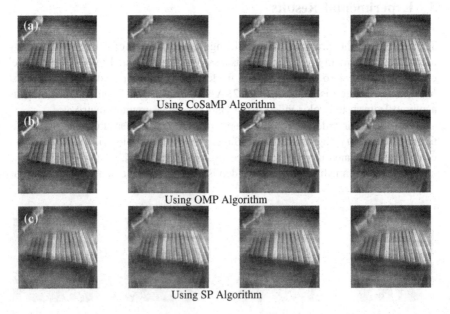

Using CoSaMP Algorithm

Using OMP Algorithm

Using SP Algorithm

Fig. 3 Reconstructed video frames using FFT basis matrix

Table 1 Quality measure for reconstructed video frames using FFT basis matrix

Greedy algorithm	Average PSNR (dB)	Average SSIM	Total processing time (s)
CoSaMP	38.97	0.9933	666.23
OMP	40.73	0.9972	121.11
SP	38.43	0.9918	351.25

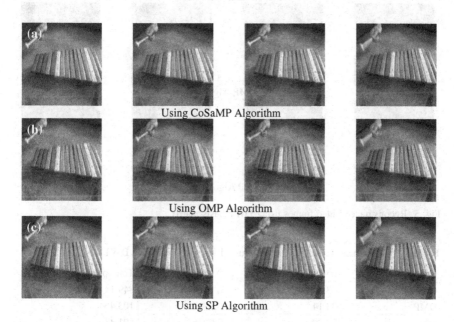

Using CoSaMP Algorithm

Using OMP Algorithm

Using SP Algorithm

Fig. 4 Reconstructed video frames using DCT basis matrix

Table 2 Quality measure for reconstructed video frames using DCT basis matrix

Greedy algorithm	Average PSNR (dB)	Average SSIM	Total processing time (s)
CoSaMP	38.43	0.9893	827.33
OMP	38.25	0.9887	101.59
SP	38.39	0.9892	489.39

using MATLAB command wavmat [16, 17] with equal size of the video frame. Table 3 shows quality measures value using DWT basis matrix and different greedy algorithms.

After obtaining results, comparative analysis of the proposed system for different greedy algorithms is done. Figures 6, 7, and 8 show comparative graph for PSNR values, SSIM value, and processing time for different transform basis and greedy

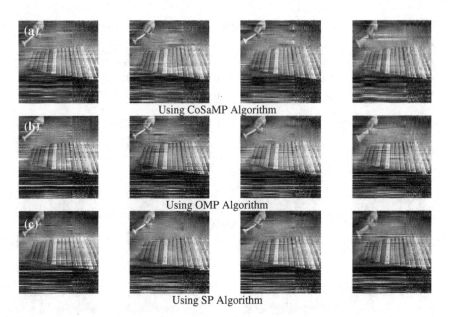

Fig. 5 Reconstructed video frames using DWT basis matrix

Table 3 Quality measure value for reconstructed video frames using DWT basis matrix

Greedy algorithm	Average PSNR (dB)	Average SSIM	Total processing time (s)
CoSaMP	33.21	0.9084	849.14
OMP	33.14	0.9048	103.46
SP	33.39	0.9135	501.43

algorithms. The graph in Fig. 6 shows that PSNR value of OMP algorithm for FFT basis is better than PSNR values of other two bases. The graph in Fig. 7 shows that SSIM value of an OMP algorithm for FFT basis is better than SSIM values of other two bases. The graph in Fig. 8 shows that processing time of OMP algorithm for DCT basis is better than processing time of other two bases.

The processing time of the proposed system is also compared with other approaches which are given for grayscale images. Table 4 shows the processing time of various approaches using CS theory for digital video sensing. The comparison shows that 2D DCT-OMP algorithm performs faster than other algorithms for digital video reconstruction using CS theory.

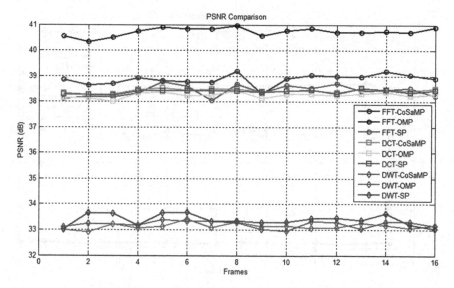

Fig. 6 PSNR comparison graph for different transform basis and greedy algorithms

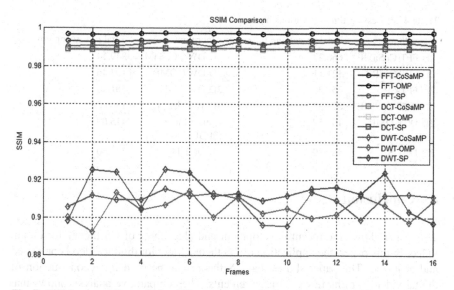

Fig. 7 SSIM comparison graph for different transform basis and greedy algorithms

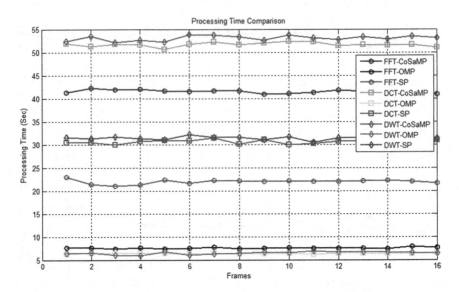

Fig. 8 Time comparison graph for different transform basis and greedy algorithms

Table 4 Processing time for various algorithms

Algorithms	Total processing time (s)	Algorithms	Total processing time (s)
2D FFT-CoSaMP	666.23	2D DWT-CoSaMP	849.14
2D FFT-OMP	121.11	2D DWT-OMP	103.46
2D FFT-SP	351.25	2D DWT-SP	501.43
2D DCT-CoSaMP	827.33	3D Wavelet [4]	134.00
2D DCT-OMP	101.59	Optical Flow (SPGL1) [4]	415.00
2D DCT-SP	489.39	3D DCT [4]	332.00

4 Conclusion

This paper shows the recent development and algorithms of CS theory for digital video. In this paper, the application of CS theory for color digital video is proposed and analyzed. The paper shows that CS theory can be used for reconstruction of digital video from its few CS measurements. The comparative analysis shows that FFT-OMP algorithm is performed better than other algorithms with different transform bases. The processing time of the 2D DCT-OMP algorithm is higher than other algorithms which are used for reconstruction of digital video. In this paper, few fast CS recovery approaches are proposed and analyzed for color digital video. The proposed approaches are getting fast results compared to existed approaches for grayscale digital video available in the literature.

References

1. Candes, E.: Compressive sampling. In: Proceedings of the International Congress of Mathematicians, Madrid, Spain, pp. 1433–1452, June 2006
2. Donoho, D.: Compressed sensing. IEEE Trans. Inf. Theory **52**(4), 1289–1306 (2006)
3. Baraniuk, R.: Lecture notes "Compressive Sensing". IEEE Signal Process. Mag. **24**(4), 118–124 (2007)
4. Baraniuk, R., Goldstein, T., Sankaranarayanan, A., Studer, C., Veeraraghavan, A., Wakin, M.: Compressive video sensing. IEEE Signal Process. Mag. 52–66 (2017)
5. Wakin, M., Laska, J., Duarte, M., Baron, D., Sarvotham, S., Takhar, D., Kelly, K., Baraniuk, R.: Compressive imaging for video representation and coding. In: Picture Coding Symposium, vol. 1, no. 13 (2006)
6. Park, J., Wakin, M.: A multiscale framework for compressive sensing of video. In: Picture Coding Symposium, pp. 1–4 (2009)
7. Stankovic, V., Stankovic, L., Cheng, S.: Compressive video sampling. In: 16th IEEE European Signal Processing Conference, pp. 1–5 (2008)
8. Fowler, J.: Block-Based Compressed Sensing of Images and Video, Mississippi State University, Mar 2010
9. Yang, J., Yuan, X., Liao, X., Llull, P., Brady, D., Sapiro, G., Carin, L.: Video compressive sensing using Gaussian mixture models. IEEE Trans. Image Process. **23**(11), 4863–4878 (2014)
10. Tropp, J., Gilbert, A.: Signal recovery from random measurements via orthogonal matching pursuit. IEEE Trans. Inf. Theory **53**(12), 4655–4666 (2007)
11. Needell, D., Tropp, J.: CoSaMP: iterative signal recovery from incomplete and inaccurate samples. Appl. Comput. Harmon. Anal. **26**(3), 301–321 (2009)
12. Wei, D., Milenkovic, O.: Subspace pursuit for compressive sensing signal reconstruction. IEEE Trans. Inf. Theory **55**(5), 2230–2249 (2009)
13. Mark, M., Grgic, S., Grgic, M.: Picture quality measures in image compression systems. In: EUROCON 2003, Ljubljana, Slovenia, 233-2-7 (2003)
14. Wang, Z., Bovik, A.: A universal image quality index. J. IEEE Signal Process. Lett. 9(3), 84–88 (2004)
15. Jain, A.: Fundamental of Digital Image Processing, pp. 150–153. Prentice Hall Inc., New Jersey (1999)
16. Yan, J.: Wavelet Matrix, Department of Electrical and Computer Engineering, University of Victoria, Victoria, BC, Canada (2009)
17. Vidakovic, B.: Statistical Modelling by Wavelets, pp. 115–116. Wiley (1999)

An Efficient Approach for Detecting Helmets on Motorcyclists Using Machine Learning Techniques

Abhijeet S. Talaulikar, Sanjay Sanathanan and Chirag N. Modi

Abstract In this paper, we propose a system to automatically detect motorcycle riders without helmets. It extracts moving vehicles from the captured images of the road traffic using background subtraction technique. Then, it filters out the blobs of vehicles other than motorcycles. It extracts heads of the motorcycle riders as region of interest and identifies the presence or absence of helmet. We have tested different machine learning techniques with weighted averaging method to identify a head as wearing a helmet or not, and analyzed the results.

Keywords Computer vision · Helmet detection · Machine learning

1 Introduction

Many of the vehicle users do not care to wear the necessary safety gear (helmets, seat belts, airbags, etc.) and thus, the frequency and severity of road accidents have been elevated. As per the survey, a whopping 25% of the road accident deaths involve motorcyclists, which puts riders at the greatest risk of injury and death [1]. Therefore, the safety of motorcyclists is a major concern in the country. In 2013, a project was launched in the city of Surat which had 550 CCTV cameras installed in 282 locations. The aim was to observe the traffic through these live cameras, identify the causes of accidents and take steps to prevent them. Although such an administration could be used for detecting riders without helmets, the footages were manually monitored by the crew in the control room. This was a sluggish process

A. S. Talaulikar (✉) · S. Sanathanan · C. N. Modi
Department of Computer Science and Engineering, National Institute of Technology Goa,
Ponda, India
e-mail: abhijeetstalaulikar@gmail.com

S. Sanathanan
e-mail: sanjay.sanathanan@gmail.com

C. N. Modi
e-mail: cnmodi@nitgoa.ac.in

© Springer Nature Singapore Pte Ltd. 2018   437
P. K. Sa et al. (eds.), *Recent Findings in Intelligent Computing Techniques*,
Advances in Intelligent Systems and Computing 709,
https://doi.org/10.1007/978-981-10-8633-5_43

and required manpower to monitor the streams. In the absence of the crew, the video footages were stored and watched later, but this entailed a large amount of memory for storage.

To address the above issues, we propose a system to automatically detect motorcycle riders sans helmets. It performs image processing to process frames of the captured video streams and extracts a set of features. These features are applied to different machine learning techniques with a weighted averaging method to predict whether a helmet is present or not. This system can rev up the process of identifying the violators while dodging the need to use intensive manpower.

Rest of the paper is organized as follows. Section 2 discusses the existing works on identifying helmets from images. Section 3 presents the proposed system. The experimental results and a comparative analysis of the proposed work are given in Sect. 4. Finally, Sect. 5 concludes our work with references at the end.

2 Related Work

Chiu et al. [2, 3] have suggested algorithms to detect occluded motorcycles using visual length, visual width, and pixel ratio. They assume that motorcycle riders wear helmets and this fact is used to detect motorcycles in a scene. Silva et al. [4] have proposed a system to segregate motorcycles from other vehicles in live video streams. They used Adaptive Mixture of Gaussians (AMG) to update the background with object segmentation to detect moving vehicles. Then, they extracted Histogram of Oriented Gradients (HOG), HAAR and Linear Binary Patterns (LBP) descriptors and classified the images using Multi-layer Perceptron, SVM, and Radial Basis Function Networks. Leelasantitham and Wongseree [5] have used traffic engineering knowledge for detecting and classifying moving vehicles. It has detected moving vehicles using a tracking method. With the help of features such as width, length, and the width to length ratio vehicles were classified using a decision tree method. Marayatr and Kumhom [6] have used Circle Hough Transform to detect the presence of helmet in a scene. Wen et al. [7] have identified people wearing a helmet to evade the surveillance system of the ATM room. Liu et al. [8] have presented a method to detect a full-face helmet in an image by finding circles in its Canny edges. These methods work effectively only on a full-face helmet from which a circle or an arc can be extracted. Chiverton [9] has proposed an algorithm for detecting helmets by taking advantage of their reflective properties. From the moving objects detected, it identifies motorcycles through the head to motorcycle size ratio and the aspect ratio of the bounded rectangle. Then, the approximate head region was isolated. A SVM classifier is trained over HOG features and the classification outputs of more than one frame are used to predict whether a helmet is present or not. Silva et al. [10] have used AMG to update the background. A tracking algorithm was implemented to detect moving vehicles. A feature vector of Linear Binary Patterns (LBP) was used along with the SVM classifier to distinguish motorcycles from non-motorcycles. Waranusast et al. [11] have performed background subtraction using an improved AMG model and thereafter use contour area, aspect ratio, and the standard deviation of hue around the center of the blob to distinguish between motorcycle objects and other motor vehicles with the help of a KNN Classifier [12].

In the existing research efforts, the geometric features such as circular shape were not able to singlehandedly differentiate a helmet head from a bare head. Using the reflective property of helmets involves assumptions on the environmental conditions.

3 Proposed System

The proposed system aims to detect motorcyclists without helmet from the captured images, while considering following design goals: It should accurately predict the presence or absence of helmet on motorcyclists. The execution time of the system should be minimal. False Negatives (FN) should be reduced. It should be accurate in different environmental conditions. As shown in Fig. 1, the proposed system includes modules viz; background subtraction, motorcycles detection, and helmet detection.

3.1 Background Subtraction

In the proposed system, we use adaptive background/foreground model [13] which calculates pixel intensities and a cumulative average of previous pixel values (μ_t). If the difference of the current pixel's intensity (I_t) and the cumulative average is greater than a threshold, then it is classified as a foreground. Therefore at each t time frame, pixel in I_t can be classified as foreground if $|I_t - \mu_t| > k$ holds, where k is a threshold. $\mu_{t+1} = \mu_t * I_t + (1 - \alpha) * \mu_t$, where α is the learning rate and is typically 0.5.

Fig. 1 Design of the proposed system

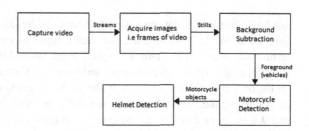

3.2   Motorcycle Detection

Among the blobs that belong to the foreground, only the blobs of motorcycles are retained. The following features are used to detect motorcycles: *Aspect ratio of bounded rectangle* and *Standard deviation of Hue* in the rectangular blob.

3.3   Helmet Detection

From the motorcycle blobs, the top 25% is chosen as the region of interest (ROI). *Features 1 and 2—Arc circularities*—the arc circularity measure [14] of the left half and right half of the head region are the first two features. The arc circularity, C, provides a measure of the arc's resemblance to the circle as given by $C = \mu_r / \sigma_r$, where μ_r is the mean and σ_r is the standard deviation of the distance r from the centroid O to the boundary of the head region.

Features 3–6—Average intensities—these features are the average pixel intensities of each quadrant. The average intensities are calculated as $\mu_I = \frac{1}{N} \sum_{i=0}^{N-1} I_i$, where I_i is the intensity of the *ith* pixel in the quadrant and N is the total number of pixels in the quadrant. Here, a quadrant has different average intensity range, if a helmet is worn. This also captures the luminosity of the helmet.

Features 7 and 8—Average hues—these features are the average hue of the third and the fourth quadrant. $\mu_H = \frac{1}{N} \sum_{i=0}^{N-1} H_i$, Where H_i is the hue of the *ith* pixel in the quadrant and N is the total number of pixels in the quadrant.

Histogram of Oriented Gradients (HOG)—The last 1980 features are the HOG features of the top half of the head region.

Above features quantitatively describe the meaningful characteristics of the image as required for its classification. A classifier, then, maps each of these feature vectors (images) to discrete classes.

4   Results and Discussion

We have prepared a dataset with 300 images of heads wearing a helmet and 300 bare heads. We have applied the extracted features to well-known machine learning techniques as shown in Table 1. Logistic Regression, MLP, and SVM work well in identifying the helmet. We have applied Majority Voting and Weighted Averaging on the output of these classifiers. It shows that Weighted Averaging performs well in identifying the helmet from the road traffic images. From Figs. 2, 3, 4, 5, 6, 7, 8, and 9 it is concluded that the weighted average on Logistic Regression, MLP and SVM have 9.8893 AUC which is very encouraging.

Table 1 Results of different classifiers in identifying helmets from the images

Classifier	Accuracy	Precision	Recall	F score	AUC	FN	Time (s)
Logistic regression (LR)	0.94	0.94	0.94	0.94	0.98	3.00	1.23
MLP	0.93	0.93	0.94	0.93	0.97	2.66	8.99
SVM	0.94	0.94	0.94	0.94	0.98	2.83	18.21
Decision tree	0.83	0.83	0.84	0.83	0.84	7.50	3.80
Random forests	0.91	0.91	0.91	0.91	0.96	4.83	1.75
KNN	0.80	0.81	0.80	0.79	0.89	4.33	5.35
Majority voting on LR, MLP, SVM	0.92	0.92	0.92	0.92	0.98	1.33	42.74
Weighted average on LR, MLP, SVM	0.94	0.94	0.93	0.93	0.98	0.83	24.08

Fig. 2 ROC curve for logistic regression in detecting helmets

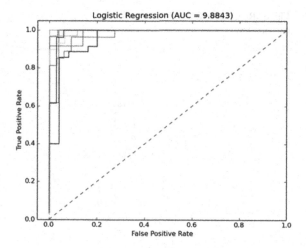

Fig. 3 ROC curve for multi-layer perceptron in detecting helmets

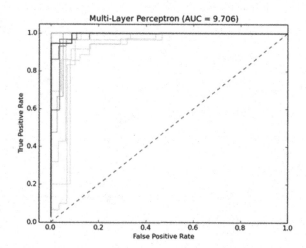

Fig. 4 ROC curve for SVM

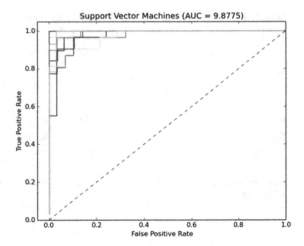

Fig. 5 ROC curve for decision tree

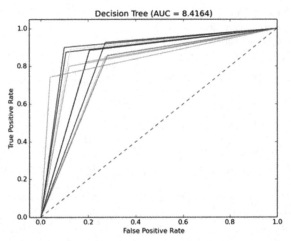

Fig. 6 ROC curve for random forests

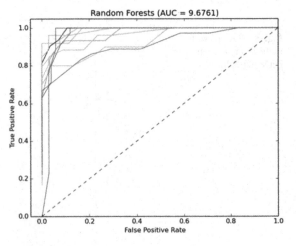

Fig. 7 ROC curve for KNN

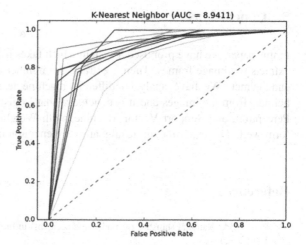

Fig. 8 ROC curve for majority voting over the selected classifiers

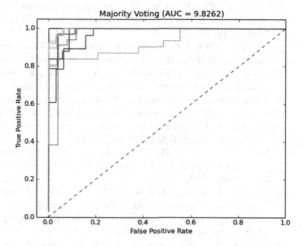

Fig. 9 ROC curve for weighted average over the selected classifiers

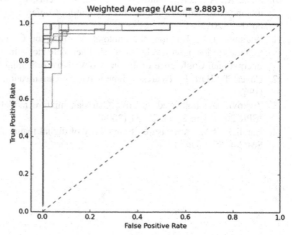

5 Conclusions

In this paper, we have proposed a system which takes road traffic video streams and extracts the image frames. Then it extracts the number of features relevant to head and helmet. We have analyzed different machine learning techniques to detect helmets from the images and it is concluded that Logistic Regression, Multi-Layer Perceptron, and Support Vector Machines with Weighted Averaging method perform well. The experimental results are very encouraging.

References

1. Ruikar, M.: National statistics of road traffic accidents in India. J. Orthop Traumatol Rehabil 1–6 (2013)
2. Chiu, C.C., Ku, M.Y., Chen, H.T.: Motorcycle detection and tracking system with occlusion segmentation. In: 8th International Workshop on Image Analysis for Multimedia Interactive Services, pp. 32–32 (2007)
3. Ku, M.Y., Chin, C.C., Chen, H.T., Hong, S.H.: Visual motorcycle detection and tracking algorithms. WSEAS Trans. Electron 5(4), pp. 121–131 (2008)
4. Silva, R., Aires, K., Veras, R., Santos, T., Lima, K., Soares, A.: Automatic motorcycle detection on public roads. CLEI Electron. J. 16(3) (2013)
5. Leelasantitham, A., Wongseree, W.: Detection and classification of moving thai vehicles based on traffic engineering knowledge. In: ITST, pp. 439–442 (2008)
6. Marayatr, T., Kumhom, P.: Motorcyclist's Helmet wearing detection using image processing. Adv. Mater. Res. 931–932, 588–592 (2014)
7. Wen, C-Y., Chiu, S.-H., Liaw, J.-J., Lu, C.-P.: The safety Helmet detection for ATM's surveillance system via the modified hough transform. In: IEEE 37th Annual ICCST, pp. 364–369 (2003)
8. Liu, C.C., Liao, J.S., Chen, W.Y., Chen, J.H.: The full motorcycle helmet detection scheme using canny detection. In: 18th IPPR Conference on CVGIP, pp. 1104–1110 (2005)
9. Chiverton, J.: Helmet presence classification with motorcycle detection and tracking. IET Intell. Trans. Syst. 6(3), pp. 259–269 (2012)
10. Silva, R., Aires, K., Santos, T., Abdala, K., Veras, R., Soares, A.: Automatic detection of motorcyclists without helmet. In: Computing Conference (CLEI), XXXIX Latin American, pp. 1–7 (2013)
11. Waranusast, R., Bundon, N., Timtong, V., Tangnoi, C., Pattanathaburt, P.: Machine vision techniques for motorcycle safety helmet detection. In: Proceedings of the 2013 28th International Conference on Image and Vision Computing New Zealand, pp. 41–46 (2013)
12. Cover, T., Hart, P.: Nearest-neighbor pattern classification. IEEE Trans. Inf. Theory 21–27 (1967)
13. Zivkovic, Z.: Improved adaptive Gaussian mixture model for background subtraction. In: ICPR 2004, vol. 2, pp. 28–31 (2004)
14. Haralick, R.M.: A measure of circularity of digital features. IEEE Trans. Syst. Man Cybern. SMC-4, 394–396 (1974)

Image Classification Using an Ensemble-Based Deep CNN

Aloysius Neena and M. Geetha

Abstract For the customary classification algorithms, performance depends on feature extraction methods. However, it is challenging to extract such unique features. With the advancement of Convolutional Neural Networks (CNN), which is the widely used Deep Learning Framework, there seems to be a substantial improvement in classification performance combined with implicit feature extraction process. But, training a CNN is an intensive process that often needs high computing machines (GPU) and may take hours or even days. This may confine its application in a few situations. Considering these factors, an ensemble architecture is modelled, that is trained on a subset of mutually exclusive classes, grouped by Hierarchical Agglomerative Clustering based on similarity. A new Probabilistic Ensemble-Based Classifier is designed for classifying an image. This new model is trained in comparatively lesser time with classification accuracy comparable to the traditional ensemble model. Also, GPUs are not necessary for training this model, even for large datasets.

Keywords Convolutional neural networks · Deep learning · Computer vision · Image classification

1 Introduction

Convolutional Neural Network is the widely used deep learning framework which was inspired by the visual cortex of animals [1]. Initially, it had been widely used for object recognition tasks but now it is being examined in other domains as well [2]. The neocognitron in 1980 [3] is considered as the predecessor of ConvNets. LeNet was the pioneering work in Convolutional Neural Networks by Jackel et al. [4]

A. Neena (✉) · M. Geetha
Department of Computer Science and Engineering, Amrita School of Engineering,
Amritapuri, Amrita Vishwa Vidyapeetham, Amrita University, Coimbatore, India
e-mail: neenaloysius@ymail.com

M. Geetha
e-mail: geetham@am.amrita.edu

© Springer Nature Singapore Pte Ltd. 2018
P. K. Sa et al. (eds.), *Recent Findings in Intelligent Computing Techniques*,
Advances in Intelligent Systems and Computing 709,
https://doi.org/10.1007/978-981-10-8633-5_44

445

in 1990. It was specifically designed to classify handwritten digits and was successful in recognizing visual patterns directly from the input image without any preprocessing. But, due to lack of sufficient training data and computing power, this architecture failed to perform well in complex problems. Later in 2012, with the rise of GPU computing, Krizhevsky et al. [5] had come up with a CNN model that succeeded in drastically bringing down the error rate on ImageNet 2012 Large-Scale Visual Recognition Challenge (ILSVRC-2012) [6]. Over the years later, their work has become one of the most influential one in the field of computer vision and used by many for trying out variations in CNN architecture. But initially their results also daunted many in the area of computer vision due to the fact that the high-capacity classification of CNN is owed to huge labelled training dataset like ImageNet and it is obviously difficult in practice to have such large labelled datasets in different domains.

The aforementioned problem is addressed using Transfer Learning. The mid-level feature representations learned by a ConvNet on a large dataset are transferred to other object recognition tasks with limited training data. The main challenge while transferring knowledge is that it should produce positive learning in the target task. There is a high chance of negative transfer learning when the source and target tasks are less related. The datasets chosen as source is ImageNet and the target is Caltech101. Chances of negative transfer learning are less in our case since the source and target dataset are not totally unrelated.

The specific contributions of this work are as follows: we have trained a new ensemble model of convolutional neural network on Caltech101 and achieved the best results in terms of time complexity on this dataset, without compromise on accuracy. The subset of classes fed to each network of the ensemble (conveniently called pipeline) is grouped by Hierarchical Clustering algorithm based on single linkage to suit our need for grouping similar classes. The subsets chosen are mutually exclusive. Also, the concept of transfer learning is applied by retraining a trained AlexNet model which has significantly reduced the training time and also improved learning. Further to improve the learning process, visual saliency maps of all training images are generated to identify the salient portion of images and the network is trained using this. Much of the unnecessary background details are eliminated in the process.

2 Related Works

Even though Convolutional Neural Networks were introduced in 1990 by LeCun et al. [7], the architecture developed by Alex Krizhevsky et al. [5] is credited as the first work in CNN to popularize it in the field of computer vision. It has a total of 8 learned layers—5 convolution layers and 3 fully connected layers. The network was similar to LeNet but instead of alternating convolution layers and pooling layers, AlexNet had all the convolution layers stacked together. And compared to LeNet, this network is much bigger and deeper.

An improvement over AlexNet was the CNN architecture by Zeiler and Fergus [8]. They have presented a novel way to visualize the activity within the ConvNets using a multi-layered Deconvolutional Network (DeConvNet) [9]. A DeConvNet is also a ConvNet that operates in reverse direction, mapping features to input pixel space. So the visualization of a ConvNet is done by attaching a DeConvNet to each of it layers. This architecture can be used to observe the evolution of features during training and also to troubleshoot the network in case of any issues. They have used these tools to analyse the components of AlexNet and did a tweaking of the network by reducing the filter size and stride in first layer and expanded the size of the middle convolutional layers, resulting in an improved version over AlexNet.

Szegedy et al. [10] from Google, later in 2012 proposed an architecture called GoogleLeNet with a new module, Inception(v1), that gives more utilization of the computing resources in the network. GoogLeNet is a particular incarnation that has 22 layers of Inception module but with less parameters compared to AlexNet. This module has multiple convolution filters applied on the input image along with pooling and then combining the results. This leads to multi-level feature extraction from each input and also abstract features from different scales simultaneously.

Another famous architecture is VGGNet by K. Simonyan and A. Zisserman [11]. They have done a thorough analysis of the depth factor in a ConvNet, keeping all other parameters fixed. This try could have led to huge number of parameters in the network but it was efficiently controlled by using very small 3×3 convolution filters in all layers. This study has led to the development of a more accurate ConvNet architecture called, VGGNet.

Szegedy et al., in 2015 proposed an architecture [12], which is an improvement over GoogLeNet where the training of Inception modules (by Szegedy et al. [10]) are accelerated when trained with residual connections (introduced by He et al. [13]). The network has yielded state-of-the-art performance in the 2015 ILSVRC challenge and has won the contest.

A residual learning framework was presented by Kaiming et al. [13], where the layers learn residual functions with respect to the inputs received instead of learning unreferenced functions. The main drawback of this network is that it is much expensive to evaluate due to the huge number of parameters.

The work by Soman et al. [14] does grouping of misclassified characters together to improve accuracy which is in line with our work, but the performance is found to be dropping down when the number of classes exceeds the range of 40.

3 Proposed Ensemble Architecture

The proposed architecture can be viewed as a ConvNet which is replicated more than once (called as pipelines), each trained on a subset of class labels with different parameter settings. Here, subset of dataset refers to subset of classes or labels. This inherently means that the training subsets formed are mutually exclusive. The advantage of training on a subset of classes are analysed to be multifold, i.e. training

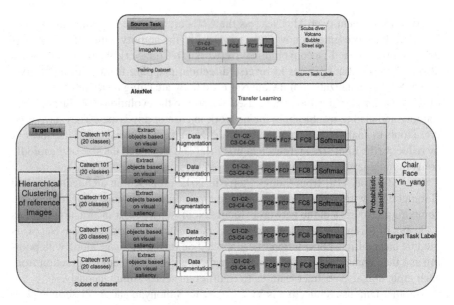

Fig. 1 Proposed Ensemble Architecture

time is expected to be reduced significantly; eliminate the need for high-end GPUs for the training of ConvNets on huge datasets.

Figure 1 clearly shows all the components involved, starting from the process of transfer learning whereby the new model gets initial weights from AlexNet trained on ImageNet. From the 101 classes, reference images are selected for each class and it is subjected to Hierarchical Agglomerative Clustering which results in group of similar images. Based on this grouping, mutually exclusive subsets are formed which is fed to network after preprocessing steps (training). The following sections detail all the steps involved.

3.1 Transfer Learning

The mid-level feature representations learned by AlexNet model on ImageNet are efficiently transferred for training the new network on Caltech101. Mid-level features or generalized features are captured in the first seven layers, i.e. from first conv layer to second fully connected layer (FC7). The learned weights of these layers are used in our model as well and these are kept constant and not updated during training. The final fully connected layer FC8 and classifier of source task are more specific to ImageNet hence we ignore them and add new FC8 and softmax classifier, which are retrained.

3.2 Clustering

Each pipeline in the new ensemble architecture is to be trained on images that belong to similar set of classes. And the grouping of similar classes is done by hierarchical clustering. Initially, reference images are selected for each class, the one with minimum noise. Based on the similarity matrix computed, Hierarchical Agglomerative Clustering (HAC) of the reference images is done. HAC follows a bottom-up approach, i.e. hierarchy of clusters are formed by recursively merging, starting from individual elements. Maximum similarity metric is considered for merging process, called as single-linkage clustering. Classes belonging to a cluster are considered for training a pipeline of the proposed ensemble, and thereby expecting a model that can be trained in lesser time without the need of GPUs, compared to the existing one.

3.3 Preprocessing Steps

The bottom-up method of computing saliency maps, Graph-Based Visual Saliency (GBVS), proposed by Koch et al. [15] is used to detect the objects (Fig. 2). The method is particularly useful when images have multiple objects and background. Based on the saliency maps, bounding box is drawn and objects are cropped from the original image, thereby removing much of the background information.

The customary procedure of random cropping is replaced by resizing the visual saliency-based detected object image to the standard input size required for AlexNet model. In addition to this, another data augmentation applied is horizontal flipping of the images. This is done based on the requirement that objects should be equally recognizable even if it is its mirror image. Applying more relevant transformations, the model is exposed to additional variations without the need of more labelled training images. Also, the problem of overfitting can be solved and thereby improving the model's ability to generalize.

Fig. 2 Starting from left, original image followed by the saliency map and original image with bounding box

3.4   Ensemble Training

Based on the results of hierarchical clustering, classes in each cluster is given as input to each pipeline of the ensemble model. ConvNets are usually trained on GPUs. But we have trained the new model without GPU, with parameter settings like 50 training data and 10 testing data, trained for a total of 15 epochs with batch size as 10 and 0.03 as learning rate. The activation function used is Rectified Linear Units (ReLU), i.e.

$$f(x) = max(0, x) \tag{1}$$

If $f(x, y)$ is the input image and $w[s, t]$ is the filter, then the basic convolution operation is given by

$$g[x, y] = \sum_{s=-a}^{a} \sum_{t=-b}^{b} w[s, t] \cdot f[x + s, y + t] \tag{2}$$

Existing ensemble architecture, with equal number of pipelines, is also trained on full dataset by varying parameters to do a comparative study on performance. Top-1 and top-5 errors are computed in the process. Both the metrics decrease progressively over the training phase.

Algorithm 1 Proposed Ensemble

INPUT: Set of images *img*
OUTPUT: A trained network *t*
1: **procedure** PROPOSED–ENSEMBLE(*img*)
2:      Initialize weights with that of pretrained AlexNet.
3:      Select reference image from each class.
4:      Compute similarity score matrix *M*.
5:      set of clusters $C \leftarrow Hierarchical\_Clustering(M)$
6:      **for** each dataset $i \in C(i)$ **do**
7:           Saliency_Extraction(*i*);
8:            $t \leftarrow$ Train(*i*).
9:      **end for**
10:     Return trained model *t*
11: **end procedure**

3.5   Probabilistic Classifier

Testing of ensemble model involves feature computation and softmax classification (with scores) with each pipeline model (as given in Algorithm 2). The state-of-art ensemble networks does prediction by averaging the softmax classifier's score values. We have come up with a probabilistic classifier where we select the maximum score of softmax classifier from each pipeline and again a maxima of all the maximum scores. This is based on the presumption that given a test image, the pipeline which has learned the features accurately will recognize it with a very high probability compared to other incorrect classification scores of other pipelines.

Algorithm 2 CNN Testing

INPUT: Any image from Caltech-101 or of similar data distribution
OUTPUT: Object label with predicted score
1: **procedure** CNN–TEST
2:    Load the saved models of each pipeline.
3:    Replace the last softmax loss layer with softmax classifier.
4:    Each pipeline computes score for the given image using the same convolution and pooling operations done during training.
5:    Find maximum of scores from each pipeline. Let it be $score(i)$, where i represents pipeline.
6:    $final\_score \leftarrow max(score(i))$
7:    Return associated label, $final\_score$.
8: **end procedure**

---

4   Results and Performance Evaluation

The new ensemble model is trained on Caltech101 dataset using initialization weights from AlexNet trained on ImageNet.

4.1   Dataset

The source dataset for transfer learning of mid-level representations is chosen as ImageNet. The images have center-focused objects with less background clutter. The AlexNet model trained on ImageNet is chosen as the source task. The main advantage of selecting AlexNet as the source model over other models is that, since it is trained on the largest image database available, the mid-level representations learned will be more accurate and can be easily adapted to any other challenging datasets of different data distributions. The target dataset chosen for studying the impacts of transfer learning is Caltech101. It contains a total of 9,146 images distributed across 102 categories.

4.2   Testing

Testing is done on Caltech101 dataset by considering 8 classes, 25 classes, and full dataset. This incremental testing approach has ultimately proved useful in understanding the correlation between the number of classes, number of pipelines in the ensemble and classification accuracy.

Table 1 Caltech-101 classification accuracy for our ConvNet model trained on 8 classes, against the alternate approach

Models	Acc%	Train time
New ensemble	80	approx. 20 mins
Score-averaging ensemble	79	approx. 45 mins
Single-nonpipelined	78	approx. 20 mins

Table 2 Class-wise accuracy

Class	Acc% (new)	Acc% (existing)
Airplanes	70	60
Beaver	80	80
Car side	30	20
Dalmatian	100	100
Elephant	100	90
Helicopter	100	100
Kangaroo	90	100
Motorbikes	70	80

Table 3 Caltech-101 classification accuracy for our ConvNet model trained on 25 classes, against the alternate approach

Models	Acc%	Train time
New ensemble	84	approx. 40 mins
Score-averaging ensemble	83.66	approx. 1.5 h
Single-non-pipelined	83	approx. 40 mins

Case 1: 8 Classes and 2 Pipelines—We have trained an ensemble model of two pipelines, for a total of 8 classes, i.e. 4 classes per pipeline. Also, a score-averaging ensemble of comparable size (two pipelines), 8 classes per pipeline is also modelled. And the results are given in Tables 1 and 2.

Case 2: 25 Classes and 2 Pipelines—Next, the number of classes are increased and trained an ensemble model of two pipelines, for a total of 25 classes, 12 in one pipeline and 13 in the other. In this case as well a score-averaging ensemble of comparable size (two pipelines), 25 classes per pipeline is modelled. The test results are shown in Tables 3 and 4.

Case 3: 101 Classes and 5 Pipelines - Having seen the good results in above two scenarios, we have trained the ensemble on the whole Caltech-101 dataset, for 101 classes. Since we have more number of classes in this case, the ensemble is designed to have 5 pipelines with 20 classes per pipeline except one having 21 classes. The score averaging ensemble as well has 5 pipelines, each trained on full dataset.

Table 4 Class-wise accuracy

Class	Acc% (new)	Acc% (existing)
Airplanes	50	60
Beaver	80	80
Binocular	30	40
Bonsai	100	100
Brontosaurus	90	100
Camera	100	100
Cellphone	100	80
Chair	80	90
Dalmatian	100	70
Elephant	50	60
Ferry	100	100
Garfield	100	100
Gerenuk	100	90
Helicopter	90	90
Joshua tree	90	100
Kangaroo	100	90
Leopards	60	50
Llama	60	70
Okapi	80	90
Panda	100	90
Rhino	90	90
Stegosaurus	70	80
wheelchair	100	100
Wild cat	100	90
Windsor chair	80	70

Table 5 Caltech-101 classification accuracy for our ConvNet model, against the alternate approach

Models	Acc%	Train time
New ensemble	68	approx. 3
Score-averaging ensemble	78.48	approx. 15 h
Single-non-pipelined	78	approx. 3 h

Classification accuracies for the model as such as well as for per-class are detailed in Tables 5 and 6.

Figure 3 shows the top 3 classes with high classification accuracies and Fig. 4 shows top 3 classes with low classification accuracies, compared to the state-of-the-art model. Incorrectly classified are highlighted in red and those in green are correctly

Table 6 Class-wise accuracy

Class	Acc% (new)	Acc% (existing)
Airplanes	50	40
Beaver	70	80
Binocular	50	20
Bonsai	100	100
Brontosaurus	90	60
Camera	90	50
Cellphone	20	0
Chair	90	80
Dalmatian	70	60
Elephant	50	50
Ferry	100	100
Garfield	100	90
Gerenuk	90	80
Helicopter	90	100
Joshua tree	100	100
Kangaroo	80	60
Leopards	50	50
Llama	60	50
Okapi	80	80
Panda	90	90
Rhino	80	80
Stegosaurus	80	80
Wheelchair	90	100
Wild cat	80	50
Windsor chair	80	70

Fig. 3 Top 3 classes for which our method has performed well compared to alternate approach

Fig. 4 Classes for which our method has very low classification results compared to the alternate approach

Fig. 5 Sample predicted output

airplanes (2), score 1.000

classified. Figure 5 shows a sample prediction, where the given image (airplanes) is predicted with the highest score of 1.

4.3 Evaluation

The result clearly shows improved accuracy with the new model when compared to the existing architecture, for less number of classes, with a reduction in training time. Thus, the new method is particularly useful when a new convolutional neural network is to be trained on large datasets like ImageNet where training complexity is a critical factor. However, performance drops as the number of training classes increases. Since we have increased the number of pipelines proportional to the number of classes, the probabilistic classification is severely impacted when maximal is chosen from the set of maximals. From an analysis of the score and class predictions it is found that in most of the cases, individual pipelines have correctly predicted the desired class. But with probabilistic classification we miss the desired result. This has led to a significant reduction in the classification accuracy of the new model. But the conventional requirement of high computing machines for training convents on huge datasets can be eliminated with the proposed ensemble architecture.

5 Conclusion

Various aspects of CNN have been analysed, starting from transfer learning of feature representations from a pretrained model and the new model is actually found to be well adapted to the target dataset. With accuracies comparable to the existing model, we were able to bring about a decrease in the training time, thus reducing the time complexity of network. Our testing is limited to only one dataset in this work. We plan to have more rigorous testing of the model on challenging datasets like Caltech-256 and Pascal-VOC, in our future work.

References

1. Hubel, D.H., Wiesel, T.N.: Receptive elds and functional architecture of monkey striate cortex. J. Physiol. **195**(1), 215 (1968)
2. Nithin, D.K., Sivakumar, P.B.: Generic feature learning in computer vision. Procedia Comput. Sci. **58**, 202–209 (2015)
3. Fukushima, K.: Neocognitron: a self-organizing neural network model for a mechanism of pattern recognition unaffected by shift in position. Biol. Cybern. **36**(4), 193–202 (1980)
4. John, S., Henderson, D., Howard, R.E., Hubbard, W., Jackel, L.C., Boser enker, B., Lawrence, D.: Handwritten digit recognition with a back-propagation network. In: Advances in Neural Information Processing Systems. Citeseer (1990)
5. Krizhevsky, A., Sutskever, I., Hinton, G.E.: Imagenet classification with deep convolutional neural networks. In: Advances in Neural Information Processing Systems, pp. 1097–1105 (2012)
6. Fei-Fei, L., Berg, A., Deng, J.: Large Scale Visual Recognition Challenge (2010)
7. LeCun, Y., Bottou, L., Bengio, Y., Haffner, P.: Gradient-based learning applied to document recognition. Proc. IEEE **86**(11), 2278–2324 (1998)
8. Zeiler, M.D., Fergus, R.: Visualizing and understanding convolutional networks. In: ECCV, pp. 818–833. Springer (2014)
9. Matthew D Zeiler, Graham W Taylor, and Rob Fergus. Adaptive deconvolutional networks for mid and high level feature learning. In: 2011 ICCV, pp. 2018–2025. IEEE (2011)
10. Szegedy, C., et al.: Going deeper with convolutions. In: Proceedings of the IEEE Conference on Computer Vision and Pattern Recognition, pp. 1–9 (2015)
11. Simonyan, K., Zisserman, A.: Very deep convolutional networks for large-scale image recognition (2014). arXiv:1409.1556
12. Szegedy, C., Ioffe, S., Vanhoucke, V.: Inception-v4, inception-resnet and the impact of residual connections on learning (2016). arXiv:1602.07261
13. He, K., Zhang, X., Ren, S., Sun, J.: Deep residual learning for image recognition (2015). arXiv:1512.03385
14. Kumar, K., Sachin, S., Anil, R.M., Soman, K.P.: Convolutional neural networks for the recognition of malayalam characters. In: FICTA 2014, pp. 493–500. Springer (2015)
15. Harel, J., Koch, C., Perona, P.: Graph-based visual saliency. In: Advances in Neural Information Processing Systems, pp. 545–552 (2006)

Performance Metric Evaluation of Segmentation Algorithms for Gold Standard Medical Images

S. N. Kumar, A. Lenin Fred, H. Ajay Kumar
and P. Sebastin Varghese

Abstract Image segmentation plays a vital role in medical image processing for the delineation of anatomical organs and analysis of anomalies. The evaluation of segmentation algorithms is vital to select the appropriate algorithm and parameters for optimum performance. In this paper, we are describing various metrics for evaluating the quality of segmentation algorithms with respect to ground truth images. The analysis of metrics has been carried out on real-time data sets of abdomen and retina. The variants of active contour algorithms are employed for the abdomen CT images, Kirsch and Wavelet algorithm were used for the retinal fundus images. This paper presents performance evaluation parameters that can be used to analyze efficiency of segmentation algorithms.

Keywords Segmentation · Metrics · Evaluation · Success rates
Error rates · Distance measures

S. N. Kumar (✉)
Department of Electronics and Communication Engineering, Sathyabama Institute of Science and Technology, Chennai 600119, Tamil Nadu, India
e-mail: appu123kumar@gmail.com

A. Lenin Fred
School of Computer Science Engineering, Elavuvilai, Mar Ephraem College of Engineering and Technology, Marthandam 629171, Tamil Nadu, India
e-mail: leninfred.a@gmail.com

H. Ajay Kumar
School of Electronics and Communication Engineering, Elavuvilai, Mar Ephraem College of Engineering and Technology, Marthandam 629171, Tamil Nadu, India
e-mail: ajayhakkumar@gmail.com

P. Sebastin Varghese
Metro Scans and Laboratory, Consultant Radiologist, Trivandrum 695011, Kerala, India
e-mail: sebastin464@gmail.com

© Springer Nature Singapore Pte Ltd. 2018   457
P. K. Sa et al. (eds.), *Recent Findings in Intelligent Computing Techniques*,
Advances in Intelligent Systems and Computing 709,
https://doi.org/10.1007/978-981-10-8633-5_45

1   Introduction

The role of segmentation is vital in medical image processing in the analysis of anomalies like tumor and cyst. An efficient segmentation algorithm output will drive the classifier efficiently in determining the stages of anomalies like a tumor. Prior to the extraction of ROI, a robust preprocessing stage will be there in aiding the segmentation algorithm to produce the accurate result. A wide variety of segmentation algorithms are there for the extraction of desired area of interest in medical image processing. The evaluation of segmentation algorithms is vital in choosing the appropriate algorithm for the application. The analysis of existing algorithms is also needed for the evaluation of newly developed algorithms by the researchers [1]. The metrics for evaluation of segmentation algorithms are mainly classified into two types supervised and unsupervised. The supervised evaluation metrics need a reference image for analyzing the efficiency of the segmentation algorithm. The reference image is also called ground truth image or gold standard image. The ground truth image is generated by an expert physician (radiologist) by carefully tracing the region of interest in the medical image [2]. The reference image will be compared with the segmentation algorithm output to determine the efficiency. In some cases, it is not possible to create a reference image and in that case, the unsupervised evaluation metrics are used. The texture descriptors such as color uniformity, inter-region contrast, intra-region uniformity, entropy, and shape of the desired ROI are used as unsupervised evaluation metrics [3]. In medical image processing, computer-aided segmentation algorithms play a vital role in the extraction of ROI (tumor, cyst, anatomical organ, etc.) [4]. The determination of accuracy is required to validate a novel algorithm with the existing techniques. The optimum selection of parameters of the algorithm to produce accurate result was influenced by the proper segmentation metrics [5]. In the comparison of segmentation algorithms, the metrics play a vital role for the evaluation of the algorithm. The tuning of the parameters is done to improve the performance of the segmentation algorithm for target applications [1]. The ground truth or gold standard image is the manually segmented ROI structures and it is compared with the result produced by the segmentation algorithm to evaluate the efficiency. Vikram et al. framed statistical approach for the evaluation of medical image segmentation algorithm; the only information available is boundaries traced by multiple expert observers [6]. Daniel et al. used regional mutual information (RMI) to evaluate the efficiency of the segmentation algorithm. RMI takes into account neighboring pixel characteristics and produces a robust result than mutual information, which is calculated on a pixel-by-pixel basis [7].

Jaime et al. proposed symmetric (dsym) and asymmetric (dasym) distance metrics based on the distance between segmentation partitions. The reference or ground truth image is required and the efficient results are produced for complex applications [8]. Fenster et al. proposed that accuracy, precision, and efficiency must be measured and evaluated in the analysis of segmentation algorithms for medical organs delineation. The accuracy refers to the degree of closeness of segmentation result with the gold standard image. The precision was determined using the appropriate metric (distance, area, and volume) estimated from the test image, ground truth image and it provides the information on the repeatability of the segmentation algorithm. In general, the efficiency was determined from the execution time, however, other parameters such as user interaction and applicability of segmentation algorithm to different medical images were also considered [9]. Ruben et al. developed novel segmentation metrics based on the pixel characteristics such as local intensity, connectivity, and boundary of the segmented data [10]. F. C. Monteiro et al. take into account of the over segmentation and under segmentation and proposed a new discrepancy measure metric for evaluation [11].

The widely used segmentation metrics were analyzed and a new metric was proposed that encompasses inherent ambiguity in segmentation, partition region sensitivity, and concern about the segmentation boundary labeling error constraints. D.J. Withey et al. performed an analysis of various segmentation algorithms in medical images and discussed about segmentation software's and validation databases [12]. K. O. Babalola et al. used Williams' Index in addition to conventional metrics like Hausdorff distance, Dice coefficient in the evaluation of four algorithms (Classifier Fusion and Labelling (CFL), Profile Active Appearance Models (PAM), Bayesian Appearance Models (BAM), Expectation–maximization-based segmentation (EMS)) for the segmentation of subcortical structures in the brain MR images; the CFL model produces efficient result [13]. Xiang Li et al. estimated the ground truth for skin lesion segmentation by three methods voting policy, variation based, maximum a posterior probability based, and voting policy method is simple, robust, and produces slightly better results than other two methods [14]. T. Kholberger proposed a classifier-based approach for evaluating the segmentation result in the absence of ground truth image. The widely used energy-based graph cut segmentation algorithms were analyzed and 42 shape and appearance features were extracted from the basic building blocks of the algorithm [15]. G. Safarzadeh Khooshabi used Simultaneous Truth and Performance-Level Estimation (STAPLE) method to estimate the ground truth image from the manual expert's segmentation result for the validation of automatic segmentation algorithms [16]. R. Agrawal et al. used Dice similarity coefficient, overlap ratio, and Jaccard coefficient in the validation of segmentation algorithms (adaptive spatial fuzzy c-means (ASFCM), Markov random field, fuzzy connectedness method, and atlas-based re-fuzzy connectedness) for the analysis of brain tissues on MR images; the ASFCM produces efficient result [17]. J Herttuainen used Dice coefficient (DC), Jaccard index (JI), and relative absolute area difference for the validation of Naive Bayes (NB) and Gaussian Mixture Model (GMM) classifier in the extraction of exudates from retinal images; GMM classifier produces superior results [18]. Z. F. Khan et al. used segmentation accuracy derived from region properties of images (solidity, area, and perimeter) for the evaluation of fuzzy bit plane thresholding algorithm on lung and retinal images [19]. Abdel et al. evaluated for 3D medical image segmentation various segmentation metrics and framed a rank table that comprises of the result produced by metrics and manual ranking by experts [20]. A. Ahirwar et al. used success and error rates for analyzing the segmentation algorithms on MR images of the brain [21].

2 Segmentation Metrics

The ground truth image generated manually by experts is represented by G and the machine-generated image by segmentation algorithm is represented by S. Figure 1, depicts the ground truth image having two regionss of interest a and b [22]. The segmentation result is depicted in Fig. 1f, when there is a perfect matching between the region of interest in G and S.

The over segmentation results, when there is matching between more than one region in S corresponds to one region in G and is depicted in Fig. 1b. The under segmentation occurs, when more than one region in G matches to single region in S and is depicted in Fig. 1c. The segmentation result is prone to noise as shown in Fig. 1d and when a region in S does not corresponds to a region in G and sometimes desired region in the G will be missed in the S Fig. 1e.

Let X is the set of all voxels in the image, G is the set of all voxels labeled as a segmented object in ground truth image, and S is the set of all voxels labeled as a segmented object by a proposed segmentation algorithm. Figure 2 represents success and error rates with respect to ground truth image.

The supervised segmentation evaluation metrics relies on ground truth image and segmented image. Some of the widely used segmentation metrics based on the segmented and ground truth image are depicted in Table 1.

Fig. 1 An example of segmentation metrics

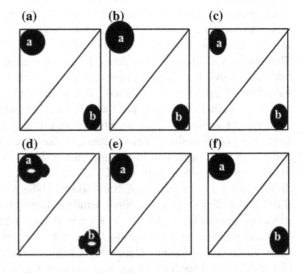

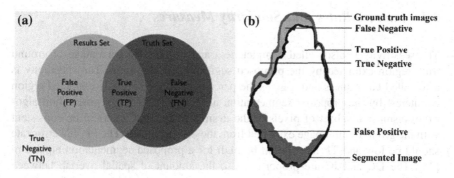

Fig. 2 **a** Success and error rates in segmentation evaluation, **b** Gold standard image (Red) and segmented image (Black)

Table 1 Metrics for the evaluation of segmentation algorithms

Success and Error rates		Similarity and Distance measures	
Sensitivity	$\frac{|TP|}{|TP|+|FN|}$	Dice coefficient (DC)	$\frac{2|TP|}{2|TP|+|FN|+|FP|}$
Specificity	$\frac{|TN|}{|TN|+|FP|}$	Jaccard coefficient (JC)	$\frac{|TP|}{|TP|+|FN|+|FP|}$
Precision	$\frac{|TP|}{|TP|+|FP|}$	Overlap ratio (OR)	$\frac{DC}{2-DC}$
TP rate	$\frac{|TP|}{|TP|+|FP|}$	Volume similarity (VS)	$\left(\frac{V_s-V_g}{V_g}\right)\times 100$
TN rate	$\frac{|TN|}{|FN|+|TN|}$	Variation of Information (VI)	$VI(S,G)=H(S)+H(G)+I(S,G)$
FP rate	$\frac{|FP|}{|TN|+|FP|}$	Williams index	$W_g=\frac{(n-1)\sum_i^n D_{pg}}{2\sum_{i=2}^n \sum_{j=1}^{i=1} D_{pq}}$
FN rate	$\frac{|FN|}{|TP|+|FN|}$	Hausdorff distance	$H=max(H_{SG},H_{GS})$
Classification error rate	$\frac{|FP+|FN|}{|FP|+|TP|+|FN|+|TN|}$	Rand index (RI)	$\frac{|TP|+|TN|}{|TP|+|FN|+|TN|+|FP|}$
		Adjusted rand index	$\frac{RI-R_{exp}}{1-R_{exp}}$
Likelihood ratio positive	$\frac{sensitivity}{|1-specificity|}$	Hamming distance	$D_H(S\Rightarrow R)=\sum\limits_{r_i\epsilon RS_k\neq S_jS_k\,\cap\,r_i\neq 0}\sum |r_i\cap S_k|$
		Region-based hamming distance	$p=1-\frac{D_H(S\Rightarrow R)+D_H(R\Rightarrow S)}{2\times|S|}$
Likelihood ratio negative	$\frac{|1-sensitivity|}{specificity}$	Mean absolute difference	$MAD_J=\frac{1}{K}\sum\limits_{i=1}^{K} d(b_i,T)$
Accuracy	$\frac{|TP|+|TN|}{|TP|+|FN|+|TN|+|FP|}$	Maximum difference	$MAXD_j=max_{i\in[1,K]}\{d(b_1,T)\}$

2.1 Success Rates and Similarity Measures

The sensitivity is also called completeness and it gives the percentage of ground truth region extracted by the proposed segmentation algorithm. The specificity is also called correctness and it gives the percentage of the correctly extracted region of interest by the proposed segmentation algorithm [22]. The segmentation algorithm result is a subset of pixels in the desired image and it should be consistent with the ground truth image extracted from the desired image. The FP, FN error rate should be low and TP rate should be high for a proposed segmentation algorithm [23]. The DC and JC are proportional to the amount of spatial overlap between machine segmented and ground truth binary image and the value ranges from 0 (no overlap) to 1 (perfect matching) [24].

The relationship between DC and JC is represented as follows.

$$JC = \frac{DC}{2 - DC}. \tag{1}$$

The false positive Dice coefficient is a measure of over segmentation; false negative Dice coefficient is a measure of under segmentation and they are represented by the following formula

$$FPD = \frac{2|S \cap \bar{G}|}{|S| + |G|} \times 100. \tag{2}$$

$$FND = \frac{2|\bar{S} \cap G|}{|S| + |G|} \times 100 \tag{3}$$

The rand index examines the consistency of pixels in the segmented and ground truth image. The value of rand index ranges from 0 to 1 [25]. The value '0' indicates the dissimilarity between S and G and the value of '1' indicates the perfect similarity between S and G.

The Williams index is also an efficient metric to determine the efficiency of segmentation algorithm and it is widely used to compare multiple approaches. In the case of formula in Table 1, the n represents the number of methods, D_{pg} is the Dice coefficient of method p result with respect to the ground truth image and D_{pq} is the Dice coefficient of method p result with respect to the method q result. The William index value falls into the any one of the categories

Case i   When W_g value is greater than 1, the result of proposed segmentation algorithms are matching with gold standard than to each other.

Case ii   When W_g value is equal to 1, the result of the proposed segmentation algorithms and the gold standard are matching.

Case iii   When W_g value is less than 1, the result of proposed segmentation algorithms are matching to each other than each to the gold standard.

2.2 Error Rates

The local consistency error takes into account of the measurement of error in both direction, while the global consistency error (GCE) assumes that both directions are same. The GCE should be zero for efficient segmentation and its value increases when there is an inconsistency between the segmentation algorithm output and ground truth image [26]. The validation of LCE is similar to that of GCE.

The classes in S and G that contains the given pixel P_i are represented by C(S, P_i) and C (G, P_i). The local refinement error is represented as follows.

$$LRE(S, G, Pi) = \frac{|C(S, P_i)/C(G, P_i)|}{|C(S, P_i)|} \tag{4}$$

Similarly, the expression for LRE is also represented as follows.

$$LRE(G, S, P_i) = \frac{C(G, P_i)/C(S, P_i)}{|C(G, P_i)|}. \tag{5}$$

Global consistency error (GCE) forces all local refinements in the same direction and it is represented as follows.

$$GCE(S, G) = \frac{1}{N} \min\left\{ \sum_i LRE(S, G, x_i) \sum_i LRE(G, S, x_i) \right\}. \tag{6}$$

2.3 Distance Measures

The distance-based metrics are used for the evaluation of segmentation algorithms for boundary extraction [27, 28]. They measure the distance between the segmented boundary and true boundary. The segmented boundary by the machine segmentation algorithm S comprises of set of points $S = \{s_1, s_2, s_3, \ldots, s_m\}$ and the gold standard image comprises of set of points $G = \{g_1, g_2, g_3, \ldots, g_n\}$. The variation of information is distance measure and its computation is based on entropy and mutual information. It is a measure of randomness and lowers the value better is the segmentation algorithm. In the expression shown in Table 1, the H (S) and H (G) are the entropy values and I(S, Y) is the mutual information between S and G. The Hausdorff distance is the maximum distance between the true segmented boundary and gold standard boundary to the closest point between them.

The expression for Hausdorff distance (H_{SG}) is written as follows:

$$H_{SG} = \max(d_i^{sg}); \quad i = 1, 2, 3, \ldots, n_s \tag{7}$$

where d_i^{sg} is the minimum distance for the ith surface voxel in S to the set of surface voxels in G.

The mean absolute distance uses the same distance metric as in Hausdorff distance and is calculated as the average of the absolute of the distances for all points on curve S (n_s). The mean absolute difference (MAD) is the average of error in segmentation and maximum difference (MAXD) is the maximum error in the segmentation.

3 Results and Discussion

This paper explores some of the metrics for the evaluation of segmentation approaches by the implementation of algorithms [29–31]. Two case studies based on abdominal organs, anomalies extraction, and retinal blood vessel extraction are considered in this paper.

The active contour algorithms are validated on real-time abdomen CT data sets. In the evaluation of active contour models apart from similarity measures, the distance measures are also widely used to determine the proficiency of the algorithm. The CREASEG software based on MATLAB comprises of a collection of active contour models like Caselles, Chan and Vese, Chimming Li, Bernard, Shi output, and Lankton algorithm. The medical images were obtained from Metro Sans and Research Laboratory, Trivandrum. For the evaluation of algorithms, three data sets are used here. Each data set comprises of plain and contrast-enhanced CT images of 0.6 mm slice thickness. The results of typical slices are depicted here. The first data set is normal liver CT data, second data set comprises of malignant renal cell carcinoma CT data and third data set comprises of benign liver tumor data.

The localized region-based Lankton active contour model produces superior results than other active contour models in terms of DC and HD for the segmentation of abdominal organs and anomalies. The results are depicted in Table 2. The ground truth image was obtained from the expert physician who carefully traces the

Table 2 Performance metrics of active contour segmentation algorithms

Active contour algorithms	Dice coefficient (DC)			Hausdorff distance (HD)			Mean sum of square distance (MSSD)		
	S1	S2	S3	S1	S2	S3	S1	S2	S3
Caselles	0.94	0.92	0.88	10.63	10.2	10.61	9.35	7.46	8.99
Chan and Vese	0.85	0.41	0.49	16.32	13.1	18.01	14.57	16.57	12.97
Chiming Li	0.58	0.64	0.61	35.87	38.88	36.46	18.88	20.63	28.13
Bernard	0.95	0.89	0.93	37.21	37.07	32.83	14.06	11.56	3.15
Shi	0.62	0.61	0.63	18.23	20.6	26.19	8.84	8.95	11.08
Lankton	0.96	0.94	0.91	9.87	9.12	9.88	3.46	4.62	4.09

abdominal organs and anomalies. From the table, it is clear that the Lankton algorithm result is superior when compared with other approaches, since DC $\geq$ 0.9, HD $\leq$ 15, and MSSD $\leq$ 5. In Fig. 3a–c represents Caselles algorithm results, (d, e, f) represents Chan and Vese results, (g, h, i) represents Chiming Li algorithm results, (j, k, l) represents Bernard algorithm results, (m, n, o) represents Shi algorithm results, (p, q, r) represents Lankton algorithm results.

The high-resolution publically available fundus images from strive database is used in this paper for retinal blood vessel segmentation. The retinal blood vessel segmentation was evaluated by Kirsch edge detection operator and wavelet

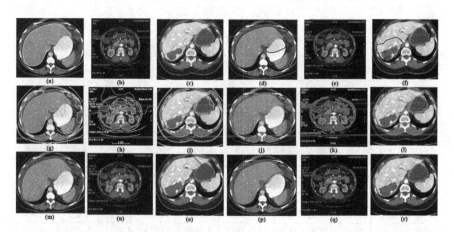

Fig. 3 Active contour algorithms segmentation results

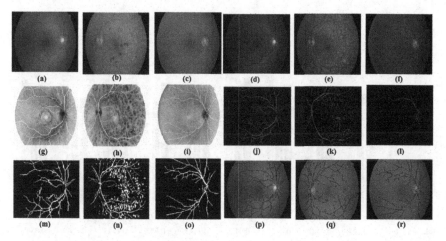

Fig. 4 **a–c** Retinal input images from strive database, **d–f** green component extraction, **g–i** Decision based median filter and Histogram equalization output, **j–l** optical disc removed output, **m–o** Retinal blood vessel segmentation by wavelet transform, **p–r** Blood vessel traced on input fundus images

transform-based approach. The wavelet approach produces better results than Kirsch edge detection operator and was quantitatively evaluated by success and error rates, similarity measures. The algorithms were evaluated on healthy, diabetic retinopathy, and glaucoma cases, the ground truth images were provided in the database for verification of result.

The retinal blood vessel segmentation results of typical slices are depicted above. In Fig. 4a depicts the healthy eye, (b) depicts the diabetic retinopathy case, (c) depicts the glaucoma case. The input images are preprocessed and wavelet

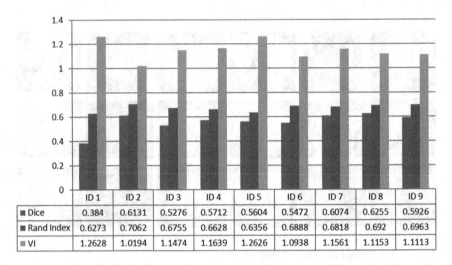

Fig. 5 Performance metrics plot of Kirsch algorithm for retinal blood vessel extraction

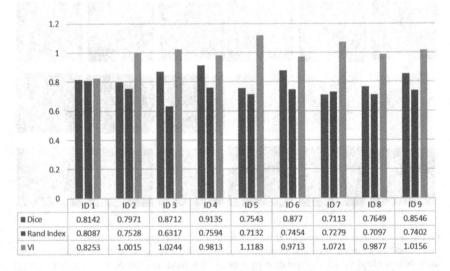

Fig. 6 Performance metrics plot of wavelet algorithm for retinal blood vessel extraction

transform is applied to the blood vessel extraction. The plots below show the performance metrics for Kirsch and Wavelet segmentation approach. The ID1 to ID3 are the normal retina cases, ID4 to ID6 are the diabetic retinopathy cases and ID7 to ID9 are the glaucoma cases.

From the plot Figs. 5 and 6, it is clear that, wavelet transform produces better segmentation results since DC and RI values are greater than 0.8 and VI value is low when compared with Kirsch approach. From the success and error rates plot of Figs. 7 and 8 also, it is clear that wavelet produces superior results. The FP, FN rate is low and TP rate is high for wavelet approach and GCE is low when compared with kirsch approach.

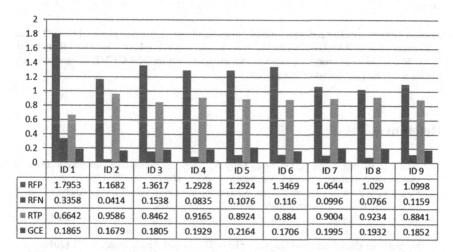

Fig. 7 Success and error rates of Kirsch algorithm for retinal blood vessel extraction

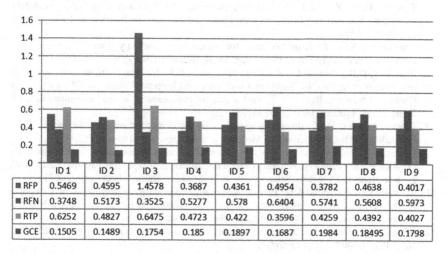

Fig. 8 Success and error rates of wavelet algorithm for retinal blood vessel extraction

4   Conclusion

This paper proposes various metrics that can be used in medical image processing to qualify the segmentation algorithms with gold standard images. The segmentation metrics will help the researchers in ranking algorithms with respect to their efficiency and proper selection of parameters for satisfactory performance. In this work, the variants of active contour algorithms are analyzed on abdomen CT images and wavelet transform, Kirsch algorithm are analyzed on retinal fundus images. The metrics will be vital in the case of evaluation of novel segmentation algorithm and quantitative comparison of existing algorithms with their modified ones.

Acknowledgements The authors would like to acknowledge the support provided by DST under IDP scheme (No: IDP/MED/03/2015). We thank Dr. Sebastian Varghese (Consultant Radiologist, Metro Scans and Laboratory, Trivandrum) for providing the medical CT/MR images and supporting us in the preparation of manuscript.

References

1. Zhang, Hui, Fritts, Jason E., Goldman, Sally A.: Image segmentation evaluation: a survey of unsupervised methods. Comput. Vis. Image Underst. **110**(2), 260–280 (2008)
2. Suri, J.S., Wilson, D., Laxminarayan, S.: Handbook of Biomedical Image Analysis, vol. 2. Springer Science & Business Media, Berlin, Germany (2005)
3. Ilea, D.E., Whelan, P.F., Ghita, O.: Unsupervised image segmentation based on the multi-resolution integration of adaptive local texture descriptors. In: Proceedings of the Fifth International Conference on Computer Vision Theory and Applications, vol. 2 (2010)
4. Pham, D.L., Xu, C., Prince, J.L.: Current methods in medical image segmentation. Annu. Rev. Biomed. Eng. **2**, 1–754 (2000)
5. Tian, W., Geng, Y., Liu, J. and Ai, L.: Optimal parameter algorithm for image segmentation. In: Second International Conference on Future Information Technology and Management Engineering (2009)
6. Chalana, V., Kim, Y.: A methodology for evaluation of boundary detection algorithms on medical images. IEEE Trans. Med. Imag. **16**(5), 642–652 (1997)
7. Russakoff, D.B., Tomasi, C., Rohlfing, T., Maurer, C.R. Jr.: Image similarity using mutual information of regions. In: European Conference on Computer Vision, pp. 596–607 (2004)
8. Cardoso, J.S., Corte-Real, L.: Toward a generic evaluation of image segmentation. IEEE Trans. Image Process. **14**(11), 1773–1782 (2005)
9. Fenster, A., Chiu, B.: Evaluation of segmentation algorithms for medical imaging. In: IEEE Proceedings of the Engineering in Medicine and Biology, pp. 7186–7189 (2005)
10. Cárdenes, R. et al.: Multimodal evaluation for medical image segmentation. In: Kropatsch, W. G., Kampel, M., Hanbury, A. (eds.) Computer Analysis of Images and Patterns. CAIP 2007, Lecture Notes in Computer Science, vol. 4673. Springer, Berlin, Heidelberg (2007)
11. Monteiro F.C., Campilho A.C.: Performance evaluation of image segmentation. In: Campilho, A., Kamel, M.S. (eds.) Image Analysis and Recognition, ICIAR 2006, Lecture Notes in Computer Science, vol 4141. Springer, Berlin, Heidelberg (2006)
12. Withey, D.J., Koles, Z.J.: Medical image segmentation: methods and software. IEEE Proc. NFSI ICFBI **2007**, 140–143 (2007)

13. Babalola, K.O., Patenaude, B., Aljabar, P., Schnabel, J., Kennedy, D., Crum, W., Smith, S., Cootes, T., Jenkinson, M., Rueckert, D.: An evaluation of four automatic methods of segmenting the subcortical structures in the brain. Neuro Image **47**, 1435–1447 (2009)
14. Li, X., Aldridge, B., Rees, J., Fisher, R.: Estimating the ground truth from multiple individual segmentations with application to skin lesion segmentation. In: Proceedings of the Medical Image Understanding and Analysis Conference, UK, vol. 1, pp. 101–106 (2010)
15. Kohlberger, T., Singh V., Alvino C., Bahlmann C., Grady L.: Evaluating segmentation error without ground truth. In: Ayache, N., Delingette, H., Golland, P., Mori, K. (eds.) Medical Image Computing and Computer-Assisted Intervention—MICCAI 2012. MICCAI 2012, Lecture Notes in Computer Science, vol. 7510. Springer, Berlin, Heidelberg (2012)
16. Safarzadeh Khooshabi, G.: Segmentation Validation Framework, CMIV, Linköping University, Department of Biomedical Engineering (2013)
17. Agrawal, R., Sharma, M.: Review of segmentation methods for brain tissue with magnetic resonance images. Int. J. Comput. Netw. Inf. Secur. **6**(4), 55 (2014)
18. Khan, Z.F., Kannan, A.: Intelligent segmentation of medical images using fuzzy bitplane thresholding. Meas. Sci. Rev. **14**(2), 94–101 (2014)
19. Taha, A.A., Hanbury, A., Del Toro, O.A.J.: A formal method for selecting evaluation metrics for image segmentation. In: IEEE ICIP, pp.932–936 (2014)
20. Ahirwar, A.: Study of techniques used for medical image segmentation and computation of statistical test for region classification of brain MRI. I. J. Inf. Technol. Comput. Sci. **05**, 44–53 (2013)
21. Gajanayake, G.M.N.R., Yapa, R.D., Hewawithana, B.: Comparison of standard image segmentation methods for segmentation of brain tumors from 2D MR images. IEEE International Conference on Industrial and Information Systems, pp. 301–305 (2009)
22. Sinthanayothin, C., Boyce, J.F., Williamson, T.H., Cook, H.L., Mensah, E., Lal, S., Usher, D.: Automated detection of diabetic retinopathy on digital fundus images. Diabet. Med. **19**(2), 105–112 (2002)
23. Wang, Z., Bovik, A.C., Sheikh, H.R., Simoncelli, E.P.: Image quality assessment: from error visibility to structural similarity. IEEE Trans. Image Process. **13**(4), 600–612 (2004)
24. Hamamci, A., Kucuk, N., Karaman, K., Engin, K., Unal, G.: Tumor-cut: segmentation of brain tumors on contrast enhanced MR images for radiosurgery applications. IEEE Trans. Med. Imaging **31**(3), 790–804 (2012)
25. Coelho, L.P., Shariff, A., Murphy, R.F.: Nuclear segmentation in microscope cell images: a hand-segmented dataset and comparison of algorithms. In: Proceedings of the IEEE International Symposium on Biomed Imaging, vol. 5193098, pp. 518–521 (2009)
26. Xess, M., Agnes, S.A.: Analysis of image segmentation methods based on performance evaluation parameters. Int. J. Comput. Eng. Res. **4**(3), 68–75 (2014)
27. Chen, S., Ma, B., Zhang, K.: On the similarity metric and the distance metric. Theor. Comput. Sci. **410**(24–25), 2365–2376 (2009)
28. Zhang, Y., Huang, D., Ji, M., Xie, F.: Image segmentation using PSO and PCM with Mahalanobis distance. Expert Syst. Appl. **38**(7), 9036–9040 (2011)
29. Kumar, S.N, Lenin Fred, A., Ajay Kumar, H., Jonisha Miriam, L.R., Asha, M.R.: Retinal blood vessel extraction using wavelet transform and morphological operations. Res. J. Pharm. Biol. Chem. Sci. **7**(5) (2016)
30. Kumar, S.N., Fred, A.L., Kumari, L.S., Varghese, P.S.: Localized region based active contour algorithm for segmentation of abdominal organs and tumors in computer tomography images. Asian J. Inf. Technol. **15**(23), 4783–4789 (2016)
31. Kumar, S.N., Fred, A.L., Kumari, S.L., Anchalo Bensiger, S.M.: Feed forward neural network based automatic detection of liver in computer tomography images. Int. J. PharmTech Res. **9** (5) 231–239 (2016)

An Advanced Ripplet Transform Based Medical Image Compression Method

Shrinivas D. Desai and R. P. Neha

Abstract Diagnostic medical imaging is tremendously increased over past few years. It basically involves multiple images or sequence of images taken at different cross sections leading to very high volume of data. Hence, there exists a need for compression of these images for effective storage and retrieval purposes. Current compression techniques provide a very high compression rate, but with a considerable loss of image quality. In medicine, it is necessary to preserve high image quality in diagnostically important regions. In this paper, a better compression algorithm for medical images is presented to address the aforementioned challenge. The proposed is method that uses a new transformation called "Ripplet transform" that preserves inconsistencies along curves and provides high-quality decompressed image by representing images at different scales and directions. The proposed method is compared with conventional as well as current compression techniques. Better performance is observed for proposed method when assessed by qualitative and quantitative measures.

Keywords Medical image · Compression · Ripplet transform · Multi-resolution
Huffman coding

1 Introduction

Achieving highest compression ratio as well as preserving the minute diagnostic details of medical images (DICOM images) is current research trend in the community. The healthcare services historically have generated huge volume of data and are subjected to degradation in image quality during storage and retrieval. Such huge

S. D. Desai (✉) · R. P. Neha
School of Computer Science and Engineering, K L E Technological University,
Hubballi 580031, India
e-mail: sd_desai@bvb.edu
URL: http://www.kletech.ac.in/

R. P. Neha
e-mail: nehapud.np@gmail.com

© Springer Nature Singapore Pte Ltd. 2018
P. K. Sa et al. (eds.), *Recent Findings in Intelligent Computing Techniques*,
Advances in Intelligent Systems and Computing 709,
https://doi.org/10.1007/978-981-10-8633-5_46

volume of data requires large amount storage space. Hence, there exists a need for better compression algorithm that yields good compression ratio while preserving image quality.

Image compression shall be broadly classified into two major categories: 1— Lossy and 2—Lossless. Lossy compression techniques do not preserve image quality after decompression. Hence, these techniques are preferably not used in medical imaging which may yield diagnostically incorrect results. However, standard lossless compression algorithms generate identical results compared with original images. Conventional transform such as Fourier transform aids to represent smooth images but is not able to preserve edges in images. In contrast, transforms based on wavelets resolve 1D singularities but are not able to preserve 2D singularities. Thus, conventional transforms could not preserve curve discontinuities to a greater extent. The proposed method is designed to resolve the limitations of conventional transforms. The proposed algorithm uses ripplet transform to preserve ripple-like structures across the curve and Huffman encoder to compress the resulting coefficients. The prime objective of proposed algorithm is to yield high-quality decompressed image and achieve a significant compression ratio [1].

1.1 Literature Survey

1. Thomas et al. [2] proposed an algorithm for medical image compression based on automated ROI selection using fast discrete curvelet transform with adaptive arithmetic coding, lifting wavelet transform with set partitioning embedded block coding. This hybrid approach increases compression ratio and reduces information loss [2].
2. Bruckmann and Uhl [3] proposed selective medical image compression using wavelet techniques. Main objective was here to compress explicitly defined region of interest in a lossless manner using wavelets algorithm and non-region of interest in lossy manner for efficient storage [3].
3. Sanchez et al. [4] proposed a compression algorithm compression based on HEVC intra-coding. This algorithm uses DCT and improves image compression in picture archiving and communication systems [4].
4. Mallat [5] proposed a wavelet-based algorithm which can resolve c1 singularities [5].
5. Anandan et al. [6] proposed an image compression algorithm using fast discrete curvelet transform. This algorithm achieves high compression ratio but could not curve discontinuities to a greater extent [6].

Research gap noticed through survey is that, in conventional methods such as Discrete Cosine Transform (DCT), Discrete Wavelet transform (DWT), curvelet, and JPEG transforms during compression and decompression of images, discontinuities across a curve (or 2D singularities) could not be preserved. Due to this, high-frequency components of images get affected. Thus, conventional transforms

Table 1 Performance of fast discrete curvelet transform

Quality parameter	In relative to DWT	
	Modality	Fast discrete curvelet transform Anandan et al. [6] (%)
PSNR	MRI	93.80
	CT	76.00
	USG	101.00
CR	MRI	207.02
	CT	212.40
	USG	469.00

could not preserve curve discontinuities to a greater extent. Table 1 presents performance of fast discrete curvelet transform based method proposed by Anandan et al. [6]. These performances are considered as benchmark and proposed method will be assessed against this method.

1.2 Medical Data Acquisition

Digital Imaging and Communication in Medicine (DICOM) datasets from various diagnostic centers are collected. They include Computed Tomography (CT), Magnetic Resonance Imaging (MRI), X-ray, Ultrasound Sonography (USG), Positron Emission Tomography (PET), and PET-CT datasets. These datasets will be used for experimentation. Table 2 presents an overview of datasets collected.

Table 2 Medical data acquisition

Modality	Avg. image size (KB)	Resolution	Avg. HU at ROI	Avg. count of images	Gender	Age
CT	200	(512×512)	21.81	3000 (10 cases)	M-10 F-4	41
MRI	332	(512×512)	259.49	348 (1 case)	F-2	41
X-RAY	29427	(512×512)	485.73	15 (10 cases)	F-2 M-8	37
USG	226	(512×512)	—	100 (10 cases)	F-10	22
PET	58	(512×512)	595.17	198 (1 case)	M	65
PET-CT	516	(512×512)	71.75	2633 (1 case)	M	65

Table 3 Quality metrics

S. no.	Type	Description	Formula
1	Mean square error (MSE)	MSE is the difference between original and compressed image	$MSE = \frac{1}{mn} \sum_{i=0}^{m-1} \sum_{j=0}^{n-1} [I(i,j) - k(i,j)]$ (1)
2	Peak signal-to-noise ratio (PSNR)	PSNR is the ratio between highest signal power to the noise power	$PSNR = 10 * \log_{10}\left(\frac{255^2}{MSE}\right)$ (2)
3	Compression ratio (CR)	CR is ratio of original image to the compressed image	$CompressionRatio = \frac{originalSize}{CompressedSize}$ (3)
4	Response time	It is the total time taken to respond to a request	Calculated using Matlab inbuilt functions TIC and TOC

2 Performance Measuring Parameters

In this section, performance of the proposed method is assessed by the following parameters. Table 3 presents the quality metrics where $I(i,j)$ represents original image and $K(i,j)$ represents transformed image.

3 Proposed Methodology

Ripplet transform proposed by Jun Xu et al. is an extension of curvelet transform which is designed to resolve the limitations of traditional transforms which was discussed in Sect. 1.2 [7]. Proposed method is capable of resolving C2 singularities by preserving edges at different scales and directions.

The proposed method generalizes curvelet transform by adding two additional parameters such as Support "c" and Degree "d". For c = 1 and d = 1, anisotropy property is not preserved since it has linear scaling. For d = 2 and d = 3, ripples have parabolic and cubic scaling. These new parameters provide with anisotropy property of preserving edges at different scales and directions. Ripplet transform is given by

$$\rho_{\vec{ab}\theta} = \rho_{\vec{a00}}\left(R_\theta\left(\vec{x} - \vec{b}\right)\right) \tag{4}$$

where the $R_\theta = \begin{bmatrix} cos\theta & sin\theta \\ -sin\theta & cos\theta \end{bmatrix}$ is used for rotation. $\vec{x}$ and $\vec{b}$ are 2D vectors. $\rho_{\vec{a00}}(.)$ is the main function. $\rho_{\vec{ab}\theta}$ defines the ripplet functions because in spatial domain it produces ripple-like shapes [1, 7].

Algorithm and mathematical background for compression and decompression are presented below.

3.1 Algorithm for Compression

1. Medical image $g(i,j)$ of size 512×512 is made as a input to the algorithm.
2. Decomposed medical image into set of low- and high-frequency components using 9/7 wavelet filter. $g(i,j) = (p_o g(i,j), \Delta_1 g(i,j), \Delta_2 g(i,j) \dots)$ where $p_o g(i,j)$ is the low-frequency component and $(\Delta_1 g(i,j), \Delta_2 g(i,j) \dots)$ is high-frequency component.
3. Normalize the resulting components.
4. Allow resulting high-level components to ripplet domain. Continuous ripplet function is defined as [1, 7]:

$$R_{(a,\vec{b},\theta)} = \left\langle f, \rho_{\vec{a00}} \right\rangle = \int f(\vec{x}) \overline{\rho_{\vec{a00}}(\vec{x})} d\vec{x} \tag{5}$$

where $R_{(a,\vec{b},\theta)}$ are ripplet coefficients and $f(\vec{x})$ represents high-level component.
5. Apply Huffman encoder to high- and low-level components.
6. Assess the performance of algorithm using the parameters mentioned in Sect. 2 [1, 7].

3.2 Algorithm Decompression

1. Decode the resulted coefficients using Huffman decoder.
2. Apply inverse ripplet transform and is given by [1, 7]:

$$R_{\vec{j},k,l} = \sum_{i=0}^{M-1} \sum_{j=0}^{N-1} \overline{\rho_{\vec{j},k,l}(i,j)} \tag{6}$$

Our proposal is different from recently reported ripplet transform based technique [7] in terms of simplified normalization and smoothening by median filter.

4 Results and Discussions

In this section, we present experimental results that demonstrate the better ability of ripplet transform.

A. Spatial Filtering

It is a technique used for enhancement of image quality as well as to remove noise in spatial domain. Here, original image of size 512×512 is filtered to get smoothed image. Filtered image presents edge enhancement and smoothing as shown in Fig. 1.

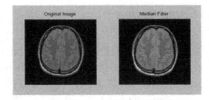

Fig. 1 Filtered image

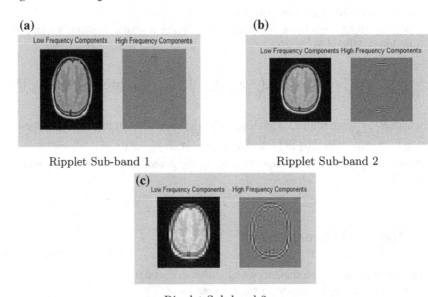

Ripplet Sub-band 1   Ripplet Sub-band 2

Ripplet Sub-band 3

Fig. 2 Ripplet transformed images

B. Ripplet transform applied at different levels of images

Multi-resolution analysis of images is very important concept to preserve singularities along arbitrarily shaped curves. Here, Laplacian transform is applied on high-level components of images to preserve edges with low energy. Image of 512×512 is divided into three levels each level consisting of reduced image size compared to the original size. Figure 2a represents ripplet sub-band 1 which consists of low- and high-level components of image. To get ripplet sub-band 2, low-level component of previous sub-band is again divided into low- and high-level components by applying transform. Figure 2b depicts ripplet sub-band 2. Similar step is applied to obtain ripplet sub-band 3 and is shown in Fig. 2c.

Fig. 3 Inverse transformed image

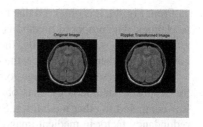

After multi-resolution analysis of image, low- and high-level components of image are sent to Huffman encoder to improve the compression performance resulting in compressed size of 8 kb. Figure 3 represents reconstructed image of size 128 kb (original size) after applying inverse transformations.

4.1 Experiment 1: Performance Analysis

In this section, we present performance analysis of proposed method considering medical images acquired over different modalities. All medical images as discussed in Sect. 2, are subjected to compression and decompression by the proposed method. In each run, response time, compression ratio as well as quality assessing parameter like PSNR are recorded. Figure 4a presents the average response time consumed for both compression and decompression.

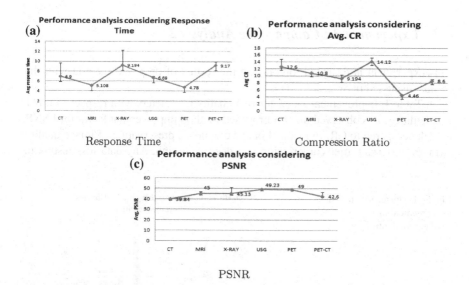

Fig. 4 Performance analysis

From Fig. 4a, it is observed that response time for PET is least compared to other modalities. The proposed method is edge oriented, and this modality has very less number of edges. Consequently, the ripplet transform performs less iterations as compared to other cases. Hence, the response time is less in this case.

Figure 4b presents the average compression ratio achieved for different modalities for proposed methodology. It is observed that compression ratio for PET is very low compared to other modalities. Since the extent of compression depends on the redundancy factor in medical image, the CR in PET images is less. In case of PET images, pixel values are distinct and of varying in nature.

Figure 4c presents the average PSNR achieved for different modalities for proposed methodology. It is observed that PSNR values range from 40–50 dB. This is due to the fact that the proposed algorithm performs multi-resolution analysis of images and is capable of capturing singularities across multiple directions.

4.2   Experiment 2: Comparative Analysis

This experiment is designed to evaluate the proposed method against conventional methods, using various medical images, which are acquired by different modalities. From Fig. 5 it is observed that compression ratio for proposed methodology is outperformed as compared to other conventional methods. Because the proposed methodology performs in-depth analysis and normalization, it achieves high CR compared to other conventional methods.

4.3   Experiment 3 : Comparative Analysis 2

This experiment is designed to compare the results of proposed method with the results discussed in Sect. 2.

In Table 4, improvement factor for two quality parameters with respect to DWT is calculated. It is observed that for proposed method improvement factor for PSNR is high compared to CR. In medical image diagnosis, preservation of image quality is more important after compression and decompression. Any data loss results in

Fig. 5 Performance analysis using medical images of different modalities

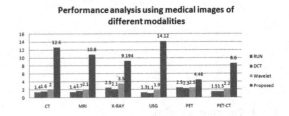

Performance analysis using medical images of different modalities

incorrect diagnosis. Since the proposed method is edge oriented and preserves edges to a greater extent, the quality of image is preserved.

4.4 Experiment 4: Qualitative Analysis (Validation)

To evaluate quality of reconstructed image, subjective analysis by domain experts is carried out and assessment metric is presented in Table 5.

Figure 6 presents the score with respect to sharpness, noise presence, artifacts, and diagnostic acceptability reconstructed images for proposed methodology. It is observed that subjective evaluation is highest for CT and PET-CT images and least for PET images.

Table 4 Comparative analysis

Quality parameter	In relative to DWT		
	Modality	Fast discrete curvelet transform Anandan et al. [6] (%)	Proposed method (%)
PSNR	MRI	93.80	76.12
	CT	76.00	**106.74**
	USG	101.00	**106.07**
CR	MRI	207.02	**342.00**
	CT	212.40	200.30
	USG	469.00	259.30

Table 5 Assessment metric

Qualitative grading scale	Image quality			
	Sharpness	Noise	Artifacts	Diagnostic acceptability
1	Blurry	Unacceptable	Highly present	Unacceptable
2	Lesser than average	Lesser than average	Lesser than average	Suboptimal
3	Average	Average	Average	Average
4	Better than average	Better than average	Better than average	Better than average
5	Sharp	Minimum or no noise	Absent	Superior

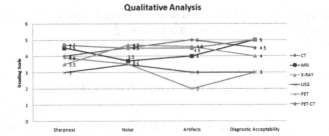

Fig. 6 Assessment metric

5 Conclusions and Future Scope

In this paper, an attempt is made to compress medical images, which were acquired by various modalities and achieve highest compression ratio as well as preserve diagnostically important information. The objective is met through systematic design and development of proposed algorithm. The response time of proposed algorithm for compression and decompression is quite appreciable.

Experimental results revealed that proposed methodology provides more efficient presentation of images while preserving 2D singularities along the curves by doing multi-resolution analysis of the image. It also provides an appreciable compression ratio compared to conventional transforms through normalization.

To further improve the performance of ripplet-based image compression, one can experiment with different transformations and filters to preserve edges of smooth areas (i.e., low-level components) and also levels for multi-resolution shall be based on the intensity of edges in image.

References

1. Juliet, S., Rajsingh, E.B., Ezra, K.: A novel medical image compression using ripplet transform. J. Real-Time Image Process. **11**(2), 401–412 (2016)
2. Thomas, D.S., Moorthi, M., Muthalagu, R.: Medical image compression based on automated ROI selection for tele medicine application. Int. J. Eng. Comput. Sci **3**, 3638–3642 (2014)
3. Bruckmann, A., Uhl, A.: Selective medical image compression using wavelet techniques. CIT J. Comput. Inf. Technol. **6**(2) 203–213 (2015)
4. Sanchez, V., Bartrina-Rapesta, J.: Lossless compression of medical images based on HEVC intra coding. In: IEEE International Conference on Acoustics, Speech and Signal Processing (ICASSP). IEEE (2014)
5. Mallat, S.: A Wavelet Tour of Signal Processing, 2nd edn. Academic Press, New York (1999)
6. Anandan, P., Sabeenian, R.S.: Medical image compression using wrapping based fast discrete curvelet transform and arithmetic coding. In: Circuits and Systems vol. 7, no. 08, p. 20–59 (2016)
7. Xu, J., Yang, L., Wu, D.: Ripplet: a new transform for image processing. J. Vis. Commun. Image Represent. **21**(7), 627–639 (2010)

Fast Feature Extraction on Graphic Processing Unit for a Video Sequence

Krunal Randive and M. Sridevi

Abstract This paper presents an efficient feature extraction method which uses Graphic Processing Unit (GPU) for digital three-dimensional (3D) reconstruction for the captured real-time video sequence. 3D reconstruction system is a time-consuming process, because it has to perform various intermediate processes such as feature extraction, feature correspondence, projective reconstruction, and image registration. The feature extraction is the vital step in the 3D reconstruction process. It is demonstrated that the use of GPU-based stream processing model can significantly accelerate the process of feature extraction of 3D reconstruction on the set of frames. Frames in the video are processed, and features are extracted from each frame using various feature extraction methods. The techniques are implemented on both CPU and GPU. The computation time for feature extraction is calculated. The experimental results are demonstrated to show the performance in terms of accuracy and speed.

Keywords Three-dimensional reconstruction · Feature extraction · Graphic processing unit · Computing time

1 Introduction

Three-dimensional reconstruction is a very hot topic for research in engineering. The procedure for generating a computerized or digital model of an object or any terrain that graphically represents buildings and other objects is called as 3D reconstructed models. These reconstructed scenes provide the experience of virtual reality, where

K. Randive (✉)
Department of Computer Science and Engineering, Indian Institute
of Information Technology, Tiruchirappalli, India
e-mail: krunalrandive@gmail.com

M. Sridevi
Department of Computer Science and Engineering, National Institute of Technology
Tiruchirappalli, Tiruchirappalli 620015, Tamil Nadu, India
e-mail: msridevi@nitt.edu

© Springer Nature Singapore Pte Ltd. 2018
P. K. Sa et al. (eds.), *Recent Findings in Intelligent Computing Techniques*,
Advances in Intelligent Systems and Computing 709,
https://doi.org/10.1007/978-981-10-8633-5_47

481

users feel involved with the scene because of the perception of depth in 3D images which makes reconstructed scene interactive. The process of 3D reconstruction can be divided into four stages, namely feature extraction, feature correspondence, projective reconstruction, and image registration. In the first stage, from a set of images, candidate features are extracted. The second stage involves feature correspondence by using the extracted features [1]. The third stage uses projective reconstruction which retrieves the camera position and 3D position of the matched feature points up to a projective transformation. The final stage applies image registration to get the 3D-reconstructed model.

In the process of 3D reconstruction, feature extraction has the major role. Feature extraction is applied for dimensionality reduction. Features present in image include blobs, edges, ridges, and corners. Feature extraction from multiple view images may obtain redundant data. Hence, candidate point features are chosen. The process of feature extraction becomes cumbersome and takes more time with the increase in a number of images. The problem can be solved by a fast feature extraction technique. GPUs are used in arithmetic-intensive operations having significant parallelism [2]. Hence, the burden of CPU can be mitigated to GPU by extracting parallel arithmetic-intensive computation.

This paper focuses on the accelerated feature extraction using the stream processing model based on GPU. The proposed method reduces the computation time of feature extraction process from the given set of images or a video sequence.

2 Related Work

Feature extraction plays influential role in image processing applications such as object recognition, 3D reconstruction, analysis and classification, artificial intelligence, medical imaging, agriculture and computer vision, etc. and explored by many researchers [1–7]. In [5], different feature extraction techniques were adopted to obtain a comparison protocol of various classification techniques. The result conveys that the more precisely the features are extracted, the more is the accuracy rate of the classification technique.

Image is divided into optimal subareas using percentile-based approximation for faster processing [8]. In [9], the implementation of Gabor filter for the texture feature extraction with separable filters provides the improvement in the performance and minimizing the error. Majority of existing approaches cannot distinguish similar actions in chaotic data. Hence, for feature extraction, the Gabor filter was used and it is also sensitive to distortion and noise.

In [7], the author proposes a modified SIFT algorithm, Affine-SIFT to extract 3D point cloud of the image and achieved the real shape of the object using triangulation algorithm. The time complexity of this algorithm is same as the order of magnitude as the SIFT algorithm uses. In [10], the author proposes a new feature extraction method to address the limitations of the features extracted from of the low-resolution image and that cannot adjust to changes in image direction and came up with a multiple extracted feature based new global optimization framework.

To overcome the problems, this paper proposes faster feature extraction method using GPU instead of CPU. The remainder of the paper is organized as follows: Sect. 3 discusses the proposed method. Experimental results and performance analysis are discussed in Sect. 4. Section 5 concludes the paper.

3   Proposed Method

The methods used in the literature are very much sensitive to the slight distinction between the images. Local features can enable feature-based algorithms to handle scale changes, rotation, and occlusion in a better manner. FAST, Harris, and BRISK methods are applied for detecting corner features. SURF and SIFT methods are used to detect blob features, and Histogram of Oriented Gradients (HOG) method is used for object detection. Various feature extraction methods are explained below.

Harris corner detector algorithm detects corners, and a local correlation function is used to detect changes in shifted patterns. Harris uses a formula to calculate the change in pixel values in any direction, rather than just eight fixed directions [6]. Speed Up Robust Feature (SURF) features are used in 3D reconstruction, object recognition, and classification. The interest points in SURF are detected by using the determinant of Hessian blob detector. Its feature descriptor depends on the point of interest which is the sum of the Haar wavelet response. In three feature detection steps, i.e., matching, description, and detection, SURF was believed to ensure high speed. SIFT and SURF algorithms employ a slight difference in the way of detecting its features. SIFT builds an image pyramid, filtering each layer with Gaussians of increasing sigma values and taking their difference as mentioned in [11].

Binary Robust Invariant Scalable Keypoints (BRISK) uses Hamming distance instead of Euclidean distance. In a specific sampling pattern, a limited number of points are used by the descriptor bitstream built by BRISK [4]. Features from Accelerated Segment Test (FAST) method detects corner features used for tracking and mapping objects. This method can be used for real-time video processing due to its high-speed performance [4]. FAST algorithm is based on the Smallest Univalue Segment Assimilating Nucleus (SUSAN) corner criterion. Corner detection is prioritized over edge detection [11]. The Histogram of Oriented Gradients (HOG) is used for object detection. HOG is a dense feature extraction method as opposed to SIFT which is restricted to local neighborhood of key points.

3.1   Parallel Feature Extraction on GPU

The process of 3D reconstruction itself is time-consuming and not all the stages in reconstruction are parallelizable. Parallelizing the process of feature extraction will

Fig. 1 GPU-based SIMD
streaming of feature
extraction techniques

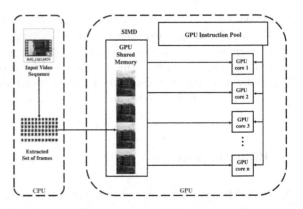

reduce the time taken for extracting the candidate feature points. The first step in constructing the parallel feature extraction is to upload a set of input images to the GPU. The GPU shared memory is based on Single Instruction Multiple Data (SIMD) parallel hardware architecture. The preprocessing of data is performed on a per-frame level in the form of intensive arithmetic operations. It can be handled by the GPU. The feature extraction process discussed in the above sections has the operations like applying Gaussian filters on the grayscale images, applying convolution, transformations, etc. When implementing these feature-based descriptors on GPU, several GPU operations are clump together. That is, once all the image frames are uploaded to the GPU, as many tasks as possible are performed in parallel on the GPU before transferring back to CPU. Because there is significant overhead involved in copying data from CPU to the GPU and vice versa, images in the GPU are stored in the GPU array. These GPU arrays are given as an argument to the feature-based descriptors like Harris, SURF, BRISK, FAST, and HOG. Steps involved in the proposed fast feature extraction from the video using GPU are shown in Fig. 1.

Video frames are created using frame extraction process. Frame Rate or Frames Per Second (FPS) is the characteristic feature of any video camera. Frames are the unique consecutive images produced by the camera device at a particular frequency (or rate). These set of frames are used for extracting the corner features which will be used for 3D reconstruction. The implementation is performed in two scenarios, CPU sequential computing of the features and GPU parallel computing to extract candidate features. In both these scenarios, the computation time for feature extraction is calculated and their performance is analyzed.

4 Results and Discussion

The experiment is conducted to examine the computational speedup of the various feature extraction methods for a video sequence using GPU. The experimental results of the proposed method and the performance analysis are explained in Sects. 4.1 and 4.2, respectively.

4.1 Experimental Result

The proposed method is implemented and the experiment was carried out with the following setup. (1) The real-time video is captured using mobile camera (iPhone SE) of 12 megapixel resolution for image 1080p HD at 30 fps. (2) It is implemented on intel core i5 processor with 4 GB RAM and Nvidia GeForce 410M GPU chip. (3) The experimental setup includes CUDA Toolkit 6.5 and MATLAB R2016b. The recorded video of the scene was given as an input to the proposed method to perform the frame extraction process. The video recording of the scene is given as the input to the frame extraction, and set of frame is extracted from the video sequence instead of all the frames in video. The consecutive image frames have more overlapping areas. Hence, the image frame after some interval was selected. The candidate point feature of from every image is extracted using various methods like Harris, SURF, BRISK, FAST, and HOG feature descriptor. This step of frame processing is implemented on both CPU and GPU to demonstrate the performance of the proposed method. The result of frame processing is shown in Fig. 2.

4.2 Performance Analysis

Two scenarios are explored for comparison of computational speed feature extraction methods. The computational time of each scenario is measured, namely GPU parallel computing and CPU sequential computing. The execution time is considered as the measure for computation time for the extraction of candidate point feature on CPU, in the CPU sequential computing as well as in GPU parallel computing on 48 cores of Nvidia GPU.

Table 1 shows the speedup of GPU parallel computing over CPU sequential computing, measured using the computation time for extraction of candidate point fea-

Fig. 2 Extracted candidate point features for **a** frame no. 48, **b** frame no. 55

Table 1 Speedup of GPU parallel processing over CPU sequential processing

Sr. no.	n	No. of frames	Speedup (approx.)				
			Harris	SURF	BRISK	FAST	HOG
1	14	61	1.61	1.40	1.51	1.70	1.24
2	12	71	2.25	1.30	2.03	1.40	2.85
3	10	85	2.80	1.58	1.98	1.80	3.82
4	8	107	2.85	3.25	5.00	3.53	4.34
5	6	142	4.74	5.25	7.70	3.91	5.24

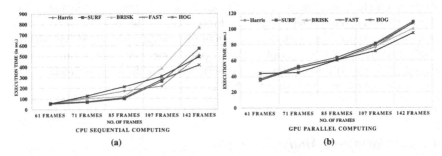

Fig. 3 **a** Performance of the feature extraction techniques on CPU, **b** Performance of the feature extraction techniques on GPU

tures from the set of frames using GPU parallel and CPU sequential computing, respectively. The frames are varied from 6 to 14 in this experiment. The speedup based on the execution time for all the different feature descriptor methods is calculated as shown in Eq. 1.

$$Speedup = \frac{Execution\ time\ for\ CPU\ sequential\ processing\ (L_{old})}{Execution\ time\ for\ GPU\ parallel\ processing\ (L_{new})} \qquad (1)$$

To help with visualization of the relationship between numbers of frames and execution time, Fig. 3a, b presents a comparison diagram for both the CPU sequential and the GPU parallel computing, respectively. From Fig. 3a, it is inferred that for all feature extraction methods sequential computing time increases with increase in number of frames. It is observed that the HOG, Harris, and FAST descriptors give better performance on both CPU and GPU from all the feature extraction techniques. But, HOG feature descriptor is used for object detection. Hence, it may not give best corner features. On the other hand, Harris corner and FAST feature descriptors are mainly used for corner feature extraction. It is observed from Fig. 3b that GPU parallel computing performs significantly better than CPU sequential computing.

The performance analysis is done based on the CPU sequential computation time and the GPU parallel computation time for Harris and FAST as shown in Fig. 4.

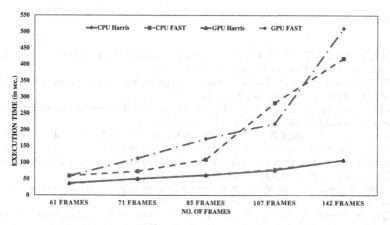

Fig. 4 Performance comparison of Harris and FAST method on CPU and GPU

It is inferred that the Harris corner feature descriptor takes more time on CPU for computation when compared to FAST on CPU. But, it is also observed that the GPU computation time for Harris and FAST feature descriptor is almost same and low with respect to the CPU sequential time for both Harris and FAST feature descriptors as shown in Fig. 4.

5 Conclusion

A parallel computing method based on GPU for fast feature extraction in the 3D reconstruction was proposed. The streaming version of the feature extraction for 3D reconstruction for video sequence is implemented and executed on GPU chip. The method is examined and compared with various feature extraction techniques, namely Harris, SURF, BRISK, FAST, and HOG. The execution times for sequential- and parallel-based feature extractor are calculated by varying the number of frames. It is observed that for Harris and FAST feature descriptor, performance can be significantly improved by using parallel computing method than sequential method. The proposed fast feature extraction method is applicable in real-time 3D reconstruction and various applications in image processing. However, the proposed method based on GPU has significant overhead involved in copying data from CPU to GPU and from GPU to CPU. In future, the parallel method can be adapted to other stages such as feature correspondence and image registration.

References

1. Han, B., Paulson, C., Wu, D.: Depth-based image registration via three-dimensional geometric segmentation. IET Comput. Vis. **6**(5), 397–406 (2012)
2. Yi, F., Moon, I., Lee, J.-A., Javidi, B.: Fast 3D computational integral imaging using graphics processing unit. J. Disp. Technol. **8**(12), 714–722 (2012)
3. Li, C., Chen, Q., Hua, H., Mao, C., Shao, A.: Digital three-dimensional reconstruction based on integral imaging. Opt. Rev. **22**(3), 427–433 (2015)
4. Verina, A.B., Yuvaraju, B.N.: Feature extraction techniques for video processing in MATLAB. Int. J. Innov. Res. Comput. Commun. Eng. **4**(4), 5292–5296 (2016)
5. Medjahed, S.A.: A comparative study of feature extraction methods in images classification. Int. J. Image Graph. Signal Process. (IJIGSP) **7**(3), 16–23 (2015)
6. Bheda, D., Joshi, M., Agrawal, V.: A study on feature extraction techniques for image mosaicing. Int. J. Innov. Res. Comput. Commun. Eng. (IJRICCE) **2**(3), 3432–3437 (2014)
7. Qun, W., Zhaohe, X., Jue, W.: Research and implementation of 3D reconstruction technique based on images. In: 2nd International Conference IEEE Information Science and Control Engineering (ICISCE), pp. 408–411 (2015)
8. Eriksson, E., Dn, G., Fodor, V.: Real-time distributed visual feature extraction from video in sensor networks. Proc. IEEE Int. Conf. Distrib. Comput. Sensor Syst. 152–161 (2014)
9. Pang, W.-M., Choi, K.-S., Qin, J.: Fast gabor texture feature extraction with separable filters using GPU. J. Real-Time Image Process. **12**(1), 5–13 (2013)
10. Zhang, L., Wang, X., Li, C., Li, Q., Li, X.: One new method on image super-resolution reconstruction. In: 2015 IEEE International Conference IEEE Information and Automation, pp. 79–82 (2015)
11. El-gayar, M., Soliman, H.: A comparative study of image low level feature extraction algorithms. Egypt. Inf. J. **14**(2), 175–181 (2013)

Prediction of Human Mobility Using Mobile Traffic Dataset with HMM

Anshika Rawal and Pravati Swain

Abstract The connectivity and usability of cellular communications influencing to infer the information about user mobility based on the mobile traffic data. Thus, mobile traffic data can be used to analyze human location histories and prediction of human mobility. This paper proposes a framework for predicting human mobility which is based on Hidden Markov Models (HMMs). First, locations are clustered according to their characteristics such the highest traffic generated in a particular location, in certain time period. Then, the proposed HMM will be trained by the generated clusters. The usage of HMMs empowers to deal with spatiotemporal data, location characteristics, and possible visited states which are called the observable states.

Keywords Human mobility prediction · Hidden Markov Model · Mobile traffic dataset

1 Introduction

The research on human mobility patterns and prediction of human mobility trajectories are important in traffic planning, epidemic modeling, and mobile computing. Human mobility patterns produce some fundamental rules that govern the prediction of next movement. Here, one simulative or mathematical model needs to be constructed to reproduce such patterns, so that prediction can be done with adequate accuracy. There are some crucial factors which derive the fundamental laws for prediction of human mobility such as *visited locations*, *travel distance*, *spatiotemporal regularity*, result of mobile data analysis determines for how long a person been

A. Rawal (✉) · P. Swain
Department of Computer Science and Engineering, National Institute
of Technology Goa, Ponda, India
e-mail: parulrawal47@gmail.com

P. Swain
e-mail: pravati@nitgoa.ac.in

© Springer Nature Singapore Pte Ltd. 2018
P. K. Sa et al. (eds.), *Recent Findings in Intelligent Computing Techniques*,
Advances in Intelligent Systems and Computing 709,
https://doi.org/10.1007/978-981-10-8633-5_48

in a spatial locality and helps in the predictability, and *Predictability*, the strong regularity of human mobility patterns helps to inferred that how easy to predict individual movements and how many times an individual is going to visit that place and when? There are several methodologies that exist to analyze the location history. All the methods are categories into three different parts, namely state-space models, template matching, and data mining technique. Hidden Markov Model (HMM) is one state-space approach which is based on the spatiotemporal data model. In the case of human mobility prediction, HMM supports the estimation of transition probability to possible future visited locations. Hence, HMM is a better approach to predicting human mobility. This paper proposes a framework for predicting human mobility which is based on HMMs using the large-scale mobile traffic dataset.

2 Related Work

In this section, we review the quite extensive literature on the dependability of mobile traffic data as a source of mobility information. Mobility models are derived from the mobile data which contributes to a various number of domains such as extracting users' location patterns from location history, urban planning, and traffic engineering [1]. In the following, we review most relevant works that leverage mobile data to study human mobility. In Paper [2, 3], GSM dataset has been used for predicting human mobility on the basis of Hidden Markov Models (HMMs). They have stated that keeping track of location by GPS is not feasible because of its battery consumption and spurious signal problem. However, the impact of time parameter on the prediction model is missing. The paper [8] proposes a social relationship-based mobile node location prediction algorithm using daily routines. After considering users' dynamic behavior resulting from their different daily schedules, they have proposed algorithm to predict users' mobility in different daily time periods. Then, prediction results are amended using other users' location information which has strong relationship to that particular user. In paper [4], Sebastien et al. have addressed the issue of predicting the next location of an individual based on the observations of his mobility behavior over some period of time and the recent locations that he has visited using Mixed Markov Chains (MMC). This work has several potential applications such as the evaluation of geo-privacy mechanisms, the development of location-based services anticipating the next movement of a user, and the design of location-aware proactive resource migration.

3 Problem Statement

This paper presents a probabilistic model to predict human mobility using large-scale mobile traffic data to estimate the transition probability of an individual or aggregate of people from the current location to future location by capturing the mobile traffic data. The proposed framework is divided into two phases where in the first phase, the geographical locations are clustered according to their characteristics such as the highest traffic generated in a particular time period. In the second phase, the proposed HMM will be trained by the generated clusters. The usage of HMMs empowers to deal with spatiotemporal data, location characteristics, and possible visited states which are the observable states.

4 Proposed Framework

In the proposed framework, the mobile traffic data are classified into some optimal number of clusters based on the method proposed by Marco et al. [5] with some required modification. The cellular users are classified into different clusters based on their cellular traffic generation patterns, i.e., use of cellular data or call-in and call-out activities. The resultant clusters are the input to the prediction model and are considered as hidden states, and the possible visited locations are the observable states of the proposed HMM.

4.1 Classifying Call Profiles

In the given dataset, the attribute called square ID represents the geographical location. One particular city is fitted in a grid, and each cell is referred to a particular location associated with unique square ID. To make the proposed framework more robust, the low traffic generated locations are merged with adjacent location. If the traffic generated by the adjacent location is below the threshold value, then repeat the cluster merging process called as the location filtering. After location filtering, the resultant locations are identified by the unique square ID. The resultant dataset is transformed into a graph $G = (V, E)$ where each location (identified by square ID) is represented by vertices $V = s_1, s_2, s_3 \ldots s_n$, s_i represents the ith square ID. Each pair of (s_i, s_j) is connected by an edge e_{ij}. Each edge e_{ij} associated with a weight w_{ij}. Here, w_{ij} represents the similarity between the vertices s_i and s_j. There are two different types of traffic volume similarity measure exist [5] such as traffic volume similarity in equation and traffic distribution similarity in equation. The weight w_{ij} is calculated using the traffic volume similarity Algorithm 1.

Here, Z is the set of locations and V_i^t is the volume of traffic at ith location at time period t.

Clustering: The generated weighted graph in Fig. 1 is the input to the hierarchical clustering algorithm which gives the optimal number of clusters by using Calinski and Harabasz index or Beale index [5]. Initially, each vertex is assigned to a different cluster C_i and, at each step, we calculate the distance between two clusters C_i and C_j, i.e., d_{ij}. Then, merge the pair of cluster which possesses the minimum distance. Here, the distance is calculated as

Algorithm 1 Traffic volume similarity:

1: To measure the similarity between two distinct locations s_i and s_j at a particular time t. The weight function is
2: **if** (i == j) **then**
3:    $w_{ij} = 1$;
4: **else**
5:    $w_{i,j} = \dfrac{1}{\sqrt{\sum_{i,j \in Z} (v_i^t - v_{j}^t)^2}}$
6: **end if**

(a)   **(b)**

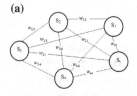   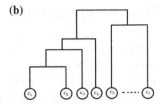

Fig. 1 **a** Generated weighted graph. **b** Dendrogram of clusters

Fig. 2 Discrete HMM with
n states and n possible
outputs

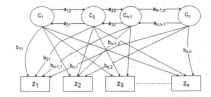

$$d_{ij} = \frac{1}{|C_i||C_j|} \sum_{i \in C_i, j \in C_j} \frac{1}{w_{ij}} \qquad (1)$$

Until we get optimal number of clusters, the above merging process is repeated which is presented in Fig. 1.

4.2 Prediction Model

The last phase of framework is the prediction model. Here, the locations are clustered based on the highest traffic generation among all other locations at a certain time period (t). It is noted that a particular location can be considered in more than one cluster based on the traffic generation at different time periods. In this proposed prediction model, the whole day is divided into different time intervals because of user's current and future visits are highly related to on time intervals based on their daily routines. Unlike the papers [6, 7], only the location has been considered as the parameter for the prediction model. The behavior in a particular time period explains 50–70% of human movements [8] such as the same set of user's behavior in morning time are different in evening time. Hence, the various time periods should be considered in the prediction model to predict the most possible visited state which is more realistic results. This spatiotemporal dataset can be modeled using HMM, efficiently.

HMM is most suited approach for the analysis of time series data, in which the sequences of observation states are assumed to be generated by a Markov process with unobserved or hidden states [9]. In the proposed HMM, clusters which are gained from classification are hidden states and the locations that can be visited are considered as visible states. Each hidden state has probability distribution for transition from one hidden state to another hidden state as well as over the possible location to be visited.

Figure 2 depicts the architecture of the proposed hidden Markov model. The random variable $C(t)$ represents the cluster, i.e., the set of locations generates the highest traffic at a particular time t, $C(t) = \{C_1, C_2, \ldots, C_n\}$ and $t = 1, \ldots n$. Similarly, the random variable $Z(t)$ represents the observed state or possible visited locations at time t, $Z(t) = \{Z_1, Z_2, \ldots Z_n\}$ and $t = 1, \ldots, n$. The proposed prediction model satisfies the Markov property, i.e., the process will be in state C_t at time period t which only based on the probability that the process in the state C_{t-1} at time period $t - 1$, because the mobile user's mobility is based on their daily routines with corresponding time periods. Similarly, $Z(t)$ depends only on $C(t)$ at time period t.

$$P(Location_{future}) = \alpha \pi_t^{t+1} P(Z_{t+1}|C_{t+1}) \cdot P(C_{t+1}|C_t) \cdot P(C_t|Z_t) \qquad (2)$$

Here, $Location_{future}$ represents the most probable visited location at time instant $t + 1$ where $P(C_t \mid Z_t)$ is a *decoding* problem that can be stated as given a present location Z_t (visible symbol) how do we choose a corresponding hidden state C_t. The term $P(Z_{t+1}|C_{t+1}) \cdot P(C_{t+1}|C_t) \cdot P(C_t|Z_t)$ is an evaluation problem, where present hidden state is given and we have to compute efficiently the next most probable location (visible symbol). $P(C_{t+1} \mid C_t)$ is probability of transition between hidden state and $P(C_t \mid Z_t)$ is the probability that process will be in hidden state C_t, while the process emits the visible symbol Z_t at time instant t. The $P(Z_{t+1} \mid C_{t+1})$ gives the probability that the process being emitted Z_{t+1} while in the hidden state C_{t+1} at time instance $t + 1$. The α is a constant. Suppose there are n number of clusters; then, the probability of transitions is shown in the following matrix:

The initial probability distribution of Markov chain can be given as $P(0) = (P_1, P_2, P_3, \ldots P_n)$. If C_3 is initial state, then $P(0) = (0, 0, 1, 0 \ldots 0)$ and at time $t + 1$ and the probability distribution will be $P(t + 1) = (P_1^{t+1}, P_2^{t+1}, P_3^{t+1}, \ldots, P_n^{t+1})$.

The whole day has been divided into four periods such as morning, noon, evening, and night. In the proposed HMM, there are four hidden states based on the traffic generated for different locations at a particular time period, i.e., C_1, C_2, C_3, and C_4, respectively. If the visible state Z_t is given, at present time t, then probabilities of hidden state can be calculated as $P(C_i \mid Z_i) = \frac{load_t}{Total_{load}}$. The load t represents that the traffic load generates at that particular hidden state, for all $i \in \{1, 2, 3, 4\}$, and $Total_{load}$ is the total traffic generated in the whole day. By using the forward algorithm, select the $\max\{ P(C_i \mid Z_t) \}$ for all $i \in \{1, 2, 3, 4\}$. After getting most probable hidden state at time t+1, probability of possible locations can be calculated as $P(Z_{i+1} \mid C_{i+1}) = \frac{load_{i+1} - load_i}{Total_{load}}$. Here, load t + 1 is traffic generated at the hidden state C_{i+1} and load t is traffic generated in cluster C_i, by using forward algorithm for evaluation problem in which location possesses the highest probability that would be considered as most probable future location Z_{t+1}.

5 Conclusion

This paper proposed a framework with hidden Markov model to predict the human mobility using the mobile traffic data. The literature survey on mobile traffic analysis has been presented and captured the relationship between human mobility with mobile traffic data. The location flittering has been embedded in the existing classification of the call profiles, which gives the number of clusters from the large-scale mobile traffic data. After using of hierarchical algorithm, the optimal numbers of clusters are input for the human mobility prediction model. In the proposed framework, HMMs is going to serve as prediction model.

References

1. Furno, A., Stanica, R., Fiore, M.: A comparative evaluation of urban fabric detection techniques based on mobile traffic data. In: Proceedings of the 2015 IEEE/ACM International Conference on Advances in Social Networks Analysis and Mining 2015, pp. 689–696. ACM (2015)
2. Mathew, W., Raposo, R., Martins, B.: Predicting future locations with Hidden Markov Models. In: Proceedings of the 2012 ACM Conference on Ubiquitous Computing, pp. 911–918. ACM (2012)
3. Qiu, D., Papotti, P., Blanco, L.: Future locations prediction with uncertain data. In: Joint European Conference on Machine Learning and Knowledge Discovery in Databases, pp. 417–432. Springer, Berlin, Heidelberg (2013)
4. Gambs, S., Killijian, M.-O., del Prado Cortez, M.N.: Next place prediction using mobility Markov chains. In: Proceedings of the First Workshop on Measurement, Privacy, and Mobility, p. 3. ACM (2012)
5. Naboulsi, D., Stanica, R., Fiore, M.: Classifying call profiles in large-scale mobile traffic datasets. In: IEEE INFOCOM 2014-IEEE Conference on Computer Communications, pp. 1806–1814. IEEE (2014)
6. Gomes, J.B., Phua, C., Krishnaswamy, S.: Where will you go? mobile data mining for next place prediction. In: International Conference on Data Warehousing and Knowledge Discovery, pp. 146–158. Springer, Berlin, Heidelberg (2013)
7. Monreale, A., Pinelli, F., Trasarti, R., Giannotti, F.: Wherenext: a location predictor on trajectory pattern mining. In: Proceedings of the 15th ACM SIGKDD International Conference on Knowledge Discovery and Data Mining, pp. 637–646. ACM (2009)

8. Yu, R., Xia, X., Liao, S., Wang, X.: A location prediction algorithm with daily routines in location-based participatory sensing systems. Int. J. Distrib. Sens. Netw. (2015)
9. Ossama, O., Mokhtar, H.M.O.: Similarity search in moving object trajectories. In: Proceedings of the 15th International Conference on Management of Data, pp. 1–6 (2009)

Reckoning the Hatch Rate of Multivoltine Silkworm Eggs by Differentiating Yellow Grains from White Shells Using Blob Analysis Technique

R. N. Nikitha, R. G. Srinidhi, R. Harshith, T. Amar
and C. G. Raghavendra

Abstract There is an increasing need to concentrate on healthy production of silkworm eggs in order to cater to the worldwide demand for silk. Accurately predicting the hatch rate helps in this direction. To fulfil this requirement, we have proposed a novel image processing algorithm based on blob analysis that determines the hatch rate. The algorithm depends on detecting unhealthy eggs that are yellow in colour and counting them. This is achieved by extraction of only those pixel values pertaining to yellow colour and their subsequent enumeration. The proposed method gives satisfactorily accurate results. Knowing the hatch rate at an early stage will help in predicting the yield rate, which is a measure of the total silk produced.

Keywords Colour image processing · Colour extraction · Silkworm eggs
Sericulture and hatch rate

1 Introduction

Silk, one of the finest fibres used in the textile industry, is gaining popularity today in other fields. Constant research is taking place in the field of medicine, to use silk as a suture material in healing wounds and silk protein sericin for regenerative medicine

R. N. Nikitha (✉) · R. G. Srinidhi · R. Harshith · T. Amar · C. G. Raghavendra
Department of Electronics & Communication Engineering, M S Ramaiah Institute
of Technology, Bengaluru 560054, Karnataka, India
e-mail: nikitharn235@gmail.com

R. G. Srinidhi
e-mail: srinidhirg@gmail.com

R. Harshith
e-mail: harshithr10.01@gmail.com

T. Amar
e-mail: amart080@gmail.com

C. G. Raghavendra
e-mail: cgraagu@msrit.edu

© Springer Nature Singapore Pte Ltd. 2018
P. K. Sa et al. (eds.), *Recent Findings in Intelligent Computing Techniques*,
Advances in Intelligent Systems and Computing 709,
https://doi.org/10.1007/978-981-10-8633-5_49

497

[1], and in electronic devices, antennas, lasers, etc. [2] as silk exhibits the properties of an insulating material. But silk is produced only in few countries in the world because of the climatic conditions. To meet the demand for silk, these countries have to increase their production. India, being fifth largest producer of silk, needs to focus on early developmental stages of silkworm to increase its silk production. Silk as a cash crop will be deemed successful depending on its vitality and disease-free factor.

The health of the silkworm plays an important role in the production of silk. To provide a healthy growing environment and adequate feed for the worms, it becomes necessary to know the accurate count of silkworm eggs. Knowing the count, the farmer pays accordingly. In this regard, counting of eggs becomes a very crucial step. There are a few methods described for counting of silkworm eggs in [3, 4]. Mathematical morphology-based erosion and connected-component algorithm forms the basis for these counting algorithms.

Robust silkworms develop from healthy eggs. The egg phase is a very delicate stage in the life cycle of a silkworm. Proper environment should be provided for the development of eggs, such as ambient temperature, humidity, light, etc. which determines the development of the silkworms at a later stage [5]. Also, as only healthy eggs hatch and develop into silkworms, determination of hatch rate predicts the yield rate at an early stage.

In the sericulture industry, yield rate refers to the number of silkworms which are capable of spinning cocoon. This directly translates to the total amount of silk produced. Hatch rate is defined as the number of eggs that hatch, to the total number of eggs laid. By determining the hatch rate, final yield rate can be easily predicted well in advance.

Also, hatch rate can act as a feedback mechanism, i.e. if hatch rate reduces rapidly, it will alert the farmers and scientists to take corrective measures and improve the yield in the next cycle. This is particularly useful in preventing the spread of diseases to subsequent cycles, which is a possibility as the silk moth that lays eggs for the next cycle is most likely to be affected. Without the knowledge of hatch rate, there is no way to detect most types of diseases which could have been easily prevented from propagating. This is indication enough for the usefulness of determining hatch rate.

This is accomplished by differentiating yellow grains (eggs) from empty white shells (by image analysis) in a silkworm egg sheet, which is used in the experiment after removing the worms.

2 Proposed Methodology

Colour image analysis of silkworm eggs gives us valuable information regarding its nature and characteristics. This analysis can be used to build algorithms for determination of important parameters such as hatch rate, type of eggs, infection, etc.

We have developed an algorithm to determine the hatch rate of silkworm eggs by applying image processing and colour filtering techniques to the captured images.

2.1 Fundamental Concepts

The fundamental concepts that help in clearer understanding of the proposed methodology are briefly explained.

It is difficult to describe colours in RGB model to suit practical human interpretation, because we do not view any colour as a combination of the three primary colours; rather, we view it in terms of its hue, saturation and brightness. In this regard, HSV colour space provides a more comprehensive and practical approach for image analysis [6].

Hence, RGB image is converted into HSV colour space, where H denotes Hue which describes a pure colour, S denotes Saturation which gives a degree to which white light dilutes the pure colour and V denotes Value which is the grey-level intensity. HSV colour space is preferred because it separates colour information from intensity information which helps in our analysis [7].

The conversion of RGB to HSV [6] is governed by the following equations:

$$H = \begin{cases} \theta, & B < G \\ 360 - \theta, & B > G \end{cases} \tag{1}$$

where

$$cos\theta = \frac{0.5[(R - G) + (R - B)]}{\sqrt{[(R - G)^2 + (R - B)(G - B)^{0.5}]}} \tag{2}$$

$$V = max\{R, G, B\} \tag{3}$$

$$S = \frac{max(R, G, B) - min(R, G, B)}{V} \tag{4}$$

After experimentation, green channel of RGB colour space was found to provide good results and hence it is extracted.

Thresholding is used to distinguish the pixel values of the image to range of pixels above and below the specified threshold. Threshold can be specified by knowing histogram of the image which determines the pixel value that can be used to separate foreground and background of the image [8]. Adaptive thresholding methods can also be used to find the threshold of the image.

If I(p, q) is the intensity of any pixel (p, q) in the image, and level denotes the specified threshold value, then the image is thresholded to separate foreground and background based on the following equation:

$$I(p,q) = \begin{cases} 1, if I(p,q) > level \\ 0, if I(p,q) < level \end{cases} \tag{5}$$

Pixels that have same intensity value and differ from neighbouring pixels with respect to brightness or colour are considered as blobs [9]. Blob detection is done based on determining the connectivity of any pixel (e.g. 8-connectivity) to the pixels having same properties. By blob analysis, regions of interest are extracted and used in discerning the hatch rate. Blob analysis is a tool of choice in various machine vision applications where object detection is the primary concern, such as analysis of electronic circuit boards and IC chips [9], tracking of blood cells in microfluidic channels [10], defect inspection, etc.

2.2 Algorithm

The egg sheet consists of several disease-free layings (DFLs) which are nothing but the layings of the moths. For further processing, image of a single laying after the eggs have hatched is captured. This is preferable because before hatching, it is not possible to accurately determine which eggs will hatch and which will not, as different eggs mature at different rates.

After image acquisition, extracting yellow eggs from the input image is a major step towards calculating hatch rate. This is achieved by blob analysis and subsequently counting the unhatched yellow eggs, which is accomplished by the following algorithm:

Step 1: After the silkworm eggs have hatched, a colour image of a single laying in the sheet is taken using high-resolution camera under suitable lighting conditions.

Step 2: The RGB image is converted into HSV space and H-, S- and V-planes are extracted from it.

Step 3: Green channel (G) is extracted from the RGB image (Eq. 6) and top-hat transform is applied in order to easily distinguish background from the foreground (Eqs. 7–10).

$$G = \frac{G}{R+G+B} \tag{6}$$

where R, G and B denote Red, Green and Blue channels of an RGB image.

Top-hat transform is governed by the following set of equations:

$$T_w = f - f \Psi b \tag{7}$$

where Ψ denotes opening operation.

Opening operation performed on binary image A with structuring element S is given as

$$\Psi = (A \ominus S) \oplus \bar{S} \tag{8}$$

Erosion of A by structuring element S:

$$A \ominus S = \{(i,j) : S(i,j) \subset A\} \tag{9}$$

where $\oplus$ is defined by the equation:

$$A \oplus S = \bigcup_{(i,j) \in A} \overline{S}(i,j) \tag{10}$$

where $\overline{S}$ is the reflection of S in (i,j).

Step 4: The green channel image is binarized using a suitable threshold value (Eq. 5), noise is removed from the binary image and pixel coordinates in the regions of interest are obtained.

Step 5: The pixel coordinates are mapped to the HSV model of the original image.

Fig. 1 Flowchart depicting the algorithm

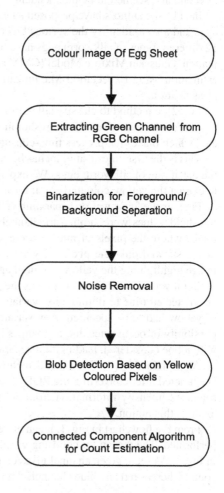

The pixels whose H, S and V values fall in the range of the colour yellow are separated, thus leading to the extraction of the yellow-coloured areas within the region of interest.

Step 6: The extracted yellow-coloured areas are counted using connected-component algorithm to obtain the number of yellow eggs.

The flowchart depicting the proposed algorithm is shown in Fig. 1.

3 Results and Observation

The proposed algorithm has been implemented using MATLAB 2016a on a PC with Intel Core i7-6500U processor of 2.5 GHz clock speed, 8 GB RAM loaded with Windows 7 (64-bit) operating system. The results were obtained within a fraction of a second and seemed to be independent of the system.

In [11], the authors have proposed a technique to detect the quantity of silkworm eggs and also to identify the various breeds with diverse regional heritage. Around 12 different types of silkworm eggs have been classified based on their colour information. Gaussian Mixture Model (GMM) based on the Maximum Likelihood (ML) estimation using Expectation Maximization (EM) algorithm has been adopted for classification.

In [12], a method to classify silkworm eggs in the last incubation stage has been proposed which involves Otsu thresholding, feature extraction, K-means clustering, etc., which makes the process time-consuming and cumbersome to implement.

This is the first time an attempt has been made to accurately and swiftly determine the hatch rate of silkworm eggs. We experimented with various techniques in order to extract the yellow-coloured egg information.

First, as the focus was on separating yellow-coloured pixels from others, different threshold values were applied to channels of RGB colour space to form a binary image where the pixels of interest were set to 1 and all others to 0. The resultant image showed only the pixels of yellow-coloured eggs. Figure 2 shows the input image highlighting the yellow-coloured eggs.

But it was difficult to enumerate as the pixels which are not a part of an egg were also detected (due to illumination variation). Also, RGB values for various shades of yellow had to be specified. Any variations beyond the given range, which has a possibility of occurrence due to changes in illumination and acquisition conditions will not be considered, leading to inaccurate count.

As explained in [7], for object detection from colour images, the required colour is extracted after converting the RGB image into HSV image, as HSV colour space separates intensity information from colour information. Our proposed algorithm exploits this notion.

From the flowchart in Fig. 1, it can be seen that green channel extraction has been performed. The reason for choosing green channel is apparent from the plot shown in Fig. 3. As seen, green channel gives comparable results with respect to the actual count, whereas red and blue channels deviate considerably. Also, the count obtained

Fig. 2 Original image with yellow eggs being highlighted

Fig. 3 Comparison of results obtained using different channels of RGB

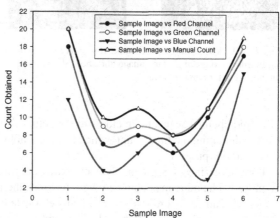

Table 1 Tabulation of results

Image	Manual count	Count obtained by algorithm	Accuracy (%)
1	22	20	90.9
2	12	10	80
3	9	9	100
4	18	17	94
5	11	10	90.9
6	9	8	88.88
7	7	7	100
8	2	2	100
9	13	12	92.3
10	6	6	100

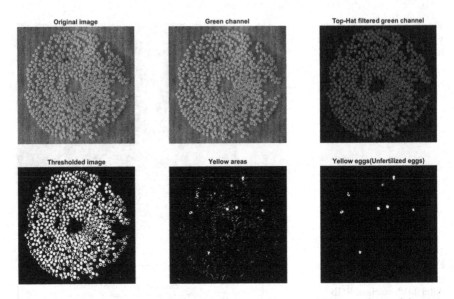

Fig. 4 (Clockwise from top): Original image, green channel extracted image, after applying top-hat transform to green channel image, thresholded image, yellow egg regions after pixel mapping, final image after noise removal

using red channel seems to be closer to the actual count because it considers noise while calculating the final count, which is undesirable. Hence, green channel is best suited for our purpose.

For binarizing the green channel extracted image, Otsu thresholding was employed. The outcomes obtained were far from the actual count and were subsequently discarded. In order to automatically estimate the threshold value, histogram method was employed where the valleys provide apparently suitable threshold levels. However, when the image was thresholded using these values, the accuracy drastically reduced. Hence, experimentally obtained threshold value of 65 (in the range 0–255) was adopted.

For separating out yellow egg regions, H, S and V channels were thresholded using the following values—$(0.067 < H < 0.120), (0.344 < S < 0.664)$ and $(0.412 < V < 0.984)$ (in the range 0–1), which were established with the help of colour thresholder application in MATLAB.

The pixel coordinates corresponding to yellow eggs, i.e. those whose H, S and V values lie in the above-specified range, were identified and set to 1, others to 0. This gave us the final image containing only yellow egg regions, to which connected-component algorithm was applied to obtain the count of unhatched eggs. Table 1 tabulates the results obtained for 10 sample images, which gives an average accuracy of 94%. With the knowledge of the total egg count that can be calculated using any of the methods described in [3, 4, 11], hatch rate can be determined.

Figure 4 shows different stages in the progress of the algorithm. The proposed algorithm gives an average accuracy of 94% under practical conditions. Accuracy can be improved by eliminating illumination variation.

One limitation of the proposed method is the variation of the count with respect to changes in illumination. Illumination-invariant property is not completely satisfied which leads to difference in the actual count and the count obtained by implementing the proposed algorithm. Very light yellow-coloured eggs may be interpreted as egg shells. There may be a chance of not detecting those eggs. In the same way, the eggs shells may also be interpreted as yellow eggs due to shadowing effect. This may give a count which is more than the actual number of unhatched eggs.

The removal of unwanted areas may sometimes lead to the removal of egg areas which may be tackled by using appropriate noise removal techniques.

4 Conclusion and Future Work

A novel method for calculating hatch rate of silkworm eggs by visual inspection using blob analysis technique has been proposed. Determination of hatch rate is preeminent in the sericulture industry as it helps in early prediction of yield rate and also acts as a barometer for developing disease-free worms. Unhatched eggs which are yellow in colour are detected from the egg sheet by colour extraction using HSV colour space. After successful detection, these unhatched egg regions are computed using connected-component algorithm. An average accuracy of 94% has been obtained.

In some cases, unhatched eggs may be bluish black in colour. This occurs when eggs do not hatch even after developing post-larva stage. The larva would have died due to inadequate nutrition available inside the egg, and hence failed to hatch. The proposed algorithm can be further extended to detect bluish-black eggs along with yellow eggs. This will improve the accuracy of hatch rate in such scenarios.

References

1. Wang, Z., Zhang, Y., Zhang, J., Huang, L., Liu, J., Li, Y., Zhang, G., Kundu, S.C., Wang, L.: Exploring natural silk protein sericin for regenerative medicine: an injectable, photoluminescent, cell-adhesive 3D hydrogel. In: Scientific reports (nature.com), vol. 4, p. 7064 (2014)
2. Omenetto, F. G.: Silk-based materials for technology at the micro-and nanoscale. In: Conference on Lasers and Electro-Optics (CLEO), San Jose, CA, pp. 1–1 (2016)
3. Pandit, A., Rangole, J., Shastri, R., Deosarkar, S.: Vision system for automatic counting of silkworm eggs. In: International Conference on Information Communication and Embedded Systems (ICICES2014), Chennai, pp. 1–5 (2014)
4. Kawade, R., Sadalage, J., Shastri, R., Deosarkar, S.: Automatic silkworm egg counting mechanism for sericulture. In: Proceedings of International Conference on Internet Computing and Information Communications, pp. 121–128. Springer, India (2014)

5. Reddy, G.V., Rao, V., Kamble, C.K.: Fundamentals of Silkworm Egg Bombyx Mori L, 1st edn. (2003)
6. Gonzalez, R.C., Woods, R.E.: Digital Image Processing, 3rd edn. (2007)
7. Khule, A., Nagmode, M.S., Komati, R.D.: Automated object counting for visual inspection applications. In: IEEE International Conference on Image Processing (ICIP), pp. 801–806 (2015)
8. Kurban, T., Civicioglu, P., Kurban, R., Besdok, E.: Comparison of evolutionary and swarm based computational techniques for multilevel color image thresholding. In: Applied Soft Computing 23 , pp. 128–143, Elsevier (2014)
9. Zhong, Q., Chen, Z., Zhang, X., Hu, G.: Feature-based object location of IC pins by using fast run length encoding BLOB analysis. IEEE Trans. Compon. Packag. Manuf. Technol. $4(11)$ (2014)
10. Arvind, B.C., Nagaraj, S.K., Seelamantula, C.S., Gorthi, S.S.: Active-disc-based Kalman filter technique for tracking of blood cells in microfluidic channels. In: IEEE International Conference on Image Processing (ICIP), pp. 3394–3398. Phoenix, AZ (2016)
11. Kiratiratanapruk, K., Methasate, I., Watcharapinchai, N., Sinthupinyo, W.: Silkworm eggs detection and classification using image analysis. In: IEEE International Computer Science and Engineering Conference (ICSEC) pp. 340–345 (2014)
12. Kiratiratanapruk, K., Sinthupinyo, W.: Silkworm egg image analysis using different color information for improving quality inspection. In: International Symposium on Intelligent Signal Processing and Communication Systems (ISPACS), pp. 1–6. IEEE (2016)

Role of Intensity of Emotions for Effective Personalized Video Recommendation: A Reinforcement Learning Approach

Abhishek Tripathi, D. G. Manasa, K. Rakshitha, T. S. Ashwin
and G. Ram Mohan Reddy

Abstract Development of artificially intelligent agents in video recommendation systems over past decade has been an active research area. In this paper, we have presented a novel hybrid approach (combining collaborative as well as content-based filtering) to create an agent which targets the intensity of emotional content present in a video for recommendation. Since cognitive preferences of a user in real world are always in a dynamic state, tracking user behavior in real time as well as the general cognitive preferences of the users toward different emotions is a key parameter for recommendation. The proposed system monitors the user interactions with the recommended video from its user interface and web camera to learn the criterion of decision-making in real time through reinforcement learning. To evaluate the proposed system, we have created our own UI, collected videos from YouTube, and applied Q-learning to train our system to effectively adapt user preferences.

Keywords Reinforcement learning · Q-learning · Emotional intensities
Cognitive preferences · Affectiva

1 Introduction

Today, video recommendation is one of the extensive problems being considered by researchers. The main reason being extensive and diverse distribution of videos

A. Tripathi (✉) · D. G. Manasa · K. Rakshitha · T. S. Ashwin · G. Ram Mohan Reddy
National Institute of Technology Karnataka, Surathkal, Mangalore, India
e-mail: abhishek.tripathi2421@gmail.com

D. G. Manasa
e-mail: dgmanasa1@gmail.com

K. Rakshitha
e-mail: rakshi.chandu@gmail.com

T. S. Ashwin
e-mail: ashwindixit9@gmail.com

G. Ram Mohan Reddy
e-mail: profgrmreddy@gmail.com

© Springer Nature Singapore Pte Ltd. 2018
P. K. Sa et al. (eds.), *Recent Findings in Intelligent Computing Techniques*,
Advances in Intelligent Systems and Computing 709,
https://doi.org/10.1007/978-981-10-8633-5_50

with constant increase in the volume shared by people over massive scale. There are a lot of video recommendation systems till date. YouTube, a famous video sharing website, offers many advanced metrics apart from commenting, rating, favorites, subscribing, appropriate social filters for different age groups, etc. to help a user find best available stuff. The core and non-core metrics defined by YouTube [1], play an important role but limitations do exist when real-time decision making for user's mood is considered. It is observed that users still have to redefine search criteria whenever there is a change in their mood which takes a fair amount of time, thus making the entire process time-consuming and sometimes frustrating.

Consider the situation when the user is in a very good mood and is continuously watching comedy videos. YouTube might provide the user with all the relevant comedy content belonging to the class or channel the user is digging into, along with other popular videos which are related to user's query. But there is no guarantee that the user finds them funny at that point of time. Perspective may change with mood. In fact, some videos of the returned list can contain items which have already been watched. Here, the recommendation is not personalized. Rather, it is just based on the context of the keywords typed in by the user. Moreover, the same list will be returned to every user who has typed similar keywords. Search engines do not yet know the emotional preferences at different points of time and how to adapt according to present emotional state. Emotional preferences play an important role in a recommendation system and the user should always be considered in a dynamic state with continuous transitions involved from one state to another over the time and the system understand them and adapt accordingly.

This paper focuses on exploring the impact of intensities of emotions in a video to model users' emotional interests and emotional state transition preferences. We have created a hybrid framework incorporating effects of both collaborative and content-based filtering and consider this problem as continuous optimization problem. Considering personalized recommendation to be a crucial factor to make users feel that the system was created just for them, keeping close watch on user activity is needed. To serve the purpose, user's facial behavior is continuously monitored through a webcam toward the recommended video, simultaneously logging user's feedback, e.g., liking/disliking, rating, skipping, and fast forwarding are done. A web browser-based interface is created for the user, like YouTube, hosted over a local server. Evaluation of the proposed methodology reveals that Q-learning is effectively able to learn user behavior and converges within seven sessions to get zero negative feedbacks.

2 Related Work

The idea of the model created is based on the work of Chi et al. [2] who proposed an automatic playlist generation based on song emotions using reinforcement learning. Here, songs are considered as states and rewards are generated based on the feedback. Simpkins et al. [3] mentioned a reinforcement model to build user personality model. This personality model can be used to model states and give rewards before the reinforcement learning is applied. That is, the video set is being annotated before we start recommending videos to the user. Chen et al. [4] designed a model to detect change in emotional intensity before and after proposing each video but does not describe initial annotation of videos. Therefore, to achieve annotation of videos, the model described by Das et al. [5] was followed. The emotional tagging is done based on the context, emotional expressions. Moreover, fresh emotion-based insight has been recognized as a substantial construct for language education by Pishghadam et al. [6]. After the annotation of videos, research on reinforcement learning has been considered. Bozinovski et al. [7] constructed a model where the agent learns in an environment which gives the reinforcement not after an action but several steps later. Therefore, we can consider immediate action for a certain class of feedback, while we can consider feedback after a set of actions for a different class of feedback. By considering the feedback, Marinier et al. [8] proposed a way to determine whether a stimulus is relevant to the current goal. The goal could be driving the user to an active state. To suggest next video, we could suggest a random video. But the miss rate in this case might be high. Instead, we could use the similarity index to suggest the next video as defined by Aucouturier et al. [9]. Then, research on linking emotional intensity values with the recommendation has been considered. The main objective is to improve the human mood score which can be verified using the model proposed by Addo et al. [10]. Broekens et al. [11] showed that results support the idea that the function of emotion is to provide a complex feedback signal for an organism to adapt its behavior. Chi et al. [2] used feedback such as forwarding the song, unliking the song, etc. But we could consider the user's facial expressions as well while watching the video as a feedback. Broekens et al. [12] considered facial expressions to facilitate robot learning. This type of feedback can be considered after a set of sessions. To check whether the user has made an emotion-based decision in experience-based paradigm and whether the impact of regret depends on its degree of unexpectedness as suggested by the current regret theory, we can refer to Ahn et al. [13]. Therefore, the combination of both emotional intensity-driven approach to recommend videos and reinforcement learning has been applied to make a learned decision as to which video is to be recommended next and to calculate the misrates. The base work in this aspect comes from Feldmaier et al. [14] where the framework to incorporate emotional model into decision-making process of a machine learning agent is proposed.

3   Methodology

3.1   Modeling Emotion Vector

According to Robert Plutchik's Wheel of Emotions, Fig. 1, proposed by Robert Plutchik's psychoevolutionary theory of emotion [15], there are eight primary bipolar emotions of which we are considering only 4 in our experiment: Joy versus sad-

Fig. 1 Wheel of emotions

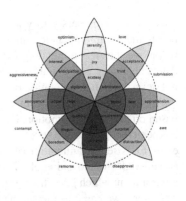

Fig. 2 Valence-arousal
model

ness and anger versus fear. The location of these emotions according to valence-arousal model is shown in Fig. 2. We annotate each video of the dataset with an emotion vector (E-Vector) comprised of these four emotions. Each value (E-Value) denotes the intensity of that particular emotion in the entire video.

$$E = \{Joy, Sad, Anger, Fear\}$$

3.2 Dataset Annotation

For initializing the emotion vectors, we analyze the videos throughout their lengths. Since nearby frames would be similar, we extract frames after an interval of 2 or 3 sec and then analyze the type of emotions on the human faces. Since many such tools are already developed, here we are using Affectiva [16], an emotion measurement technology company that grew out of MIT's media lab. Their classifier has ROC (Receiver Operating Characteristic) score of more than 0.8 for happy emotions. Other emotions tend to be more nuanced and subtle and are therefore harder to detect, and therefore we verify the annotated data and improve them by manual annotation. The initial annotation might be or might not be accurate but later on, it is modified based on user feedback.

3.3 System Monitors

To keep a close watch on user behavior, system is built with two monitors: (1) The frontend user interface and (2) Live Web Camera. User interaction with the system is through a web browser-based frontend as shown in Fig. 3. In the beginning, user selects a video from the collection which will act as the seed video. The next recommendation will happen dynamically based on the intensity values of the present video and user actions (like fast forwarding, replying, skipping, rating, liking/disliking) toward the played video which is logged constantly along with a web camera focusing on viewers face to record emotions throughout video length in real time. The logged actions and user's facial analysis are used to make decisions by the system about the user preferences and to identify the video segments with high emotion intensities, respectively. If user is just running through multiple videos, these dominating segments can be shown first to give the major impact of the video in starting itself. It further helps to evolve the dataset with more accurate E-values (Joy, Sad, Anger, and Fear).

3.4 Mapping Dynamic Recommendation to Reinforcement Learning

In the diverse area of machine learning, reinforcement learning, inspired by behaviorist psychology, is concerned with how agents should take actions ideally in an environment to create a policy which maximizes some notion of cumulative reward.

Fig. 3 User interface

Cognitive behavior of humans is dynamic and is largely affected by the situations around them. Their mood preferences change over time. If recommendation systems are to incorporate such dynamic behavior in real time, they should change their policies according, to adapt the present demand or the usual behavior over a period. Adapting to such a changing environment takes time and is a continuous process of policy optimization to get synchronized with the user needs for long-term benefits. The faster it happens, more intelligent the system is considered. Clearly, it is proven that reinforcement learning is an ideal solution for recommendation in dynamic environment over supervised learning and other paradigms of machine learning.

3.5 Defining States and Actions

Reinforcement learning considers a system into a set of states and actions are associated with each state to move from one state to another. There could be many ways to formulate the format of a state and actions. Since each video is tagged with unique E-vector, here we consider each video to be a unique state and action denotes transition of user from one state to another. When user watches a video, the present video's E-vector is considered to be possible emotional state of the user which upon user feedback is decided to be right or wrong. The next video to be recommended will be among the next three videos with E-vector closest to the current video, chosen randomly, representing three possible actions.

3.6 Shaping Reward Function

Reward function is used in decision-making by the agent to find optimal paths. Usually, the reward function is handcoded prior to learning and must accurately assign reward values to any state the agent might reach. But here a human generates the rewards interactively: the human observes the state, takes action, and returns a scalar to the agent in the form of reward. In our system, we have three components of reward function. Feedbacks from frontend user interface are classified as (1) implicit feedback, comprised of {replay, skipping, backwarding, forwarding, listening time}, and (2) explicit feedback, comprising liking, disliking, and rating. Feedback from web camera forms the third contribution parameter. The final reward function is shown below:

$$r_t = \alpha R_{1(t)} + \beta R_{2(t)} + (1 - \alpha - \beta)R_{3(t)}, \quad 0 < \alpha, \beta < 1 \qquad (1)$$

where $R_{1(t)}$, $R_{2(t)}$, and $R_{3(t)}$ are the three components of the total reward (r_t) at time t, respectively. Each one of them represents the cumulative reward which is equal to the weighted summation of all the action in the respective sets.

Table 1 Rewards Description

Positive		Negative	
Action	Weight (%)	Action	Weight (%)
Replay	20	Skipping	−20
Backwarding	10	Forwarding	−10
Like	50	Dislike	−50
Rating (4, 5)	10, 20	Rating (1, 2)	−20, −10
Listening Time (Full length)	10		

$$R_{x(t)} = \sum_i w_i a_i, \forall x \in \{1,2,3\}$$

Table 1 summarizes the contribution(weights) of each action in reward components one ($R_{1(t)}$) and two ($R_{1(t)}$), respectively. We have chosen these values by trial and error to ultimately get the correct user feedback and nullify the effect of each other in case user mistakenly takes an action which was not meant to be.

The third component is calculated on the basis of difference between the usual trend shown by all the other users who have watched the video and the trend shown by the user. Since the system monitors user's face, each video has the cumulative trend shown by all the users who have watched the videos so far, the extent to which they are matched, form the weight of this component.

3.7 Applying Temporal Difference Learning Techniques

Temporal difference (TD) learning, which is primarily used in reinforcement learning, is considered to be a combination of Monte Carlo ideas and dynamic programming (DP) ideas and is best suited for learning in dynamic environments. In our problem, the world is composed of states but does not have predefined actions. They are formulated by user along with their weights (rewards) as described in Table 1 when they start interacting with the system. So to learn such a new world through temporal learning, we have used Q-learning (an off-policy control strategy).

For a walk containing an alternating sequence of states and state-action pairs, as shown in Fig. 4, in simplest form, the algorithm, Q-Learning, is described by Eq. 2 and complete algorithm by Algorithm 1.

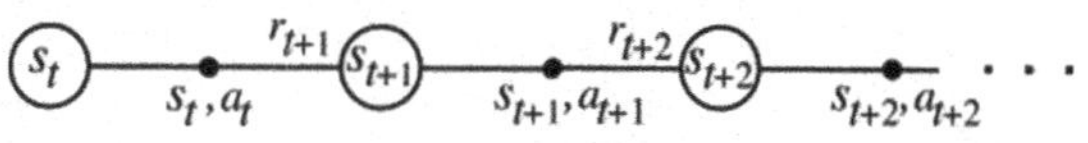

Fig. 4 Sequence of states and actions

$$Q(s_t, a_t) \leftarrow Q(s_t, a_t) + \alpha[r_{t+1} + \gamma \max_a Q(s_{t+1}, a) - Q(s_t, a_t)] \tag{2}$$

where s_t, a_t, and r_t represent the state, action, and reward at time t, respectively. r_{t+1} is the reward obtained after performing action a_t in state s_t and s_{t+1} represents the next state. Parameter α is the learning rate and γ is the discount factor. In our experiment, we have kept their values to be 0.1 and 0.9, respectively. With small learning rate, Q-learning converges slowly but less chances of overfitting. Keeping discount factor value approaching 1 will make it strive for a long-term high reward.

Algorithm 1: Q:Learning

```
1 Initialize the Q-matrix, Q(s,a) arbitrarily;
    /*for each session/episode/walk*/
2 repeat{
    Initialize state s;
    /*for each transition in a session/episode/walk*/
    repeat{
        Choose action (a(t)) from state (s(t))
                using (epsilon-greedy) policy;
        Recommend next video using f($a_{t}$);
        Observe reward $r_{t+1}$ and new state $s_{t+1}$;
        Update Q(s(t),a(t)) from (2);
        s(t)<-s(t+1)
    }
}
```

Considering ϵ-greedy policy, in our experiment, we have kept it constant. System considers exploring videos of other classes of emotions by taking a random action after recommending three videos from the same class. To explore other emotions, the random action first recommends videos with less E-values, i.e., on change of quadrant, videos which are close to origin are recommended first. Upon getting consecutive 3–4 negative feedback from other classes of emotion, exploration is decreased in that class. Talking about function F(a), it tells the system about next recommendation. Since videos should be recommended in increasing order of their emotional intensities, F(a) randomly chooses one of the next three videos closest to the E-vector of the present video being watched by the user.

4 Experiment and Evaluation

4.1 Dataset

To evaluate the performance of the recommendation system using proposed method-ology, we manually collected videos from one of the most popular video sharing websites, YouTube. There are 20 videos in total, 7 belonging to quadrant-I (+,+), 6 from quadrant-II (−,+), and 7 from quadrant-III (−,−) of the valence-arousal model shown in Fig. 2. The collected videos are then annotated and tagged with E-vector using Affectiva. For classes of emotions with less accuracy, annotations are verified manually.

4.2 Experimental Setup

We tried to put in decent efforts to manage complexity for implementing our archi-tecture and chose web browser-based frontend user interface connected to backend via AJAX calls once the page gets loaded completely. The experiment was conducted with three users. Since our dataset is small, whenever users log in to the system, we considered using entire dataset at once for each session. A session is defined as the duration for which users watch videos and interact with the system, until they log out. Since in every session mood of the user can be different, system considers mod-eled emotional behavior and preferences from past sessions as well as strives to learn present behavior using system monitors and other methods as explained in method-ology in detail. Due to low accuracy of the tools to identify sad emotions, α and β are tuned to give more weights to $R_{1(t)}$ and $R_{2(t)}$.

After logging user feedbacks from initial sessions, the logged data is used to train system using Q-learning algorithm to adapt the new behavior and preference. During a session also, after new interactions logged data is used to train the system using the two algorithms.

4.3 Performance Results

To analyze the results of both the algorithms, we consider negative feedback as error in prediction and define the following evaluation metrics:

$$Error\ Rate = \frac{Count\ of\ Negative\ Feedbacks\ (i)}{n(i)} \qquad (3)$$

where n(i) is the total number of states visited in ith session i.e., number of videos watched.

Fig. 5 Q-Learning curve

As shown in Fig. 5, in initial sessions, system recommends videos to a user based on increasing intensities of emotions simultaneously exploring other classes which reduces the count of negative feedbacks by the end of session 2. Later on, after applying Q-learning by the end of session 3, system learns the user preferences and after that recommendation happens in more optimized way favoring user's preferences and ultimately the negative feedback count reaches to zero by the end of seven sessions. The proposed system works effectively and preferred over random recommendation from shuffled list of available videos.

5 Conclusion

Considering a major role of moods and emotions, we have focused on personalized recommendation. Exploring tools which can effectively identify human facial emotions, we ended up with Affectiva. We describe how the problem can be solved using reinforcement learning. We have highlighted how the present video database can efficiently evolve based on live user feedback over the entire video length. Through this approach, we are able to extract segments of dominant emotions in videos during which users have responded very actively which further increases performance of recommendation system. The temporal learning technique: Q-learning is used to train the system in real time and their results are highlighted.

Our approach to incorporate emotional intensities for effective video recommendation can also be extended to incorporate the audio and text (meta-data) components of videos, thus pushing the limits further to create a more generic mood-based content recommendation system.

References

1. YouTube Core and Non-Core Metrics to track individual user. https://developers.google.com/youtube/analytics/v1/dimsmets/mets
2. Chi, C.-Y., Tsai, R.T.-H., Lai, J.-Y., Hsu, J.Y.: A reinforcement learning approach to emotion-based automatic playlist generation. In: 2010 International Conference on Technologies and

Applications of Artificial Intelligence (TAAI), pp. 60–65. IEEE (2010)

3. Simkins, C., Isbell, C., Marquez, N.: Deriving behavior from personality: a reinforcement learning approach. In: Proceedings of the International Conference on Cognitive Modeling, Philadelphia, PA, pp. 229–234 (2010)

4. Chen, X.Y., Segall, Z.: XV-Pod: an emotion aware, affective mobile video player. In: 2009 WRI World Congress on Computer Science and Information Engineering, vol. 3, pp. 277–281. IEEE (2009)

5. Das, D., Bandyopadhyay, S.: Identifying emotional expressions, intensities and sentence level emotion tags using a supervised framework. PACLIC **24**, 95–104 (2010)

6. Pishghadam, R., Adamson, B., Shayesteh, S.: Emotion-based language instruction (EBLI) as a new perspective in bilingual education. Multilin. Educ. **3**(1), 9 (2013)

7. Bozinovski, S., Bozinovska, L.: Emotion and learning: solving delayed reinforcement learning problem using emotionally reinforced connectionist network. In: Proc AAAI Fall Symposium on Emotional and Intelligent: The Tangled Knot of Cognition, pp. 29–30 (1998)

8. Marinier, R.P.: A computational unification of cognitive control, emotion and learning. (Unpublished doctoral dissertation). University of Michigan. Ann Arbor, MI (2008)

9. Aucouturier, J.-J., Pachet, F.: Finding songs that sound the same. In: Proceedings of IEEE Benelux Workshop on Model based Processing and Coding of Audio, pp. 1–8 (2002)

10. Addo, I.D., Ahamed, S.I.: Applying affective feedback to reinforcement learning in ZOEI, a comic humanoid robot. In: 2014 RO-MAN: The 23rd IEEE International Symposium on Robot and Human Interactive Communication, pp. 423–428. IEEE (2014)

11. Broekens, J., Jacobs, E., Jonker, C.M.: A reinforcement learning model of joy, distress, hope and fear. Connect. Sci. **27**(3), 215–233 (2015)

12. Broekens, J.: Emotion and reinforcement: affective facial expressions facilitate robot learning. In: Artifical Intelligence for Human Computing, pp. 113–132. Springer, Berlin, Heidelberg, 2007

13. Ahn, W.-Y., Rass, O., Shin, Y., Busemeyer, J.R., Brown, J.W., O'Donnell, B.F.: Emotion-based reinforcement learning. In: CogSci (2012)

14. Feldmaier, J., Diepold, K.: Path-finding using reinforcement learning and affective states. In: 2014 RO-MAN: The 23rd IEEE International Symposium on Robot and Human Interactive Communication, pp. 543–548. IEEE (2014)

15. Robert Plutchik's psychoevolutionary theory of emotion. https://en.wikipedia.org/wiki/Robert_Plutchik

16. McDuff, D., Kaliouby, R., Senechal, T., Amr, M., Cohn, J., Picard, R.: Affectiva-mit facial expression dataset (am-fed): naturalistic and spontaneous facial expressions collected. In: Proceedings of the IEEE Conference on Computer Vision and Pattern Recognition Workshops, pp. 881–888 (2013)

Performance Analysis of Classifiers and Future Directions for Image Analysis Based Leaf Disease Detection

Manisha Goswami, Saurabh Maheshwari and Amarjeet Poonia

Abstract Plants play a very important role in the environment to maintain ecosystem, so this is our responsibility to protect it by detected disease which appears in it. In the plant disease, most symptoms appear on leaf, so by performing some image analysis we can detect these diseases very fast and accurately. This paper includes survey of different techniques which are used in leaf disease detection. To detect plant disease color conversion, Canny and Sobel edge detectors are used initially and then some segmentation techniques, i.e., Otsu and k-means, are used; after then, feature extraction takes place and is classified with classification techniques.

Keywords Edge detection · K-means segmentation · GLCM and classification technique

1 Introduction

Increase growing rate is the demand of world population and we all connected with plant directly or indirectly. Leaf disease detection is the initial way to detect plant diseases. In the early stage, it can be detected by symptoms which are showing on leaf but this is not correct way because in some diseases symptom similarities can occur. Image processing provides an easy and correct way to detect these by performing some steps. It takes two steps training and testing. Training is used to create the database of collecting leaf images. Testing is used to detect and classify the diseased leaf [1, 2].

M. Goswami (✉) · S. Maheshwari · A. Poonia
Govt Women Engineering College, Ajmer, India
e-mail: manishagswmi@gmail.com

S. Maheshwari
e-mail: dr.masurabh@gmail.com

A. Poonia
e-mail: amar.gweca@gmail.com

© Springer Nature Singapore Pte Ltd. 2018
P. K. Sa et al. (eds.), *Recent Findings in Intelligent Computing Techniques*,
Advances in Intelligent Systems and Computing 709,
https://doi.org/10.1007/978-981-10-8633-5_51

519

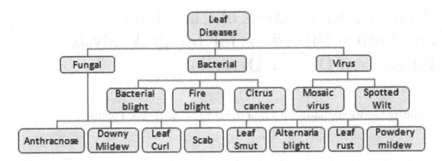

Fig. 1 Classification of leaf diseases

Plant diseases occur because of fungicides, bacteria, and viruses. Here, we present categorization of these diseases. Fungal disease appears as yellow and brown spots on leaves, and these spots are dark day by day. Bacterial diseases appear as water-soaked yellow spot that turns into brown in old age and virus diseases appear as yellow and green spots [3–6].

1. **Fungal Diseases**

   Anthracnose: It initially appears as irregular, small yellow, pink, or brown spot. These spots darken day by day and expand covering the leaf [3].

   Downy mildew: Downy mildew affects as white to yellow spot on underside of leaves and turns into black with old age [1, 7, 8].

   Leaf curl: It appears as reddish leaf area on new leaf. In the old age, area becomes thick; leaves are puckered, curl, and reduce fruit growth [4].

   Scab: In early stage, it appears as water-soaked pale yellow and olive-green spot on the upper surface and dark spot on lower surface turns brown [9].

   Leaf smut: Smut appears as small whitish-gray galls which increase with time and convert into black, hard, and dry [4].

   Alternaria blight: It found in tomatoes, potato, and pepper plant. It first appears as water-soaked, black-brown spot. In old edge, spots are dry and crack and then leaf dropout from plant [10].

   Leaf rust: In the early stage, white spot found undersides of leaves. In the later stage, these spots turn into reddish-orange spore masses and then yellow-green and black, after then leaf dropout from plant [4].

   Powdery mildew: Infected leaf starts with white to gray powdery, turns into brown, and drops it. Flower bud with mildew may never open [1, 7, 8].

2. **Bacterial Diseases**

   Bacterial blight: It appears as small, angular water-soaked spot which turns into yellow, after then brown, and infected portion dies. In the old age, these spots darken and covered by yellowish-green and the dead portion fall out [9, 11, 12].

Fire blight: It affects new growth first, so upper part of plant is affected. Most infected leaves wilt rapidly and turn down brown or black; leaves die but not drop off [4].

Citrus canker: Citrus canker lesion shows a water-soaked, yellow concentric circle leaf with 2–10mm diameter. It changes into brown color with time and shows as pimple on both sides of plant [3].

3. **Virus Diseases**

Mosaic virus: Symptoms of mosaic virus disease are yellow, white, or green spot, wrinkled, curled, or small leaves, and affect growth rate [5].

Spotted wilt: It causes leaves to turn into yellow or purplish-bronze [13] (Fig. 1).

2 Literature Review

S. Arivazhagan et al. proposed a system to automatically detect leaf disease. Here, the green pixel is masked and then extracted texture features passed to SVM classifier with 94% accuracy [14]. The work done by Sanjeev S Sannakki et al. is used to diagnose disease using image processing and artificial intelligence technique. Image is enhanced by anisotropic diffusion and then k-means segmentation is applied; back-propagation neural network classifier is used with 100% accuracy [7]. The method proposed by Auzi Asfarian et al. tries to detect disease when two diseases of same color occur. The diseased lesion is cropped and then s component is used for classification using PNN with 83% accuracy [12]. Kiran R. Gavhale et al. proposed a method to detect unhealthy region of citrus leaf, and performed resizing and color transformation and then k-means. Here, texture and color features are extracted and then SVM is used [3]. The method proposed by Umapathy Eaganathan et al. is used to identify late scorch disease. In preprocessing thresholding, median filter and morphological operation are used and then k-means segmentation is applied. KNN classifies disease with 95% accuracy [15]. The work done by P. Revathi and Hemalatha to detect leaf disease utilizes feature selection method. Edge is detected by cany edge detector and then genetic algorithm is performed. Feature is extracted by EPSO, and then comparison of SVM, BPNN, and fuzzy classification is performed [16]. The method is proposed by Suman and Dhruvakumar to classify paddy disease. Group of homogenous pixel is used for detection, and then block-wise features are extracted and identified the disease with 70% accuracy [11]. The method proposed by Prakash M. Mainkar et al. is used to detect and classify plant disease. K-means segmentation is used, and then the green pixel is masked; texture features are extracted, and then BPNN-based classification is performed with 94–96% accuracy [17]. The work done by Aakanksha Rastogi aims to detect leaf disease using fuzzy logic and a computer vision technique to grade. First, the system followed by ANN is trained and then the affected cluster is detected using k-means segmentation and ANN-based classifica-

Table 1 Yearwise classification of literature review

Reference paper	Technique used	Future direction
[14] Detection of unhealthy region of plant leaves and classification of plant leaf diseases using texture features	Color transformation, Otsu, GLCM, and SVM	Features in training phase can be increased, color and shape feature with optimal feature use for disease detection
[7] Diagnosis and classification of grape leaf diseases using neural networks	Anisotropic diffusion, k-means segmentation, and PNN	Other than hue component can be used and result can be compared
[12] Paddy diseases identification with texture analysis using fractal descriptors based on Fourier spectrum	Lesion crop, Laplacian filter and PNN	Other segmentation and classification technique can be used to increase recognition rate
[10] Classification of cotton leaf spot disease using support vector machine	Otsu, blob analysis, statistical analysis, and SVM	All types of images can be taken and k-means can be used
[3] Unhealthy region of citrus leaf detection using image processing techniques	Resize, enhance, k-means, GLCM, and SVM classifier	We can apply better image analysis to increase accuracy on other plants
[15] Identification of sugarcane leaf scorch diseases using k-means clustering segmentation and KNN-based classification	Median filter, Otsu, k-means segmentation, block-wise feature extraction, and KNN	Neural network can be used to increase recognition rate
[16] Cotton leaf spot diseases detection utilizing feature selection with skew divergence method	Resize, Canny edge detector, genetic algorithm, EPSO, SVM, BPNN, and fuzzy classifier	Recognition rate can be increased
[9] Grading & identification of disease in pomegranate leaf and fruit	Resize, morphological operation, and k-means segmentation	Classifier can be used to detect type of disease
[11] Classification of paddy leaf diseases using shape and color features	smoothing, binarization, group of homogenous pixel, block-wise feature extraction and SVM	Can be used for all types of leaf and better classification technique can be used
[23] Detection of unhealthy region of plant leaves using image processing and genetic algorithm	Genetic algorithm	To improve recognition rate ANN, fuzzy, hybrid algorithm can be used
[8] An implementation of grape plant disease detection	Threshold, GLCM, and SVM	K-means can be used for segmentation
[17] Plant leaf disease detection and classification using image processing techniques	Color transformation, k-means, GLCM, and BPNN	Accuracy can be increased by using segmented image analysis
[2] Leaf disease detection and grading using computer vision technology & Fuzzy logic	Resize, filter, k-means segmentation, and ANN	Better image analysis method and other classifier can be used to increase accuracy

(continued)

Table 1 (continued)

Reference paper	Technique used	Future direction
[18] Potato leaf diseases detection and classification system	Resize, filtering, color pixel conversion, k-means segmentation, GLCM, and BPNN	Recognition rate can be increased
[19] Detection of plant leaf diseases using image segmentation and soft computing techniques	Threshold, genetic algorithm, GLCM, and SVM	To improve recognition rate ANN, fuzzy, hybrid algorithm can be used
[20] Plant leaf and disease detection by using HSV features and SVM classifier	HSV color conversion, genetic algorithm, GLCM, NN, and SVM	Accuracy can be increased by using better color component for feature extraction
[24] Leaf disease recognition system	Resizing, thresholding, GLCM, and SVM	Other segmentation technique can be used, more features can be extracted
[1] SVM classifier-based grape leaf disease detection	Gaussian filtering, k-means, GLCM, and SVM	Accuracy can be increased by using BPNN classifier.
[21] Developing an algorithm for tomato leaf disease detection and classification	Thresholding, k-means, GLCM, and SVM	More features can be extracted to increase recognition rate
[22]Identification of leaf diseases in pepper plants using soft computing techniques	Color transformation, threshold, binarization, GLCM, and ANN	K-means can be used for segmentation
[6]Image processing based leaf rot disease, detection of betel vine (Piper BetleL.)	Resizing, cropping, and Otsu	Other segmentation techniques can be used
[25] Leaf disease detection using image processing	Contrast enhancement, k-means, and SVM	We can use other segmentation and classification technique

tion [2]. Girish and Priti proposed a method to detect and classify potato leaf disease using k-means segmentation, GLCM, and BPNN classifier with 92% accuracy [18]. Vijai and Misra et al. proposed a soft computing approach to detect plant leaf diseases. The green pixel is masked using threshold and then genetic algorithm is used to detect unhealthy portion; then, features are extracted and SVM classified the disease with 95.71% accuracy [19]. The method proposed by M. Ravindra Naik et al. tries to detect leaf disease with HSV feature and SVN classifier. Genetic algorithm is used for segmentation, and present comparison of SVM and NN classifier shows that SVM gives better classification accuracy [20]. The work done by Pranjali and Anjali tries to detect grape disease using SVM classifier. Image is preprocessed and then k-means segmentation and block-wise feature extraction are used which are classified with 88.89% accuracy [1]. Vidyaraj and Priya proposed an algorithm to detect tomato leaf disease. K-means is used for segmentation and then GLCM-based feature extraction and SVM based classification are performed [21]. Jobin Francis et al. proposed a soft computing technique to detect disease in pepper plant. Thresholding method is used to detect diseased area and then GLCM-extracted feature is sent to neural network [22] (Table 1).

3 Conclusion and Future Work

This paper presents a review of recent techniques used to detect leaf disease and classify it. The purpose of this paper is to introduce some existing techniques and future direction to improve result. By changing segmentation, feature extraction, and classification techniques, we can increase accuracy and processing time. To improve disease detection and classification accuracy, we can use proper image preprocessing steps, k-means segmentation and then edge detector to detect a closed portion of disease, better color component for feature extraction, and SVM or BPNN classifier.

References

1. Padol, P.B., Yadav, A.A.: SVM classifier based grape leaf disease detection. In: Conference on Advances in Signal Processing (CASP), pp. 175–179 (2016)
2. Rastogi, A., Arora, R., Sharma, S.: Leaf disease detection and grading using computer vision technology & fuzzy logic. In: 2nd International Conference on Signal Processing and Integrated Networks (SPIN), pp. 500–505 (2015)
3. Gavhale, K.R., Gawande, U., Hajari, K.O.: Unhealthy region of citrus leaf detection using image processing techniques. In: International Conference for Convergence of Technology (2014)
4. Planet natural research center. https://www.planetnatural.com
5. Agriculture: university of Minnesota. http://www.extension.umn.edu/agriculture/
6. Deya, A.K., Sharma, M., Meshram, M.R.: Image processing based leaf rot disease, detection of betel vine (Piper BetelL.). Procedia Comput. Sci. 85, 748–754 (2016). https://doi.org/10.1016/j.procs.2016.05.262
7. Sannakki, S.S., Rajpurohit, V.S., Nargund, V.B., Kulkarni, P.: Diagnosis and classification of grape leaf diseases using neural networks. In: 4th ICCCNT (2013). https://doi.org/10.1109/ICCCNT.2013.6726616
8. Kakade, N.R., Ahire, D.D.: An implementaion of grape leaf plant disease detection. IJARIIE, 527–535 (2015)
9. Deshpande, T., Sengupta, S., Raghuvanshi, K.S.: Grading & identification of disease in pomegranate leaf and fruit. IJCSIT, 4638–4635 (2014)
10. Zambre, R.S., Patil, S.P.: Classification of cotton leaf spot disease using support vector machine. Int. J. Eng. Res. Appl., 92–97 (2014)
11. Suman, D.: Classification of paddy leaf diseases using shape and color features. IJEEE, 239–250 (2015)
12. Asfarian, A., Herdiyeni, Y., Rauf, A., Mutaqin, K.H.: Paddy diseases identification with texture analysis using fractal descriptors based on Fourier spectrum. In: International Conference on Computer Control Informatics and Its Application, pp. 77–81 (2013). https://doi.org/10.1109/IC3INA.2013.6819152
13. A service of SDSU extension. https://igrow.org/gardens/gardening/
14. Arivazhagan, S., Newlin Shebiah, R., Ananthi, S., Vishnu Varthini, S.: Detection of unhealthy region of plant leaves and classification of plant leaf diseases using texture features. Agric. Eng. Int. CIGR J., 211–217 (2013)

15. Eaganathan, U., Sophia, J., Luckose, V., Benjamin, F.J.: Identification of sugarcane leaf scorch diseases using K-means clustering segmentation and K-NN based classification. IJACST, pp. 11–16 (2014)
16. Revathi, P., Hemalatha, M.: Cotton leaf spot diseases detection utilizing feature selection with skew divergence method. Int. J. Sci. Eng. Technol., 22–30 (2014)
17. Mainkar, P.M., Ghorpade, S., Adawadkar, M.: Plant leaf disease detection and classification using image processing techniques. Int. J. Innov. Emerg. Res. Eng., 139–144 (2015)
18. Athanikar, G., Badar, P.: Potato leaf diseases detection and classification system. Int. J. Comput. Sci. Mob. Comput., 76–78 (2016)
19. Singh, V., Misra, A.K.: Detection of plant leaf diseases using image segmentation and soft computing techniques. Inf. Process. Agric. (2016)
20. Ravindra Naik, M., Sivappagari, C.M.R.: Plant leaf and disease detection by using HSV features and SVM classifier. Int. J. Eng. Sci. Comput., 3794–3797 (2016)
21. Vidyaraj, K., Priya, S.: Developing an algorithm for Tomato leaf disease detection and classification. IJIREEICE, 105–108 (2016)
22. Francis, J., Anto Sahaya Dhas, D., Anoop, B.K.: Identification of leaf disease in pepper plant using soft computing. In: ICEDSS (2016)
23. Vijai Singh, V., Misra, A.K.: Detection of unhealthy region of plant leaves using image processing and genetic algorithm. In: ICACEA, pp. 1028–1032 (2015)
24. Pachore, K.A., Kishore, R., Bhawar, S.G.: Leaf disease recognition system. Int. J. Comput. Appl., 31–34 (2016)
25. Sujatha R., Sravan Kumar, Y., Akhil, G.U.: Leaf disease detection using image processing. J. Chem. Pharm. Sci., 670–672 (2017)

Pixel Position Based Efficient Image Secret Sharing Scheme

P. Reshma Sagar, J. Pruthvi Sri Chakra and B. R. Purushothama

Abstract Every image irrespective of whether it is a color image or grayscale image can be mapped to a collection of pixels whose intensity values lie between 0 and 255. Construction of any image whose pixel count is greater than 256 involves the recurrence of any of these 256 pixels. In other words, we can say a pixel of a particular intensity is bound to occur in multiple places inside the image. The paper introduces to the reader an algorithm that exploits the fact that a pixel of particular grayscale value occurs in multiple positions in the image. The algorithm extends the basic idea of image secret sharing which enables a secret image to be shared through n shadow images and retrieved only when the retriever has any r images out of these n shadow images. It ensures that the size of the shadow image remains constant irrespective of the size of the original image. A constant shadow image size would allow the sender to transmit any extra information to the receiver in cases where there is a need to do so. The receiver can easily pinpoint the extra information present inside the shadow image when it has a size greater than the usual size used. This property provides benefits during further processing of shadow images such as polynomial computation and storage.

Keywords Secret sharing · Shadow images · Image sharing · Spatial domain

P. Reshma Sagar (✉) · J. Pruthvi Sri Chakra · B. R. Purushothama
Department of Computer Science, National Institute of Technology Goa,
Farmagudi, India
e-mail: prsreshma@gmail.com

J. Pruthvi Sri Chakra
e-mail: pruthvisrichakra@gmail.com

B. R. Purushothama
e-mail: puru@nitgoa.ac.in

© Springer Nature Singapore Pte Ltd. 2018
P. K. Sa et al. (eds.), *Recent Findings in Intelligent Computing Techniques*,
Advances in Intelligent Systems and Computing 709,
https://doi.org/10.1007/978-981-10-8633-5_52

527

1   Introduction

Digital image sharing has played an prominent role in image transmission due to the increasing demand in the need for transmission and the uses of image resources. The security involved in such transmissions is an inevitable concern especially when they are used in situations where transmission and secrecy go hand in hand. Military transmissions are good examples of such situations.

The basic idea behind image secret sharing is to convert an image into n shadow images which on mere observation do not reveal any information about the original image. The (r, n) threshold scheme is one such scheme that uses n shadow images for sharing any r images out of these n images for retrieval. The idea of using a subset of the n shadow images makes sure that the retrieval is not highly dependent on any single share which provides an edge over the loss of the share. Yet, another key aspect of this scheme is that users cannot obtain the original image even if they have complete knowledge of any $r - 1$ shares.

The main objective of secret sharing methods is to achieve resistance to various image tampering and forgery methods. Many different algorithms and techniques [1, 4–6] have been proposed for secret image sharing in order to satisfy these objectives. In this paper, we use an image as an information carrier and secret sharing as a technique to hide information. It has also been ensured that the image is not tampered during transmission. The rest of paper is organized as follows. Section 2 describes the literature survey for secret sharing schemes. Section 3 describes proposed scheme. Section 4 presents the comparative study of implemented scheme with few existing schemes. Section 5 concludes about the proposed scheme.

2   Literature Survey

There are different categories of secret sharing schemes like secret sharing scheme in the spatial domain [7], secret sharing in frequency domain, visual secret sharing [8], etc. To understand the importance of secret sharing, let us look into a real-life application. Imagine an organization with confidential information. In order to access this confidential information, mutual agreements from all the members of the company are mandatory. Loss of confidentiality will occur only when all members lose their share of the secret the chances of which are low. We base our method on (r, n) threshold scheme proposed by Shamir [2] and image secret sharing method proposed by Thien et al. [3]. Before we proceed to our algorithm, we roughly introduce this scheme.

A. Shamir's (r, n) **Threshold Scheme**: In 1979, Shamir [2] proposed the theory of secret sharing scheme based on Lagrange's polynomial interpolation. In his method, a particular data D is divided into n pieces. In order to reconstruct the data D, any r out of the n pieces can be used. Shamir's method is designed in such a way that it denies the liberty to retrieve D to even authorized people who hold the knowledge of any $r - 1$ pieces. The key aspect used in the retrieval of data is

Lagrangian polynomial interpolation. It is based on the fact that given any r coordinates in the 2-dimensional space $(x_1, y_1,) \dots (x_r, y_r)$ with distinct $x'_j s$, there is one and only one polynomial $p(x)$ of degree $r - 1$ such that $p(x) = y_j$ for all j. Without loss of generality, we can assume that the data D is an integer, if not we convert into integer. Polynomial $p(x)$ of degree $r - 1$ is constructed as follows.

$$p(x) = b_o + b_1 x + \dots b_{r-1} x^{r-1} mod P \tag{1}$$

where P is a prime chosen at random. b_o is initialized to D, and the following values are computed: $D_1 = p(1) \dots D_i = p(i) \dots D_n = p(n)$. Lagrangian interpolation [2] is used to determine coefficients of $p(x)$ if any r of these D_i values are available. The data D is then obtained by evaluating $D = p(0)$. The direct application of Shamir's (r, n) threshold scheme to images is ineffective because we need to replace b_0 by grayscale value of all pixels in the image and obtain corresponding output values. The resulting shadow image of each is of size $n \times n$, if the original image size is $n \times n$.

B. Thiens's Secret Image Sharing:

In 2002, Thien and Lin proposed a method based on Shamir's method to carry out secret sharing of images. Thien et al. [3] improved upon Shamir's method by devising an algorithm in which each shadow image size is less than that of the original image unlike those obtained in Shamir's method. This method has an upper hand over Shamir's method during storage and transmission. The algorithm is segregated into two phases, namely the sharing phase and the reveal phase. Sharing phase is as follows:

1. Pixels with intensity values greater than 250 are converted to 250 so as to allow all grayscale intensity values to fall in the range of 0–250.
2. The pixels are then reordered, using a permutation key, within the image in order to increase security.
3. The image is sectioned so that each section consists of r pixels taken sequentially of the permuted image. These r pixels are used as coefficients $b_0 \dots b_{r-1}$.
4. $p(1), \dots, p(n)$ are computed and placed in the n shadow images.
5. Steps three and four are repeated until all sections are covered. The structure of the equation is as follows:

$$p(x) = (b_0 + b_1 x + \dots + b_{r-1} x^{r-1}) mod 251 \tag{2}$$

In the reveal phase, the coefficients b_0 to b_{r-1} of $b(x)$ are determined using Lagrangian interpolation [2]. Each shadow image size is reduced to $1/r$ times the original image size. This saves storage space and transmission time. The conversion of pixel intensity to 250 from values greater than 250 may prove to be a problem while handling light images as pixel values greater than 250 are lost. In this scheme, the permutation key can lead to single point failure because the key can be corrupted or lost. Our proposed method mainly focuses on the improvement of the Thien and Lin's scheme by using position values instead of pixel intensity.

3 Proposed Approach

We propose a new method for image secret sharing scheme where the size of the shadow image is constant. This method exploits the fact that a pixel of particular grayscale value occurs in multiple positions in the image. The basic idea is to apply Shamir's secret sharing scheme to the positions of a pixel with a particular grayscale value instead of applying it to all or a section of the pixels present in the image.

3.1 Methodology

- Let I be a grayscale image of size $n \times n$
- The grayscale entries in the image lie in the range $[0, 255]$
- We begin by extracting the positions in which a pixel of a particular grayscale value occurs and storing these value.
- Thus let $S_p = \{(i,j)|p$ is the value at position (i,j) in $I\}$, where $0 \le p \le 255$.
- Let $n_p = |S_p|$. That is n_p is the total number of pixels with value p in I.
- We associate polynomials to each S_p as below

  - Let $n_p = k$
  - Let $S_p = \{(i_1,j_1), (i_2,j_2), \dots, (i_k,j_k)\}$
  - Let $S_x = \{i_1, i_2, \dots, i_k\}$
  - Let $S_y = \{j_1, j_2, \dots, j_k\}$
  - Execute the algorithm for **Polynomial Construction** of the polynomial $P_{p_i}(x)$ of degree $k - 1$
  - Execute the algorithm for **Polynomial Construction** of the polynomial $P_{p_j}(y)$ of degree $k - 1$
  - Execute the algorithm for **Image Recovery.**

We observe that in the above method the total number of polynomials to be computed is only 256. Even though a drastic decrease in the number of polynomials is observed, the degree of these polynomials is large. The evaluation of such high powers is found to incur large computation overheads. Thus, in order to reduce the degree of the polynomials, we propose a method to combine two or more positions of a pixel of particular grayscale into one coefficient of the polynomial.

3.2 Polynomial Construction Algorithm

- Let us set the number of terms to be present in the polynomial to a constant value and denote this value by r

- Setting a constant r implies that the number of available coefficients is also constant. Thus, in order to determine the number of positions to be combined into one coefficient, we compute $n_p/r = t$
- Let $p_1, p_2, \ldots, p_t$ be t prime numbers greater than maximum position value of pixel in the image($=n$). We illustrate our method by taking a 256×256 image. Thus, we take prime greater than 256
- We construct the coefficients of polynomials using x coordinates of positions as follows:

$$C_0 = p_1 \times x_1 + p_1 \times p_2 \times x_2 + p_1 \times p_2 \times p_3 \times x_3 \ldots + p_1 \times p_2 \times \cdots p_k \times x_k \quad (3)$$

- Similarly, the coefficients $C_1, C_2, \ldots C_{r-1}$ can be calculated. Therefore,

$$P_{p_i}(x) = C_0 + C_1 \times x + C_2 \times x^2 + \cdots + C_{r-1} \times x^{r-1} \quad (4)$$

- Similarly, we construct the polynomial $P_{p_j}(y)$ using the y coordinates of positions.

We compute $P_{p_i}(1), P_{p_i}(2), \ldots, P_{p_i}(n)$ and $P_{p_j}(1), P_{p_j}(2), \ldots, P_{p_j}(n)$ for all grayscale values p and share these values in the form of shadow images. Each shadow image contains at least 512 pixels as p in P_{p_i} and P_{p_j} takes values from 0 to 255. In an attempt to place these pixels in a shadow image of equal dimensions, we compute the square root of 512 which is approximately equal to 23. Thus, the shadow images are of size 23×23.

3.3 Image Recovery Algorithm

In order to retrieve the position values, the receiver must perform the following step:

- Use Lagrangian interpolation to determine the polynomial.
- Let C_i be ith coefficient where $0 \le i < r$.
- Obtain the position values by executing the following steps to each of the coefficients

1. Initialize $j = 1$.
2. Perform steps 3 to 6 till $C_i > p_j$.
3. Divide C_i by p_j and store result in C_i.
4. Perform modulus operation on C_i with p_{j+1}. The value obtained will be x_j.
5. Subtract x_j from C_i and store result in C_i.
6. Increment j.

- Perform the above process for all r coefficients and get the position values of a pixel with grayscale value p.
- Reconstruct the original image using p and the positions in which these pixels occur.

Table 1 Comparison study for 1024 × 1024 image

Parameters	Naor and Shamir method	Thien's method	Proposed method
No. of. polynomials	(1024 × 1024)	(1024 × 1024)/r	512 (always constant)
Shadow image size	1024 × 1024	1024 × 4	23 × 23
Number of pixels	1024 × 1024 × 127	1024 × 4 × 127	23 × 23 × 127
Total pixels	133169152	520192	67183

We know that all position values lie between 0 and 255 if we are taking a grayscale image of size 256 × 256. Suppose we choose a prime number less than 256; there are chances that they will be factors of any of the position values. Thus, while performing modulus operation for obtaining these positions in Step 3 to 6 (explained previously), the remainder would be zero and the position value would be lost. So we choose primes $p_1, p_2, \ldots, p_t$ greater than 256. The primes are to be chosen according to the size of the original image. If size equals $n \times n$, all primes must be greater than n.

4 Comparison Study

The encapsulation of multiple position values into a single coefficient proposed in our method allows us the liberty of choosing the number of terms in the polynomials constructed. Table 1 is a comparison study of our method, Naor and Shamir method, and Thien's method performed on a grayscale image of size 1024 × 1024 taking the number of terms in the polynomial to be 127.

The ratio of pixels between Naor and Shamir method (lossless image recovery) and proposed scheme $= 133169152/67183 = 1982.19$ and the ratio of pixels between Thien's method and proposed scheme $= 520192/67183 = 7.743$. Thus, our method performs 7.743 times better than Thien's method theoretically. The total number of polynomial computation is always a constant and equals to 512. The number of terms in the polynomial is independent of the size of the original image.

5 Conclusion

This proposed method approaches image secret sharing from the perspective of the pixel position. The concept proposed in our method uses prime numbers greater than 256 (in general greater than the size of the image) to encapsulate multiple position values, which aids us in the computation of large powers. The main highlights of our method are as follows: (i) It produces shadow images whose size is always constant and independent of the size of the original image. This aids during storage, accessing, and transmission of the shadow images. (ii) The method also reduces the number of polynomials computed by making the number a constant irrespective of the number of terms and the size of the original image. The constant size would also allow the sender to attach another information about the shadow images before transmitting to the receiver. The receiver would have no trouble detecting any extra information present within the shadow image.

References

1. Saleh, S., Balafar, M.A.: An investigation on image secret sharing. Int. J. Secur. Its Appl. **9**(3), 163–190 (2015)
2. Shamir, A.: How to share a secret. Commun. ACM **22**(11), 612–613 (1979)
3. Thien, C.-C., Lin, J.-C.: Secret image sharing. Comput. Graph. **26**(5), 765–770 (2002)
4. Rishiwal, V., Gupta, A.: An efficient secret image sharing scheme. In: World Applied Programming, vol. 2, Issue (1), pp. 42–48, Jan 2012
5. Wang, K., Zou, X., Sui, Y.: A multiple secret sharing scheme based on matrix projection. In: Annual IEEE International Computer Software and Applications Conference (2009)
6. Patil, S., Deshmukh, P.: Enhancing security in secret sharing with embedding of shares in cover images. Int. J. Adv. Res. Comput. Commun. Eng. **3**(5) (2014)
7. Mohamed, F.P., Arockia Jansi Rani, P.: (N, N) Secret color image sharing scheme with dynamic group. Int. J. Comput. Netw. Inf. Secur. **7**, 46–52 (2015)
8. John Justin, M., Alagendran, B., Manimurugan, S.: A survey on various visual secret sharing schemes with an application. Int. J. Comput. Appl. (0975 8887) **41**(18) (2012)

Optimal Approach for Image Recognition Using Deep Convolutional Architecture

Parth Shah, Vishvajit Bakrola and Supriya Pati

Abstract In the recent time, deep learning has achieved huge popularity due to its performance in various machine learning algorithms. Deep learning as hierarchical or structured learning attempts to model high-level abstractions in data by using a group of processing layers. The foundation of deep learning architectures is inspired by the understanding of information processing and neural responses in human brain. The architectures are created by stacking multiple linear or nonlinear operations. The article mainly focuses on the state-of-the-art deep learning models and various real-world application-specific training methods. Selecting optimal architecture for specific problem is a challenging task; at a closing stage of the article, we proposed optimal approach to deep convolutional architecture for the application of image recognition.

Keywords Deep learning · Image recognition · Transfer learning · Deep neural networks · Image processing · Convolutional neural networks

1 Introduction

In any artificial intelligence problem, we require two main things. First, we need to identify and extract right set of features that represent the problem. Second, we need to have an algorithm that takes these extracted features and provides predicted outputs. Identifying right set of features itself is a challenging task especially when we are dealing with the images. The solution to this problem is to allow machines to learn from their own experience instead of fixed rules and to understand concepts

P. Shah (✉) · V. Bakrola · S. Pati
C.G. Patel Institute of Technology, Uka Tarsadia University, Bardoli, India
e-mail: parthpunita@yahoo.in

V. Bakrola
e-mail: vishvajit.bakrola@utu.ac.in

S. Pati
e-mail: supriya.pati@utu.ac.in

© Springer Nature Singapore Pte Ltd. 2018
P. K. Sa et al. (eds.), *Recent Findings in Intelligent Computing Techniques*,
Advances in Intelligent Systems and Computing 709,
https://doi.org/10.1007/978-981-10-8633-5_53

in terms of hierarchy of simpler concepts. If we create a graph that shows how these concepts are stacked on each other, then that resulting graph becomes deep with high number of layers. This is why we call this approach deep learning. In deep learning, we normally use deep neural networks where each neuron in same layer maps different features and each layer will combine features of previous layer to learn new shapes. But this brings new challenge of choosing the perfect strategy for implementing deep learning architecture as accuracy and time required for training depend on it.

Recently, very deep convolutional neural networks are in main focus for image or object recognition. For the task of image recognition, several different models like LeNet, AlexNet, GoogLeNet, ResNet, Inception-ResNet, etc. are available. Most of these models are result of ImageNet Large-Scale Visual Recognition Challenge (LSVRC) and MSCOCO competition, which is a yearly competition where teams from around the globe compete for achieving best accuracy in image recognition task. In ImageNet LSVRC, evaluation criteria are top-1 and top-5 error rate, where the top-N error rate is the fraction of test images for which the correct label is not among the N labels considered most probable by the model. In addition to that, we can judge any deep neural network based on computation cost, memory it requires to execute, etc. Selecting best appropriate model from that is tricky task. It depends on size of input, type of input, as well available resources. In this paper, we have analyzed effect of different training methods on these models and evaluated performance on different sizes of dataset. We have also discussed the benefits and trade-offs of increasing number of layers.

In Sect. 2 of this paper, we have presented literature review of different deep learning models for image recognition in detail. In Sect. 3, we have described different training methodologies for deep neural network architectures. Section 4 describes implementation environment used for implementing various models described in literature review. Comparison of these model under various scenarios is presented in Sect. 5. In Sect. 6, we conclude the finding of this paper.

2 Literature Review

First concept of deep learning was introduced way back in early 80s when computers were not in even day-to-day usage. In 1989, Yann LeCun successfully demonstrated deep convolutional neural network called LeNet for task of handwritten character recognition. But it was not further developed because there was not enough data and high computation power available at that time which was required by deep neural networks. This slowed down the research in area of deep learning. The new wave of research started only after Alex Krizhevsky successfully demonstrated use of deep convolutional neural networks by beating traditional object recognition methods in ImageNet LSVRC-2012 by large margin in 2012. After that, year-on-year new deep neural networks were introduced like GoogLeNet, ResNet, Inception-ResNet, etc. Each architecture was designed to have more accuracy than its precedent. We have covered all these models in brief in this section.

2.1 LeNet

LeNet consists of five convolution layers for feature extraction and object detection. Before LeNet, people used methods that require feature vector to be provided to algorithm. These feature vectors need to be handcrafted from knowledge about the task to be solved. This problem is solved in LeNet by using convolutional layers as a feature extractor [1]. These convolutional layer's weights are learnable parameters which we can use during training process. Due to usage of convolution layers in LeNet, it requires high computation power. In order to tune weight of these convolutions, higher number of training dataset is required. These two were the main limitations of LeNet when it was introduced. This prevented further development in deep learning in early 80s.

2.2 AlexNet

One of the major breakthroughs in the area of deep learning was achieved when Alex Krizhevsky successfully demonstrated use of convolutional neural networks by beating traditional object recognition methods in ImageNet LSVRC by large margin in 2012. AlexNet's architecture was based on concept established by LeNet. AlexNet uses total eight different hidden layers. From this eight layers, first five layers are convolution layers and other three layers are fully connected layers. Here, these convolution layers are used for feature extraction task. Lower layer will extract basic features like edges. As we go to higher level, it combines shapes from lower layer and identifies shapes. Output of last fully connected layer is fed as input to 1000 way softmax layer which acts as output layer which represents 1000 class of ImageNet dataset [2]. Softmax layer will output probability of each output class between 0 and 1, where 0 means object is not present in an image, while 1 means object is present in an image. Demonstrated AlexNet model at ILSVRC-2012 used GPU for meeting computation needs of these convolution layers. AlexNet had directly reduced top-5 error rate of 26% of 2011 ImageNet winner to 16% which was more than 10% improvement in single year. One of the main reasons of this huge performance improvement was instead of hard coding of features; it had extracted featured automatically using deep convolutional layers like LeNet [3].

2.3 GoogLeNet

Although AlexNet improves accuracy greatly compared to traditional architectures, it was not able to provide human-like accuracy because only eight layers were not able to extract all features needed for identifying all 1000 classes in ImageNet dataset. Based on architecture of AlexNet, Szegedy et al. developed new deep convolutional neural network architecture called GoogLeNet in 2014 [4]. GoogLeNet took concept of building hierarchy of feature identifier from AlexNet and stacked layers in form of inception modules. It uses the concept of network-in-network strategy [5] where the whole network is composed of multiple local networks called "inception" module. These inception modules consist of 1×1, 3×3 and 5×5 convolutions. All convolutional layers in GoogLeNet are activated by use of rectified linear unit. GoogLeNet has 3 times more layers compared to AlexNet. Architecture of GoogLeNet is 27 layers deep (more than 100 layers if we count layers in inception module separately). Inception module is designed such that it provides better result as compared to directly stacking layer on one another like in AlexNet. The network was designed with computational efficiency and practicality in mind, so that inference can be run on any devices including those with limited computational resources and low-memory footprint. GoogLeNet has achieved around 6.67% top-5 error rate and won the ImageNet LSVRC-2014 competition [6].

2.4 ResNet

Even with the optimized architecture of GoogLeNet, deep neural network remained difficult to train. In order to increase the ease of training and accuracy, the concept of residual connection was added to deep neural network. In 2015, ResNet architecture proposed by He et al. achieved superhuman accuracy of just 3.57% error rate using residual connections in ImageNet LSVRC-2015 competition and MSCOCO competition 2015 [7]. The reason for adding residual connection was that when deeper network starts converging, problem of degradation occurs where accuracy degrades rapidly after some point in training. In order to solve this problem, instead of hoping that each stacked layer directly fits a desired underlying mapping, it explicitly lets these layers fit a residual mapping using newly added residual connections. Architecture of ResNet was based on VGGNet [8] which was the runners-up of ImageNet LSVRC-2013 competition. VGGNet was composed of fixed size convolution layer with varying number of deepness of architecture. In ResNet, between each pair of 3×3 convolutional layer, a shortcut residual connection was added. Authors have proposed different architectures of ResNet having 20, 32, 44, 56, 110, 152, and 1202 layers. But as we increase number of layers computation, complexity increases due to higher number of convolution operations.

2.5 Inception-ResNet

In order to optimize architecture of deep neural networks, Szegedy et al. introduced residual connection's concept from ResNet into inception module in their GoogLeNet architecture and proposed Inception-ResNet architecture in 2016. It helps to keep performance of network while accelerating training of network using residual connection. Inception-ResNet was able to achieve error rate of just 3.08% over ILSVRC dataset [9]. In Inception-ResNet, shortcut connection was added between each inception module. Inception-ResNet used two simple convolution models of ResNet with single inception module. Authors have proved that in addition to increasing model size, residual connection also increases training.

3 Training Methodologies of Deep Neural Networks

Once architecture of neural network is defined, we need to train it so that it can learn the given problem. Training can be done in various ways, but its main goal is to map given input data to its appropriate given output value. Once training is done, we can save final updated mapping and use it while performing inference on test data. Two of the approaches that are highly used in training for deep neural networks are as follows.

3.1 Training from Scratch

This is the most common method for training neural networks. In this method at the start of training, correct class for each test data is known. Then, we will initialize weights of all layers randomly. After initialization, we will try to map these class labels with actual input. We will adjust weight of each neuron such that model will learn to predict actual class label. Using this approach, we will train all the available neurons in networks such that model will learn to output correct label. As this approach requires to update each neuron's weight, it requires large dataset for providing higher accuracy.

3.2 Training Using Transfer Learning

Generally, traditional algorithms were developed to train on specific task as it is required to extract features manually, but with the introduction of deep learning, process of feature extraction become part of neural network itself. This introduces the new window of opportunity for generalized architecture that can deal with more than one type of problem that is where the concept of transfer learning is introduced. Transfer learning provides the advance way of learning in machine learning algorithms. Transfer learning is the method of improvement of knowledge in a new task using knowledge transfer from previously learned similar task [10]. The most common method for applying transfer learning is to only train neuron of final layer keeping neurons of all other layers fixed. This approach will greatly speedup process of learning. Generally, in deep learning, due to higher number of layers, there are larger number of trainable weights are there. But in most of the cases only weight in last layers is the deciding factor for output generation. This makes transfer learning best suitable training method when we want to train model with higher number of layers but we get only small dataset due to some limitations.

4 Implementation

For evaluating performance of different models of deep neural network architecture, we have implemented some of the best models based on accuracy for object recognition and compared its performance based on different dataset sizes, different numbers of labels, and different training methods.

Dataset and implementation environment were used for implementing, and these image recognition models are discussed in detail in this section.

4.1 Datasets

Different datasets we used for testing are as follows:

MNIST Handwritten Digit Dataset: The MNIST database is a collection of handwritten digits that is created by National Institute of Standard and Technology. It contains 28×28 pixel grayscale image for English numerical digits [1]. The standard MNIST database contains 60,000 training images and 10,000 testing images. We referred it to MNIST full dataset throughout this paper. In addition to this, we have created two another datasets, from which one dataset contains subset of around 12,000 image called MNIST small and another contains total 1,40,000 images called MNIST inverted dataset.

Flowers Dataset: Flower dataset consists of different flower images that are collected from the Internet having creative commons license. It has only five different labels and each label contains around 650 images.

Yale Face Dataset: Yale face dataset is collection of face images captured under different lighting conditions and different angles for multiple persons. It has around 40 different faces where each face has around 60 different images [11].

4.2 Implementation Environment

Implementation and training of these models are done using tensorflow deep learning framework using two different machine setups. They are as follows:

Setup 1: First machine is an AWS GPU instance with Xeon E5 processor with 8 core and NVIDIA K520 GPU having 3072 CUDA cores and 8 GB GPU memory and 15 GB RAM.

Setup 2: Second machine is dedicated hardware with Intel Xeon E3 processor with 12 cores and 32 GB RAM.

5 Comparison

5.1 Training with MNIST Dataset

In the first experiment, we have trained LeNet, AlexNet, and GoogLeNet using MNIST dataset over hardware setup 1. Before performing actual training of the network, we need to preprocess them. In preprocessing, we have converted all image files to single file such that it can be used for batch processing.

As we can see from Fig. 1, time required for preprocessing is dependent on number of images in dataset as well as number of channels in image.

After data is preprocessed, we apply it to train our models. Each model is trained for 10 epoch with base learning rate of 0.01 and learning rate will decrease with every three steps in factor of 0.01 to increase the efficiency of training. We have measured Top-1 accuracy, Top-5 accuracy, and training loss for our training in Table 1. As Top-5 accuracy was introduced as a criterion for ImageNet LSVRC competition in 2013,

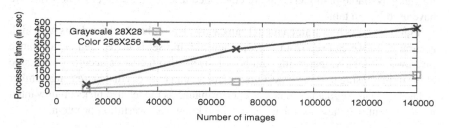

Fig. 1 Preprocessing cost for different sizes of dataset

Table 1 Performance of different models on MNIST

Dataset	MNIST small	MNIST full	MNIST small	MNIST full	MNIST small	MNIST full
Model	LeNet	LeNet	AlexNet	AlexNet	GoogLeNet	GoogLeNet
Layers	5	5	8	8	22	22
Accuracy in top-1 (%)	96.80	98.86	92.22	99.24	85.50	98.26
Accuracy in top-5	NA	NA	NA	NA	98.5372%	99.96%
Loss	0.105	0.040	0.256	0.024	0.438	0.061
Time (s)	50	45	454	2010	1167	9300

(a)   **(b)**

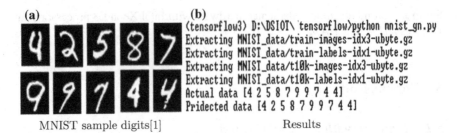

MNIST sample digits[1]   Results

Fig. 2 Handwritten digit recognition example

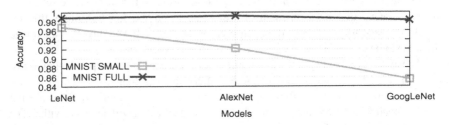

Fig. 3 Accuracy of different models

it is only available in models that are developed after that. So for other network, we have used NA in table.

Figure 2 shows how trained model can easily identify handwritten digits with high accuracy. As we can see from Fig. 3, when training set is large, then performance improves. Similarly, increasing number of layers also increases accuracy with side effect of increasing training time. Other thing we can notice is that increasing number of layers on smaller dataset decreases performance as model also picks up unwanted noise present in dataset as a feature. It is also known as overfitting problem.

5.2 Training with Flowers Dataset

As we can see from previous experiment, for smaller dataset, performance degrades compared to larger dataset. In order to solve this degradation problem, transfer learning method is used. In this experiment, we have trained GoogLeNet model from scratch and also trained GoogLeNet and Inception-ResNet using transfer learning on Flowers dataset. For training all three models, we have used hardware setup 2 which is described in previous section.

Simple training is done for GoogLeNet using 40,000 epoch in case of learning from scratch, while for transferred leaning on GoogLeNet and Inception-ResNet, 10,000 training epoch is used with base learning rate of 0.01 and learning rate will

(a)   **(b)**

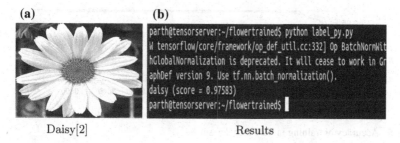

Daisy[2]   Results

Fig. 4 Daisy identification example

Table 2 Comparison of learning from scratch versus transfer learning

Training method	From scratch	Transfer learning	Transfer learning
Model	GoogLeNet	GoogLeNet	Inception-ResNet
Accuracy (top-1) (Number of epoch	40,000	10,000	10,000
Training time (h)	245	12	14

decrease every 30 step with factor of 0.16 for optimizing learning process. Simple example of trained model using transfer learning is shown in Fig. 4.

As seen from results in Table 2, using transfer learning, we can achieve better performance compared to training from scratch as in transfer learning, it preserves the features extracted when it was previously trained on larger dataset. This greatly decreases the training time required for training maintaining accuracy.

5.3 Training with Yale Face Dataset

Normally, in real-world usage scenario of any image recognition problem, inputs are taken from different image capturing devices which are fixed at particular place like security cameras. It is not always possible to have full object captured by camera every time. Our model should be capable of identifying an object even when only partial object is present in image. In order to check how model performs in case of incomplete or partial images, we have first trained model with Yale face dataset and then tested it with image containing partial objects. This experiment was performed on hardware setup 2 using transferred learning approach. Base learning rate of 0.01 is used for initialization of training. Learning rate will decrease every 30 step with factor of 0.16.

Figure 5 shows overall training accuracy and validation accuracy for Yale face dataset. As you can see after around 1000 epoch, improvement of accuracy is minimal.

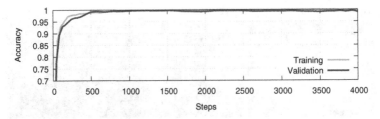

Fig. 5 Accuracy of training in Yale face dataset

(a)   **(b)**

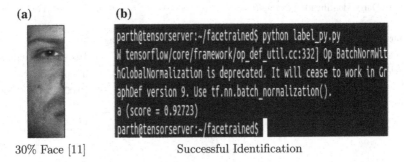

30% Face [11]   Successful Identification

Fig. 6 Face identification in case of partial image

Table 3 Results in case of partial object in image

Amount of object in image (%)	Average output of Softmax layer
10	0.92642
20	0.84098
30	0.92730
40	0.92361
50	0.81139
60	0.86707

As seen from Fig. 6, that model provides correct results even when there is only 30% of face present in image. This is due to nature of deep learning that extracted minute features that were previously not possible with hardcoded features of traditional approaches.

In Table 3, result of Softmax output layer is given for sample set of images, which shows that model provides correct results in most case even when objects are incomplete in an image. This proves the ability of deep learning to tolerate incompleteness of input.

6 Conclusion

In the presented work, we have evaluated performance of different models for image recognition. Based on the derived performance evaluations, we found Inception-ResNet with highest accuracy, while keeping moderate computation requirements. Model with higher number of hidden layers improves accuracy but causes overfitting with small dataset. In order to prevent overfitting, we have used transfer learning method that efficiently reduced training time as well as overfitting without affecting accuracy of model.

Acknowledgements We would like to thank Department of Computer Engineering, C. G. Patel Institute of Technology for providing us computer resources as and when needed for training and implementing models presented in this paper.

References

1. Lecun, Y., Bottou, L., Bengio, Y., Haffner, P.: Gradient-based learning applied to document recognition. Proc. IEEE **86**(11), 2278–2324 (1998)
2. Deng, J., Dong, W., Socher, R., Li, L.J., Li, K., Fei-Fei, L.: Imagenet: a large-scale hierarchical image database. In: IEEE Conference on Computer Vision and Pattern Recognition, 2009. CVPR 2009, pp. 248–255. IEEE (2009)
3. Krizhevsky, A., Sutskever, I., Hinton, G.E.: Imagenet classification with deep convolutional neural networks. In: Advances in Neural Information Processing Systems, pp. 1097–1105 (2012)
4. Szegedy, C., Liu, W., Jia, Y., Sermanet, P., Reed, S., Anguelov, D., Erhan, D., Vanhoucke, V., Rabinovich, A.: Going deeper with convolutions. In: Proceedings of the IEEE Conference on Computer Vision and Pattern Recognition, pp. 1–9 (2015)
5. Lin, M., Chen, Q., Yan, S.: Network in network. CoRR. arXiv:abs/1312.4400 (2013)
6. Szegedy, C., Vanhoucke, V., Ioffe, S., Shlens, J., Wojna, Z.: Rethinking the inception architecture for computer vision. arXiv:1512.00567 (2015)
7. He, K., Zhang, X., Ren, S., Sun, J.: Delving deep into rectifiers: surpassing human-level performance on imagenet classification. CoRR. arXiv:abs/1502.01852 (2015)
8. Simonyan, K., Zisserman, A.: Very deep convolutional networks for large-scale image recognition. CoRR. arXiv:abs/1409.1556 (2014)
9. Szegedy, C., Ioffe, S., Vanhoucke, V.: Inception-v4, inception-resnet and the impact of residual connections on learning. arXiv:1602.07261 (2016)
10. Torrey, L., Shavlik, J.: Transfer learning. In: Handbook of Research on Machine Learning Applications and Trends: Algorithms, Methods, and Techniques, vol. 1. Information Science Reference (2009)
11. Georghiades, A.S., Belhumeur, P.N., Kriegman, D.J.: From few to many: illumination cone models for face recognition under variable lighting and pose. IEEE Trans. Pattern Anal. Mach. Intell. **23**(6), 643–660 (2001)

A Hybrid Approach Optical Character Recognition for Mizo Using Artificial Neural Network

J. Hussain and Vanlalruata

Abstract The paper presents an overview of a new hybrid approach for Mizo optical character recognition. The preprocessing enhances the accuracy of recognition by removing noise. After this image enhancement, a hybrid approach character segmentation using bounding box and morphological dilation is performed which merges the isolated blobs of Mizo character into a single entity. In feature extraction, another hybrid method is derived using zoning and topological feature which enhances multi-font and multi-size character recognition. The segmented characters are subdivided into nine equal zones and from each zone, four topological features are extracted, forming a total of $9 \times 4 = 36$ features for each character. These features are used for classification and recognition. An experiment is carried out in 24 different types of fonts using four different types of artificial neural network architecture. Each architecture is compared and analysed. The backpropagation neural network has the highest accuracy, with above 99% rate of recognition.

Keywords Mizo OCR · Pattern recognition · Hybrid OCR · Hybrid segmentation technique for isolated character · Hybrid feature extraction Artificial neural network · Mizo image to text · Character image processing

1 Introduction

The Mizo Optical Character Recognition (OCR) is the process of translating images of printed text into a format understood by machines for the purpose of editing, indexing/searching, preserving and reduction in storage size [1]. The Mizo language is the main language in the state of Mizoram; it is also used in the

J. Hussain (✉) · Vanlalruata
Department of Mathematics and Computer Science, Mizoram University,
Aizawl 796009, Mizoram, India
e-mail: jamal.mzu@gmail.com

Vanlalruata
e-mail: ruata.mzu@gmail.com

© Springer Nature Singapore Pte Ltd. 2018
P. K. Sa et al. (eds.), *Recent Findings in Intelligent Computing Techniques*,
Advances in Intelligent Systems and Computing 709,
https://doi.org/10.1007/978-981-10-8633-5_54

547

neighbouring states including Manipur, Nagaland, Tripura and even in some part of Bangladesh and Burma. Therefore, there is a great need of accurate OCR for Mizo character as lots of vital older information have been held captive in the form of hard copy documents. The training dataset consists of all the Mizo characters, including vowels, consonants, special characters and number forming a total of 93 characters for one type of font. In this experiment, 24 different types of font were used. Therefore, in a total of $93 \times 24 = 2232$, individual character sets are used for training. Testing data are image with low contrast, image without noise and image with noise (evenly distributed noise and unevenly distributed noise).

1.1 Properties of Mizo Script

Basic Mizo character sets are comprised of 25 alphabets, 6 vowels and 10 numerical. Based on the variation of vowel character, the tone of the language changes which in turn change the meaning. Therefore, these vowel characters play a vital role in the language. The incorporated vowels are Â, â, Ê, ê, Î, î, Ô, ô, Û, û, Ṭ and ṭ.

2 Methodology

This Mizo OCR system can be divided into several stages as depicted in Fig. 1.

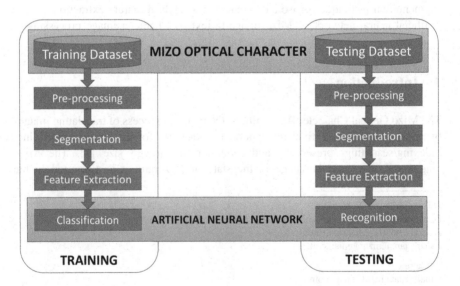

Fig. 1 Stages of Mizo OCR

2.1 Preprocessing

The recognition accuracy is greatly affected by the quality of the actual text and the presence of noise [2]. In this experiment, two types of noise known as Gaussian noise and salt pepper noise [3] are taken into consideration. Padding is applied in median filter [4], average filter [5] and Wiener filter [6]. Padding preserves the edge of each connected component pixel. In this finding, the average filter is better for Gaussian noise, whereas median filter is better for salt-and-pepper noise. This evaluation is performed by taking Root-Mean-Square Error (RMSE) as in Eq. (1).

$$RMSE = \sqrt{\frac{\sum (f(i,j) - g(i,j))^2}{mn}} \tag{1}$$

Here, $f(i,j)$ is the original image with noise, $g(i,j)$ is the enhanced image and m and n are the total number of pixels in the horizontal and vertical dimensions of the image. The low value of RMSE indicates the better enhancement. Comparison of the filter under consideration is given in Tables 1 and 2.

2.2 Segmentation

In segmentation, each individual character is segmented. However, it is difficult to segment some of the Mizo characters, as the Mizo vowel characters consist of two separate isolated blobs (collection of a connected pixel). Therefore, a new hybrid approach using bounding box and morphological dilation is implemented to connect the isolated blobs which form one character. The dataset image is duplicated; the original image is named as foreground and duplicated image as background image. Image processing known as morphological dilation [7] is performed in the

Table 1 RMSE for 1024 * 800 with 300 dpi of Gaussian noise range from 10 to 50%

Filter	10%	20%	30%	40%	50%
Average	26.267	31.520	40.976	69.660	97.524
Median	27.851	33.422	43.448	73.862	103.407
Winner	26.987	32.385	42.100	71.570	100.199

Table 2 RMSE for 1024 * 800 with 300 dpi of salt-and-pepper noise range from 10 to 50%

Filter	10%	20%	30%	40%	50%
Average	13.9821	18.1767	23.6297	30.7186	39.9342
Median	6.3423	8.2449	10.7187	13.9340	18.1142
Winner	21.9425	28.5252	37.0828	48.2076	62.6699

Fig. 2 Problem of direct
segmentation using projection
profile

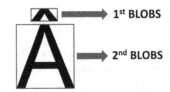

background image by merging separate blobs as represented in Fig. 2 into a single blob so that the character represented above could be combined into a single entity. Subsequently, binarization and invert operations are applied. For some characters such as A B D O P R, there will be isolated pixel after this operation. This isolated pixel is removed using bounding box [8]; the pixel inside the inner bounding box is removed. This stage of segmentation is depicted in Fig. 3.

2.3 Feature Extraction

Feature extraction is a critical part of OCR in neural network; the challenging factor is to extract a unique feature to represent each segmented character. A unique feature enhances the performance of an OCR in terms of recognition [9]. In this process, we derived a hybrid algorithm which fuses together a technique known as zoning [10] and topological feature [11]. To maintain uniformity with the multi-font size, we resized the segmented image into an equivalent scale of size 12×9 pixel, and then it is divided into nine equal zones. From each zone, topological feature is extracted which gives detailed pattern of information for each character. The zoning is depicted in below Fig. 4a.

After zoning, a process known as thinning [12] is performed. This simplifies the topological feature extraction process, by reducing the computation cost, as the number of pixels to compute is reduced. After the process of zoning and thinning, topological features are extracted. To achieve this, a 3×3 matrix is used as represented in Fig. 4b. In this matrix, the centre pixel is represented by C. The neighbouring pixel is numbered starting from pixel below the centre pixel with 1 till 8 in a clockwise direction. To extract topological vector, the algorithm travels through the entire pixels by overlapping the 3×3 matrix over each pixel from each zone. When the pixel is aligned with the centre 'C' of this 3×3 matrix, a number label is assigned to the aligned pixel base on the position of its next

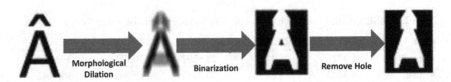

Fig. 3 Stages of segmentation

Fig. 4 Graphical representation of **a** zoning and **b** 3 × 3 matric

(a)

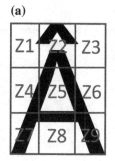

(b)

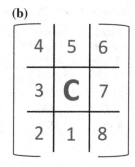

corresponding pixel. If the corresponding pixel is present in the top left corner, then 4 is given as a label, 5 is given if, in the top horizontal, 6 is given if top right corner and so on. However, topological feature is not exposed by this labelling using number. Therefore, rules are defined to generate information about the topology of this labelling using number. If the maximum occurring label is 2 or 6, the type of topology from this zone is identified as a right diagonal. If the maximum occurring label is 4 or 8, then it is identified as left diagonal. If the maximum occurring label is 1 or 5, then it is identified as a vertical line. If the maximum occurring label is 3 or 7, then it is identified as a horizontal line.

Based on this information, four topological feature vectors are generated. (1) Identified number of horizontal lines, (2) identified number of vertical lines, (3) identified number of right diagonal lines and (4) identified number of left diagonal lines. These feature vectors are extracted individually from each zone. For example, if there are N zones, there will be 4 × N feature vector. So in our case, there will be nine zones where each zone will be having four features, forming a total of 9 × 4 = 36 features which will be extracted for each segmented character.

2.4 Classification and Recognition

In classification and recognition, 36 feature vectors are obtained from feature extraction and 93 possible target values which represent the Mizo character are feed to the artificial neural network during training. In this experiment, an attempt is made to train using different neural network architectures such as BBNN (Back-propagation neural network), RNN (Recurrent neural network), RBF (Radial basis function) and LVQ (Learning vector quantization). After training, the weight of the network is saved. In recognition stage using the saved weight, a test data is simulated. The testing dataset is a data consisting of actually Mizo character on which an optical character recognition is performed.

Table 3 Mizo OCR rate of recognition

Neural network classifier	No. of character tested	Correctly recognize character	Misrecognize character	Accuracy	MSE	Time taken in second
BPNN	2320	2312	8	99.52	0.98	73.17
RBF	2320	1309	11	99.12	0.98	74.83
LVQ	2320	2260	60	97.22	0.97	77.53
RNN	2320	2297	23	98.68	0.98	77.53

Table 4 Misrecognized character using backpropagation neural network

Document image	Misrecognize character	Recognized as	No of occurrence
Doc 1	0	û	2
Doc 2	"	,,	2
Doc 3	!	1	3
Doc 4	1	I	1
Total			8

3 Experiment and Results

The proposed Mizo OCR is trained with a training dataset which is comprised of 29 lower case, 29 upper case, 10 numeric and 25 special characters with 24 different types of font. The total number of character $(29 + 29 + 10 + 25) \times 24 = 2232$ is used for training the network. The testing dataset consists of four types of font including Arial, Cambria, Tahoma and Times New Roman, which are extracted from real-life documents such as a Mizo character laser-printed document, Vanglaini daily local newspaper, Scan Mizo Bible and Scan Mizo Christian Songbook. This total consists of 2320 characters for testing. The recognition accuracy is given in Table 3.

The backpropagation neural network has the highest rate of recognition achieving 99.52%. Out of 2320 characters, only 8 characters are misrecognized. The misrecognized characters using BPNN are given in Table 4.

4 Conclusion

The misrecognized characters from the listed table above clearly indicate that the pattern (feature) of these misrecognized characters exhibits very similar feature when taken in a pixel depth orientation. However, an improvement may be achieved by refining the feature extraction algorithm. In future, various advancements could be made to the system like adding text to speech to enable blind or visually impaired users to read the Mizo text that is displayed on the computer screen with a speech synthesizer. In fact, this work is a milestone with a promising application, as the Mizo text remains mostly untapped in this field of computing.

References

1. Mori, S., Nishida, H., Yamada, H.: Optical Character Recognition. Wiley (1999)
2. Chang, F.: Retrieving information from document images: problems and solutions. Int. J. Doc. Anal. Recogn. **4**(1), 46–55 (2001)
3. da Silva, A.R.F.: Wavelet denoising with evolutionary algorithms. Digit. Sig. Proc. **15**(4), 382–399 (2005)
4. Hwang, H., Haddad, R.A.: Adaptive median filters: new algorithms and results. IEEE Trans. Image Process. **4**(4), 499–502 (1995)
5. Wang, Z., Zhang, D.: Progressive switching median filter for the removal of impulse noise from highly corrupted images. IEEE Trans. Circ. Syst. II Analog Digit. Sig. Process. **46**(1), 78–80 (1999)
6. Kumar, S., Kumar, P., Gupta, M., Nagawat, A.K.: Performance comparison of median and wiener filter in image de-noising. Int. J. Comput. Appl. **12**, 0975–8887 (1999)
7. Haralick, R.M., Sternberg, S.R., Zhuang, X.: Image analysis using mathematical morphology. IEEE Trans. Pattern Anal. Mach. Intell. **4**, 532–550 (1999)
8. Lempitsky, V., Kohli, P., Rother, C., Sharp, T.: Image segmentation with a bounding box prior. In: 2009 IEEE 12th International Conference on Computer Vision, pp. 277–284 (2009)
9. Alam, M.M., Kashem, D.M.A.: A complete Bangla OCR system for printed characters. JCIT **1**(01), 30–35 (2010)
10. Murthy, O.R., Hanmandlu, M.: Zoning based Devanagari character recognition. Int. J. Comput. Appl. **27**(4) (2011)
11. Tou, J.T., Gonzalez, R.C.: Recognition of handwritten characters by topological feature extraction and multilevel categorization. IEEE Trans. Comput. **100**(7), 776–785 (1972)
12. Wang, P.S.P., Zhang, Y.Y.: A fast and flexible thinning algorithm. IEEE Trans. Comput. **38**(5), 741–745 (1989)

Optimized Image Compression Method for Portable Devices

Maganti Syamala, Lakshmana Phaneendra Maguluri, Amarasalg Gopala Gupta, T. Akhila and T. Bhargav

Abstract The rapid advancements in technology have led to an increase in the exchange of data (image, video, audio, etc.) between portable devices. This invokes the necessity of building the algorithms which consume low power with no compromise in the performance. In this paper, the abovementioned issue is taken into account and accordingly an image compression technique using Repetitive Iteration CORDIC (RICO) architecture has been proposed. The proposed method is power efficient as it uses RICO for Discrete Cosine Transform (DCT) coefficient generation and performs equally well as compared to JPEG. Results have been obtained via MATLAB and they show that our proposed technique performs equally well and consumes less power compared with the other.

Keywords Repetitive iteration CORDIC · Discrete cosine transform

1 Introduction

Unlimited power!!! Do we really have it or can generate it? The answer is obvious, we do not and we cannot. The number of multimedia devices in user hand is exponentially increasing day by day which is consuming a lot of power and there is no end to this. So, it has become the need of the day to come up with solutions that consume less power and are equally efficient in performance. It is also essential to maintain the Speed–Power–Area–Accuracy (SPAA) tradeoff for which we need to balance some parameters at the cost of others.

M. Syamala
Department of Computer Science and Engineering, Dhanekula Institute
of Engineering and Technology, Ganguru, Vijayawada 521139, India

L. P. Maguluri (✉) · A. G. Gupta · T. Akhila · T. Bhargav
Department of Computer Science and Engineering, K L University, Green Fields,
Vaddeswaram, Guntur 522502, Andhra Pradesh, India
e-mail: phanendra51@gmail.com

© Springer Nature Singapore Pte Ltd. 2018   555
P. K. Sa et al. (eds.), *Recent Findings in Intelligent Computing Techniques*,
Advances in Intelligent Systems and Computing 709,
https://doi.org/10.1007/978-981-10-8633-5_55

Compressing the data and then using it for communication is way efficient than using the big-sized original data. This technique can be made more sophisticated and easy to use on multimedia devices by making it power efficient. Discrete Cosine Transform (DCT) [1] and Discrete Wavelet Transform (DWT) [2] are the best available transforms to achieve good image compression. For efficient power utilization, choosing DCT over DWT would be a better option as DCT provides more energy compaction [3], the ability to pack most information in fewest coefficients [4], and requires less computational resources to transform the image.

Much architecture has been proposed in the recent days to achieve energy-efficient DCT transform. Some of them being using Coordinate Rotation Digital Computer (CORDIC) [5, 6], Loffler-based DCT [7], fast 2D-DCT using CORDIC [8], etc. But the conventional CORDIC [5] suffers from two drawbacks: (i) it needs to be scaled after final computation and (ii) there is data dependency due to its iterative nature. These are overcome by using scale-free CORDIC [9] and Repetitive Iteration CORDIC (RICO) [10], respectively. These architectures can be used in JPEG compression to make it power efficient and performance effective and in this paper, we are using RICO in compression.

Further discussion of the used techniques is held as follows: Sect. 2 presents the proposed compression technique which is followed by Sect. 3 that gives the simulation environment and the results obtained via the existing and the proposed techniques. Finally, Sect. 4 concludes the paper and references are mentioned at the end.

2 Proposed Method

Here, we propose an energy-efficient compression technique that makes use of a hardware-efficient CORDIC architecture [11]. The CORDIC unit is employed to compute the DCT coefficients which make our proposed technique an energy-efficient one. The block diagrams of the compression technique and CORDIC-based 2D-DCT block are given in Figs. 1 and 2, respectively.

2.1 CORDIC-Based 2D-DCT/IDCT

Computation of sine/cosine values is a very tedious and hardware expensive task. Out of various methods available (Power series/Taylor series expansion, arc-line

Fig. 1 Proposed compression block

Fig. 2 CORDIC-based
2D-DCT block

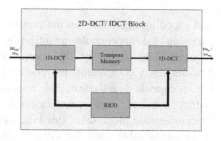

method, CORDIC, etc.), CORDIC turns out to be effective both in terms of hardware and computational complexity. Thus, here we make use of CORDIC architecture for computing the DCT coefficients. The 2D-DCT/IDCT that employs CORDIC unit is shown in Fig. 2.

Input to the 2D-DCT block is the level-shifted version of the original image (Im_{org}), in which first 1D-DCT (given by Eq. 1) is performed using the coefficients obtained using Repetitive Iteration CORDIC architecture (RICO). The output generated by the 1D-DCT block is transposed and further given to another 1D-DCT block which uses the same coefficients to evaluate the final 2D-DCT (Im_{tr}).

$$w_r = \frac{c_r}{2} \sum_{i=0}^{7} Im_{org_i} \cos \frac{(2i+1)r\pi}{16}, r = 0, 1, 2, \ldots, 7 \qquad (1)$$

where

w_r is the 1D-DCT output of 8 × 8 block

Here, RICO is nothing but an area and power-efficient CORDIC unit that is used for computing the seven DCT coefficients (a, b, ..., g) which is proposed in [10]. It is used here for obtaining energy efficiency and low complexity. For reconstruction of the transformed image, 2D-IDCT block is used which is the replica of 2D-DCT block as shown in Figs. 1, 2, and 3.

2.2 Compression/Decompression Block

The output obtained from the 2D-DCT block (Im_{tr}) is fed to the compression block to give the compressed bitstream as the output. This block performs quantitation

Fig. 3 Quantization table for
50% reconstruction quality

$$Q = \begin{bmatrix} 16 & 11 & 10 & 16 & 24 & 40 & 51 & 61 \\ 12 & 12 & 14 & 19 & 26 & 58 & 60 & 55 \\ 14 & 13 & 16 & 24 & 40 & 57 & 69 & 56 \\ 14 & 17 & 22 & 29 & 51 & 87 & 80 & 62 \\ 18 & 22 & 37 & 56 & 68 & 109 & 103 & 77 \\ 24 & 35 & 55 & 64 & 81 & 104 & 113 & 92 \\ 49 & 64 & 78 & 87 & 103 & 121 & 120 & 101 \\ 72 & 92 & 95 & 98 & 112 & 100 & 103 & 99 \end{bmatrix}$$

and encoding operations on the transformed image. Quantization is performed by choosing the appropriate quantization table per the amount of compression required. The quality of the compressed image is decided by the amount of low-frequency components preserved by the quantization table. Im_{tr} is divided by the appropriate quantization table to suppress the high-frequency components. The output obtained after quantization is coded into bitstream and stored/sent. Decompression block takes this bitstream as input and multiplies it with the same quantization table used for compressing. This operation gives the transformed image (Im_{tr}) to which 2D-IDCT is applied for image reconstruction.

3 Results and Discussions

In this section, the results of the proposed algorithm are presented. The software platform used for evaluating the proposed algorithm is MATLAB. The algorithm has been tested on 200 images of SIPI image database [12] consisting of different textures and resolutions. For evaluating the performance of the algorithm, it has been compared with the existing algorithms

$$PSNR = 10 \log_{10} \left(\frac{R^2}{MSE} \right) \tag{2}$$

$$CR = \frac{size\ of\ compressed\ data}{size\ of\ original\ raw\ image} \tag{3}$$

Volder [5], Aggarwal [11], and the normal JPEG [13] without CORDIC architecture. The algorithm has been compared on the basis of Peak Signal-to-Noise Ratio (PSNR) and Compression Ratio (CR) which are given by Eqs. 2 and 3, respectively. For presenting the results, a sample database consisting of 20 images with different resolutions and textures is presented. Results have been compared on the basis of CR and the results show that the PSNR values of all the methods are almost same which indicates that the proposed method is performing efficiently despite consuming low power. Average consolidated results for 200 images are presented.

The efficacy of proposed technique is evaluated by processing some benchmark images taken from SIPI database (200 images have been passed through the proposed method). The results have been compared with the existing techniques which are also used for processing these images. Figures 4 and 5 show the original image along with the processed ones for quantization table Q_{50}. The average PSNR obtained for all the images for different CR values using the various techniques is given in Table 1. A comparison of average PSNR obtained via the techniques for different CR values is also plotted as shown in Fig. 6.

Fig. 4 **a** Original image,
b without CORDIC [13],
c conventional CORDIC [5],
d scale-free CORDIC [11],
e proposed method

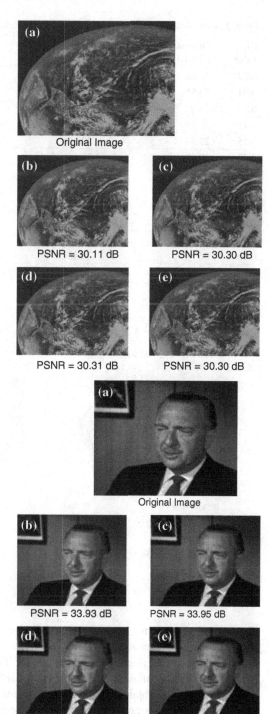

Fig. 5 **a** Original image,
b without CORDIC [13],
c conventional CORDIC [5],
d scale-free CORDIC [11],
e proposed method

Table 1 Consolidated results of SIPI database for different values of Q

CR value	Original (without CORDI) [13]	Conventional CORDIC [5]	Scale-free CORDIC [11]	Proposed method
0.02	21.64	21.63	21.66	21.64
0.04	23.55	23.54	23.56	23.54
0.07	26.02	26.02	26.03	26.01
0.11	28.10	28.09	28.10	28.08
0.15	29.34	29.33	29.34	29.32
0.22	30.88	30.86	30.86	30.85
0.25	31.25	31.24	31.24	31.22
0.29	32.03	32.01	32.01	31.99
0.34	32.96	32.93	32.92	32.90
0.42	34.06	34.02	34.01	33.99
0.43	34.25	34.21	34.20	34.17
0.63	36.89	36.83	36.81	36.73

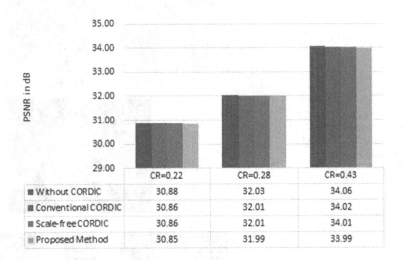

Fig. 6 Average PSNR for Q_{70} table

4 Conclusion

This paper has presented a power-efficient image compression technique that makes use of Repetitive Iteration CORDIC (RICO) architecture for the computation of Discrete Cosine Transform (DCT) coefficients. The proposed approach is modeled using MATLAB simulation environment on SIPI database consisting of 200 images. The results are obtained and have been compared with the existing technology used for compression and with methods that employ conventional CORDIC and scale-free CORDIC architectures. The basis for comparison is the CR, and the results show that the PSNR values of all the methods are almost same which indicates that the proposed method is performing efficiently.

References

1. Ahmed, N., Natarajan, T., Rao, K.R.: Discrete cosine transform. IEEE Trans. Comput. **100**(1), 90–93 (1974)
2. Madanayake, A., Cintra, R.J., Dimitrov, V., Bayer, F., Wahid, K.A., Kulasekera, S., Edirisuriya, A., Potluri, U., Madishetty, S., Rajapaksha, N.: Low-power VLSI architectures for dct/dwt: precision vs approximation for HD video, biomedical, and smart antenna applications. IEEE Circ. Syst. Mag. **15**, 25–47 (First quarter) (2015)
3. Roy, A.B., Dey, D., Mohanty, B., Banerjee, D.: Comparison of FFT, DCT, DWT, WHT compression techniques on electrocardiogram and photoplethysmography signals. In: IJCA Special Issue on International Conference on Computing, Communication and Sensor Network CCSN, pp. 6–11 (2012)
4. Kaur, A., Kaur, J.: Comparison of DCT and DWT of image compression techniques. Int. J. Eng. Res. Dev. **1**(4), 49–52 (2012)
5. Volder, J.E.: The CORDIC trigonometric computing technique. IRE Trans. Electron. Comput. **3**, 330–334 (1959)
6. Volder, J.E.: The birth of CORDIC. J. VLSI Sig. Process. Syst. Sig. Image Video Technol. **25**(2), 101–105 (2000)
7. Sun, C.C., Heyne, B., Ruan, S.J., Goetze, J.: A low-power and high-quality cordic based loeffler DCT. In: 2006 International Symposium on VLSI Design, Automation and Test, pp. 1–4. IEEE (2006)
8. Huang, H., Xiao, L.: Cordic based fast radix-2 dct algorithm. IEEE Sig. Process. Lett. **20**(5), 483–486 (2013)
9. Aggarwal, S., Meher, P.K., Khare, K.: Area-time efficient scaling-free cordic using generalized micro-rotation selection. IEEE Trans. Very Large Scale Integr. Syst. (VLSI) **20**(8), 1542–1546 (2012)
10. Nawandar, N.K., Garg, B., Sharma, G.: Rico: a low power repetitive iteration CORDIC for DSP applications in portable devices. J. Syst. Archit. (2016)
11. Aggarwal, S., Khare, K.: Hardware efficient architecture for generating sine/cosine waves. In: 2012 25th International Conference on VLSI Design, pp. 57–61. IEEE (2012)
12. Internet database. http://sipi.usc.edu/database/
13. Wallace, G.K.: The jpeg still picture compression standard. IEEE Trans. Consum. Electron. **38**, xviii–xxxiv (1992)

An Efficient Image Enhancement Method for Dark Images

V. Naga Bushanam and C. H. Satyananda Reddy

Abstract Image enhancement is a technique to give a better quality of an image in terms of its clarity, brightness, and to give the human eye comfortable to look at. There are different types of techniques to give good quality to an image. Global image contrast enhancement is one of the most commonly used techniques to enhance the quality of an image, but it has some disadvantages with the fact that it does not consider the local details of an image. By local detail, it means that the detailed texture of an image. In local contrast enhancement, it addresses the local details of an image and preserves the local details of the image. Local details of an image are very important while analyzing an image, which is that of the scientific study of an image like the image taken from planetary bodies, satellite image, and also in medical images. Local details of an image are very important for diagnosing a particular ailment. When we used either local contrast enhancement or global contrast enhancement alone, we faced the loss of brightness of the image. In order to address and reduce this discrepancy of individual enhancement methods, a new proposal is used for both these methods on the same image. First, the image is locally enhanced and the output is again processed by the global enhancement method, thereby giving a properly enhanced image without losing the brightness of the image. This enhancement method is simulated in MATLAB, and results are verified on the parameters of image.

Keywords Image enhancement · Image sharpening · Unsharp masking
Global contrast stretching · Local contrast stretching

V. Naga Bushanam (✉) · C. H. Satyananda Reddy
Department of Computer Science and Engineering, Andhra University,
Waltair Junction, Visakhapatnam 530003, Andhra Pradesh, India
e-mail: phanendra51@gmail.com

© Springer Nature Singapore Pte Ltd. 2018
P. K. Sa et al. (eds.), *Recent Findings in Intelligent Computing Techniques*,
Advances in Intelligent Systems and Computing 709,
https://doi.org/10.1007/978-981-10-8633-5_56

1   Introduction

The human visual perception of an image can be greatly affected by the contrast of an image. Contrast has defined as the difference in the pixel intensity value of a particular pixel to its neighboring pixels. If the difference in the intensity is more, then we can say that the contrast is more of that image. The more contrast gives us better clarity of an image in terms of local details. The more details in the local information of an image are very important if the image is of medical or astronomical for analyzing it and extracting information of that image and proper diagnosing of the ailments based upon the image of a cell. So with the advancement of science and technology, especially in the field of signal processing, the quality of an image can be enhanced so that it gives clear and detailed information about the image [1]. There are various techniques for enhancing the image quality, and many techniques have been proposed to address various aspects of an image over the years.

Since long back many scholars have developed techniques for enhancing an image, the equalization of the histogram of an image for enhancement of the image is very common and effective. One single technique cannot be used as a universal technique that can be applied to all types of images. One technique may give a very good result to a specific type of image but may not give a satisfactory result to another image.

Various contrast stretching methods have been proposed to enhance the image of leukemia, a medical image in [2]. When a dark stretch is performed, the bright portions of the image or the bright pixels are more brighten. A better way to address such problem is to enhance the dark regions by keeping the bright regions untouched [3]. They have shown the effects of various contrast stretching techniques like global stretching, local contrast stretching, partial contrast stretching, etc., and which one is best suitable for which type of image is studied. The problem of a blurred image, which is caused by the motion of the object while taking the image, and how to avoid are presented in [4]. It also used local edge detection to deblur the original image. In [5], the effect of application of both global and local contrast enhancements is studied on grayscale image and only the brightness parameter of the image has been studied. This method is being used in this paper on the dark color image, and image enhancement parameters like mean and measure of enhancement factor are calculated and the output image is compared with the existing image enhancement techniques.

This paper is organized as follows: In Sect. 2, the image enhancement techniques are presented. Section 3 describes the methodology of the algorithm. In Sect. 4, the implementation of methodology and results are explained and Sect. 5 concludes the paper.

2 Image Enhancement Techniques

Image enhancement techniques have been widely used to get a good quality of an image for the human interpretation. Image enhancement techniques may be broadly classified as local image enhancement and global image enhancement.

2.1 Local Enhancement of the Image

The local enhancement is employed to get the minute details of an image. It enhances the local details in terms of the gradient of the image which gives useful information to the analyzer of the image. It addresses those pixels which would be ignored by the global method. The local enhancement method employed [6] here is the unsharp masking. In this method, the image is sharpened by subtracting an unsharp that is a blurred or smoothed image from the original image; so, the name unsharp masking is derived. In this method, the following steps are involved:

1. Blurring of the image,
2. Subtracting the blurred image from the original image to make the mask, and
3. Adding the mask to the original image.

If the blurred image is denoted as $b(i, j)$ and the image as $p(i, j)$, then the mask $m(i, j)$ is given according to Eq. (1)

$$m(i,j) = p(i,j) - b(i,j) \tag{1}$$

Then, a weighted portion of the mask is added to the original image to get the sharpened image $s(i, j)$ as given by equation

$$s(i,j) = p(i,j) + w * m(i,j) \tag{2}$$

where "w" is the weight, generally greater than zero. When the weight is equal to 1, it is the unsharp masking and when greater than 1 then it is called high boost filtering.

2.2 Global Enhancement of the Image

The global enhancement of the image is used to increase the contrast of the image. In this process, each pixel of the image is adjusted so that it gives a better visualization of the image. In spatial contrast enhancement, the operation is done on the pixel directly. Different methods can be used to improve the image quality. In HE, for the discrete image, the probabilities [7] of the pixel value are taken. To take the probabilities, first, the corresponding number of pixels should have a pixel intensity value; it is calculated and divided by the total number of the pixels present in the image. The probability of occurrence of pixel intensity level "k" in the digital image is given by Eq. (3)

$$p(r_k) = \frac{n_k}{N * M} \tag{3}$$

where N * M is the total number of pixels in the image and n_k is the total number of pixels having intensity level "k". The pixels are transformed as per the following transformation equation in discrete form [7]:

$$t_k = L(r_k) = (G-1) \sum_{i=0}^{k} p(r_i) = \frac{G-!}{N * M} \sum_{i=0}^{k} n_i \tag{4}$$

where G is the highest intensity level or value, L (.) is the transform function, and k = 0, 1, 2, 3, ..., G − 1. So the output image pixel is obtained by mapping each input pixel r_i to the newly transformed value t_k. The processed output value may have fractional value, so a rounding function to the nearest integer value is needed. While doing so, some of the image pixels may go to the new value and some of the intensity pixel values may not be present in the transformed image.

3 Methodology

Figure 1 shows the methodology to be incorporated into the paper to get a good quality image and algorithm of combining both local enhancement and global enhancement of a color image. It mainly consists of the following four steps.

Step 1: Get the color image and convert it into HSV (Hue saturation and value) color space and take the luminance of that image.
Step 2: Apply the local enhancement method to enhance the local details of image.
Step 3: The local output is again given as global input and performs global image enhancement.
Step 4: Recombine the components and reconvert it back to color image.

Here in order to enhance the local gradients or the local details, an existing local enhancement method has been used. Here, the unsharp masking is used as local detail enhancement method. As the name suggests, it uses the blurred image to make the mask and enhances the local details in the form of edge sharpening. The sharpened image is used as the input to the global enhancement method. The global enhancement method uses one of the global contrast stretching methods.

At first, a color image to be enhanced is taken and it is converted to the HSV color space. From that color space, the luminance portion is taken. The enhancement in the hue and saturation is not done. The enhancement of image is performed

Fig. 1 Flow graph of the algorithm

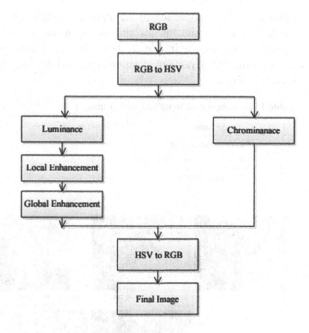

only in the luminance plane of the image. The local details of an image can be accessed or addressed through the luminance only.

4 Implementation and Results

To see the effect of the combination of local and global enhancement methods to a given image, the abovementioned algorithm is applied to the image. The color image or digital color image to be enhanced is taken and converted to the HSV color space in order to apply the algorithm. The image plane slicing is performed, and the image is divided into three different planes each of hue, saturation, and value.

In order to enhance the edges which are considered as the local features of an image, the local contrast stretching process that is unsharp masking is applied. This is the first step of the enhancement method. At the end of this step, a locally enhanced image is obtained. It gives a clear picture of the local information of the image but deficient in the overall brightness of the image. The working of the algorithm can be verified with the help of image quality parameters. One of the very common parameters is the measure of enhancement and measure of enhancement factor (MEF). In order to find the MEF, the measures of enhancement of the input and output have been calculated individually. MEF is the ratio of the measure of enhancement of output image to the measure of enhancement of the input image. A better value of MEF implies that the visual quality of the enhanced image is good. The mean of the input original image and enhanced output image is also calculated. The comparison of the input and output image is done and is shown in Table 1. The original images and its enhanced images by performing proposed algorithm are shown in Fig. 2. It is also compared with some of the existing methods like histogram equalization (HE) and discrete shearlet transform (DST) as shown in Fig. 3.

Table 1 Comparison of input and output images

SI no	Image name	Input mean value	Output mean value	MEF
1	Low light image	0.37	0.42	1.75
2	Shadow afternoon image	0.38	0.4	3.05
3	Evening image	0.37	0.38	2.95

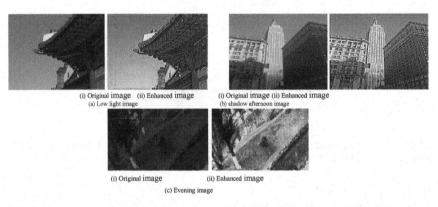

(i) Original image   (ii) Enhanced image
(a) Low light image

(i) Original image (ii) Enhanced image
(b) shadow afternoon image

(i) Original image   (ii) Enhanced image
(c) Evening image

Fig. 2 Input and output images of three different images

Fig. 3 Comparison between different methods

5 Conclusion

The results are successfully carried out in MATLAB R2014a. A combination of both local and global contrast enhancement techniques, where a local enhancement method is applied first to enhance the local details of the image, is not taken care and usually neglected in the global contrast enhancement. The locally enhanced image is given to the input of global enhancement for better visual perceptions and increases the brightness to a level which gives pleasant sensation to the human eye. This method works fine in most of the dark images. It has more significance to those images where we need local minute gradient information such as the image of planetary and heavenly bodies, satellite images, and medical images. The comparison is done with a couple of the existing methods. The different local and global methods have been used and tested their effectiveness of the different combinations of the local and global methods.

References

1. Celik, T., Tjahjadi, T.: Contextual and variational contrast enhancement. IEEE Trans. Image Process. **20**(12), 3431–3441 (2011)
2. Premkumar, S., Parthasarathi, K.A.: An efficient approach for colour image enhancement using discrete shearlet transform. In: IEEE International Conference on Current Trends in Engineering and Technology (ICCTET) (2014)
3. Lal, S., Chandra, M.: Efficient algorithm for contrast enhancement of natural images. Int. Arab J. Inf. Technol. **11**(1), 95–102 (2014)
4. Chen, S., Suleiman, A.: Scalable global histogram equalization with selective enhancement for photo processing. In: Proceedings of the 4th International Conference on Information Technology and Multimedia, Malaysia, pp. 744–752 (2008)
5. Pathak, S.S., Dahiwale, P., Padole, G.: A combined effect of local and global method for contrast image enhancement. In: IEEE International Conference on Engineering and Technology (ICETECH) (2015)
6. Chen, S., Ramli, A.: Preserving brightness in histogram equalization based contrast enhancement techniques. Digit. Sig. Proc. **14**(5), 413–428 (2004)
7. Gonzalez, R.C., Woods, R.E.: Digital Image Processing, 3rd edn. Pearson Publication

Basic Operations on the 2-D Mesh Topology with Optical Interlinks

Amritanjali

Abstract The square mesh topology is often used for interconnecting nodes in parallel computers. However, the large diameter of the basic mesh topology makes it suitable for solving problems where communication takes place mostly among neighboring nodes. The new 2-D mesh topology with optical interlinks reduces the diameter of the mesh topology, by interlinking nodes that are separated by half the number of nodes in each dimension. In this paper, parallel algorithms have been developed to demonstrate the effectiveness of this new variant over the basic topology. The presence of additional optical interlinks helps in designing faster solutions to various problems without increasing design complexity. The problems that have been addressed are data communication, matrix multiplication, and solving linear systems.

Keywords Parallel algorithms · Interconnection networks · Mesh topology

1 Introduction

The execution time of a parallel program not only depends on the number of computational steps that each processor should execute but also on the time spent in exchanging data between the processors. The interconnection network used in the parallel computer plays a key role in determining the overhead incurred on running a parallel algorithm on it. Mesh topology is quite popular and several of the commercial available parallel computers are based on it because of its regularity and low hardware complexity. In the basic mesh topology, nodes are organized in the form of q-dimensional lattice. The nodes are connected by links which can be unidirectional or bidirectional. Additionally, wrap-around links can be used to connect the border nodes at the opposite end. The mesh network with wrap-around connections is called

Amritanjali (✉)
Department of Computer Science and Engineering, Birla Institute of Technology, Mesra, Ranchi 835215, India
e-mail: amritanjali@bitmesra.ac.in

© Springer Nature Singapore Pte Ltd. 2018
P. K. Sa et al. (eds.), *Recent Findings in Intelligent Computing Techniques*,
Advances in Intelligent Systems and Computing 709,
https://doi.org/10.1007/978-981-10-8633-5_57

a torus. The large value of diameter is the main drawback of this topology. Various mesh-based hybrid networks were introduced for better performance, like mesh of trees, multi-mesh, and multi-mesh of trees [1–3]. Recently, optical links have become quite popular as an interconnection medium in parallel systems for performing high-speed computing [4]. OTIS-MOT [5] and OMULT [6] are some the hybrid topologies using optical links. Some of the advantages of optical links are increased communication speed, reduced power consumption, etc. The new mesh topology with optical interlinks [7] is a variant of the basic square mesh topology to improve its communication efficiency. This paper is extension of this work to show its efficacy in implementing basic parallel algorithms. The topology of the mesh interconnection network and its properties are described in Sect. 2 of the paper. In Sect. 3, we implement some of the basic algorithms on this topology and show their efficiency. Finally, Sect. 4 concludes the paper.

2   2-D Square Mesh with Optical Interlinks

The nodes in the network are organized in the form of n × n 2-D array, where n is the power of 2 [7]. The processors are numbered row-wise such that P_{1n} is at the top right corner, P_{n1} at the bottom left corner, and P_{nn} at the bottom right corner. In addition to the normal electronic links between adjacent processors, optical links are used to provide more connectivity and reduce diameter. In each row i, where $1 \leq i \leq n$, there are horizontal optical links between processor P_{ij} and $P_{i(j+n/2)}$, where $1 \leq j \leq n/2$. For each column j, where $1 \leq j \leq n$, there are vertical optical links between processor P_{ij} and $P_{(i+n/2)j}$, where $1 \leq i \leq n/2$. Figure 1 shows the topology of an 8 × 8 mesh with optical interlinks. The diameter of the two-dimensional n × n mesh is n, bisection width is $n^2/2 + n$, and node degree is 6. The length of the optical link increases with increase in the size of the network.

3   Extending Mesh-Based Parallel Algorithms

We present implementation of parallel algorithms for some elementary problems on the proposed mesh topology.

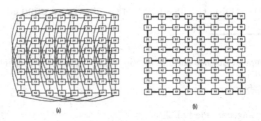

(a)   (b)

Fig. 1 **a** 8 × 8 mesh with optical interlinks. Horizontal optical links are shown only for first and last row for clarity. **b** Bold lines show logical grouping of processors into four blocks

3.1 Communication Algorithms

Unicast Routing. It is used to find optimal path to send a message from a source node to a destination node. As we have additional inter-block optical links, we can use this to reach the destination block. If the destination processor is in adjacent block, use either horizontal or vertical optical links to reach destination block. Otherwise, if destination is in diagonally opposite block, then two optical moves will be required to reach the destination block, one horizontal and one vertical. Once the destination block is reached, we can use any unicast routing method like the odd–even turn algorithm to reach the destination processor in that block. Thus, we can reach any node in maximum n steps.

Broadcasting. To broadcast a data from any node of a block (shown in Fig. 1b) to all the nodes in the network, it is first broadcasted to all the nodes in its block using electronic links in maximum of $2(n/2 - 1)$ steps. Then, using horizontal optical links, the data is sent to the nodes in the block adjacent to it horizontally. Finally, using the vertical optical links it is sent to the nodes in the remaining two blocks. Therefore, one to all broadcasting can be done in n times. Also, broadcast in a single row (or column) is achieved in n/2 steps: first broadcasting to all the nodes in the same row (column) of its block, and then one optical horizontal (vertical) move.

Data Aggregation. Aggregation (or Reduction) involves merging of values stored in each processor using some function to result in a single value. For example, finding sum, average, maximum, or minimum of n^2 data elements stored in the nodes of the mesh network. In the first phase, parallel aggregation is done in each block, row-wise, bringing the partial result to the first column of the block, using the electronic links in $(n/2 - 1)$ communication steps. Next, the horizontal optical links are used to bring the new partial result in the first column of the mesh. Then, using the vertical optical links, the partial results are brought to the first column of the first block. In the last phase, the final result is produced at the processor $P(1,1)$ using the $(n/2 - 1)$ electronic links. Hence, the parallel aggregation is generated in n communication steps. And aggregation over a single row or column can be achieved in n/2 communication steps.

3.2 Matrix Multiplication

The blocks shown in Fig. 1b are numbered row-wise from 1 to 4. To multiply two square matrices X and Y of dimension n x n, total n^3 product terms need to be computed. As we have n^2 processors, each processor is responsible for generating n product terms and finally compute one element of the result matrix. By partitioning matrices, we can design-efficient algorithm for matrix multiplication [8]. The two matrices to be multiplied are divided into sub-matrices and distributed across processors in each block in such a manner to minimize the communication overhead.

The assignment of elements of matrices X and Y to arrays a and b, respectively, of the processor P_{ijk} in the kth block, where $1 \leq i, j \leq n/2$, is defined as follows:

a[1...4] ← {X[i][2k-1], X[i][2k], X[i + 4][2k-1], X[i + 4][2k]}
b[1...4] ← {Y[2k-1][j], Y[2k][j], Y[2k-1][j + 4], Y[2k][j + 4]}

The initial block-wise distributions of elements of the input matrices X and Y are shown in Fig. 2 and Fig. 3, respectively.

Algorithm 1

for all P_{ij}, where $1 \leq i, j \leq n$, do

  {calculate the product terms}
  c[1] ← a[1].b[1] + a[2].b[2]
  c[2] ← a[1].b[3] + a[2].b[4]
  c[3] ← a[3].b[1] + a[4].b[2]
  c[4] ← a[3].b[3] + a[4].b[4]

  {Exchange terms using horizontal optical links}

  if (k%2 = 0) //even numbered blocks
      $(P_{i(j-4)(k-1)})$temp1 ← c[1]
      $(P_{i(j-4)(k-1)})$temp2 ← c[3]

      temp1 ←$(P_{i(j-4)(k-1)})$c[2]
      temp2 ←$(P_{i(j-4)(k-1)})$c[4]

      c[2] ← c[2] + temp1
      c[4] ← c[4] + temp2

  else
      $(P_{i(j+4)(k+1)})$temp1 ← c[2]
      $(P_{i(j+4)(k+1)})$temp2 ← c[4]

      temp1 ← $(P_{i(j+4)(k+1)})$c[1]
      temp2 ← $(P_{i(j+4)(k+1)})$c[3]

      c[1] ← c[1] + temp1
      c[3] ← c[3] + temp2

  {Exchange using vertical optical link and calculate the final term}

  if k ≤ 2
      $(P_{(i+4)(j)(k+2)})$temp ← c[k+2]
      temp ← $(P_{(i+4)(j)(k+2)})$c[k]
  else
      $(P_{(i-4)(j)(k-2)})$temp ← c[k-2]
      temp ← $(P_{(i-4)(j)(k-2)})$c[k]

  term ← c[k] + temp

end for

After the initial distribution of elements, the multiplication takes place in three phases described by Algorithm 1. In the first phase, each processor computes n

$X_{11} X_{12}$ **$X_{51} X_{52}$**	$X_{11} X_{12}$ $X_{51} X_{52}$	$X_{11} X_{12}$ $X_{51} X_{52}$	$X_{11} X_{12}$ $X_{51} X_{52}$	**$X_{13} X_{14}$** **$X_{53} X_{54}$**	$X_{13} X_{14}$ $X_{53} X_{54}$	$X_{13} X_{14}$ $X_{53} X_{54}$	$X_{13} X_{14}$ $X_{53} X_{54}$
$X_{21} X_{22}$ $X_{61} X_{62}$	$X_{21} X_{22}$ $X_{61} X_{62}$	$X_{21} X_{22}$ $X_{61} X_{62}$	$X_{21} X_{22}$ $X_{61} X_{62}$	$X_{23} X_{24}$ $X_{63} X_{64}$	$X_{23} X_{24}$ $X_{63} X_{64}$	$X_{23} X_{24}$ $X_{63} X_{64}$	$X_{23} X_{24}$ $X_{63} X_{64}$
$X_{31} X_{32}$ $X_{71} X_{72}$	$X_{31} X_{32}$ $X_{71} X_{72}$	$X_{31} X_{32}$ $X_{71} X_{72}$	$X_{31} X_{32}$ $X_{71} X_{72}$	$X_{33} X_{34}$ $X_{73} X_{74}$	$X_{33} X_{34}$ $X_{73} X_{74}$	$X_{33} X_{34}$ $X_{73} X_{74}$	$X_{33} X_{34}$ $X_{73} X_{74}$
$X_{41} X_{42}$ $X_{81} X_{82}$	$X_{41} X_{42}$ $X_{81} X_{82}$	$X_{41} X_{42}$ $X_{81} X_{82}$	$X_{41} X_{42}$ $X_{81} X_{82}$	$X_{43} X_{44}$ $X_{83} X_{84}$	$X_{43} X_{44}$ $X_{83} X_{84}$	$X_{43} X_{44}$ $X_{83} X_{84}$	$X_{43} X_{44}$ $X_{83} X_{84}$
$X_{15} X_{16}$ **$X_{55} X_{56}$**	$X_{15} X_{16}$ $X_{55} X_{56}$	$X_{15} X_{16}$ $X_{55} X_{56}$	$X_{15} X_{16}$ $X_{55} X_{56}$	**$X_{17} X_{18}$** **$X_{57} X_{58}$**	$X_{17} X_{18}$ $X_{57} X_{58}$	$X_{17} X_{18}$ $X_{57} X_{58}$	$X_{17} X_{18}$ $X_{57} X_{58}$
$X_{25} X_{26}$ $X_{65} X_{66}$	$X_{25} X_{26}$ $X_{65} X_{66}$	$X_{25} X_{26}$ $X_{65} X_{66}$	$X_{25} X_{26}$ $X_{65} X_{66}$	$X_{27} X_{28}$ $X_{67} X_{68}$	$X_{27} X_{28}$ $X_{67} X_{68}$	$X_{27} X_{28}$ $X_{67} X_{68}$	$X_{27} X_{28}$ $X_{67} X_{68}$
$X_{35} X_{36}$ $X_{75} X_{76}$	$X_{35} X_{36}$ $X_{75} X_{76}$	$X_{35} X_{36}$ $X_{75} X_{76}$	$X_{35} X_{36}$ $X_{75} X_{76}$	$X_{37} X_{38}$ $X_{77} X_{78}$	$X_{37} X_{38}$ $X_{77} X_{78}$	$X_{37} X_{38}$ $X_{77} X_{78}$	$X_{37} X_{38}$ $X_{77} X_{78}$
$X_{45} X_{46}$ $X_{85} X_{86}$	$X_{45} X_{46}$ $X_{85} X_{86}$	$X_{45} X_{46}$ $X_{85} X_{86}$	$X_{45} X_{46}$ $X_{85} X_{86}$	$X_{47} X_{48}$ $X_{87} X_{88}$	$X_{47} X_{48}$ $X_{87} X_{88}$	$X_{47} X_{48}$ $X_{87} X_{88}$	$X_{47} X_{48}$ $X_{87} X_{88}$

Fig. 2 Block-wise distribution of elements matrix X elements. Elements assigned to the first processor of each block are shown in bold

$Y_{11} Y_{21}$ **$Y_{15} Y_{25}$**	$Y_{12} Y_{22}$ $Y_{16} Y_{26}$	$Y_{13} Y_{27}$ $Y_{13} Y_{27}$	$Y_{14} Y_{28}$ $Y_{14} Y_{28}$	**$Y_{31} Y_{41}$** **$Y_{35} Y_{45}$**	$Y_{32} Y_{42}$ $Y_{36} Y_{46}$	$Y_{33} Y_{43}$ $Y_{37} Y_{47}$	$Y_{34} Y_{44}$ $Y_{38} Y_{48}$
$Y_{11} Y_{21}$ $Y_{15} Y_{25}$	$Y_{12} Y_{22}$ $Y_{16} Y_{26}$	$Y_{13} Y_{27}$ $Y_{13} Y_{27}$	$Y_{14} Y_{28}$ $Y_{14} Y_{28}$	$Y_{31} Y_{41}$ $Y_{35} Y_{45}$	$Y_{32} Y_{42}$ $Y_{36} Y_{46}$	$Y_{33} Y_{43}$ $Y_{37} Y_{47}$	$Y_{34} Y_{44}$ $Y_{38} Y_{48}$
$Y_{11} Y_{21}$ $Y_{15} Y_{25}$	$Y_{12} Y_{22}$ $Y_{16} Y_{26}$	$Y_{13} Y_{27}$ $Y_{13} Y_{27}$	$Y_{14} Y_{28}$ $Y_{14} Y_{28}$	$Y_{31} Y_{41}$ $Y_{35} Y_{45}$	$Y_{32} Y_{42}$ $Y_{36} Y_{46}$	$Y_{33} Y_{43}$ $Y_{37} Y_{47}$	$Y_{34} Y_{44}$ $Y_{38} Y_{48}$
$Y_{11} Y_{21}$ $Y_{15} Y_{25}$	$Y_{12} Y_{22}$ $Y_{16} Y_{26}$	$Y_{13} Y_{27}$ $Y_{13} Y_{27}$	$Y_{14} Y_{28}$ $Y_{14} Y_{28}$	$Y_{31} Y_{41}$ $Y_{35} Y_{45}$	$Y_{32} Y_{42}$ $Y_{36} Y_{46}$	$Y_{33} Y_{43}$ $Y_{37} Y_{47}$	$Y_{34} Y_{44}$ $Y_{38} Y_{48}$
$Y_{51} Y_{61}$ **$Y_{55} Y_{65}$**	$Y_{52} Y_{62}$ $Y_{56} Y_{66}$	$Y_{53} Y_{63}$ $Y_{57} Y_{67}$	$Y_{58} Y_{68}$ $Y_{58} Y_{68}$	**$Y_{71} Y_{81}$** **$Y_{75} Y_{85}$**	$Y_{72} Y_{82}$ $Y_{76} Y_{86}$	$Y_{73} Y_{83}$ $Y_{77} Y_{87}$	$Y_{74} Y_{84}$ $Y_{78} Y_{88}$
$Y_{51} Y_{61}$ $Y_{55} Y_{65}$	$Y_{52} Y_{62}$ $Y_{56} Y_{66}$	$Y_{53} Y_{63}$ $Y_{57} Y_{67}$	$Y_{58} Y_{68}$ $Y_{58} Y_{68}$	$Y_{71} Y_{81}$ $Y_{75} Y_{85}$	$Y_{72} Y_{82}$ $Y_{76} Y_{86}$	$Y_{73} Y_{83}$ $Y_{77} Y_{87}$	$Y_{74} Y_{84}$ $Y_{78} Y_{88}$
$Y_{51} Y_{61}$ $Y_{55} Y_{65}$	$Y_{52} Y_{62}$ $Y_{56} Y_{66}$	$Y_{53} Y_{63}$ $Y_{57} Y_{67}$	$Y_{58} Y_{68}$ $Y_{58} Y_{68}$	$Y_{71} Y_{81}$ $Y_{75} Y_{85}$	$Y_{72} Y_{82}$ $Y_{76} Y_{86}$	$Y_{73} Y_{83}$ $Y_{77} Y_{87}$	$Y_{74} Y_{84}$ $Y_{78} Y_{88}$
$Y_{51} Y_{61}$ $Y_{55} Y_{65}$	$Y_{52} Y_{62}$ $Y_{56} Y_{66}$	$Y_{53} Y_{63}$ $Y_{57} Y_{67}$	$Y_{58} Y_{68}$ $Y_{58} Y_{68}$	$Y_{71} Y_{81}$ $Y_{75} Y_{85}$	$Y_{72} Y_{82}$ $Y_{76} Y_{86}$	$Y_{73} Y_{83}$ $Y_{77} Y_{87}$	$Y_{74} Y_{84}$ $Y_{78} Y_{88}$

Fig. 3 Block-wise distribution of elements of matrix Y

different product terms and they are added to get four sum terms, each being sum of $n/4$ product terms. This step will take $O(n)$ time assuming that the internal computation of each processor is sequential. In the next phase, each processor exchanges two of the computed partial sums with the processors connected through horizontal optical links. When added, we get new partial sum of n/2 product terms. In the last phase, the partial sums are exchanged through vertical optical links and added to get the final sum of n product terms. Processor P_{ij} of the network holds the Element[i][j] of the product matrix. Overall, there are two horizontal moves and one vertical move. Thus, the parallel algorithm computes the product of two n × n matrices in O(n) time.

3.3 Solving Linear System of Equations

The Jacobi algorithm for solving linear system of equations is amenable to parallelization [9]. The coefficients of the linear system in n variables are distributed row-wise over processing elements, one on each. The constants are stored on the diagonal processors. Each diagonal processor will be responsible for generating one element of the solution. Algorithm 2 describes the steps in the parallel implementation the Jacobi's method.

Algorithm 2

```
for all Pᵢⱼ, where 1 ≤ i, j ≤ n, do

    if(i = j)
            x ←b / a
            columnBroadcast(x)
    endif

    repeat
            if(i ≠j)
                        receive(x)
                        term ← x * a
            endif

            rowReduce(term, Pᵢᵢ, sum)

            if(i = j)
                        newx← (b – term) / a
                        diff ← Ix – newxI
                        x ← newx
                        columnBroadcast(x)
            endif

            allReduce(diff, P₁₁, max)   [aggregating maximum difference on P₁₁]

            if( i = j = 1)
                        if(diff > ε) [convergence criterion]
                                    flag ← 0
                        else
                                    flag ← 1
                        endif

                        broadcast(flag)
            endif
    until(flag)
endfor
```

To achieve maximum parallelism, each coefficient is stored separately. It does increase communication steps; however, communication overhead is reduced because of inter-block optical links. The procedure column broadcast and row reduce takes n/2 steps, while all reduce and broadcast procedures require n steps. For fast convergence, Jacobi over-relaxation variant can be used for updating each intermediate solution.

4 Conclusions

By designing various parallel algorithms for the 2-D mesh network with optical interconnects, we have shown that it is capable of effectively solving various computational problems of interest. As a result of optical interlinking of far apart nodes in the network topology, communications between any pair of nodes as well as data distribution in the network are faster in comparison to the simple 2-D mesh topology. The existing parallel algorithms can be easily extended to take the advantage of the additional links. Therefore, it has the potential to be used as an alternative mesh topology for parallel computers.

References

1. Das, D., Sinha, B.P.: A new network topologies with multiple meshes. IEEE Trans. Comput. **44**(5), 536–551 (1999)
2. Chen, W.M., Chen, G.H., Hsu, D.F.: Combinatorial properties of mesh of trees,. In: International Symposium on Parallel, Architectures, Algorithms and Networks, pp. 134–139 (2000)
3. Jana, P.K.: Multi-mesh of trees with its parallel algorithms. J. Syst. Archit. **50**, 193–206 (2004)
4. Marsden, G.C., Marchand, P.J., Harvey, P., Esener, S.C.: Optical transpose interconnection system architecture. Opt. Lett. **18**(3), 1083–1085 (1993)
5. Wang, C.F., Sahani, S.: Basic operation on OTIS-mesh optoelectronics computer. IEEE Trans. Parallel Distrib. Syst. **19**(12), 1226–1233 (1998)
6. Sinha, B.P., Banyopadhyay, S.: OMULT: an optical interconnection system for parallel computing. LNCS **3149**, 302–312 (2004)
7. Amritanjali: A new 2-D mesh topology with optical interlinks. In: International Conference on Advanced Computing, Networking and Informatics (2016)
8. Samantray, B.S, Kanhar, D.: Implementation of dense matrix multiplication on 2D Mesh. In: International Conference on High Performance Computing & Applications (2014)
9. Quinn, M.J.: Parallel Computing: Theory and Practice. McGraw Hill, New York (1994)

Multiple Target Tracking

Kiran Phalke and Ravindra Hegadi

Abstract We have proposed a tracker which is based on multiple hypothesis model and iterative closest point algorithm. The algorithm successfully tracks multiple targets by using multiple hypothesis model matching approach. Hypothesis models are formed by using track information; the models are updated periodically whenever a new track is added. Although there are number of algorithms proposed so far in this area, our approach addresses number of issues, which others fail to do. The algorithm works well in variable number of targets, as soon as target appears in the track; it is automatically added and when the target leaves the track is dropped. The detection accuracy observed on the video tracks is noticeable.

Keywords Background model · Tracking · ICP

1 Introduction

Our approach is based on a multiple hypothesis algorithm that groups target detections. First, we found locations of the target in the previous frame to match against the location in next frame. The video of multiple targets is used to evaluate our approach. The general idea of this paper is to describe detection-based tracker. The tracker takes frames in which detections of target object are found as input and produces tracked frames as output. In the first step, the object detections are found by using pixel information, i.e., the target pixel is added to the detected list. To accomplish this task, algorithm used is tracked by using movement proposed by Rosin and Ellis in 1995. The objects are dark and the background is light so the frame differencing is used to find detections. Also, the average color values are detected as the input frame. In this way first, video data is converted into binary detections of potential targets and detections which do not represent targets to be tracked are eliminated by using background subtraction. The detections that remain after background

K. Phalke · R. Hegadi (✉)
School of Computational Sciences, Solapur University, Solapur, India
e-mail: rshegadi@gmail.com

© Springer Nature Singapore Pte Ltd. 2018
P. K. Sa et al. (eds.), *Recent Findings in Intelligent Computing Techniques*,
Advances in Intelligent Systems and Computing 709,
https://doi.org/10.1007/978-981-10-8633-5_58

subtraction are used for tracking. The model is created from sample data and the target frames are detected based on this information. In case if any object does not have a corresponding track, then they are treated as untracked frames or missing tracks.

2 Related Works

There has been number of researchers working in the same area. Khan et al. [1] proposed a particle filter-based tracker which tracks objects in occluded background also. Bruce [2] presented color segmentation which can be used to study various computer vision problems. These approaches should accompany a good problem-solving approach to tackle unknown types and number of targets. Vision tracking-based algorithms like Balch et al. [3] need additional steps for preprocessing. Due to number of difficulties in such type of tracking, many researchers like Fod et al. [4], Gutmann et al. [5], Hahnel et al. [6], Prassler et al. [7], Panangadan et al. [8], and Prassler et al. [7] use linear occupancy maps to estimate object trajectories. Fod et al. [4] use real-time tracking algorithm that uses multiple lasers in the environment. In the field of robotics vision, Jung and Sukhatme [9] used different types of sensors in order to track multiple objects. Stroupe and Balch [10] and Powers et al. [11] implemented a probabilistic model-based approach for multiple targets. Schulz et al. [12] used both Rao-Blackwellized particle filters and joint probabilistic data association filters (SJPDAFs) to track moving multiple objects. Prassler et al. [7] and Liao et al. [13] proposed a tracker to track activities of people around one building. Our work is different from others in the following ways: it includes (1) varying number of targets and (2) real-time performance.

3 Proposed Approach

The proposed approach accurately tracks unknown number of interacting and moving targets over time. Tracks are generated in real time. Background subtraction is used to remove uninteresting objects (i.e., the background). The result generated is represented in terms of series of tracks of positions of targets in the frame over time, creating tracks representing an individual object location over time. The proposed approach uses the well-known multiple hypotheses and iterative closest point (ICP) algorithm. The algorithm is used to check whether a track can be placed in the model or not that representing the targets being tracked. The calculated locations of targets in previous frames are used to initialize model placement activity in the next frame. The process of tracking objects goes through number of phases. The phases are data gathering, registration of locations, background subtraction, and tracking.

There are two major phases in the tracking process (Fig. 1).

Algorithm 1 Tracking algorithm

1: **procedure** MUTITRACK
2:    **for** $i = 1$ to n_1 **do**
3:       *Segment the image frame using background subtraction technique.*
4:       *Compute their likelihood of belonging to target or background.*
5:       *Compute model depending upon likelihood.*
6:       *Compute the target background probability distribution.*
7:       *Generate new tracks and update model accordingly.*
8:       *Use hypothesis model to optimize model localization.*
9:       *Track the locations of objects in the target frame.*
10:      *Update the model for every sequence to be tracked.*
11:      *Repeat until the sequence is over.*

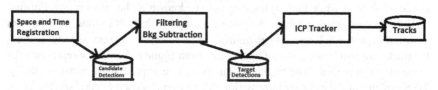

Fig. 1 Steps for proposed tracker

3.1 Preprocessing

To separate objects to be tracked from the background, the pixels corresponding to background are eliminated. Generally, the background comprises non-moving objects, e.g., wall, tables, chairs, and sometimes targets that are outside the scope of tracker. Data has been registered in space and time, once the background is over and all the non-moving background objects are removed, the data into frames is converted into Cartesian coordinates at a given time.

3.2 Model Formation

In order to estimate the location of each target, the model is created. This process is accomplished placement of an instance of a model to the data gathered. The appearance of the actual targets to be tracked is approximated by using the model consisting of various coordinate points. We can build multiple models when we have tracked multiple objects or the objects that change their shapes. The tracker automatically deals with this situation by determining which model suits to current data and fits model accordingly. The location information of a target in the track is recorded in every instance of a model. Generally, we can define a track as is a single point which stays in the middle of a model. However, it is responsibility of tracker to maintain and generate model throughout tracking. Due to this, the model also works for asymmetric model to track location of a specific part of a target to be known. Additionally, it also determines the actual orientation of the target for such models.

3.3 Tracking

We have already mentioned a track as location of a target in the frame over time: the task of determining the accurate track for object to be known and which could be challenging task. Because sometimes the target is the partially or fully occluded and the background is also not completely removed, both of these situations could create difficulties while identifying targets to be known in the given frame. Further, there is another problem while associating a target with itself in time, i.e., the data association problem. It arises when multiple targets are moving very quickly. Second, the tracks are built from group of data points from frame to frame that represents the same target over time. The tracker is supposed to complete these goals parallelly. To complete this, the cluster information in one frame is used to correspond cluster in other frame. Again, a track should represent single target at a given time. The proposed tracker has two basic components. The first component depends on pose of models in previous frames and loops, while it finds all placements for models in the current frame. The models are updated periodically to check for any remaining frames. The second component is the track splitter, which splits two tracks that are close to each other such that they can be accurately separated.

3.4 Track Assignment to model

After the preprocessing step, the tracker tracks the location of target to be known in the frame. The process has two steps. Initially, the previously created tracks are updated to reflect the new locations of the objects. The location as well as orientation is updated depending upon their previous velocity. For example, initial position of track t2 would be calculated by using a vector between its locations at t0 and t1. The t1 position is adjusted by using same vector. The vector not only denotes the magnitude and direction of location coordinates but also the rotation change between t0 and t1. The smaller distance should be required between the model and data points. It helps to prevent track jumps between one target to another while tracking multiple targets. After this, the data points with certain distance from the center of the model are found. The model point transformation is done with the help of iterative closest point (ICP). Pairing between each point and model data is done. The model is updated. Again, ICP is called to fit the model point with data. This process continues until there is no change in the pairing even after adjustment in ICP. When the model reaches to final location of the sequence, two tests are done to check if further tracks are available to a frame. First, the sum between the pairs of each data point is calculated to detect appearance of a target and the corresponding track is removed.

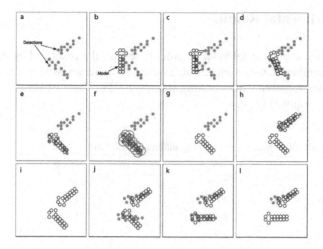

Fig. 2 Detection and model forming

The model with the best is noted as the most likely model once every model has been updated. To keep a detected track depends upon number of parameters. Abovementioned entire process is repeated for each existing track (Fig. 2).

3.5 Error Correction

There should be only one track per target in order to avoid confusion between which target belongs to which track. But in certain situations it becomes difficult to do so. In this case, the tracker should split a track into two. While splitting, the distance between the two tracks matters a lot. The higher the distance, the higher the splitting. Therefore, split distance must balance between track length and likelihood of track jumping.

3.6 Quality Check

The quality of the tracks generated by the tracker is checked by calculating average track length. The average track length is calculated by adding all the tracks generated and dividing total number of tracks. This is used to restrict the track split distance.

4 Experimental Results

We have used experimental video in order to test results generated by tracker. The video has number of frames out of which we use 3000 frames with the desired object. The accuracy achieved while tracking was about 93%. Table 1 below shows details about these results (Figs. 3, 4).

Table 1 Results of approaches proposed by different researchers

Approach	Year	Accuracy (%)
Liu and Shah	2007	83.3
Wu and Rehg	2009	83.1
Gu et al.	2011	83.7
Nanni	2012	90.3

Fig. 3 One frame of tracking results

Total Track Frames	Average Track Length	Tracks Jump	Detected Track Frames
21223	48.44 second	2	20325 (97.85%)
20440	48.43 second	2	20352 (92.57%)

Fig. 4 Tracking result statistics

5 Conclusion

There are many applications of tracking which can be used to track group of animals or people. The tracker is aimed to save time and manpower in order to track frames. The proposed algorithm showed good accuracy while tracking ants in the video. The algorithm is based on multiple hypothesis and ICP so it can deal with many issues that can arise while tracking. The tracker can be effectively used to track multiple objects in real time. Further, the research can be extended to track multiple objects parallel without using any sensors.

References

1. Khan, Z., Balch, T., Dellaert, F.: Ecient particle filter-based tracking of multiple interacting targets using an MRF-based motion model. In: Proceedings 2003 IEEE/RSJ International Conference on Intelligent Robots and Systems (IROS-2003), pp. 254–259 (2003)
2. Bruce, J., Balch, T., Veloso, M.: Fast and inexpensive color image segmentation for interactive robots. In: Proceedings 2000 IEEE/RSJ International Conference on Intelligent Robots and Systems (IROS-2000), pp. II 2061–2066 (2000)
3. Balch, T., Khan, Z., Veloso, M.: Automatically tracking and analyzing the behavior of live insect colonies. In: Proceedings of the Fifth International Conference on Autonomous Agents, ACM, pp. 521–528 (2001)
4. Fod, A., Howard, A., Matari, M.: Alaser-based peopl e tracker. In: Proceedings of the 2002 IEEE International Conference on Robotics and Automation (ICRA-2002), pp. II 3024–3029 (2002)
5. Gutmann, J., Hatzack, W., Herrmann, I., Nebel, B.: The CS Freiburg team: playing robotic soccer based on an explicit world model. In: SPIE International Symposium, AI Magazine, pp. II 21–37 (2000)
6. Hahnel, D., Burgard, W., Fox, D., Fishkin, K., Philipose, M.: Mapping and localization with RFID technology. In: Proceedings of International Conference on Robotics and Automation, IEEE, vol. 1, pp. 1015–1020 (2004)
7. Prassler, E., Scholz, J., Elfes, A.: Tracking people in a railway station during rush-hour. In: International Conference on Computer Vision Systems, pp. 162–179, Springer (1999)
8. Panangadan, A., Mataric, M., Sukhatme, G.: Detecting anomalous human interactions using laser range-nders. In: Proceedings 2004 IEEE/RSJ International Conference on Intelligent Robots and Systems, pp. 2136–2141 (2005)
9. Jung, B., Sukhatme, G.S.: Tracking targets using multiple robots: the effect of environment occlusion. Auton. Robots 13, 191–205 (2002)
10. Stroupe, A., Balch, T.: Value-based observation with robot teams (vbort) using probabilistic techniques. In: Proceedings of the 11th International Conference on Advanced Robotics (2003)
11. Powers, M., Ravichandran, R., Dellaert, F., Balch, T.: Improving multi robot multi target tracking by communicating negative information. In: Multi-Robot Systems. From swarms to intelligent automata. vol.3, pp. 107–117, Springer (2005)
12. Schulz, D., Fox, D., Hightower, J.: People tracking with anonymous and id-sensors using rao-blackwellised particle filters. In: IJCAI, pp. 921–928 (2003)
13. Liao, L., Patterson, D., Fox, D., Kautz, H.: Learning and inferring transportation routines. Artif. Intell. 311–331 (2007)

Hybrid Features of Tamura Texture and Shape-Based Image Retrieval

Naresh Pal, Aravind Kilaru, Yvon Savaria and Ahmed Lakhssassi

Abstract Search and retrieval of digital images from huge datasets has become a big problem in modern, medical, and different applications. Content-based image recovery (CBIR) is considered as the best solution for automatic retrieval of images. In such frameworks, in the ordering calculation, a few components are separated from each photo and put away as a record vector. Tamura surface features are applied on digital image and registered the low request measurements from the changed image. The separated surface components of the digital image are used for retrieval. These component mixes incorporate the pixels spatial appropriation data into numerical esteem values. The results demonstrate that this strategy is still compelling when the information scale is extensive, and it has predominant versatility than customary indexing strategies.

Keywords CBIR · Image retrieval · Image indexing · Tamura texture features

N. Pal (✉) · A. Lakhssassi
Computer Science and Engineering Department, University du Québec en Outaouais,
Gatineau, QC, Canada
e-mail: paln02@uqo.ca

A. Lakhssassi
e-mail: ahmed.lakhssassi@uqo.ca

Y. Savaria
Computer Science and Engineering Department, École Polytechnique de Montréal,
Montreal, QC, Canada
e-mail: yvon.savaria@polymtl.ca

A. Kilaru
Computer Science and Engineering Department, Manipal University Jaipur, Jaipur,
Rajasthan, India
e-mail: kilaru.aravind@gmail.com

© Springer Nature Singapore Pte Ltd. 2018
P. K. Sa et al. (eds.), *Recent Findings in Intelligent Computing Techniques*,
Advances in Intelligent Systems and Computing 709,
https://doi.org/10.1007/978-981-10-8633-5_59

1   Introduction

In the twentieth century, the new technologies like smartphones, digital cameras, and web cameras are enabling a huge growth in the amount of new digital images taken and added to Internet. Huge image databases comprising of vast knowledge are being created. It is a daunting task to store, search, and retrieve images. Image search and retrieval has become an important part of many modern-world applications. Existing image retrieval schemes are categorized into content-based image search (CIBR) and text-based image search (TIBR) [1–5]. TIBR was used in late 1970 and it relies on the textual description of the images. Keywords are mapped to each image in a huge database, and these keywords are used to search and retrieve images [6]. This may create wrong search results and adding keywords to each image in a huge database is also a tedious and never-ending task [7].

CIBR was first introduced in the year 1992 by Kato [8]. In CBIR, image search and retrieval are automated and the search is based on the features of the image like color, texture, and edge pattern. When compared with TIBR, CIBR offers an efficient way to search and retrieve images.

Mathematically, image features can be expressed using a vector of n dimensions and its components can be calculated using an image analysis technique. The most commonly used visual features are texture, shape, color, etc. For instance, a certain feature may represent the image texture information like external, internal, and region-based surfaces. In this paper, we propose a multifeature-based image search and retrieval algorithm to tackle this problem. The shape- and shading-based elements are being used for image comparison and retrieval. We concentrated on shape-based elements of digital images and pointed to both enhance recovery time and help users to express their questions productively.

Feature extraction process of an image acts as a core module in the proposed solution. Similarity measurement of an image is calculated by extracting feature

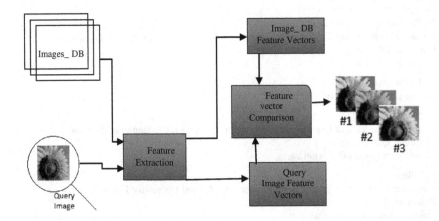

Fig. 1 Image retrieval

vectors (FV) of the image. The feature extractor process is done for both input and database images. Depending on the features extracted from the query image given by user, the database images are processed to find the similarity measurement. The database images with highest similarity measurement will be given by the system as an output set. A content-based image retrieval process is divided into two parts, which are (a) selection of an image, extraction of features, and indexing; and (b) FV-based similarity measurement calculation and image retrieval. Rather than the exact match, this algorithm works on similarity between images. The shape and shape features of an image are the most essential and fundamental features which are widely used in image retrieval (Fig. 1).

2 Related Work

Work on retrieval of images was initiated in the year 1970. In the year 1979, a first of its kind workshop on database image retrieval methods for pictorial applications was conducted in Florence. Since then, this area has attracted many scholars. In late 1970s and early 1980s, most of the work was concentrated on text-based image retrieval (TBIR) [9, 10] and extensive survey on this work. In-depth review of this area was done by Chang [11] in the year 1992. In this technique, each new image is mapped with some keywords, and some text describing the image and the image and text is saved in the database. Techniques like query evaluation and multi-dimensional indexing are made in this research direction. However, the increases in the number of images and the size of the database have led to the increase in manpower cost to maintain the system, and the results of this type of search are subjected to the particular human interpretation. The same image can be interpreted in different ways by different people. These problems have made TBIR infeasible for huge practical applications. To tackle these problems and improve the search efficiency, content-based image retrieval (CBIR) was proposed in early 1990s [3, 12–16]. CBIR is based on the image feature extraction. Image features like color, visual, and edges are used to analyze and compare the images.

Gue et al. [14] proposed an image retrieval system for medical image database which is based on CBIR and segmentation. Image features like texture, color, and shape are used for image retrieval. The system is divided into two parts, image retrieval using CBIR and image segmentation is done to retrieve a specific required part of an image. Charmi et al. [15] proposed a CBIR technique. Color co-occurrence feature and bit pattern features are introduced [15]. These are extracted from ordered-dither block truncation coding. The proposed system works for image compression. Kong et al. [16] proposed a CBIR system which works even when there are target deformation and image size variation. In the proposed system, image texture and image features are used. Songhe et al. [4] introduced a novel selective visual attention-based image retrieval algorithm which uses salient edge and region features for image retrieval. Using these two features together can represent the prominent parts of an image and characterize the human perception also.

3 Feature Extraction

CBIR system works on expression and feature extraction of a digital image. The extracted features of an image may be segregated into texture-based or visual-based feature categories. These features play a crucial role in image processing. Prior to extracting any features from an image, certain preprocessing step like resizing, normalization, and binarization are performed over the images. Image VF is used to search and extract the images from the database. The characteristics of a digital image can be conveyed in many ways from different angles that depict the nature of some of the image features. In this section, important features related to image searching and resultant expression techniques will be presented. Also, features like texture, edge, and color of an image are defined in this section.

3.1 Texture-Based Features

Texture is a basic and most important feature of an image and it can be used to describe the content of an image. It is extensively used in image searching and retrieval; this feature mainly defines the object direction in an image.

Roughness may be segregated into few calculation steps. First, the average intensity of image values of size 2k × 2k pixels is calculated, i.e., [17]

$$A_k(x, y) = \sum_{i=x-2^{k-1}}^{x+2^{k-1}-1} \sum_{j=y-2^{k-1}}^{y+2^{k-1}-1} g(i,j)/2^{2k} \tag{1}$$

For each pixel, the difference of average intensity in vertical and horizontal directions are calculated, and these may fall on an overlap window

$$\left. \begin{array}{l} E_{k,h}(x,y) = \left| A_k\left(x+2^{k-1}, y\right) - A_k\left(x-2^{k-1}, y\right) \right| \\ E_{k,v}(x,y) = \left| A_k\left(x, y+2^{k-1}\right) - A_k\left(x, y-2^{k-1}\right) \right| \end{array} \right\} \tag{2}$$

The value of E can reach extreme in either direction to set best values for k dimensions. Roughness of a digital image can be calculated by

$$F_{crs} = \frac{1}{m \times n} \sum_{i=1}^{m} \sum_{j=1}^{n} S_{best}(i,j) \tag{3}$$

The better roughness can be expressed by a variety of different texture features of an image or region, and therefore be more favorable for image retrieval. But it is more complex to calculate that with the abovementioned method, contrast of an image can be calculated by

$$F_{con} = \frac{\sigma}{\alpha_4^{1/4}}$$

$$\alpha 4 = \mu 4/\sigma 4$$

(4)

σ represents variance and the above formula is calculated for fourth iteration.

This value gives the entire image or regions in contrast global metrics.

For calculating the direction degree, each pixel's gradient vector direction has to be calculated. It is defined as

$$|\Delta G| = (|\Delta_H| + |\Delta_V|)/2$$

$$\theta = \tan^{-1}(\Delta_V/\Delta_H) + \pi/2$$

(5)

ΔH represents the horizontal direction and ΔV represents the vertical direction, and these two are a 4×4 operator variation given below:

$$
\begin{array}{cccccc}
-1 & 0 & 1 & 1 & 1 & 1 \\
-1 & 0 & 1 & 0 & 0 & 0 \\
-1 & 0 & 1 & -1 & -1 & -1
\end{array}
$$

After calculating the gradient vector, a histogram is created with $H_D(\theta)$ value in the following expression. Directional sharpness of peaks within histogram can also be expressed with

$$F_{dir} = \sum_{p}^{n_p} \sum_{\varnothing \in w_p}^{n} (\varnothing - \varnothing_p)^2 H_D(\varnothing)$$

(6)

Peak in histogram is represented by P and all peaks are represented by n_p. For a highest p, all peaks in the bin are represented by W_p. φ_p is the highest value possible for a bin.

Tamura as per Tamura [3] coarseness is considered to be the most important and fundamental feature of the six Tamura features. In an image with multiple textures of different scales, coarseness will recognize the texture with maximum scale. It is calculated by taking averages at every point with its neighborhood of linear sizes of powers of two

$$A_k(x,y) = \sum_{i=x-2^{k-1}}^{x+2^{k-1}-1} \sum_{j=y-2^{k-1}}^{y+2^{k-1}-1} f(i,j)/2^{2k}$$

(7)

In horizontal orientation at every point, averages of non-intersecting neighborhoods on both sides of the point are calculated and the difference between these averages is calculated with

$$E_{k,h}(x,y) = \left| A_k\left(x + 2^{k-1}, y\right) - A_k\left(x - 2^{k-1}, y\right) \right| \tag{8}$$

At each point, the optimum size which provides maximum output value is taken at which the K maximizes E in both ways.

$$S_{\text{best}} = 2^k \tag{9}$$

The average of S_{best} is taken as coarseness measure of the image.

3.2 Edge Directions

A feature vector of every edge of every object in an image is calculated by considering the direction of the object. This feature vector can be used for image identification. Texture data of the image is calculated by taking image's histogram of edges which are translation invariant. It is calculated by the following steps. The query image is first converted from RGB scale to Hue saturation intensity scale (HSI scale) and the HUI (H) is removed. Saturation and intensity channels are convoluted with the eight Sobel masks. Each pixel is given the maximum of the responses and the corresponding 8-quantized direction, and the gradient is threshold next. The threshold values are manually fixed to 15% and 35%, respectively, of gradient value on intensity (I) and saturation (S) channels [18]. Using OR operation, the threshold intensity image is combined with the saturation gradient image. In this operation, if the gradient directions differ between the saturation and intensity, the originally stronger gradient direction is chosen. A gray level image I(x, y) is, hence, transformed as

$$I(x,y) \rightarrow \{Ie(x,y), Id(x,y)\}$$

The binary edge image and binary direction image are represented by Ie(x, y) $\in$ {0, 1} and Id(x, y) $\in$ {0 ... 7}, respectively.

3.3 Object Texture Moment

Moment invariant method is used to define the texture of an image. It is invariant to affine transformations. For a two-dimensional function f(x, y), the moments of order p + q are defined as

$$m_{pq} = \int\limits_{-\infty}^{\infty} \int\limits_{-\infty}^{\infty} x^p y^q f(x, y) \, dx \, dy$$

For p, q = 0, 1, 2, ...

The width and height of a region are defined as the maximum value of the vertical and the horizontal projections, respectively [3].

4 Image Indexing

In CBIR, queries are processed using visual feature similarity between the query image and database images. Hence, selecting a best visual feature is very important for fast and correct search results. Euclidian distance is used to calculate the distance between feature vectors of the selected features. To calculate the equal weighting from the above-calculated distances by performing median operation, Euclidean distance (given below) is used to calculate the similarity between input image FV and database image FV

$$Sim_{FV}(IP, DI) = \sqrt{\sum_{n=1}^{m} (IP_n - DI_n)^2}$$

IP represents the input base image, the DI represents database image, and FV is the feature vector.

5 Experimental Results

In this section, we evaluate the proposed method of image retrieval based on hybrid features of texture as well as shape, as for benchmark testing and implementation, MATLAB is a platform utilized. For the testing purpose, a set of images (DB_images) is utilized which first categorizes into five different classes based on the nature. Number of images need to retrieved is being decided by the end user at runtime, and a query image is set. To search, a required number of images are indexed based on the similarity features. Search button on the system has all the processing of proposed algorithm from DB_Image feature extraction to query image feature extraction has called. System retrieved the images on the screen specified for the images based on the ranking of similarity vector from query image feature vector. If the match is found for the same index or range of the index, it returns most related image from the dataset at the very first position and so on.

The proposed system is evaluated over a standard set of images of different conditions based on the nature and features. The overall system is tested with 500

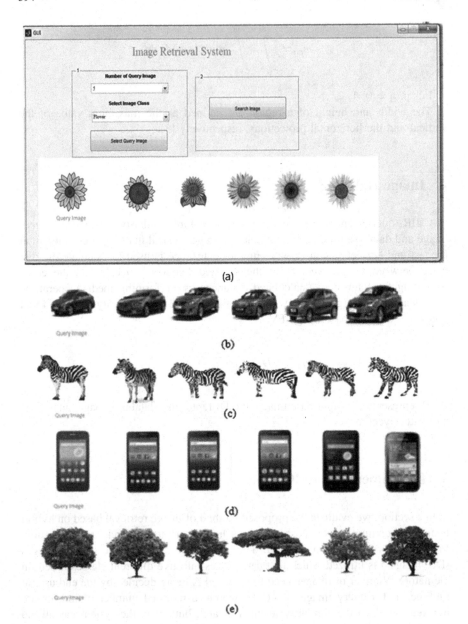

Fig. 2 Retrieved images corresponding to each class: **a** Flower, **b** Car, **c** Animal, **d** Mobile phone, **e** Tree

different images selected from the Corel database [19]. For the benchmark testing, five different classes designed based on a different nature are used. From all image databases, a set of images is defined as testing data and executed to test the

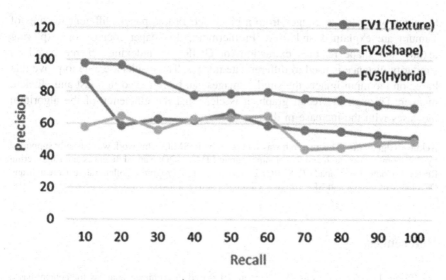

Fig. 3 Precision and recall graph

accurateness of the result retrieved. The test results are summarized in Fig. 2. Our system detected the number of images from testing database indicated in the "image detected" column and corresponding accuracy presented.

The overall system performance is calculated on the basis of these five classes. The precision and recall measure the accurateness of the proposed system with effectiveness. Precision and recall for different feature sets are evaluated for the performance measurement. At very first, we evaluated the set of 100 images which includes 20 images of each class. The corresponding precision and recall graph is shown in Fig. 3 for the FV1 as texture, FV2 as shape, and FV3 as proposed hybrid set as the combination of both.

This above graph validates that proposed hybrid feature-based method has very high accurateness. This testing result checks the effectiveness of hybrid feature-based method for image retrieval on the basis of their content. This investigation demonstrates that feature selection and their dimensionality take a major role in the feature-based information retrieval. As a selective set, our approach introduced a new set of hybrid features for efficient content-based image retrieval system.

6 Conclusion

For the present digital world, a lot of information is stored in the form of the images and the present techniques like TIBR are not providing efficient search results. Hence, there is a need for an efficient and fast content-based image retrieval system to search and retrieve images from a given set. In this paper, different features of Tamura are explained and used. Furthermore, this paper focuses on exploring texture and shape features extraction for CBIR. An indexing scheme based on similarity distance is used to differentiate images. The results are able to prove that based on the input image, the similar images can be retrieved in a fast and efficient manner. Through the recall graph, it is clear that the efficiency of the algorithm increases with the increase in dataset size.

Acknowledgements This research was supported by ReSMiQ. This work was partially supported by Collaborative Research and Development grant LIMA-UQO, Natural Sciences and Engineering Research Council of Canada (NSERC) Discovery Grants Program, Collaborative Research and Development Grants and SEP (Engage Grants for universities).

References

1. Zhang, L., Wang, L., Lin, W.: Generalized biased discriminant analysis for content-based image retrieval. IEEE Trans. Syst. Man Cybern. Part B: Cybern. **42**(1), 282–290 (2012)
2. Hu, R., Barnard, M., Collomosse, J.: Gradient field descriptor for sketch based image retrieval and localization. In: International Conference on Image Processing, pp. 1–4 (2010)
3. Kekre, H.B., Thepade, S.D.: Boosting block truncation coding using Kekre's LUV color space for image retrieval. WASET Int. J. Electr. Comput. Syst. Eng. (IJECSE) **2**(3), 172–180 (2008)
4. Feng, S., Xu, D., Yang, X.: Attention-driven salient edge(s) and region(s) extraction with application to CBIR. Signal Process. **90**(1), 1–15 (2010)
5. Murala, S., Maheshwari, R., Balasubramanian, R.: Local tetra patterns: a new feature descriptor for content-based image retrieval. IEEE Trans. Image Process. **21**(5), 2874–2886 (2012)
6. Zhang, L., Shum, H.P., Shao, L.: Discriminative semantic subspace analysis for relevance feedback. IEEE Trans. Image Process. **25**(3), 1275–1287 (2016)
7. Alkhawlani, M., Elmogy, M., El Bakry, H.: Text-based, content-based, and semantic-based image retrievals: a survey. Int. J. Comput. Inf. Technol. **4**(01) (2015)
8. Kato, T.: Database architecture for content-based image retrieval. In: SPIE/IS&T 1992 Symposium on Electronic Imaging: Science and Technology, pp. 112–123 (1992)
9. Chang, N.S., Fu, K.S.: A relational database system for images. Technical Report TR-EE 79-28, Purdue University, May 1979
10. Chang, N.S., Fu, K.S.: Query-by pictorial-example. IEEE Trans. Softw. Eng. **SE-6**(6) (1980)
11. Chang, S.-K., Hsu, A.: Image information systems: where do we go from here? IEEE Trans. Knowl. Data Eng. **4**(5) (1992)
12. Yoo, H.-W., Park, H.-S., Jangb, D.-S.: Expert System for Color Image Retrieval, vol. 28, pp. 347–357. Elsevier (2005)
13. Lee, Y.-H., Rhee, S.-B., Kim, B.: Content-based image retrieval using wavelet spatial-color and gabor normalized texture in multi resolution database. 978-0-7695-4684-1/12 © 2012 IEEE. https://doi.org/10.1109/imis.2012.98
14. Guo, J.M., Prasetyo, H.: Content-based image retrieval using features extracted from halftoning-based block truncation coding. IEEE Trans. Image Process. **24**(3), 1010–1024 (2015)
15. Charmi, V., Gudivada, N., Raghavan, J.V.: Special issue on content-based image retrieval systems. IEEE Comput. Mag. **28**(9) (1995). In: Science, Engineering and Technology, vol. 4, Issue 10, Oct 2015

16. Jin, C., Ke, S.-W.: Content-Based Image Retrieval Based on Shape Similarity Calculation
17. Das, V.V., Stephen, J. (eds.): CNC 2012, LNICST 108, pp. 54–59 (2012)
18. Brandt, S., Laaksonen, J., Oja, E.: Statistical shape features for content-based image retrieval. J. Math. Imaging Vis. **17**(2), 187–198 (2002)
19. http://sites.google.com/site/dctresearch/Home/content-based-image-retrieval
20. Tamura, H., Mori, S., Yamawaki, T.: Textural features corresponding to visual perception. IEEE Trans. Syst. Man Cybern. **8**(6), 460–473 (1978)

Codec with Neuro-Fuzzy Motion Compensation and Multi-scale Wavelets for Quality Video Frames

Prakash Jadhav and G. K. Siddesh

Abstract Virtual reality or immersive multimedia is sometimes known as the realization of real-world environment in terms of video, audio, and ambience like smell, airflow, background noise, and various ingredients that make up the real world. The combination and synchronization of audio and video with better clarity has transformed the rendition matched in quality by 3D cinema. Virtual reality still remains in research and experimental stages. The objective of this research is to explore and innovate the esoteric aspects of the virtual reality like stereo vision incorporating depth of scene, rendering of video on a spherical surface, implementing depth-based audio rendering, and applying self-modifying wavelets to compress the audio and video payload beyond levels achieved hitherto so that maximum reduction in size of transmitted payload will be achieved. Considering the finer aspects of virtual reality, we propose to implement stereo rendering of video and multichannel rendering of audio with associated back-channel activities, and the bandwidth requirements increase considerably. Against this backdrop, it becomes necessary to achieve more compression to achieve the real-time rendering of multimedia contents effortlessly.

Keywords Artificial neural networks · Peak signal-to-noise ratio (PSNR)
Root mean square error (RMS) · Quantization · Discrete cosine transform
(DCT) · Bandwidth

P. Jadhav (✉)
K. S. School of Engineering and Management, Bengaluru, India
e-mail: pcjadhav12@gmail.com

G. K. Siddesh
J.S.S. Academy of Technical Education, Bengaluru, India
e-mail: siddeshgundagatti@gmail.com

© Springer Nature Singapore Pte Ltd. 2018   599
P. K. Sa et al. (eds.), *Recent Findings in Intelligent Computing Techniques*,
Advances in Intelligent Systems and Computing 709,
https://doi.org/10.1007/978-981-10-8633-5_60

1 Introduction

Virtual reality known is the realization of real-world circumstances in terms of video and audio. This environment gives us a sense of reality as if we are living in a real world although the implementation of virtual reality is on a laboratory scale [1]. Audio has attained unimaginable clarity by splitting the spectrum into various frequency bands appropriate for rendering on a number of speakers or acoustic waveguides [2]. The combination and synchronization of audio and video with better clarity has transformed the rendition matched in quality by 3D cinema [3]. Multicasting of several channels over a single station, program menu options, parental control of channels, and various online activities like gaming, business transactions, etc. through back-channel activities have made multimedia systems truly entertaining and educative [4]. But then, there are several aspects of virtual reality that are missing from practical implementations; even today, names of the authors should be checked before the paper is sent to the volume editors.

2 Literature Survey

The existing implementations of compressor and decompressor follow the ISO Standard 13818. The standards are evolved from MPEG - 1, MPEG - 2, MPEG - 4, MPEG – 7, and MPEG - 21. There are implementations of codec based on H.264 Standard, which uses wavelets instead of DCT [5, 6]. MPEG2 implements the codec based on DCT. The shortcomings of existing codecs are not adaptive to the patterns of the pixel residents in the video frames [7]. MPEG2 uses a flat and uniform quantization, while other implementation uses vector quantization techniques. It is important to observe that vector quantization, while being marginally superior in terms of compression ratio [8], has substantial computing overheads which counterbalance the gain resulting from better compression ratio for a given visual quality metric [9]. There is an authentic need for better mechanisms of compression which will seek to achieve (1) decoding of pictures faster at the receiver end, (2) compression ratio is better, and (3) frames are constructed in better quality factor in terms of peak signal-to-noise ratio and mean square error.

3 Immersive Virtual Reality

Virtual reality it defines "virtual" and "reality". The meaning of "virtual" is near and meaning of "reality" is as human beings what we feel or experience. Therefore, the "virtual reality" term means "near-reality". This is the classical definition of virtual reality. The components that constitute virtual reality are audio, video, images, ambience, etc. To make virtual reality nearer to reality, the video or images could be immersed in a 3D world with stereo/3D vision. The rendering display unit could be on a spherical surface depicting the real-world scenario. Natural scenes are not flat but embedded in a 3D spherical world. Rendering these on a 2D flat screen removes the effect of depth of scene. Then again, the human visual system is stereo in nature. The scene is captured by both the left and right human eyes giving a true sense of depth of objects within. Audio rendering has so far reached the level of DOLBY 5.1 with the use of six different bands. Although DOLBY 5.1 is considered the most sophisticated media of audio rendering, there is still a lot left in its implementation. DOLBY standard does not generate high-frequency sounds around 20 Hz with much fidelity. Further, the depth of sound is not perceptible unless the speakers are arranged to synchronize with the acoustics of the room. This being the case, there is sufficient room for research in improving the quality with the addition of adaptive noise cancelation, faithful reproduction of high-frequency sounds (especially per-cussion instruments and piano), and selective and mild echoing to generate pleasant aural effects. Virtual (Immersive) reality seeks to combine the essential ingredients of audio, video, and back-channel activities in unobtrusive ways to generate a real "virtual world". Adding to environmental aspects will make virtual reality truly awesome. Although virtual reality has been around since a decade, the imple-mentation targeting specific areas such as education, medicine, stock markets, etc. does not see any true implementation that is worthy of being called realistic. Current implementations concentrate only on specialized.

3.1 Need for a More Efficient Transport Protocol for Payload

To form MPEG-2 transport stream, these streams are multiplexed with data source from the programs. Transport streams consist packets of 188 bytes length. To protect transport streams from interruption and noise in transmission channel, the encoder FEC takes precautionary measurement. To transmit the suitable digital symbols, the modulator converts FEC-protected transport packet into digital sym-bol, which is preferred for transmission in the global channel. Upper converter is used to convert digital symbols into required appropriate RF channel. The operation is in reverse order in the receiver section (Fig. 1).

3.2 Algorithms

Our research work so far has been related to stereo video rendering using 3D curvilinear coordinate system with left and right video frames, projecting the space-shifted frames on to the video display. Algorithms have been developed by us to give the effect of depth, and perspective projection techniques have been used.

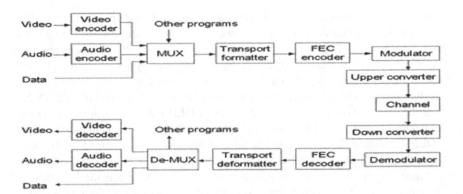

Fig. 1 DTTB (Digital Terrestrial Television Broadcasting) system. The video, audio, and data are compressed and multiplexed to get streams

Novel compression techniques based on multi-scale wavelets have been invented to generate a balance between video quality and compression levels.

The latest video quality assessment techniques based on CIDIED2000 color differences have been employed to ensure the quality of reconstructed frames.

i. The inputs to the rendering codec are the left and right frames of the stereo vision camera. The frames are in RGB (Red/Green/Blue) pixel format. Since manipulation of color images is done in the Y, Cb, and Cr, color space, the RGB pixels are converted to Y, Cb, and Cr, using the following linear equations:

$$Y' = 16 + (65.481 \cdot R' + 128.553 \cdot G' + 24.966 \cdot B')$$
$$C_B = 128 + (-37.797 \cdot R' - 74.203 \cdot G' + 112.0 \cdot B')$$
$$C_R = 128 + (112.0 \cdot R' - 93.786 \cdot G' + 18.214 \cdot B')$$

The Y,Cb,Cr image is split into three planes Y, Cb, and Cr. These three planes become input to the multi-scale wavelet algorithm.

ii. After application of the transform, we have three frequency planes corresponding to Y, Cb, and Cr. The frequency planes of both left and right images are digitally composited to get a single plane. For 3D to 2D rendering, we transform every X, Y, and Z coordinates for rendering as follows: Apply x-axis rotation to transform coordinates (x1, y1, and z1)

(a) x1 = x, (b) y1 = y * cos (Rx) + z * sin (Rx), (c) z1 = z * cos (Rx) − y * sin (Rx)

Apply y-axis rotation to (x1, y1, and z1) to get (x2, y2, and z2).

(a) x2 = x1 * cos (Ry) − z1 * sin (Ry), (b) y2 = y1, (c) z2 = z1 * cos (Ry) + x1 * sin (Ry)

Finally, apply z-axis rotation to get the point (x3, y3, and z3).

(a) $x3 = x2 * \cos(Rz) + y2 * \sin(Rz)$,   (b) $y3 = y2 * \cos(Rz) - x2 * \sin(Rz)$, (c) $z3 = z2$

iii. Apply mid-rise quantizer to the result of step 1. Each coefficient value is rounded as

$$Pixel = floor(Pixel + 0.5)$$

iv. Quantization:

$$\begin{bmatrix} 16 & 11 & 10 & 16 & 24 & 40 & 51 & 61 \\ 12 & 12 & 14 & 19 & 26 & 58 & 60 & 55 \\ 14 & 13 & 16 & 24 & 40 & 57 & 69 & 56 \\ 14 & 17 & 22 & 29 & 51 & 87 & 80 & 62 \\ 18 & 22 & 37 & 56 & 68 & 109 & 103 & 77 \\ 24 & 35 & 55 & 64 & 81 & 104 & 113 & 92 \\ 49 & 64 & 78 & 87 & 103 & 121 & 120 & 101 \\ 72 & 92 & 95 & 98 & 112 & 100 & 103 & 99 \end{bmatrix}$$

The wavelet coefficients of step 3 are quantized using the above matrix to yield a coefficient array with smaller dynamic range.

v. Rearrange coefficients in increasing frequency order:

We use the above scheme to rearrange the wavelet coefficients for subsequent compression.

vi. Temporal redundancy is achieved by adjacent frame differencing. Each pixel in the given frame is manipulated with the corresponding pixel in the next frame using:

$$Pixel(x, y) = Pixel - 1(x, y) \text{ XOR } Pixel - 2(x, y)$$

The ANN and fuzzy engine are employed at this stage.

vii. Compression is done by adaptive Huffman coding—which is the industry standard method.
viii. The picture at the receiver end is clocked to the display by retracing steps 7 to 1 in the reverse order.

4 Results

Table 1, shows the result of MPEG2 Codec with 8 × 8 Discrete Cosine Transform, by using MPEG 2 codec compressing the different frames and calculating the PSNR value of each frame.

Table 1 Results of MPEG2 codec with 8 × 8 DCT

Filename	(O) Size	(C) Size	% Comp	PSNR (dB)
frm2-00.bmp	522296	42801	91.8	1302
frm2-01.bmp	522296	51473	90.1	1290
frm2-02.bmp	522296	55790	89.3	1295
frm2-03.bmp	522296	56495	89.2	1301
frm2-04.bmp	522296	56269	89.2	1292
frm2-05.bmp	522296	58300	88.8	1291
frm2-06.bmp	522296	59766	88.6	1301
frm2-07.bmp	522296	59355	88.6	1289
frm2-08.bmp	522296	59148	88.7	1301

Table 2 Codec with neuro-fuzzy motion compensation and multi-scale wavelets

Filename	(O) Size	(C) Size	% Comp	PSNR (dB)
frm2-00.bmp	522296	7686	98.5	1603
frm2-01.bmp	522296	8071	98.5	1591
frm2-02.bmp	522296	8837	98.3	1596
frm2-03.bmp	522296	8781	98.3	1602
frm2-04.bmp	522296	9107	98.3	1593
frm2-05.bmp	522296	9165	98.2	1592
frm2-06.bmp	522296	9592	98.2	1602
frm2-07.bmp	522296	9388	98.2	1590
frm2-08.bmp	522296	9605	98.2	1602

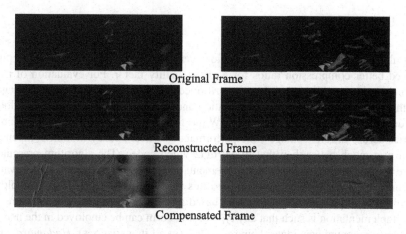

Fig. 2 Averaged frame 1 with MPEG2 codec

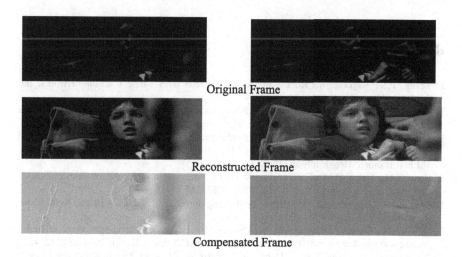

Fig. 3 Neuro-fuzzy motion compensation

The Fig. 2 shows the Averaged Frame 1 with MPEG2 Codec, compression has done by MPEG2 codec considering the 1 original frame, the compensated frame and Reconstructed frame as shown in the figure.

Table 2, shows the result of Codec with Neuro-Fuzzy Motion Compensation and Multi-Scale Wavelets, by using Codec with Neuro-Fuzzy Motion Compensation compressing the different frames and calculating the PSNR value of each frame.

The Fig. 3 shows the Neuro-Fuzzy Motion Compensation, compression has done by Neuro-Fuzzy Motion Compensation considering the 1 original frame, the compensated frame and Reconstructed frame as shown in the figure.

5 Conclusion

On all the image files without exception, neuro-fuzzy motion compensation produced better compression ratios for a given quality factor. For evaluation of performance in compression time, we performed the compression of several images both by neuro-fuzzy motion compensation and MPEG2 codec 500 times in a loop to determine the average time taken. While time for compression understandably varies across different images, the performance of neuro-fuzzy motion compensation was much better than that of MPEG2 video codec. The algorithm presented here lends itself very easily for implementation both in hardware and software. Future enhancements and extensions to this research work could include deployment of an additional layer clustering based on Markov's chains. The structure of the implementation is such that any type of transform can be employed in the neural network, potential candidates being *wavelets* (in all its variations), *Hadamard*, and *Walsh Transforms*.

References

1. Boluk, P.S., Baydere, S.: Robust image transmission over wireless sensor networks. Int. J. Mobile Netw. Appl. 149–170 (2011)
2. Lingyun, L., Haifeng, D., Liu, R.P.: CHOKeR: a novel AQM algorithm with proportional bandwidth allocation and TCP protection. IEEE Trans. Ind. Inf. **10**(1), 637–644 (2014)
3. Panayides, A., Antoniou, Z.C.: High-resolution, low-delay, and error-resilient medical ultrasound video communication using H.264/AVC over mobile WiMAX networks. IEEE J. Biomed. Health Inform. **17**(3), 619–628 (2013)
4. Aziz, S.M., Pham, D.M.: Energy efficient image transmission in wireless multimedia sensor networks. IEEE Commun. Lett. **17**(6), 1084–1087 (2013)
5. Norkin, A., Bjøntegaard, G.: HEVC deblocking filter. IEEE Trans. Circuits Syst. Video Technol. **22**(12), 1746–1754 (2012)
6. Sjöberg, R., Chen, Y., Fujibayashi, A.: Overview of HEVC high-level syntax and reference picture management. IEEE Trans. Circuits Syst. Video Technol. **22**(12), 1858–1870 (2012)
7. Nur Yilmaz, G., Arachchi, H.K., Dogan, S., Kondoz, A.: 3D video bit rate adaptation decision taking using ambient illumination context. Int. J. Eng. Sci. Technol. 01–11 (2014)
8. Ma, T., Hempel, M., Peng, D., Sharif, H.: A survey of energy-efficient compression and communication techniques for multimedia in resource constrained systems. IEEE Commun. 01–10 (2012)
9. Kandris, D., Tsagkaropoulos, M., Politis, I., Tzes, A.: Energy efficient and perceived QoS aware video routing over wireless multimedia sensor networks. Ad Hoc Netw. J. 591–607 (2011)

Author Index

© Springer Nature Singapore Pte Ltd. 2018
P. K. Sa et al. (eds.), *Recent Findings in Intelligent Computing Techniques*,
Advances in Intelligent Systems and Computing 709,
https://doi.org/10.1007/978-981-10-8633-5

Printed in the United States
By Bookmasters

Printed in the United States
By Bookmasters